AF684559

THE UNIVERSITY OF TEXAS AT AUSTIN
UNIVERSITY OF TEXAS LIBRARIES

DATE RETURNED

RETD PMA
AUG 0 6 2008

Time-Dependent Quantum Molecular Dynamics

NATO ASI Series

Advanced Science Institutes Series

A series presenting the results of activities sponsored by the NATO Science Committee, which aims at the dissemination of advanced scientific and technological knowledge, with a view to strengthening links between scientific communities.

The series is published by an international board of publishers in conjunction with the NATO Scientific Affairs Division

A Life Sciences **B Physics**	Plenum Publishing Corporation New York and London
C Mathematical and Physical Sciences **D Behavioral and Social Sciences** **E Applied Sciences**	Kluwer Academic Publishers Dordrecht, Boston, and London
F Computer and Systems Sciences **G Ecological Sciences** **H Cell Biology** **I Global Environmental Change**	Springer-Verlag Berlin, Heidelberg, New York, London, Paris, Tokyo, Hong Kong, and Barcelona

Recent Volumes in this Series

Volume 293—Methods in Computational Molecular Physics
edited by Stephen Wilson and Geerd H. F. Diercksen

Volume 294—Single Charge Tunneling: Coulomb Blockade Phenomena in Nanostructures
edited by Hermann Grabert and Michel H. Devoret

Volume 295—New Symmetry Principles in Quantum Field Theory
edited by J. Fröhlich, G.'t Hooft, A. Jaffe, G. Mack, P. K. Mitter, and R. Stora

Volume 296—Recombination of Atomic Ions
edited by W. G. Graham, W. Fritsch, Y. Hahn, and J. A. Tanis

Volume 297—Ordered and Turbulent Patterns in Taylor–Couette Flow
edited by C. David Andereck and F. Hayot

Volume 298—Chaotic Dynamics: Theory and Practice
edited by T. Bountis

Volume 299—Time-Dependent Quantum Molecular Dynamics
edited by J. Broeckhove and L. Lathouwers

Volume 300—Equilibrium Structure and Properties of Surfaces and Interfaces
edited by A. Gonis and G. M. Stocks

Series B: Physics

Time-Dependent Quantum Molecular Dynamics

Edited by

J. Broeckhove and

L. Lathouwers

Rijksuniversitair Centrum Antwerpen (RUCA)
Antwerp, Belgium

Plenum Press
New York and London
Published in cooperation with NATO Scientific Affairs Division

Proceedings of a NATO Advanced Research Workshop on
Time-Dependent Quantum Molecular Dynamics: Theory and Experiment,
held March 30–April 3, 1992,
in Snowbird, Utah

NATO-PCO-DATA BASE

The electronic index to the NATO ASI Series provides full bibliographical references (with keywords and/or abstracts) to more than 30,000 contributions from international scientists published in all sections of the NATO ASI Series. Access to the NATO-PCO-DATA BASE is possible in two ways:

—via online FILE 128 (NATO-PCO-DATA BASE) hosted by ESRIN, Via Galileo Galilei, I-00044 Frascati, Italy

—via CD-ROM "NATO-PCO-DATA BASE" with user-friendly retrieval software in English, French, and German (©WTV GmbH and DATAWARE Technologies, Inc. 1989)

The CD-ROM can be ordered through any member of the Board of Publishers or through NATO-PCO, Overijse, Belgium.

Library of Congress Cataloging-in-Publication Data

Time-dependent quantum molecular dynamics / edited by J. Broeckhove and L. Lathouwers.
p. cm. -- (NATO ASI series. Series B, Physics ; v. 299)
"Proceedings of a NATO Advanced Research Workshop on Time-dependent Quantum Molecular Dynamics, Theory and Experiment, held March 30-April 3, 1992 in Snowbird, Utah"--T.p. verso.
Includes bibliographical references and index.
ISBN 0-306-44305-8
1. Quantum theory--Congresses. 2. Molecular dynamics--Congresses. I. Broeckhove, Jan, 1955- . II. Lathouwers, Luc, 1951- . III. NATO Advanced Research Workshop on Time-dependent Quantum Molecular Dynamics: Theory and Experiment (1992 : Snowbird, Utah) IV. Series.
QC173.96.T56 1992
530.1'2--dc20 92-28749
CIP

ISBN 0-306-44305-8

© 1992 Plenum Press, New York
A Division of Plenum Publishing Corporation
233 Spring Street, New York, N.Y. 10013

All rights reserved

No part of this book may be reproduced, stored in a retrieval system, or transmitted in any form or by any means, electronic, mechanical, photocopying, microfilming, recording, or otherwise, without written permission from the Publisher

Printed in the United States of America

PREFACE

From March 30th to April 3rd, 1992, a NATO Advanced Research Workshop entitled "**Time Dependent Quantum Molecular Dynamics: Theory and Experiment**" was held at Snowbird, Utah. The organizing committee consisted of J. BROECKHOVE (Antwerp, Belgium), L. CEDERBAUM (Heidelberg, Germany), L. LATHOUWERS (Antwerp, Belgium), N. OHRN (Gainesville, Florida) and J. SIMONS (Salt Lake City, Utah). Fifty-two participants from eleven different countries attended the meeting at which thirty-three talks and one poster session were held. Twenty-eight participants submitted contributions to the Proceedings of the meeting, which are reproduced in this volume.

The workshop brought together experts in different areas of molecular quantum dynamics, all adhering to the time dependent approach. The aim was to discuss and compare methods and applications. The familiarity of the audience with the concepts of time dependent approaches greatly facilitated topical discussions and probing towards new applications. A broad area of subject matter was covered including time resolved laser chemistry, intramolecular dynamics, photodissociation dynamics, reactive and inelastic collisions as well as new time dependent methodologies. This diversity in applications is reflected in the contributions included in this volume.

The Editors

CONTENTS

VARIABLE TIME-STEP INTEGRATION FOR INTENSE FIELD EFFECTS

Charles Cerjan* and Ronnie Kosloff**

*Lawrence Livermore National Laboratory, Livermore, CA 94550
**The Fritz Haber Institute for Molecular Dynamics,
The Hebrew University, Jerusalem 91904, Israel

INTRODUCTION

In many physically interesting problems with explicit time dependence, it is advantageous to directly solve the time-dependent Schrödinger equation. For example, the interaction of a particle with a time-dependent dissipative bath, such as a solid, would find its most natural interpretation with an explicit solution, compared to say a high order perturbative decomposition. Furthermore, the time-dependent formulation more closely resembles most experimental investigations which typically prepare an initial wavepacket with a specified temporal and spatial spread. Given the desirability of using a time-dependent formulation, the question of numerical efficiency will inevitably arise when detailed calculations are attempted.

Strictly numerical attempts to solve the Schrödinger equation can be broadly classified as either time-independent or time-dependent methods. The former category typically expands the solution in an appropriate set of convenient basis functions and then solves the associated energy characteristic value problem. The difficult part of these calculations is usually the evaluation of potential matrix elements over the chosen basis set. The second category typically uses a spatial grid and an approximation to the temporal evolution operator to generate a pointwise solution to the Hamiltonian operator. This class of methods does not require potential matrix evaluations, but defers the difficulties to the representation of the Laplacian operator. It is not possible in general to simultaneously have diagonal matrix representations for both the kinetic and potential energy contributions to the Hamiltonian operator. Thus any numerical method must confront a more or less difficult calculational problem for either contribution.

Grid methods, in particular, have been increasingly applied to explicit time-dependent problems which are the situations most amenable to this type of treatment. One significant advantage to these methods is the simple form that the potential operator achieves, since this operation becomes pointwise multiplication on the grid. This ease of application compares very favorably with basis-set expansions which present the often formidable problem of complicated integral evaluations. Broadly speaking, there are two approaches to the representation of the Laplacian operator on the grid: local or global representation. A local representation uses a low-order polynomial expansion such as in finite differencing or finite element approaches. A typical global method would use an infinite order polynomial, such as sinusoidal functions, to evaluate the Laplacian operator.

Polynomial approximations are very flexible in their application, handling grid matching problems or unusual geometries easily. Since the infinite dimensional operator is reduced to a sparse, usually banded, matrix, any algebraic manipulation can be rapidly evaluated. These methods necessarily display polynomia' convergence, hence large grids are often required for suitable accuracy. Conversely, global approximations tend to be less useful for non-Cartesian geometries and multi-grid matching becomes difficult. Errors made in a localized section of the grid contaminate the entire solution quickly.

Time-Dependent Quantum Molecular Dynamics, Edited by J. Broeckhove
and L. Lathouwers, Plenum Press, New York, 1992

For some global representations, such as Fourier decomposition, the power of the Fast Fourier Transform algorithm can be employed to calculate the Laplacian efficiently. Furthermore, if the function being represented is band-limited, then the Fourier decomposition is exact at the nodal points. In practice, this means that localized functions display exponential convergence leading to significant reductions in grid size.

In addition to the spatial decomposition, the temporal evolution must be addressed. Of the many methods proposed, the applications below will concentrate on the "Crank-Nicolson" form [1] and polynomial approximations to the evolution operator[2]. The specific application will be to the relatively simple one-dimensional model of the atom-field interaction. The difficulty in these calculations arises from the large dynamic range required for numerical accuracy since the system will initially be well-described by the field-free states but subsequently will be quite removed from that description.

BACKGROUND

The particular form of the Hamiltonian operator which is the focus of the calculations presented in this work has the form

$$H(x,t) = \frac{\hbar^2}{2m}(p - \frac{e}{c}A(t))^2 + V(x). \tag{1}$$

This operator describes the motion of an electron in the presence of the applied field, $A(t)$, within the dipole approximation and with a stationary potential $V(x)$. The choice of atomic units for this operator removes several constants since $\hbar = e = m = 1$. Following Javanainen *et al.* [3], the stationary potential is the "softened" Coulombic form

$$V(x) = -\frac{1}{(1+x^2)^{1/2}}, \tag{2}$$

which retains the asymptotic form of the Coulomb potential while reducing the singularity at the origin. The applied field will consist of two parts: an envelope function, $A_0(t)$, multiplying a sinusoidal function of the laser frequency, ω_l.

The dynamics of the electron in this field can be monitored by two quantities which are experimental observables in the multi-dimensional case – the ejected photo-electron spectra and the emitted photon spectra. More specifically, if the evolving wavefunction is denoted by $\psi(x,t)$ then projection of the stationary (zero field) bound states, $\phi_n(x)$, from this wavefunction will provide a measure of the ionized electron components, $\psi_c(x,t)$, at selected times. This projection yields

$$\psi_c(x,t) \approx \psi(x,t) - \sum_{n=0}^{N}(\psi(x,t), \phi_n(x))\phi_n(x). \tag{3}$$

The amplitude of the spatial Fourier decomposition of this projection then approximately measures the probability of finding the ejected electron at different energies.

The temporal dependence of the induced atomic dipole, $\langle x(t)\rangle$, establishes the emitted photon response, where

$$\langle x(t)\rangle \equiv \int_{-\infty}^{\infty} \psi^*(x,t)\, x\, \psi(x,t)\, dx. \tag{4}$$

The choice of this form neglects atom-atom correlations [4], but the overall simplicity of the one-dimensional treatment presented here does not warrant any more sophisticated calculation. In analogy to the ejected electron spectra, the temporal Fourier transform of the induced dipole provides the energy dependence of the emitted photons.

Two different numerical schemes were used to investigate the time and space evolution of an initial wavepacket centered in the stationary potential given by (2) – the well-known Crank-Nicolson propagator with spatial finite differencing [1] and the recently introduced Residuum method using a low order Krylov subspace propagator with a pseudo-spectral spatial decomposition [5]. Krylov subspace techniques are known to have superior numerical efficiency, especially when used with the pseudo-spectral representation of the Laplacian operator [2]. These techniques supply an error estimate which can be used to control the time-step chosen, thus offering the additional advantage of variable time-stepping.

When used with a three-point formula for the spatial derivatives, time propagation of the initial wavefunction can be accomplished rapidly and stably by a unitarized approximation to the true time-ordered exponential operator

$$U(t+\Delta t) = Te^{-i\int_t^{t+\Delta t} H(s)\,ds}, \tag{5}$$

where T denotes the time-ordering operator, by the rational fraction expression

$$U(t+\Delta t) \approx (1 - \frac{i\Delta t}{2}H(t+\frac{\Delta t}{2}))\cdot(1+\frac{i\Delta t}{2}H(t+\frac{\Delta t}{2}))^{-1}. \tag{6}$$

Although this method is explicitly unitary, hence stable, its accuracy is only second-order in the time expansion. It is well-known that the Crank-Nicolson algorithm is dominated by dispersion error, which tends to slow the propagation speed of the associated Fourier components [6]. For rapidly varying temporal structure, such as that encountered near the classical turning points, the low-order of the temporal representation can become inefficient.

In contrast to the simple finite differencing plus Crank-Nicolson propagation (CN), a pseudo-spectral method with a low-order Krylov propagation (R) offers increased numerical efficiency. The spatial operation becomes

$$(H_0\psi(t))_n = -\frac{1}{2}P^2\psi_n(t) + V(x_n)\psi_n(t), \tag{7}$$

where

$$P^2\psi(x,t) = k^2\widehat{\psi(k,t)}, \tag{8}$$

where $\hat{f}(k,t)$ is the spatial Fourier transform of the function $f(x,t)$. That is, the operator is evaluated by first Fourier transforming the wavefuntion at a given time t, multiplying by k^2, and then Fourier synthesizing to the original spatial grid. The remaining differential operator is calculated in the same manner

$$(-p\cdot A\psi(t))_n = iA(t)(P\psi(x,t) = ik\widehat{\psi(k,t)})_n. \tag{9}$$

Note that it is straightforward to eliminate the dipole approximation in this representation; realistic spatial dependence for the vector potential $A(x,t)$ does not require exceptional treatment.

Time propagation can also be evaluated in a manner which has greater control of dispersion and phase errors. The approach adopted for this work entails the use of a low-order Kyrlov subspace calculation with error control supplied by the minimization of the so-called residual error in the subspace [5]. For a short time step, the full evolution operator (5) is approximated by the exponential operator which in turn can be expanded in terms of a low-order (order m) polynomial

$$e^{iH\Delta t} \approx \sum_{k=0}^{m} a_k(\Delta t)R_k(x). \tag{10}$$

As in all Krylov subspace techniques, a subspace is generated for the operation of the evolution approximation by starting with an initial guess and iterating that guess in powers of the underlying operator. Thus, the first basis vector is the initial vector $v_0 = \psi(x,t)$, and subsequent vectors are generated by $v_j = H^j v_0$ for $j = 1,\ldots m$. The minimization of the error in this projected subspace with respect to a particular operator generates the matrix equation

$$A\delta = b. \tag{11}$$

The i, jth component of the matrix A is given by the inner product of the iterated vectors, $A_{i,j} = (H^i v, H^j v)$, and the ith component of the vector b is $b_i = -(H^i v, v)$. The solution vector, δ, provides the coefficients of the residual error polynomial; the roots of this polynomial are the interpolation points of the original polynomial approximation to the exponential evolution operator. Time-step control is achieved by monitoring the l_2-norm of the residual vector – a value greater than a specified error tolerance forces a time-step reduction and an iteration of the procedure.

NUMERICAL APPLICATIONS

The methods described above were used in a variety of contexts emphasizing different choices of laser frequency, pulse duration, and temporal envelope. Since the initial-time wavefunction is taken to be the ground state of the "softened" Coulomb and the field-free bound states are needed for the projection in (3), a suitable representation of these functions on the grid is required. For the differencing method, the grid choice of Javaninen *et al.*[3] is reasonable: the total number of grid points was 32001 with a grid spacing of $\Delta x = 0.0707$. The ground state and first 30 bound states were evaluated by inverting the difference matrix representation of the Hamiltonian operator. These values reproduce previously reported values [3]. Identical accuracy (within 10^{-5}) for the pseudo-spectral representation of the *energies* of the same bound states was obtained by using a grid of 4096 points with a grid spacing of $\Delta x = 0.50$. These states were evaluated by propagating the Hamiltonian operator in imaginary time after successively filtering the previously determined lower bound states [2].

Since all the calculations were performed in the $p \cdot A$ gauge, the relation between the electric field, E, and vector potential must be used

$$A(t) = -\int_0^t E(s)\sin(\omega_l s)\,ds. \tag{12}$$

The temporal electric field which was used was the square pulse

$$E(t) = \begin{cases} E_0 & \text{if } 0 \le t \le T_c\ ; \\ 0 & \text{otherwise,} \end{cases} \tag{13}$$

and a full smooth pulse

$$E(t) = \begin{cases} E_0 \sin^2(\frac{\pi t}{T_c}) & \text{if } 0 \le t \le T_c; \\ 0 & \text{otherwise.} \end{cases} \tag{14}$$

The parameter T_c determines the envelope of the pulse and is chosen to be a multiple of the optical cycle time.

The specific application follows that of Eberly *et al.* [7] who applied the finite differencing scheme with a Crank-Nicolson time propagation to a square pulse field with an amplitude, E_0, of 0.04 a.u., a frequency of 0.07 a.u., and a pulse duration, T_c, of 16.25 field cycles. The time step, Δt, is 0.08 a.u. which corresponds to approximately 1122 time steps for each optical cycle. Since the ground state energy of the potential is -0.670 a.u., this frequency corresponds to 10-photon ionization. The generation of high-order harmonic frequencies is prominently displayed in Figure 1, especially odd harmonics of the applied field. The intensities are all scaled to the largest value and their logarithms plotted on the vertical axis.

The Residuum method was applied to this case also and the results of this calculation are plotted in Figure 2. The grid parameters were selected as noted above with variable time-stepping. A fifth-order polynomial approximation was chosen for the short-time propagation step with a specified error tolerance of 10^{-7}. The average number of time steps for each optical cycle was approximately 2278, which is an average time step of 0.039 a.u.. Most of the prominent features of the spectrum agree between the two calculations, especially for the odd-order harmonics.

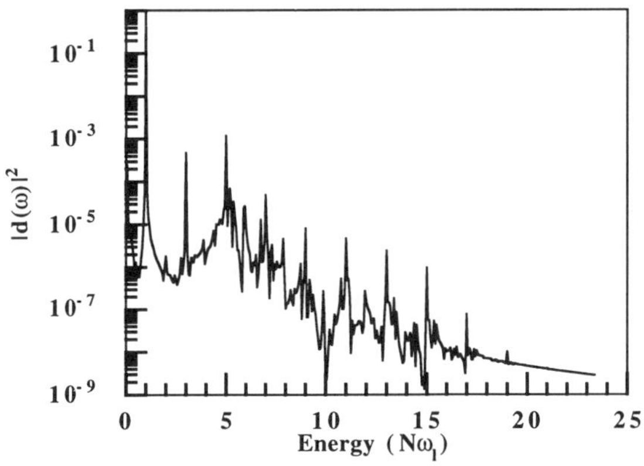

Figure 1. The absolute magnitude of the temporal Fourier transform of the dipole moment function plotted against the energy in units of the field frequency for Crank-Nicolson propagation with $E_0 = 0.04$ a.u. and $\omega_l = 0.07$ a.u. A square-pulse was applied for a total of 16.0 field cycles.

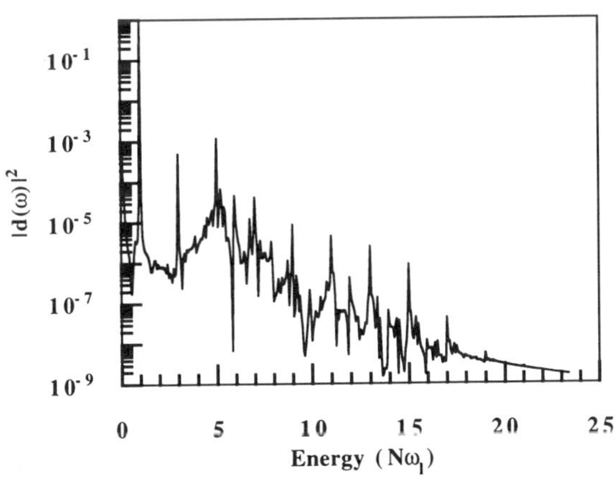

Figure 2. The same as Figure 1 for Residuum propagation.

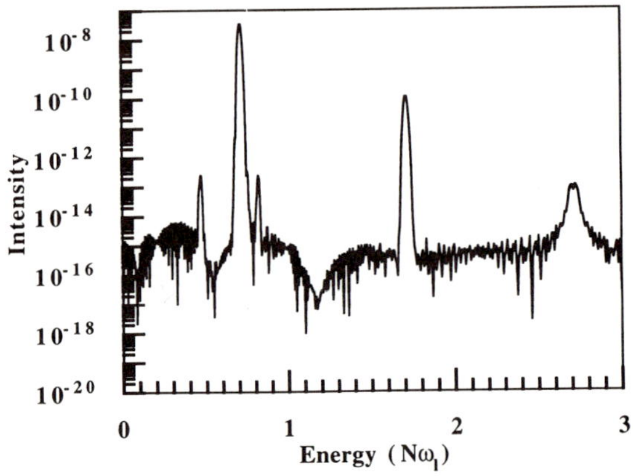

Figure 3a. ATI spectra for Crank-Nicolson propagation with $E_0 = 0.05$ a.u., $\omega_l =$ 0.52 a.u. and a 96-cycle smooth pulse turn-on. The energy is plotted in units of the applied field frequency.

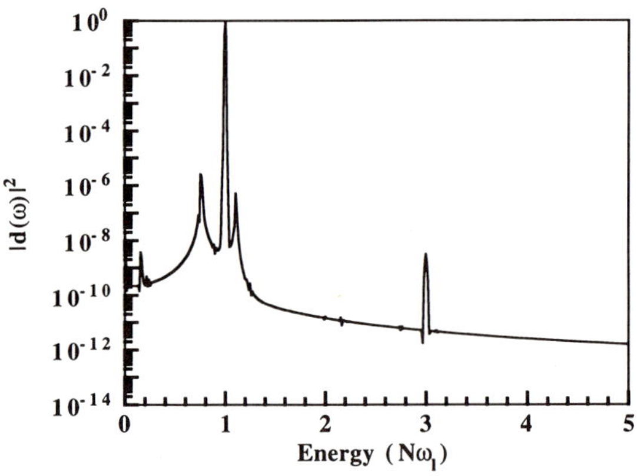

Figure 3b. Harmonic generation spectra for the conditions given in Figure 3a.

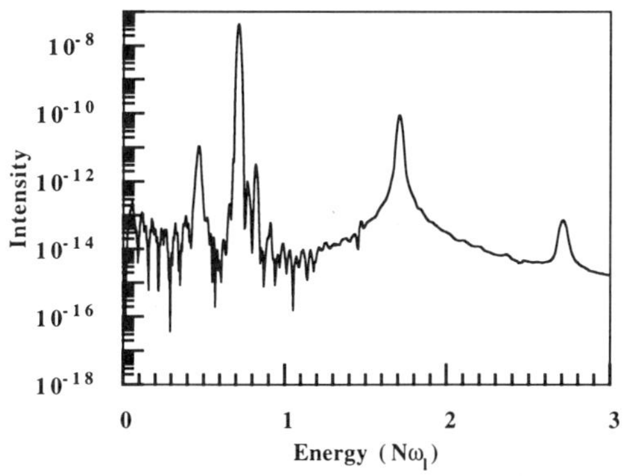

Figure 4a. Same plot as Figure 3a for Residuum propagation.

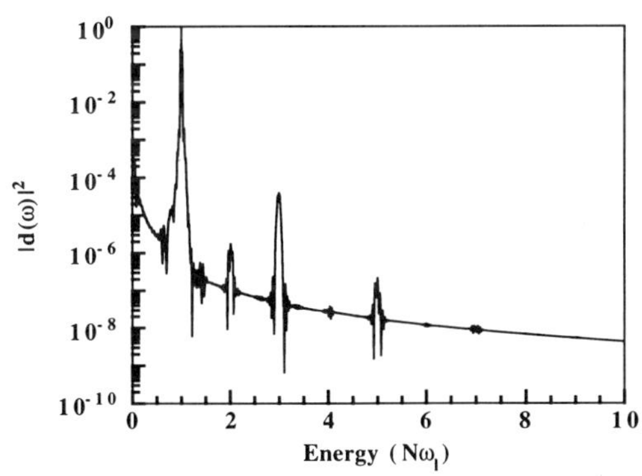

Figure 4b. Same plot as Figure 3b for Residuum propagation.

The second application represents two-photon ionization from the static potential, namely an ω_l of 0.52 a.u.. The field amplitude, E_0 was chosen to be 0.05 a.u. with a full smooth pulse of duration T_c of 96 field cycles. The predicted ATI and harmonic generation spectra are given in Figure 3a and Figure 3b, respectively, for the simple differencing algorithm. The corresponding spectra from the Residuum method are plotted in Figures 4a and 4b. The intensities in all cases were scaled to the largest value for comparison; the vertical axis is again logarithmic. An examination of the figures reveals that the ATI spectra are qualitatively similar with obvious, but minor, differences in the less prominent features. The harmonic spectra display qualitatively different structure, with the presence of even-order harmonic generation predicted by the Residuum calculation. Since the wavepacket remains localized in the potential well throughout most of the interaction time, interference effects will be more important for this case. The low-order approximation inherent in the CN treatment is incapable of following the rapid changes as the wavepacket evolves.

CONCLUSIONS

The investigation of intense field interactions with atomic systems places stringent demands on any numerical solution. The method of choice must be capable of accurately following the development of the interaction from the weak-field case when the interaction is initiated to peak intensity conditions. Variable time-stepping is clearly a major advantage in this regard, especially if many field cycles are required to simulate the true temporal laser pulse shape. Furthermore, it is essential that the grid representation selected provide an accurate evaluation of the properties of the evolving wavefunction. Since these numerical studies are then analyzed for insight into the physical mechanism of intense-field stabilization, reliable predictions of level populations must be obtained and these values are sensitive to properties calculated with the wavefunction. The applications discussed above demonstrate that the loss of phase information can seriously compromise the predicted harmonic photon spectra since various orders of emitted photons will be sensitive to the details of the initial weak-field evolution. Likewise, uncontrolled dispersion error leads to erroneously shifted ejected electron spectra.

The choice of a pseudo-spectral grid representation with variable time-step control offers a powerful alternative to other numerical grid methods. In the above applications, the grid spacing was about 7 times larger for the pseudo-spectral method than for the differencing grid with superior accuracy. This reduction in the number of grid points resulted in a factor of 35 increase in computational speed when both methods were used for a fixed number of time steps. With the additional advantage of variable time-stepping, another twofold increase in efficiency was often achieved.

The great flexibility afforded by numerical grid methods accounts for their widespread use in the study of intense field-atom interactions. The applications above emphasize the need for an accurate and computationally tractable method of solution. These requirements can be met by the Residuum method. Extensions of these one-dimensional results are required to provide more detailed comparison with experiments; the relatively small grids required in this Coulomb-like potential indicate that accurate higher dimensional calculations will be possible.

ACKNOWLEDGEMENTS

Work at the Lawrence Livermore National Laboratory is performed under the auspices of the U.S. Department of Energy, administered by the University of California under Contract No. W-7405-Eng-48. Work at the Fritz Haber Institute is supported by the Minerva gesellschaft and the U.S.-Israel Binational Science Foundation.

REFERENCES

1. K. C. Kulander, K. J. Schafer, and J. L. Krause, in "Atoms in Intense Radiation Fields", Ed. M. Gavrila (Advances in Atomic, Molecular, and Optical Physics, Supplement 1, Academic Press, New York, 1992) (in press).

2. C. Leforestier, R. Bisseling, C. Cerjan, M. Feit, R. Friesner, A. Guldberg, A. Hammerich, G. Jolicard, W. Karrlein, H. D. Meyer, N. Lipkin, O. Roncero, and R. Kosloff, J. Comp. Phys. **94**, 59(1991).
3. Javanainen, J. Eberly, and Q. Su, Phys. Rev. A**38** , 3430(1988).
4. B. Sundaram and P. W. Milonni, Phys. Rev. A**41**, 6571 (1990).
5. H. Tal-Ezer, R. Kosloff, and C. Cerjan, J. Comp. Phys. (to appear).
6. C. A. J. Fletcher, "Computational Techniques for Fluid Dynamics", vol. I, Chp. 9 (Springer-Verlag,New York, 1987).
7. J. H. Eberly, Q. Su, and J. Javanainen, Phys. Rev. Lett. **62**, 881(1989).

ADIABATIC AND DIABATIC BASIS SETS IN MOLECULAR CALCULATIONS

Renzo Cimiraglia

Dipartimento di Chimica
Università di Ferrara
Via Borsari 46, I-44100 Ferrara, Italy

Introduction

Most of today's quantum chemical methods of dealing with molecular structure determination and molecular spectra are based on Born and Oppenheimer's approximation [1] (B.O.), so that it can be stated that the notion itself of molecular structure would lose much of its meaning if this approximation did not hold. Simple order–of–magnitude arguments [2] show that, so far as the energy hypersurfaces are well separated, the contributions of the non–adiabatic couplings due to the nuclear kinetic energy operator, are negligible. The energy hypersurface for the molecular electronic ground state in the vicinity of the equilibrium geometry is usually fairly well separated from the potential energy surfaces of the excited states but, far from the equilibrium geometry and for electronically excited states there may be many regions of nuclear configuration space where the potential energy surfaces of the same symmetry approach each other closely or even come into contact.

The study of a potential energy surface usually includes the determination of critical points, such as minima, which give the possible stable conformations of the molecule and saddle points, which provide the transition states between the various conformations; in cases of close approaches of potential surfaces, the regions of nuclear configurations which bring to strong interactions in the electronic states are of great importance in the study of nuclear dynamics and have to be included among the important characteristics of the molecule in the same manner as happens with the usual (zero-gradient) critical points [3].

In cases where the B.O. approximation is judged to perform poorly because of nearly avoided crossings between the potential surfaces, one could resort to the expansion of the molecular wavefunctions over the strongly interacting states:

$$\Psi(q,Q) = \sum_{i=1}^{n} \Phi_i(q;Q)\chi_i(Q) \tag{1}$$

In this formula q and Q stand for the collection of electronic and nuclear coordinates, respectively, and the $\Phi_i(q;Q)$'s designate the adiabatic wavefunctions, i.e. the solutions to the electronic Schrödinger equation. So, even in these cases, it would seem

Time-Dependent Quantum Molecular Dynamics, Edited by J. Broeckhove and L. Lathouwers, Plenum Press, New York, 1992

that there is no reason to abandon the determination of adiabatic states, for which so vast experience has been accumulated in the course of the years, in favour of other possible non–adiabatic electronic functions. The crux with this procedure is that the successive well known elaboration of eq. (1), leading to a coupled equations scheme, necessitates the calculation of the non–adiabatic coupling functions $\left\langle \varphi_i | \frac{\partial}{\partial Q} | \varphi_j \right\rangle$ and $\left\langle \varphi_i | \frac{\partial^2}{\partial Q^2} | \varphi_j \right\rangle$; even though nowadays a great amount of experience has been gained in the practical evaluation of these coupling functions, both analytically [4] and numerically [5], the computational burden required is no trivial task; furthermore, in the case of nearly avoided crossings, the regions in which the couplings are significantly different from zero concentrate around the avoided crossing, causing the coupling functions not to be amenable to easy interpolation. Therefore the introduction of new electronic functions, no longer eigenfunctions of the electronic hamiltonian, endowed with the property of exhibiting a negligibly small value of the coupling functions, would appear as a desirable possibility; to be sure, the price to be paid for such an introduction is that the electronic hamiltonian matrix would be no longer diagonal and therefore one would have instead of the dynamical coupling due to the nuclear kinetic energy another coupling due to the off–diagonal terms in the hamiltonian; of course, the exchange would be advantageous if the new coupling had a more regular behaviour than the original dynamical coupling, thus lending itself better to interpolation. Usually one refers to those states which annihilate or minimize the dynamical couplings as "diabatic states" [6]. One can agree to call "strict" diabatic states those that actually annihilate the dynamical coupling and "quasi–diabatic" (or "nearly diabatic") the others, that simply manage to render the dynamic coupling very small. The task of building strict diabatic states, as has been shown in a by now famous paper by Mead and Truhlar [7], can only be achieved for diatomic molecules but, even in these cases, one meets with difficulties in the definition of a diabatic basis [8, 9, 10]; furthermore, from a practical point of view, before effecting the transformation from adiabatic to diabatic states, one ought to calculate the dynamic coupling functions, a task which one would like, if not completely to avoid, at least to reduce as much as possible. As a consequence of their minimizing property, the diabatic states maintain a well defined "character" through the whole range of variability of the nuclear coordinates, in contrast with the adiabatic functions which change their electronic nature when going through an avoided crossing. This property of the diabatic states is important for the understanding and characterization of the electronic states in terms of simple well–defined structures.

A requirement that should be asked of the diabatic states is that the great accuracy provided by the modern techniques of computational quantum chemistry in the calculation of adiabatic wavefunctions, should not be lost but, on the contrary, transferred to the diabatic functions. Thus, most modern methods of calculation of diabatic states are based on a unitary transformation carried out on a manifold spanned by strongly interacting adiabatic wavefunctions.

In this contribution we shall be entirely concerned with those methods of evaluating diabatic states that do not presuppose the previous calculation of the dynamic coupling functions. Most of such "diabatization" techniques were developed over the last decade; some methods, which are based on the evaluation of suitable molecular properties to build the adiabatic-to-diabatic transformation, can be termed, following Malrieu and Spiegelmann [11], "intrinsic" methods. Other methods, which we shall agree to call "extrinsic", hinge on the idea that suitable approximate diabatic states are already known in some more or less intuitive manner and strive to transfer the accuracy of the adiabatic wavefunctions to such states.

Intrinsic diabatization techniques

We shall briefly touch on a few methods which, as we said in the introduction, define a transformation matrix on the basis of suitable molecular properties (different from the $\frac{\partial}{\partial Q}$ operator), judged to assume extremal values in the diabatic states.

Werner and Meyer [12] have employed the dipole moment matrix in their study of the avoided curve crossing in the two lowest $^1\Sigma^+$ states of LiF: the unitary transformation diagonalizing the dipole matrix is used to construct the diabatic ionic and neutral states which, in this way, possess maximum and minimum dipole moment respectively. Similar methods, based on an analysis of electronic properties in the region of avoided crossings had been proposed earlier by Macìas and Riera [13]. Recently Petrongolo *et al.* [14] have taken up this method in a study of a conical intersection in the NH_2 molecule.

A different approach is followed by Hendeković [15] whose proposal is based on the assumption that diabatic states should be dominated as far as possible by a single electronic configuration; thus it is required that the diabatic states, resulting from a unitary transformation of an adiabatic set, should be determined by maximization of the sum of squared occupation probabilities of natural spin–orbitals. It should anyway be remarked that the idea of identifying the diabatic state with one maintaining a single electronic configuration, attractive though it is, is not always justified, as has been clearly shown by Malrieu [11] on the basis of the diversified nature of the various possible avoided crossings [16].

Extrinsic diabatization techniques

Lévy's method

The method of Lévy [17] is based on the theory of effective Hamiltonians, in which all the information contained in the diagonalization of the electronic Hamiltonian in an adequate space of adiabatic wavefunctions is transferred, as best as possible, in a small subset of diabatic reference states whose knowledge is assumed to be given in terms of simple electronic functions maintaining a well defined character in the course of the process under investigation. The method requires that the effective Hamiltonian should have the same spectrum as the true Hamiltonian for the (few) states of interest with eigenfunctions given as the projections of the true adiabatic wavefunctions onto the reference states; so, if we indicate with $|\boldsymbol{\chi}\rangle$ the set of reference states ($|\boldsymbol{\chi}\rangle = \{|\chi_1\rangle, |\chi_2\rangle, \ldots |\chi_n\rangle\}$) and with $\boldsymbol{\psi}$ the set of the n adiabatic states of interest (significantly interacting) we shall have: $|\boldsymbol{\psi}'\rangle = P|\boldsymbol{\psi}\rangle$ with $P = |\boldsymbol{\chi}\rangle\langle\boldsymbol{\chi}|\boldsymbol{\chi}\rangle^{-1}\langle\boldsymbol{\chi}|$ the projector onto the manifold spanned by $|\boldsymbol{\chi}\rangle$. If $E_1, E_2, \ldots, E_n$ are the adiabatic energies pertaining to the $|\boldsymbol{\psi}\rangle$ set, one could define the desired effective Hamiltonian as $\hat{H}_{eff} = \sum_{k=1}^{n} |\psi'_k\rangle E_k \langle\psi''_k|$ with $|\boldsymbol{\psi}'\rangle$ and $|\boldsymbol{\psi}''\rangle$ constituting a biorthogonal set. Alternatively -and this is the way followed by Lévy- one can orthonormalize the $|\boldsymbol{\psi}'\rangle$ set with Löwdin's $\mathbf{S}^{-\frac{1}{2}}$ transformation, obtaining a new set $|\overline{\boldsymbol{\psi}}\rangle$ which, as is well known, among all possible orthonormal sets exhibits the least deviation from the original $|\boldsymbol{\psi}'\rangle$: $|\overline{\boldsymbol{\psi}}\rangle = |\boldsymbol{\psi}'\rangle\langle\boldsymbol{\psi}'|\boldsymbol{\psi}'\rangle^{-\frac{1}{2}}$. At this point the now hermitian effective Hamiltonian takes the form $\hat{H}_{eff} = \sum_{k=1}^{n} \left|\overline{\psi}_k\right\rangle E_k \left\langle\overline{\psi}_k\right|$. In terms of the χ one has $\left|\overline{\psi}\right\rangle = |\chi\rangle \mathbf{C}$ with $\mathbf{C} = \langle\chi|\chi\rangle^{-1} \langle\chi|\psi\rangle \langle\psi'|\psi'\rangle^{-\frac{1}{2}}$ and $\mathbf{C}$ diagonalises the matrix of $\hat{H}_{eff}$ in the diabatic basis $|\chi\rangle$;

$$\mathbf{H}_{eff}\mathbf{C} = \langle\chi|\chi\rangle \, \mathbf{CE}, \qquad (\mathbf{H}_{eff})_{ij} = \langle\chi_i|H_{eff}|\chi_j\rangle$$

The calculation of the $\left\langle \psi_k|\frac{\partial}{\partial Q}|\psi_l \right\rangle$ coupling function can be approximated with $\left\langle \overline{\psi}_k|\frac{\partial}{\partial Q}|\overline{\psi}_l \right\rangle$;

$$\left\langle \overline{\psi}|\frac{\partial}{\partial Q}|\overline{\psi} \right\rangle = \mathbf{C}^+ \langle\chi|\chi\rangle \frac{\partial}{\partial Q}\mathbf{C} + \mathbf{C}^+ \left\langle \chi|\frac{\partial}{\partial Q}|\chi \right\rangle \mathbf{C} \tag{2}$$

If the χ set is nearly diabatic the last term of eq. (2) can be neglected in comparison with the first which, in turn, can be easily calculated by interpolating the regularly varying $\mathbf{H}_{eff}$ matrix. The diagonal elements of $\mathbf{H}_{eff}$, $\langle\chi_i|H_{eff}|\chi_j\rangle$ play the role of diabatic energies and are not subjected to the non–crossing rule.

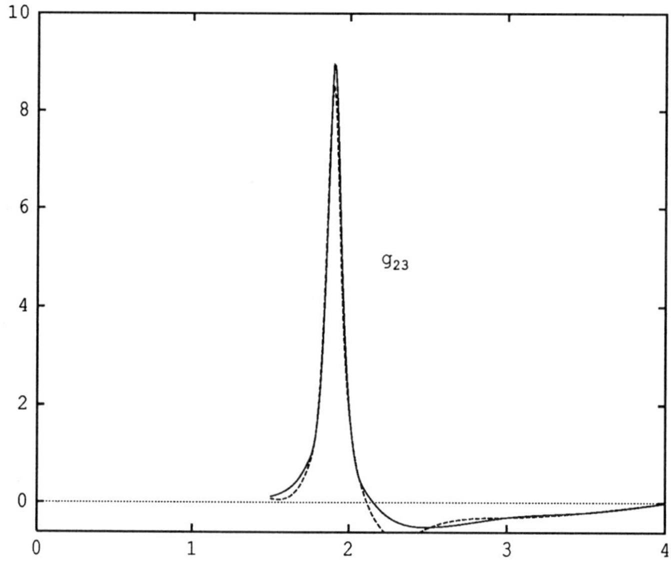

Figure 1. Dynamic coupling matrix g_{23} calculated by the finite differences technique (full line) and by Lévy's projection method (dashed line) in the $HeNe^{2+}$ molecule. All values are in atomic units.

It should be noted that in Levy's method there is no loss of information as to the eigenvalues of the $\hat{H}_{eff}$ hamiltonian (they are by construction the same as in the adiabatic space) but there is a loss as to the eigenvectors since one obtains only the projection of the adiabatic states onto the χ's; as a matter of fact the $\mathbf{C}$ matrix does not bear any relationship with the eigenvalues E_k, depending only on the $|\psi\rangle$

and $|\chi\rangle$ states given at the outset. A good measure for the adequacy of the reference basis $|\chi\rangle$ is provided by the norms of the projected wavefunctions ψ'_k which should be not too far from unity.

As an example of application we show in figure 1 the calculation of the non–adiabatic coupling function between the second and third $^1\Sigma^+$ states of the $HeNe^{+2}$ system [18], which exhibits a strictly avoided crossing; as can be seen, the difference between the values of $g_{23}(R)$ calculated with the finite difference technique and those obtained from eq. (2) with neglect of the χ's derivative term is negligible.

The method of Spiegelmann and Malrieu

Spiegelmann and Malrieu [19] proposed a method for obtaining nearly diabatic states which can be classified as of "*a priori*" type, in that it does not require a previously performed accurate adiabatic calculation. The method seems to be ideally suited to the CIPSI variation–perturbation technique [20] even though it can be adapted to other CI techniques also.

The CIPSI formalism consists in the iterative determination of a determinantal space upon which a diagonalization of the electronic Hamiltonian is carried out, followed by second order perturbation treatment on the energies of the states of interest. The iterative nature of the method consists in successive enlargements of the determinantal space S according to the overcoming in the perturbation step of certain threshold values, decreasing after each iteration. In summary, the gist of the CIPSI method consists in forming a good zero order determinantal space S which, upon diagonalization of the Hamiltonian, should provide the most important characteristics of the wavefunctions of the states under investigation; the contribution of the orthogonal space $S^{\perp}$, containing all the determinants outside S, is only taken into account through second order perturbation theory. It is the perturbation step which becomes particularly heavy from the computational point of view when the dimension of S becomes large; the present dimensions allowed by the CIPSI method in the current implementation are within a few tens of thousands of determinants for the space S, with the perturbation step carried out with efficient diagrammatic techniques [21, 22].

The idea of Spiegelmann and Malrieu consists in partitioning the determinantal space S in complementary subspaces S_A and S_B (or more if more than two diabatic states are required) in such a manner that S_A contains a given original configuration $|\varphi_A\rangle$ and all those determinants of S interacting with $|\varphi_A\rangle$ more strongly than with $|\varphi_B\rangle$, where $|\varphi_B\rangle$ is a primitive configuration for the diabatic state B. So, $S_A = \{|\varphi_A\rangle, |K\rangle, \ldots\}$ and $S_B = \{|\varphi_B\rangle, |L\rangle, \ldots\}$ with the $|K\rangle$'s satisfying $|\langle K|H|\varphi_A\rangle| > |\langle K|H|\varphi_B\rangle|$ and analogously for the $|L\rangle$'s. The choice of the primitive configurations $|\varphi_A\rangle$ and $|\varphi_B\rangle$ is done on the ground that they should not change their character with the geometry of the molecule. This desired property is ensured by using valence–bond type orbitals, judiciously orthogonalized, which undergo only small (and smooth) variations with geometrical changes of the molecular system.

Once the S_A and S_B subspaces are constructed, one can build zeroth order wavefunctions $\Psi_A^{(0)}$ and $\Psi_B^{(0)}$ through diagonalization of the electronic Hamiltonian; if we denote by P_{S_A} the projector onto the S_A subspace, we have:

$$P_{S_A} = |\varphi_A\rangle\langle\varphi_A| + \sum_{K\in S_A} |K\rangle\langle K|, \qquad P_{S_A} H_{el} P_{S_A} \Psi_A^{(0)} = E_A^{(0)} \Psi_A^{(0)} \tag{3}$$

and a similar equation for $\Psi_B^{(0)}$. Among the solutions of the eigenvalue problem (3),

$\Psi_A^{(0)}$ is obviously chosen so that its overlap with $|\varphi_A\rangle$ should be maximum:

$$\Psi_A^{(0)} = c_A\,|\varphi_A\rangle + \sum_{K\in S_A} c_K\,|K\rangle\,, \qquad c_A \approx 1$$

The two wavefunctions $\Psi_A^{(0)}$ and $\Psi_B^{(0)}$ are thus the desired zeroth order diabatic states, since they do not change their character during the course of the process, the c_A and c_B coefficients being all the time close to unity.

In order to obtain the adiabatic energies and wavefunctions one could obviously diagonalize H_{el} in the total space $S = S_A \oplus S_B$ and proceed to the perturbation treatment; it is possible, however, that this procedure should produce some artifacts [19] due to the non variational nature of the orbitals adopted. Spiegelmann and Malrieu advise therefore to build an effective Hamiltonian on the 2×2 diabatic subspace spanned by $\Psi_A^{(0)}$ and $\Psi_B^{(0)}$ by adopting Quasi Degenerate Perturbation Theory (QDPT) [23] to second order:

$$H_{AA}^{eff} = \left\langle \Psi_A^{(0)}|H_{el}|\Psi_A^{(0)}\right\rangle + \sum_{\substack{\beta\in S_B\\ \beta\neq B}} \frac{\left|\left\langle \Psi_A^{(0)}|H_{el}|\Psi_\beta^{(0)}\right\rangle\right|^2}{E_A^{(0)} - E_\beta^{(0)}} + \sum_{K\notin S} \frac{\left|\left\langle \Psi_A^{(0)}|H_{el}|K\right\rangle\right|^2}{E_A^{(0)} - E_K^{(0)}}$$

$$H_{AB}^{eff} = \left\langle \Psi_A^{(0)}|H_{el}|\Psi_B^{(0)}\right\rangle + \sum_{K\notin S} \frac{\left\langle \Psi_A^{(0)}|H_{el}|K\right\rangle \left\langle K|H_{el}|\Psi_B^{(0)}\right\rangle}{E_A^{(0)} - E_K^{(0)}}$$

and a similar definition for H_{BB}. The diagonal matrix elements H_{AA}^{eff} and H_{BB}^{eff} contain second order perturbation contributions to the zero-th order diabatic potentials and can therefore be considered as well correlated quasi–diabatic energies. Diagonalization of the 2×2 effective hamiltonian matrix produces *adiabatic* energies E_1 and E_2 which are coincident (apart from third and higher order contributions) with the energies one would have obtained by construction of an effective Hamiltonian over the basis of the eigenfunctions of $P_S H_{el} P_S$.

The numerical applications of the method (Ar_2*, NaCl in ref[19]) are very encouraging and qualify this technique as a good compromise between intuitively chosen diabatic functions and highly correlated electronic calculations; a shortcoming of the method is mainly of practical nature since, due to the valence–bond type of orbitals adopted, the dimensions of the subspaces S_A and S_B tend to increase overmuch, beyond the possibilities of our existing programs.

The method of maximum overlap

This method, originally developed by Persico [24] and further taken up in collaboration with the group of Toulouse [25], is inspired on Levy's method, already described, in that it assumes that a reference basis of approximately diabatic states should be known. Such approximate diabatic states $|R_K\rangle$ should therefore be slowly varying functions of the geometrical parameters, maintaining a well defined character throughout the process under investigation; in the case of a dissociation reaction one commonly chooses for $|R_K\rangle$ the antisymmetrized products of the wavefunctions of the fragments. We shall describe the maximum overlap method in connection with the above mentioned variation–perturbation CIPSI algorithm.

A zeroth order determinantal space is constructed and the electronic Hamiltonian matrix is diagonalized yielding the n zero order adiabatic wavefunctions of interest:

$$P_S H_{el} P_S \Psi_K^{(0)} = E_K^{(0)} \Psi_K^{(0)}, \qquad \text{with} \quad P_S = \sum_{K \in S} |K\rangle \langle K|$$

in the same notation as in Spiegelmann and Malrieu's method.

The adiabatic set $\left|\boldsymbol{\psi}^{(0)}\right\rangle$ is now subjected to a unitary transformation and a new set of states is obtained:

$$\left|\boldsymbol{\eta}^{(0)}\right\rangle = \left|\boldsymbol{\psi}^{(0)}\right\rangle \mathbf{T}^{(0)}, \qquad \mathbf{T}^{(0)+}\mathbf{T}^{(0)} = \mathbf{1}$$

It is required that the η functions should bear as much "resemblance" as possible to the n given reference states $|R_1\rangle, \ldots |R_n\rangle$. This task can be achieved by imposing the condition of maximum squared overlap between the η's and the R's:

$$\sum_{k=1}^{n} \left|\left\langle \eta_K^{(0)} | R_K \right\rangle\right|^2 = \max$$

To determine the $\mathbf{T}^{(0)}$ matrix one can adopt either an iterative procedure *à la* Jacobi [24] or, as shown by Petsalakis *et al.* [26], one can diagonalize the matrix with elements $\left\langle \eta_K^{(0)} | \Psi_i^{(0)} \right\rangle \left\langle \eta_K^{(0)} | \Psi_j^{(0)} \right\rangle$, with subsequent orthonormalization procedure, to get the k–th column of $\mathbf{T}^{(0)}$. One could alternatively choose to maximize the sum of the overlap contributions $\sum_{K=1}^{n} \left\langle \eta_i^{(0)} | R_i \right\rangle$; this procedure would then be equivalent to project the reference functions $|R_K\rangle$ onto the adiabatic subspace with subsequent symmetrical orthonormalization:

$$|\boldsymbol{\eta}'\rangle = |\boldsymbol{\psi}\rangle \langle \boldsymbol{\psi} | \mathbf{R}\rangle, \qquad |\boldsymbol{\eta}\rangle = |\boldsymbol{\eta}'\rangle \mathbf{S}^{-\frac{1}{2}} \qquad \text{with} \quad \mathbf{S} = \langle \boldsymbol{\eta}' | \boldsymbol{\eta}'\rangle$$

It can be remarked that this last procedure is in a sense complementary to Levy's method [17], where it is the adiabatic functions that are projected onto the reference set.

The two overlap maximization procedures just described, along with many more (actually infinite) that are conceivable, are not entirely equivalent, thus reflecting a certain degree of uncertainty in the concept of quasi–diabatic states defined according to their maximum similarity with the reference functions.

The zero order hamiltonian matrix in the diabatic set $\left|\boldsymbol{\eta}^{(0)}\right\rangle$ is given by

$$\mathbf{H}^{(0)} = \mathbf{T}^{+}\mathbf{E}^{(0)}\mathbf{T}$$

where $\mathbf{E}^{(0)}$ is the diagonal matrix with the adiabatic energies $E_K^{(0)}$ as elements. It is now possible to invoke QDPT (see above) to build a second–order corrected diabatic Hamiltonian exactly in the same way as indicated in Spiegelmann and Malrieu's method:

$$\mathbf{H}_{eff} = \mathbf{H}^{(0)} + \mathbf{H}^{(2)}, \quad \text{with} \quad H_{KL}^{(2)} = \sum_{I \notin S} \frac{\left\langle \eta_K^{(0)} | H_{el} | I \right\rangle \left\langle I | H_{el} | \eta_L^{(0)} \right\rangle}{H_{II}^{(0)} - E_I}$$

where E_I is the energy of determinant $|I\rangle$ in the adopted partition of the electronic Hamiltonian (usually Møller–Plesset or Epstein–Nesbet [20]).

Consistently with the above second–order corrected Hamiltonian, one will have first–order corrected diabatic wavefunctions:

$$|\eta_K\rangle = \left|\eta_K^{(0)}\right\rangle + \left|\eta_K^{(1)}\right\rangle, \qquad \left|\eta_K^{(1)}\right\rangle = \sum_{I \notin S} \frac{|I\rangle \left\langle I|H_{el}|\eta_K^{(0)}\right\rangle}{H_{II}^{(0)} - E_I}$$

The final adiabatic energies will be obtained by diagonalization of the effective Hamiltonian matrix

$$\mathbf{H}_{eff}\mathbf{T}^{+} = \mathbf{T}^{+}\mathbf{E}, \qquad |\psi\rangle = |\eta\rangle\,\mathbf{T}^{+}$$

where the $\mathbf{T}$ matrix can be regarded as a refinement brought to $\mathbf{T}^{(0)}$ through QDPT.

The evaluation of the non adiabatic coupling function g_{KL} can now be accomplished as follows:

$$\begin{aligned} g_{KL} = \left\langle \Psi_K|\frac{\partial}{\partial Q}|\Psi_L\right\rangle =& \mathbf{t}_K^{+}\frac{\partial}{\partial Q}\mathbf{t}_L + \mathbf{t}_K^{+}\left\langle \boldsymbol{\eta}^{(0)}|\frac{\partial}{\partial Q}|\boldsymbol{\eta}^{(0)}\right\rangle \mathbf{t}_L \\ &+ \mathbf{t}_K^{+}\left\{\left\langle \boldsymbol{\eta}^{(0)}|\frac{\partial}{\partial Q}|\boldsymbol{\eta}^{(1)}\right\rangle + \left\langle \boldsymbol{\eta}^{(1)}|\frac{\partial}{\partial Q}|\boldsymbol{\eta}^{(0)}\right\rangle\right\}\mathbf{t}_L \\ &+ \text{second–order terms} \end{aligned} \tag{4}$$

with $\mathbf{t}_K$ designating the k-th column of $\mathbf{T}^{+}$. The first term of the r.h.s. of eq.(4) is expected to play the dominant role in the case of avoided crossings of adiabatic potentials; such term can be best calculated by a Hellmann–Feynman type formula:

$$\mathbf{t}_K^{+}\frac{\partial}{\partial Q}\mathbf{t}_L = \mathbf{t}_K^{+}\frac{\partial \mathbf{H}_{eff}}{\partial Q}\mathbf{t}_L(E_L - E_K)^{-1} \tag{5}$$

Since the $\mathbf{H}_{eff}$ matrix is a smoothly varying function of Q by construction, it can be easily interpolated and thus the calculation of the l.h.s. of eq.(5) can be readily done. The approximate calculation of g_{KL} making use of eq. (5) will be particularly appreciated when more than one internal coordinate is considered and the direct evaluation of the dynamic coupling functions quickly becomes intractable. As to the evaluation of the matrix elements $\left\langle \eta_K^{(0)}|\frac{\partial}{\partial Q}|\eta_L^{(0)}\right\rangle$, which are expected to be very small if the η's are quasidiabatic states, one can resort to the well known finite–difference scheme:

$$\left\langle \eta_K^{(0)}|\frac{\partial}{\partial Q}|\eta_L^{(0)}\right\rangle \cong \frac{1}{\Delta}\left(\mathbf{C}_K^{+}(Q)\mathbf{C}_L(Q+\Delta) + \sum_{\alpha\beta}(\boldsymbol{\rho}_{KL}^{(0)})_{\alpha\beta}\,\langle\varphi_\alpha(Q)|\varphi_\beta(Q+\Delta)\rangle\right)$$

where $\boldsymbol{\rho}_{KL}^{(0)}$ is the one–particle transition density matrix relative to the zero–order diabatic wavefunctions $\eta_K^{(0)}$ and $\eta_L^{(0)}$, the φ's are the molecular orbitals used and the $\mathbf{C}$ coefficients are the expansion coefficients of the η's in terms of the determinants $|K\rangle$ belonging to S. Finally the third term in the r.h.s. of eq.(4) represents a still more negligible contribution which, if need be, can be calculated by use again of the finite difference technique:

$$\left\langle \eta_K^{(0)}|\frac{\partial}{\partial Q}|\eta_L^{(1)}\right\rangle + \left\langle \eta_K^{(1)}|\frac{\partial}{\partial Q}|\eta_L^{(0)}\right\rangle \cong \frac{1}{\Delta}\sum_{\alpha\beta}(\boldsymbol{\rho}_{KL}^{(1)})_{\alpha\beta}\,\langle\varphi_\alpha(Q)|\varphi_\beta(Q+\Delta)\rangle$$

where $\boldsymbol{\rho}^{(1)}_{KL}$ is the first order perturbation correction to the one–particle density matrix whose calculation is not difficult and represents a minor task in comparison with the perturbation CI; it can be noticed that there is no contribution due to the variation of the CI coefficients in this formula because the zero order and the first order CI subspaces are orthogonal. Obviously one could have applied the finite difference scheme to the adiabatic wavefunctions directly but there are noticeable advantages in applying the above described diabatization technique. Firstly, the $|\boldsymbol{\eta}\rangle$ functions are expected to vary much more smoothly than the $|\boldsymbol{\psi}\rangle$, thus eliminating most of the difficulties encountered in the direct application of the finite difference technique to the adiabatic functions, especially in the case of nearly avoided crossings. Secondly, the most relevant contribution to g_{KL} is brought by the $\mathbf{t}^+_K \frac{\partial}{\partial Q} \mathbf{t}_L$ term whose evaluation is very easy; a reasonable and often tenable approximation is simply to retain only this quantity in the evaluation of g_{KL}.

A difficulty of the maximum overlap method is usually met when the description of the n adiabatic states of interest requires more than n reference functions. A typical occurrence of such circumstance presents itself when ionic states come into play; owing to the long range nature of the Coulomb potential, a large energy interval is scanned when two ionic fragments pass from bond length distances to infinity, thus providing the possibility for many crossings to occur with underlying excited states. An obvious way to get around this difficulty would be to increase the number n of "interesting" adiabatic states, until the $|\boldsymbol{\psi}\rangle$ and the $|\eta\rangle$ sets have the same dimension. But, besides increasing the dimension and the difficulty of the electronic problem, this procedure is not automatically destined to be succesful because a growth in the $|\boldsymbol{\psi}\rangle$ set can bring about a growth in the $|\mathbf{R}\rangle$ set and so on with increasing degree of intricacy. In order to circumvent this "intruder state" problem, one can decrease the number of reference states by introducing suitable linear combinations of them with coefficients depending on the nuclear coordinates; so, for instance, we can define $R_n = \lambda R_a + \mu R_b$ where R_a and R_b are two reference functions necessary to describe the n-th adiabatic state in distinct regions of the space of nuclear coordinates. The expansion coefficients, tied to one another by the normalization condition for R_n, are determined by the same maximum squared overlap criterium already invoked to define the quasi–diabatic functions η's: $\sum_i \left| \left\langle \eta_i^{(0)} | R_i \right\rangle \right|^2 = \max$. The price one has to pay if this point of view is adopted, is that the function $\left| \eta_n^{(0)} \right\rangle$ has a lesser degree of diabaticity than the other η's because it does not have a well defined overall character; in spite of this, the matrix elements $\left\langle \eta_i^{(0)} | \frac{\partial}{\partial Q} | \eta_n^{(0)} \right\rangle$ are expected to be everywhere small since both $\lambda \left\langle R_i | \frac{\partial}{\partial Q} | R_a \right\rangle$ and $\frac{\partial \lambda}{\partial Q} \langle R_i | R_a \rangle$ are small.

As an example of application of the maximum overlap method, we shall show calculations relating to the BH molecule. The first four singlet states of BH at infinite internuclear distance correlate with the product of the H(^{2}S,1s) state and the following B states: 1) B(^{2}P,2p), the ground state 2) B(^{2}S,3s), Rydberg excitation 2p→3s 3) B(^{2}D,2s2p^2), valence excitation 2s→2p 4) B(^{2}P,3p) Rydberg excitation 2s→3p. Furthermore, the ionic state B^+H^- crosses the first three excited states together with 7 more Rydberg states lying below the B^+H^- dissociation limit. We have confined our calculations to 4 reference states, consisting of the dissociation products just described but with the last one substituted by $|R_4\rangle = \lambda[B(^2P,3p)H(^2S)] + \mu[B^+H^-]$. The atomic basis set adopted in this calculation is sufficiently rich so as to adequately describe the various Rydberg states under consideration. A total of 10s, 7p and 3d contracted gaussian orbitals were used for the B atom and a set of 6s, 3p was adopted for H. SCF–CI calculations were carried out at various internuclear distances from 1.2 a.u. to ∞. Electron correlation was dealt with by the CIPSI variation–perturbation method already described, making use of the QDPT to refine the diabatic effective Hamiltonian. We shall report a detailed description of the calculation elsewhere [27], limiting ourselves here to a brief illustration of some results.

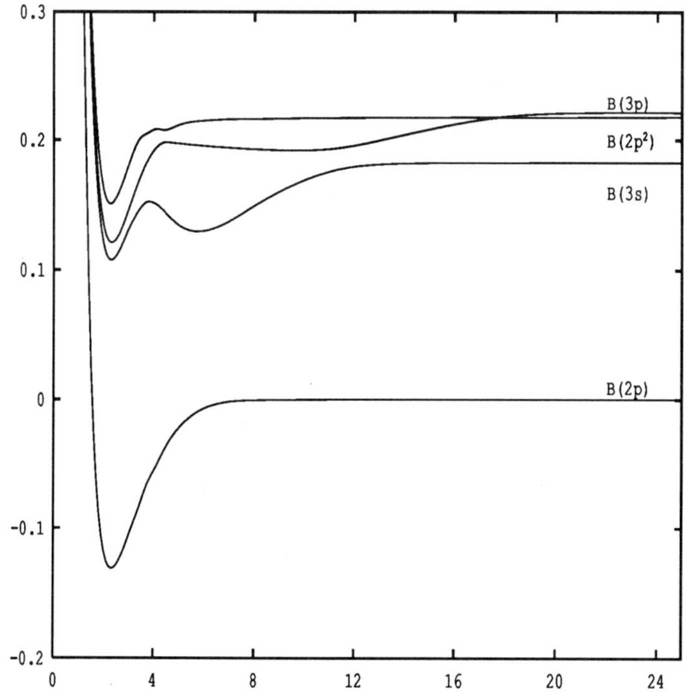

Figure 2. Adiabatic potentials for the first four $^1\Sigma^+$ states of BH. Values in atomic units.

Figure 2 shows the adiabatic potential energy curves of the first four Σ^+ singlets of BH. The adiabatic ground state has substantially [B(^{2}P,2p) H(^{2}S,1s)] character at all distances with some mixing with B(^{2}D,2s,2p^2) at short distances (we avoid specifying the H(^{2}S,1s) term in the fragment product). The first excited state is characterized by a double minimum; it dissociates into B(^{2}S,3s) and H(^{2}S) but it acquires a strong ionic character in the outer potential well; around the maximum in the curve, such ionic character disappears so that the state regains its B(^{2}S,3s) character in the first minimum region. The second excited Σ^+ singlet, dissociating into B(^{2}D,2s2p^2) + H(^{2}S), undergoes a strictly avoided crossing near 17.7 a.u., acquires a mixed character of B^+H^- and B(^{2}P,3p) and is again substantially B(^{2}D,2s2p^2) at shorter distances; this state too exhibits a double well potential with a very shallow outer minimum. Finally the third excited state, dissociating as B(^{2}P,3p) + H(^{2}S), undergoes two avoided crossings, a broad one around 19 a.u. with the ionic state and a narrow one at R=17.7 a.u. (see above) with the B(^{2}D,2s2p^2) state; the

$2p^2$ character persists down to R≅4.5 a.u. when the character changes again to B(^{2}P,3p).

All the avoided crossings just described become true crossings in the quasidiabatic potential curves set out in figure 3.

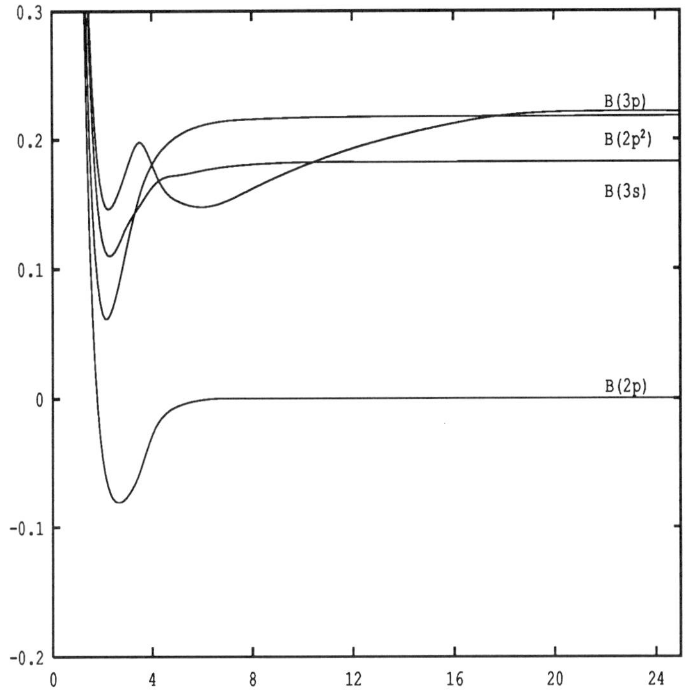

Figure 3. Diabatic potentials for the first four $^1\Sigma^+$ states of BH. Values in atomic units.

The double–minimum curve pertains to the reference function $|R_4\rangle$, the combination of the ionic state B^+H^- and the B(^{2}P,3p) state; the first minimum has 3p character, as well as the dissociation limit, whereas the wide second minimum pertains neatly to the ionic state. Such a minimum is located around R=6 a.u., in the region where the other curves are nearly flat; the reason for this large value of R_{min} is due to the large ionic radius of H^- (≈4.9 a.u.). Therefore the ionic curve is able to cross the three neutral excited states twice, both at long and at short range. Two coupling functions g_{KL} for the adiabatic states and the off–diagonal hamiltonian matrix elements H_{KL} for the diabatic ones are reported in figures 4 and 5.

A close examination of these interactions permits one to justify the differences between adiabatic and diabatic curves; in particular, the rather large H_{13} interaction between B(^{2}P,2p) and B(^{2}D,2s2p^2) causes the ground state minimum to be deeper and tighter than the corresponding diabatic one; at the same time the B(^{2}D,2s2p^2) state is displaced towards higher energies and becomes the second adiabatic excited state instead of the first. It should also be noticed that the approximate calculation of the g_{KL} coupling functions, i.e. with inclusion of only the first term in the r.h.s. of eq.(5), appears to be quite satisfactory in comparison with the exact calculation.

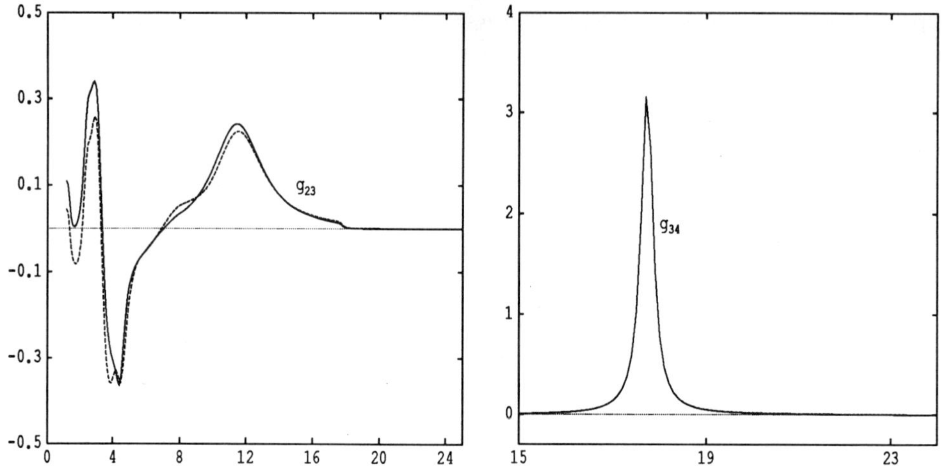

Figure 4. Non-adiabatic coupling functions g_{23} and g_{34}. Full line: finite difference calculation, dashed line: approximate calculation. Values in atomic units.

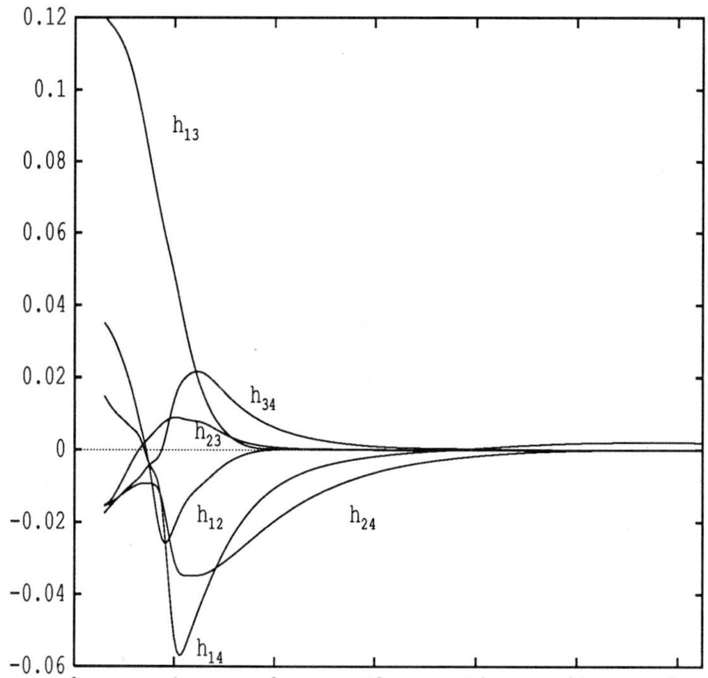

Figure 5. Off-diagonal matrix elements of the Hamiltoniam in the diabatic basis. Values in atomic units.

We would like to conclude this description of the maximum overlap method with some considerations on the choice of the reference functions $|R_i\rangle$; the identification of the adiabatic states with linear combinations of valence–bond–like (VB) functions is not exempt from a certain degree of arbitrariness which manifests itself at least in two distinct stages: a) as is the case for pure VB treatments, the reference functions, which are not an orthogonal set, cannot be attributed a perfectly defined meaning; so, in the BH case, the 3s Rydberg orbital of boron overlaps with the diffuse 1s of the H^- anion and consequently the overlap between R_2 (i.e. $B(^2S,3s)$) and R_4 (i.e. the combination of $B(^2P,3p)$ and B^+H^-) has a maximum of 0.4 at about R=6 a.u., in proximity of the largest weight of the ionic component in R_4; at short distances the $\langle R_1|R_3\rangle$ overlap reaches a value of 0.5 at R=2.3 a.u. on account of the superposition of the B(2p) and H(1s) orbitals. b) A second possible cause of arbitrariness is a lack in ability of the $|\mathbf{R}\rangle$ set to span the adiabatic subspace generated by $|\psi\rangle$; it is anyway easy to build a measure for the adequacy of the reference set: we have chosen the function

$$S = \left(\sum_{i=1}^{n} \langle R_i|\eta_i\rangle^2 / n \right)^{\frac{1}{2}}$$

the values of which, in the BH case, range from 0.96 at large distances to 0.86 at short R.

Recent developments in the construction of nearly diabatic states

Gadéa and Pelissier [28] have proposed a new transformation to propagate a given diabatic wavefunction from one nuclear configuration to another, within the frame of effective Hamiltonians. The gist of their method consists in the calculation of a modified overlap between two wavefunctions at different nuclear geometries, in such a way that, following an observation of Lorquet [29], the contribution from the translation of the basis functions with the moving nuclei should be minimized; thus a transformation can be defined, in terms of the modified overlap matrix, within the frame of De Cloizeaux effective Hamiltonian theory [30], and new diabatic references can be built at the new geometry.

The method of Gadéa and Pelissier has been taken up recently by Petsalakis *et al.* [26] who combine it with the maximum overlap method; given a set of diabatic functions at a given geometry, a new set of diabatic functions is calculated by satisfying the criterion of maximum squared overlap between the two sets. Since the two diabatic functions are expressed as linear combinations of adiabatic wavefunctions at the respective geometries, the method of Petsalakis *et al.* is based on the efficient evaluation of the overlap matrix between the two non–orthogonal adiabatic sets: in the use of CI wavefunctions, such calculations can be made without great difficulty, only requiring the evaluation of the determinant of the overlap matrix over the molecular orbitals at the two different geometries.

Pacher *et al.* [31] have proposed the construction of quasi–diabatic states from the block diagonalization of the electronic Hamiltonian matrix. These authors consider a set of strongly interacting states, well separated in energy from the other states, and subject the Hamiltonian matrix to a similarity transformation with a unitary matrix so as to achieve block–diagonalization of the hamiltonian. The uniqueness of the transformation is assured by imposing the condition that it should be as close as possible to the unit matrix in the Euclidean matricial norm. In this way the eigenvectors of the block–diagonalized hamiltonian and those of the original hamiltonian matrix are expected to be as similar as possible. Since the block–diagonalizing transformation changes the initial basis as little as possible, the authors adopt a crude adiabatic basis as initial reference set in which the Hamiltonian matrix is constructed, in order to ensure the quasi–diabaticity of the resulting states. The authors have also shown that the functions so constructed conform to a gauge theoretical definition of diabatic states, recently put forward by the same authors and Mead [32], particularly suitable, in principle, for the treatment of non adiabatic phenomena in polyatomic molecules.

Concluding remarks

In this contribution we have briefly considered a few diabatization techniques, developed over the last years, with a view to providing some insight into the various possibilities of obtaining accurate electronic functions endowed with the desired property of maintaining a well defined character throughout a given physico–chemical process. In this sense, the point of view that we have adopted is similar to the one originally formulated in the thirties by Zener [33] in the study of avoided crossing problems. The motivation for this attitude lies mainly in the awareness that intuitive "a priori" diabatic functions, given for instance in terms of simple valence-bond-like structures, can be rather easily guessed and provide a good starting point for the construction of a more refined diabatic set.

A good diabatic set should be of the same quality as the corresponding adiabatic one, without substantial loss of information. The merits of a diabatic set should essentially consist a) in providing an easy interpretation of the process under study, disentangling it from the intricacies which most of the times show up, even in simple diatomic molecules, in the occurrence of avoided crossings, b) in affording a simplified (but reliable) procedure for the evaluation of the dynamic coupling functions. These desiderata may appear rather reductive in scope, if compared with the more ambitious aim of providing a new means to describe the molecular dynamics, using a basis set where the differential couplings ($\left\langle \Phi_K | \frac{\partial}{\partial Q} | \Phi_L \right\rangle \frac{\partial}{\partial Q}$ etc.) of the adiabatic representation could be replaced by the more comfortable multiplicative couplings ($\langle \eta_K | H | \eta_L \rangle$) of the diabatic basis. As a matter of fact in many problems of molecular spectroscopy and photochemistry the initial and final molecular states are well described in terms of well defined adiabatic wavefunctions and therefore it is probable that the adiabatic basis, in which the potentials are most of the times more realistic than the diabatic ones, should be more reliable than the alternate diabatic set, even though, in the limit of completeness, both approaches should be obviously equivalent. The importance of the diabatic representation remains anyway considerable both theoretically, for its interpretative value, and practically, for its ability in providing the dynamic coupling functions in a rather inexpensive way.

Acknowledgements

I am grateful to Maurizio Persico for interesting discussions over the subject of this contribution. Financial support from the Italian M.U.R.S.T. is gratefully acknowledged.

References

[1] M. Born and R. Oppenheimer *Ann. Phys.*, vol. 84, p. 457, 1927.

[2] A. Messiah, *Quantum Mechanics.* Amsterdam: North-Holland, 1958.

[3] J. Michl *Topics in Current Chemistry*, vol. 46, p. 1, 1974.

[4] P. Saxe, B. H. L. III, and D. R. Yarkony *Chem. Phys. Letters*, vol. 113, p. 159, 1985.

[5] G. Hirsch, P. J. Bruna, R. J. Buenker, and S. D. Peyerimhoff *Chem. Phys.*, vol. 45, p. 335, 1980.

[6] W. Lichten *Phys. Rev.*, vol. 131, p. 229, 1963.

[7] C. A. Mead and D. G. Truhlar *J. Chem. Phys.*, vol. 77, p. 6090, 1982.

[8] H. Gabriel and K. Taulbjerg *Phys. Rev. A*, vol. 10, p. 741, 1974.

[9] J. B. Delos and W. R. Thorson *J. Chem. Phys.*, vol. 70, p. 1774, 1979.

[10] R. Cimiraglia and M. Persico *Mol. Phys.*, vol. 38, p. 1707, 1979.

[11] J. P. Malrieu and F. Spiegelmann in *Photophysics and Photochemistry above 6 eV* (F. Lahmani, ed.), (Amsterdam), Elsevier, 1985.

[12] H.-J. Werner and W. Meyer *J. Chem. Phys.*, vol. 74, p. 5802, 1981.

[13] A. Macìas and A. Riera *J. Phys. B: At. Mol. Phys.*, vol. 11, p. L489, 1978.

[14] C. Petrongolo *Mol. Phys.*, vol. 70, p. 825, 1990.

[15] J. Hendeković *Chem. Phys. Letters*, vol. 90, p. 193, 1982.

[16] L. Salem, C. Leforestier, G. Segal, and R. Wetmore *J. Am. Chem. Soc.*, vol. 97, p. 479, 1975.

[17] B. Lévy in *Spectral Line Shapes* (B.Wende, ed.), (Berlin-New York), Walter de Gruyter & Co., 1981.

[18] M. C. Bacchus-Montabonel, R. Cimiraglia, and M. Persico *J. Phys. B: At. Mol. Phys.*, vol. 17, p. 1931, 1984.

[19] F. Spiegelmann and J. P. Malrieu *J. Phys. B: At. Mol. Phys.*, vol. 17, p. 1259, 1984.

[20] B. Huron, P. Rancurel, and J. P. Malrieu *J. Chem. Phys.*, vol. 58, p. 5745, 1973.

[21] R. Cimiraglia *J. Chem. Phys.*, vol. 83, p. 1746, 1985.

[22] R. Cimiraglia and M. Persico *J. Comput. Chem.*, vol. 8, p. 39, 1987.

[23] F. Spiegelmann and J. P. Malrieu *J. Phys. B: At. Mol. Phys.*, vol. 17, p. 1235, 1984.

[24] M. Persico in *Spectral Line Shapes* (F. Rostas, ed.), vol. III, (Berlin-New York), Walter de Gruyter & Co., 1985.

[25] R. Cimiraglia, J.-P. Malrieu, M. Persico, and F. Spiegelmann *J. Phys. B: At. Mol. Phys.*, vol. 18, p. 3073, 1985.

[26] I. D. Petsalakis, G. Theodorakopoulos, C. A. Nicolaides, and R. J. Buenker *Chem. Phys. Letters*, vol. 185, p. 359, 1991.

[27] R. Cimiraglia, M. Persico, and F. Spiegelmann *to be published.*

[28] F. X. Gadéa and M. Pélissier *J. Chem. Phys.*, vol. 93, p. 545, 1990.

[29] M. Desouter-Lecomte, D. Dehareng, and J. C. Lorquet *J. Chem. Phys.*, vol. 86, p. 1429, 1987.

[30] J. D. Cloizeaux *Nucl. Phys.*, vol. 20, p. 321, 1960.

[31] T. Pacher, L. S. Cederbaum, and H. Köppel *J. Chem. Phys.*, vol. 89, p. 7367, 1988.

[32] T. Pacher, C. A. Mead, L. S. Cederbaum, and H. Köppel *J. Chem. Phys.*, vol. 91, p. 7057, 1989.

[33] C. Zener *Proc. R. Soc. (London) Ser. A*, vol. 137, p. 696, 1932.

CHEMICAL DYNAMICS: A PERIODIC ORBITS APPROACH

Stavros C. Farantos

Department of Chemistry, University of Crete, and
Institute of Electronic Structure and Laser, FORTH,
Iraklion, Crete 711 10, Greece

ABSTRACT

This paper surveys recent work on the correlation of periodic orbits and the dynamical behaviour of triatomic molecules. A systematic method for obtaining the principal families of periodic orbits and bifurcations of them are presented. It is shown, that periodic orbits of rotating type emanate above saddle points of the potential energy surface, and play a significant role in the dynamics of highly excited polyatomic molecules.

1. INTRODUCTION

Qualitative and quantitative studies of the dynamics and spectroscopy of highly excited molecules have been the endeavor of the experimentalists and theorists for many years [1]. To overcome the difficulties of solving the quantum equations even for triatomic molecules, the technique of classical trajectories has successfully been applied, at least for a qualitative description of the dynamics [2].

Recent advances of nonlinear mechanics invigorate the role of classical mechanics in chemical dynamics. Research on the chaotic behaviour of quantum systems revealed the importance of periodic orbits for understanding the localization of wavefunctions in configuration space. There are two major outcomes of this work. First, the semiclassical formula for calculating the density of states as a sum over periodic solutions of the classical equations of motion obtained by Gutzwiller, and improved by others [3]. Second, Heller and coworkers [4,5] pointed out that the eigenfunctions may stay localized around unstable periodic orbits, and that in spite of the strong chaotic behaviour of the classical analog of the system. Experimental manifestation of quantum localization provide the photodissociation spectra of hydrogen atom in strong magnetic fields [6–8], and the high resolution spectra of acetylene [9–11], Na_3 [12–14], HCN [15,16], CS_2 [17], SO_2 [18], O_3 [19], and H_3^+ [20].

Localized or not localized eigenstates may have serious consequences in chemistry, since that will allow the control of a chemical reaction by exciting specific states.

Time-Dependent Quantum Molecular Dynamics, Edited by J. Broeckhove
and L. Lathouwers, Plenum Press, New York, 1992

Thus, knowledge of the topology of the eigenfunctions, and not only the eigenenergies, is necessary for the interpretation of molecular dynamics.

Our work in investigating the dynamics of small molecules, is based on the assumption that there is a *correspondence between classical phase space geometries and the topology of the eigenfunctions.* The phase space structure of a dynamical system is determined by locating, i) equilibrium points, ii) periodic orbits, iii) tori, iv) reduced dimension tori, v) chaotic trajectories, and vi) stable and unstable manifolds [21].

Equilibrium points and periodic orbits located for a range of energies or any other parameter, are a good starting point to portrait the structure of phase space of small molecules. This has been demonstrated by studying in parallel the classical and quantum mechanics of model systems [22, 23] and a few triatomic molecules. In this article we present the basic theory of periodic orbits and discuss results for $ArCO$ and HCN used as prototypical systems, and comment for some other molecules. Emphasis is given for the rotating type periodic orbits which emerge above saddle points of the potential function.

2. EQUILIBRIUM POINTS AND PERIODIC ORBITS

Let q_i, $i = 1, ..., N$, be the generalized coordinates of a dynamical system of N degrees of freedom, and p_i, $i = 1, ..., N$, the conjugate momenta. If $H(\vec{q}, \vec{p})$ is the hamiltonian of the system, the equations of motion are written in Hamilton's form as,

$$\begin{aligned} dq_i/dt = \dot{q}_i\,(t) &= \quad \partial H/\partial p_i \\ dp_i/dt = \dot{p}_i\,(t) &= -\partial H/\partial q_i \quad , i = 1, ..., N. \end{aligned} \tag{1}$$

For convenience, we consider coordinates and momenta as the components of the vector $\vec{x}$,

$$\vec{x} = (q_1, ..., q_N, p_1, ..., p_N)^T, \tag{2}$$

where T denotes the transpose of the $2N$-D column vector. The equations of motion are then written,

$$\dot{\vec{x}}\,(t) = J\frac{\partial H(\vec{x})}{\partial \vec{x}} = J\nabla H(\vec{x}), \tag{3}$$

where,

$$J = \begin{pmatrix} 0_N & I_N \\ -I_N & 0_N \end{pmatrix}. \tag{4}$$

0_N and I_N are the zero and unit $N \times N$ matrices respectively. $J\nabla H(\vec{x})$ is a vector field, and J satisfies the relations,

$$J^{-1} = -J \quad \text{and} \quad J^2 = -I_{2N} \tag{5}$$

The *equilibrium points* are defined by requiring $\dot{\vec{x}} = 0$. In other words, we seek the roots of equation,

$$\nabla H(\vec{x}) = 0. \tag{6}$$

The significance of equilibrium points, which coincide with the extrema of the potential energy surface [24], stems on some important theorems which are proved in Milnor's book [25], and state that the topological type of the energy manifolds, $H = E$, change only at the equilibrium points. Therefore, substantial changes in the dynamics are expected when the total energy exceeds an equilibrium point. The stability of equilibrium points is determined by the second derivatives of the hamiltonian. Stable equilibrium points in the sense of Lyapunov (asymptotically stable) occur when the

Hessian of the hamiltonian, $\partial^2 H$, is positive definite [24]. In this case the eigenvalues of the Hessian matrix are pairs of pure imaginary numbers, $\pm i\mu_k$, $k = 1, ..., N$. For saddle points there are real eigenvalues, $\pm\mu$, and the corresponding eigenvectors point to the unstable directions.

What happens in phase space as we move away from an equilibrium point and introduce a small perturbation? For stable equilibrium points the theorems of Kolmogorov-Arnold-Moser (KAM) [26], and of Poincaré-Birkoff (PB) [27] provide an answer. For incommensurable unperturbed frequencies (i.e. there is no relation, $\sum_i^N n_i\omega_i = 0$, among the frequencies ω_i and the integers n_i) KAM theorem guarantees the existence of tori when a perturbation is introduced, whereas the PB theorem states that the commensurable unperturbed tori will break and an even number of periodic orbits will appear; half of them are stable and half are unstable. With stable we mean that nearby trajectories stay close to the periodic orbit for infinite time, and with unstable we describe the property of nearby trajectories to deviate from the periodic orbit either exponentially or with a power of time t.

After having located the equilibrium points the next geometrical objects with higher dimension, are the *periodic orbits.* One important existence theorem of periodic orbits is that of Weinstein [28]. This theorem guarantees, that arbitrarily close to a stable equilibrium point there are at least N periodic orbits whose periods are close to those of the linearized system. Generalizations of this theorem were given by Moser [29,30]. If the ratio of two eigenvalues of the equilibrium point is a rational number, $\mu_1/\mu_2 = m/n$, there are several theorems for different values of m, n which prove the existence of two more families of periodic orbits, one stable and one unstable [31–33].

Linearization of the equations of motion is a general strategy to study the behaviour of the trajectories around an equilibrium point or around a trajectory $\vec{x}(t)$. The Hartman - Grobman theorem [34] states that there is a continuous change of coordinates that relates the linearization to the vector field $(J\nabla H(\vec{x}))$. Thus, important conclusions obtained for the linearized system can be generalized to the nonlinear system.

If $\vec{x}(t)$ is a solution of Eq. (3) we want to know the behaviour of a nearby trajectory

$$\vec{x}'(t) = \vec{x}(t) + \vec{\zeta}(t). \tag{7}$$

From Eq. (3) we have,

$$\dot{\vec{x}}'(t) - \dot{\vec{x}}(t) = J\nabla H(\vec{x}') - J\nabla H(\vec{x}). \tag{8}$$

A Taylor expansion of the rhs of Eq. (8) up to the first order gives,

$$\dot{\vec{\zeta}}(t) = J\partial^2 H(\vec{x}(t))\vec{\zeta}(t). \tag{9}$$

$\partial^2 H(\vec{x}(t))$ denotes the matrix of second derivatives of the hamiltonian evaluated at $\vec{x}(t)$.

If,

$$A(t) = J\partial^2 H(\vec{x}(t)),$$

then Eq. (9) is written as,

$$\dot{\vec{\zeta}}(t) = A(t)\vec{\zeta}(t). \tag{10}$$

These are $2N$ linear differential equations with time dependent coefficients, and are called *variational equations.*

The general solution of Eq. (10) can be expressed by evaluating the fundamental matrix, $Z(t)$. This is the matrix with columns the vectors (1,0,...,0), (0,1,0,...,0),..., (0,0,...,1) at $t = 0$, i.e.

$$Z(0) = I_{2N}. \tag{11}$$

The general solution of Eq. (10) is then given by,

$$\vec{\zeta}(t) = Z(t)\vec{\zeta}(0), \tag{12}$$

where $\vec{\zeta}(0)$ describes the initial displacement from the trajectory $\vec{x}$. The fundamental matrix satisfies the variational equations as can be easily proved;

$$\dot{Z}(t) = A(t)Z(t). \tag{13}$$

If $\vec{x}(\vec{x}_0; t)$ is the reference trajectory with initial conditions $\vec{x}_0$, then we can show that the fundamental matrix has columns the vectors,

$$\vec{\zeta}_k = \frac{\partial \vec{x}(t)}{\partial x_{k0}}, \tag{14}$$

i.e. the derivatives of the trajectory $\vec{x}(t)$ with respect to the initial coordinates x_{k0}, $k = 1, ..., 2N$. Indeed, differentiating both sides of Eq. (3) we take,

$$\begin{aligned} \frac{\partial \dot{\vec{x}}}{\partial x_{k0}} &= J \frac{\partial}{\partial x_{k0}} \nabla H, \\ \frac{d}{dt}\left(\frac{\partial \vec{x}}{\partial x_{k0}}\right) &= J(\partial^2 H)\frac{\partial \vec{x}}{\partial x_{k0}} \\ &= A(t)\frac{\partial \vec{x}}{\partial x_{k0}}. \end{aligned} \tag{15}$$

Thus,

$$\dot{\vec{\zeta}}_k = A(t)\vec{\zeta}_k, \tag{16}$$

where $\vec{\zeta}_k = \partial\vec{x}/\partial x_{k0}$. Obviously, $Z(0) = I_{2N}$, and we conclude that $Z(t)$ is the fundamental matrix.

We now show, that the velocity vector, $\dot{\vec{x}}(t)$, is also solution of the variational equations. Indeed, from Eq. (3) we have,

$$\begin{aligned} \frac{d}{dt}\dot{\vec{x}}(t) &= \frac{d}{dt} J\nabla H(\vec{x}) \\ &= J\partial^2 H \, \dot{\vec{x}}(t). \end{aligned} \tag{17}$$

That is,

$$\dot{\vec{\zeta}} = A(t)\vec{\zeta}, \tag{18}$$

where $\vec{\zeta}(t) = \dot{\vec{x}}(t)$.

We are interested in further investigating the variational equations when $\vec{x}(t)$ is a periodic solution of equations of motion of period T;

$$\vec{x}(t+T) = \vec{x}(t). \tag{19}$$

In this case, the linear Eqs. (10) have periodic coefficients, and the fundamental matrix at $t = T$,

$$M = Z(T) = \frac{\partial \vec{x}(T)}{\partial \vec{x}_0}, \tag{20}$$

is called *monodromy matrix*. As we shall see, the monodromy matrix play an important role in the theory of periodic orbits and their stability, and therefore, we summarize some of its properties [35].

1.

$$
\begin{aligned}
M(nT) &= M(T)M((n-1)T), \\
M(nT) &= M(T)^n. \qquad (21)
\end{aligned}
$$

This is a consequence of Eq. (12).

2. Since $\dot{\vec{x}}$ are periodic functions of time and solutions of the variational equations (Eq. (16)), we can show that one of the eigenvalues of the monodromy matrix is equal to one. Indeed, for $\vec{\zeta} = \dot{\vec{x}}$ we have,

$$
\begin{aligned}
\vec{\zeta}(0) = \vec{\zeta}(T) &= M(T)\vec{\zeta}(0) \Rightarrow \qquad (22) \\
(M - I)\vec{\zeta}(0) &= 0. \qquad (23)
\end{aligned}
$$

Eq. (23) is true if at least one of the eigenvalues of the monodromy matrix is one. Thus, if $\vec{x}(t)$ is a periodic orbit the variational equations have also a periodic solution ($\dot{\vec{x}}$), and then, one eigenvalue of the monodromy matrix is equal to one. The inverse is also true and more important; *if the monodromy matrix has an eigenvalue equal to one, then, the variational equations have a periodic solution.*

3. It can be shown, that for every constant of motion the monodromy matrix has an eigenvalue equal to one. Let $C(\vec{x}_0; T) = const.$ is a constant of motion along the periodic orbit with initial conditions $\vec{x}_0$ and period T. This means that,

$$
C(\vec{x}; t) = C(\vec{x}_0; t = 0) \qquad (24)
$$

Differentiating the above equation with respect of the initial conditions we take;

$$
\nabla_{\vec{x}_0} C(\vec{x}; t = T) \frac{\partial \vec{x}}{\partial \vec{x}_0} = \nabla_{\vec{x}_0} C(\vec{x}_0), \qquad (25)
$$

or,

$$
(M^T - I)\nabla_{\vec{x}_0} C = 0. \qquad (26)
$$

Eq. (26) means that the monodromy matrix has one eigenvalue equal to one provided $\nabla_{\vec{x}_0} C \neq 0$.

4. For conservative hamiltonian systems Liouville's theorem results in,

$$
det(M(T)) = 1. \qquad (27)
$$

5. The matrix M satisfies the symplectic property, i.e.

$$
MJM^T = J. \qquad (28)
$$

6. For a conservative hamiltonian system if λ is an eigenvalue of the monodromy matrix, then the complex conjugate, λ^*, and the inverses, $\lambda^{-1}, (\lambda^*)^{-1}$, are also eigenvalues of M. In other words, the eigenvalues of the monodromy matrix come in pairs, (λ, λ^*) for complex eigenvalues, and (λ, λ^{-1}), for real eigenvalues. According to the properties 2 and 3, for a conservative hamiltonian system the monodromy matrix has always two eigenvalues equal to one.

Sometimes it is convenient to express the eigenvalues of the monodromy matrix of a periodic orbit as,

$$\lambda = \exp(\alpha T), \tag{29}$$

and the exponents α are called *characteristic exponents.*

3. STABILITY OF PERIODIC ORBITS AND BIFURCATION THEORY

Several important conclusions can be drawn from the eigenvalues of the monodromy matrix, which concern the behaviour of the nearby trajectories. Does a neighboring trajectory deviate or remain close to the periodic orbit? Are there other periodic orbits which bifurcate from the parent one? From Eqs. (12) and (21) we can write;

$$\vec{\zeta}(nT) = M(T)^n \vec{\zeta}(0) = \begin{pmatrix} \lambda_1^n & 0 & \dots & 0 \\ 0 & \lambda_2^n & \dots & 0 \\ \dots & \dots & \dots & \dots \\ 0 & 0 & \dots & \lambda_{2N}^n \end{pmatrix} \vec{\zeta}(0), \tag{30}$$

assuming that the coordinate system is the set of the eigenvectors of M. Therefore, the eigenvalues can describe the deviation from the periodic orbit of an initial displacement $\vec{\zeta}(0)$ after n iterations.

By excluding the two unit eigenvalues of the monodromy matrix of a conservative hamiltonian system, the following cases can arise:

i) If all the eigenvalues are on the unit complex circle, and multiple eigenvalues have independent eigenvectors which are equal to the multiplicity of the eigenvalues, then, the periodic orbit is *stable (elliptic)*, and it is surrounded by tori. The characteristic exponents are pure imaginary numbers and may be considered as the frequencies of rotation of a neighboring trajectory around the periodic orbit, $\alpha = i\sigma$.

It may happen that,

$$T/(2\pi/\sigma) = m/n, \tag{31}$$

where m, and n are integers. Then, from Eq. (30) we can see that the variational equations have a periodic solution, and there should be a new periodic orbit of period $T' = nT$ in the neighborhood of the initial one. Another way to state it is, after n iterations $\lambda^n = 1$, and then according to the property 2 of the monodromy matrix the variational equations have a periodic solution.

ii) If there are eigenvalues equal to one, then $\alpha = 0$ and a new periodic orbit of period T also exists in the neighborhood of the parent one. For eigenvalues equal to -1, $\alpha = i\pi/T$ and a new periodic orbit of double period exists nearby according to the analysis given above.

iii) If there are real eigenvalues greater than ± 1, $\vec{\zeta}(t)$ will deviate exponentially with time, and the periodic orbit is *unstable (hyperbolic)* in the directions of the corresponding eigenvectors.

iv) If there is a complex eigenvalue, λ, with norm greater than one, then λ^{-1}, λ^*, and $(\lambda^*)^{-1}$ are also eigenvalues and the periodic orbit is called *complex unstable.*

v) If there are multiple eigenvalues with independent eigenvectors less than the multiplicity of the eigenvalue, the periodic orbit is unstable with $\vec{\zeta}(t)$ deviating not exponentially but with a power of t.

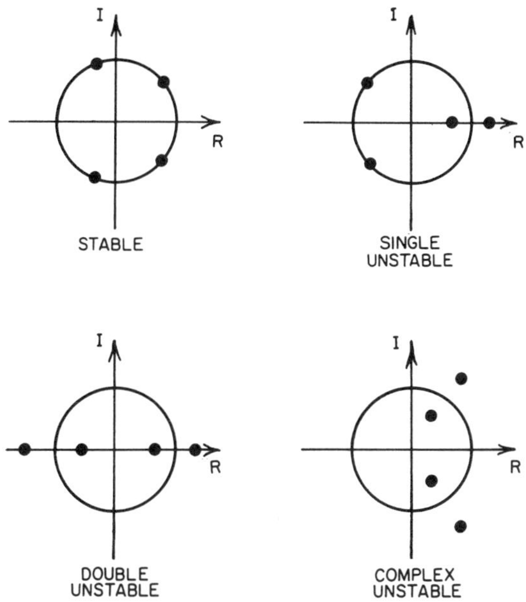

FIG. 1 Different cases of stability of periodic orbits in a system of three degrees of freedom.

In Fig. 1 we plot the eigenvalues of M on the unit complex circle for a system of three degrees of freedom. The two pairs of eigenvalues which are different than one may be: complex conjugate with norm equal to one (stable orbits), or one pair of complex conjugate with unit norm and one pair of real eigenvalues (single unstable periodic orbits), or two real pairs for double unstable periodic orbits. There are three different cases in the double instability; to have positive, negative, or one positive one negative pairs of real eigenvalues. Finally, if there is a quadruplet of complex eigenvalues the periodic orbit is complex unstable.

In investigating the phase space structure of a molecule it is important to follow the evolution of a family of periodic orbits with respect to the total energy. Then, the question about the change of stability of periodic orbits and the possibility of observing new families of periodic orbits arises. Specifically, we can ask when a stable periodic orbit will become unstable by varying a parameter of the hamiltonian. Answer to this problem is given by the *theory of Krein, Gelfand, and Lindskii* [36]. In the following we summarize the main results of this theory.

We first introduce the indefinite inner product:

$$< \vec{x}, \vec{y} >= -i(J\vec{x}, \vec{y}). \tag{32}$$

$\vec{x}, \vec{y}$ are $2N$ dimensional vectors, $(,)$ denotes the ordinary inner product, and $i = \sqrt{-1}$. An eigenvalue of the monodromy matrix with eigenvector $\vec{e}$ is of *first kind* if $< \vec{e}, \vec{e} >$ is positive, and *second kind* if $< \vec{e}, \vec{e} >$ is negative. If $< \vec{e}, \vec{e} >= 0$ the corresponding eigenvalue is called of *mixed kind.*

The necessary and sufficient condition for having stable periodic orbits is that all eigenvalues on the unit circle are of the first or second kind. It is also proved that the character of the eigenvalue (first or second) does not change by varying the total energy (or any other parameter). In other words, if two eigenvalues of the same kind collide on the unit circle, as the total energy changes, the periodic orbit remains stable. However, if two eigenvalues of different kind collide, then a multiple eigenvalue of mixed kind is formed and there exists a small perturbation which can drive the periodic orbit to become unstable.

For a system of two degrees of freedom with only one pair of eigenvalues moving on the unit circle, collision can occur only on the real axis at ± 1. In this case bifurcation of new families of periodic orbits are observed with the same period as that of the parent periodic orbit when the collision happens at 1. When the eigenvalues collide at -1 the new periodic orbits are of double period. In the latter case the periodic orbit becomes unstable and is often called *hyperbolic by reflection*. For three and higher degrees of freedom systems several combinations can happen as is explained in Fig. 1.

Can we predict the number and the stability of the bifurcating families of periodic orbits? Answer to this question is given by *degree theory*. The basic concept in this theory is the conservation of *rotation*.

We first define the *index* of an *isolated* periodic orbit $\vec{x}$ with period T [37].

$$ind(\vec{x}(t), T) = (-1)^{\beta}. \tag{33}$$

β is the sum of multiplicities of the real eigenvalues of the monodromy matrix in $(1, \infty)$. Thus, the index of a stable or complex unstable periodic orbit is 1 ($\beta = 0$), and that of a single unstable periodic orbit -1.

Similarly, the index of an equilibrium point $\vec{x_0}$ of a vector field ∇H is defined as:

$$ind(\vec{x}_0, \nabla H) = (-1)^{\beta}. \tag{34}$$

β is the sum of the multiplicities of the negative real eigenvalues of $\partial^2 H$. It is assumed that the determinant, $|\partial^2 H| \neq 0$, a condition which guarantees that the equilibrium point is *isolated.*

We conclude that stable equilibrium points have index 1, and saddles with one unstable direction have index -1.

The rotation of a vector field is defined on the boundary of a bounded open set $\Omega \subset R^{2N}$, on which the vector field is nonsingular: there are no equilibrium points.

$$\gamma = \sum_i ind(\vec{x_i}, T_i). \tag{35}$$

If E_0 is the critical energy at which bifurcation is observed conservation of the rotation means that;

$$\gamma_{E<E_0} = \gamma_{E>E_0}. \tag{36}$$

Equivalently,

$$\sum_i ind(\vec{x_i}, T_i)_{E<E_0} = \sum_i ind(\vec{x_i}, T_i)_{E>E_0}. \tag{37}$$

In Fig. 2 we show several bifurcation schemes for which the above rules of calculating the index and rotation of periodic orbits can be tested.

Plots of the initial coordinates of the periodic orbits as functions of the varying parameter (in our case the total energy) are named *diagrams of characteristics or bifurcation diagrams*, and show the continuation of the periodic orbits with the parameter and the critical bifurcation values. Usually, we plot projections of these diagrams on several planes, and as we shall see the diagrams of characteristics reveal significant information about the structure of phase space.

4. APPLICATIONS

According to the analysis of the previous sections a systematic way to portrait the phase space of a molecule for a range of energies, is first to locate the equilibrium points, and second to find those families of periodic orbits, which emanate from the equilibrium points. We call these families *principal* and their existence are predicted by Weinstein-Moser theorems. Using continuation schemes, the evolution of periodic orbits with the total energy or any other parameter (e.g. total angular momentum) is obtained. Bifurcation of new families of periodic orbits with the same or multiple periods are found by examining the eigenvalues of the monodromy matrix, and computing the index of periodic orbits. Every time the index changes a bifurcation is observed.

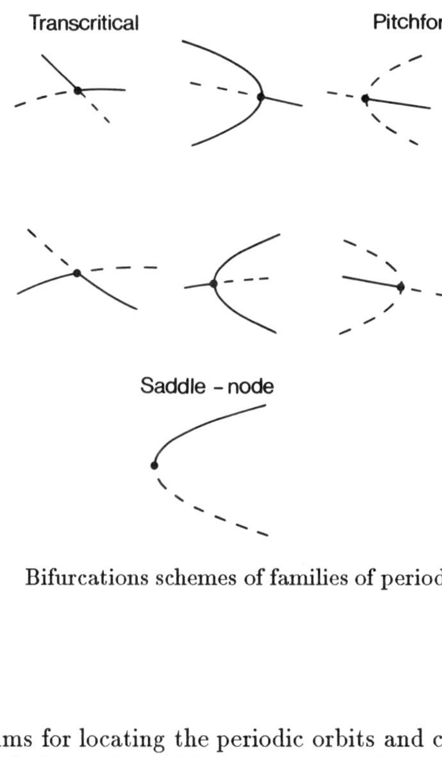

FIG. 2 Bifurcations schemes of families of periodic orbits.

Numerical algorithms for locating the periodic orbits and continuation schemes are described in reference [38]. Both equilibrium points and periodic orbits are obtained by solving the nonlinear equations (6), and (19), with a secant or Newton-Raphson method [39]. However, we have to mention that the effort of finding periodic orbits is not always successful if they are highly unstable.

More difficulties arise when we try to locate *saddle-node* type families, i.e. families of periodic orbits which emerge from a saddle-node. Contopoulos [40] has named them *irregular* since they can not be predicted as the principal or bifurcating families. The term irregular does not imply that the periodic orbit is always unstable. Sometimes the terms periodic orbits "out of nowhere" or "tangent bifurcations" are used. Saddle-node points in phase space exist for any dynamical system but in our case we shall discuss those which appear above the saddles of the potential energy surface.

Since, we want to investigate the role of periodic orbits in molecular dynamics we carry out quantum mechanical calculations in parallel to the periodic orbit search. Results are presented for two typical systems: a van der Waals molecule, $ArCO$, and a bound triatomic molecule, HCN.

$ArCO$ [41]: We keep the CO bond length fixed at its equilibrium length, r, and study the dynamics in the coordinates (R, θ); R is the distance of Ar from the center of mass of CO, and θ the angle between R, and r. The potential function is plotted in Fig. 3 and for θ between $-180^0, 180^0$. There is a minimum of -0.01368 eV at $R = 6.8729\ a_0$, and $\theta = 77^0$, and two saddle points for linear geometries, at $(r, \theta) = (7.5041\ a_0, 0^0)$ $(Ar - OC)$, and at $(8.3318\ a_0, 180^0)$ for $Ar - CO$.

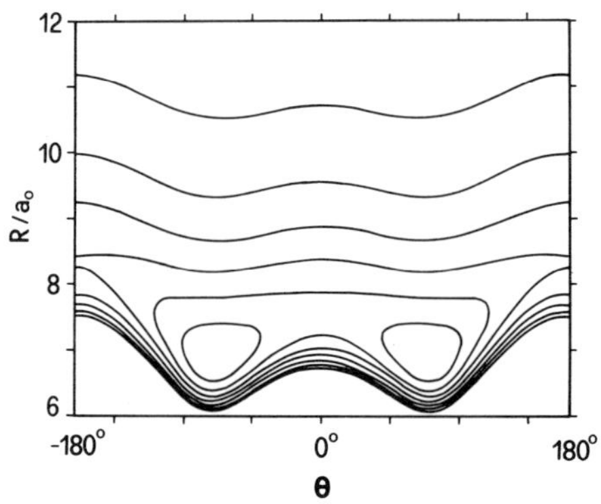

FIG. 3 Potential function for $ArCO$.

The diagram of characteristics is shown in Fig. 4. R1 and TH1 denote the principal families which emerge from the minimum, and imply excitation in the stretching and bending modes respectively. Bifurcating families are characterized with the letter A and B; i.e. TH1A, TH1B, R1A etc. Continuous curves correspond to stable periodic orbits, and dotted curves to unstable periodic orbits. Periodic orbits which appear above the barriers of linearization of the molecule are labeled as I. They are of irregular type according to our terminology, and occur in pairs, one stable and one unstable. At the saddle-node we find all the eigenvalues to be equal to one, and with equal eigenvectors, and this should be characteristic for all 2-D systems. The importance of irregular periodic orbits is that they connect remote regions of phase space.

In order to examine more globally the phase space we draw Poincaré surfaces of section. One is shown in Fig. 5 at an energy above the first barrier but below the second one ($E = -0.0087$ eV). We can see that the phase space is dominated by chaos, but a small island of stability exists which surrounds the I1A periodic orbit.

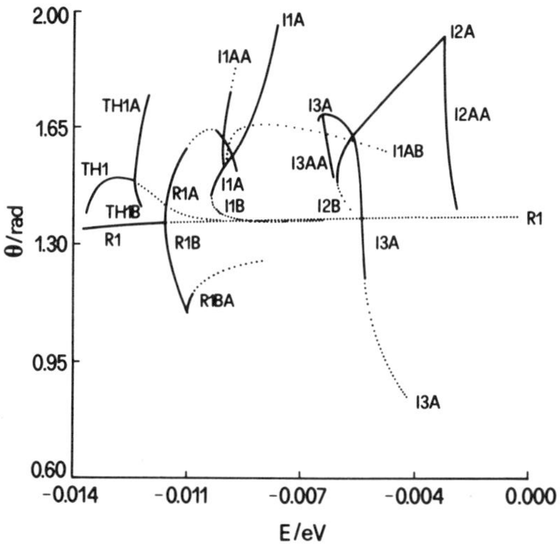

FIG. 4 Projection of the diagram of characteristics of $ArCO$ in the (E, θ) plane.

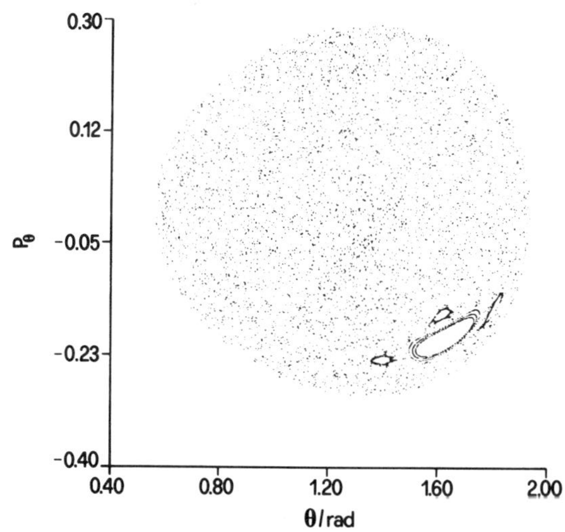

FIG. 5 A Poincaré surface of section at -0.0087 eV.

This classical structure has been used to explain the topologies of the quantum eigenfunctions. It has been found [42] that among twenty bound states diffused to all available configuration space, four are localized as can be seen in Fig. 6. The zero point energy of $ArCO$ is just above the barrier of linearization of the molecule, where as we have seen the classical chaos dominates. Therefore, the delocalization of the eigenstates is not surprising. Instead, the localized ones, which are embedded in the delocalized states, are unexpected and their existence are explained by the irregular type periodic orbits.

HCN [43]: This molecule has been studied extensively by spectroscopists who predict regular vibrational states with extended excitations in the stretching modes [15, 16, 44, 45]. A significant number of theoretical papers have also been published both classical [46] and quantum mechanical [47, 48], which explain satisfactorily the spectroscopic results. Here, we concentrate on the *rotating type* states which have been found in quantum mechanical calculations, both in 2-D with the CN bond frozen [47], and in 3-D quantum calculations [48].

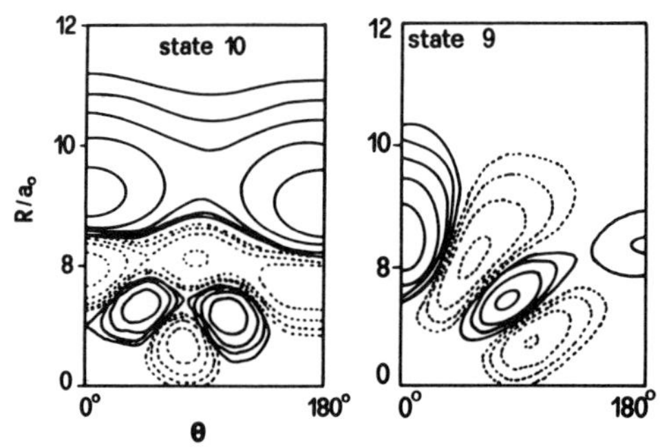

FIG. 6 The eigenfunctions of the 9th (localized) and 10th (delocalized) quantum states of $ArCO$.

The ground electronic state of HCN supports two minima with linear geometries, which correspond to HCN (absolute) and HNC (relative). With *rotating type* periodic orbits we mean those which emerge above the barrier of isomerization of the molecule and visit both minima. We could also call them "isomerizing" periodic orbits. Fig. 7 shows the diagram of characteristics of the rotating type periodic orbits found for this system. We plot the CH distance, R_{CH}, with the total energy, E.

As before stable periodic orbits are described with continuous curves, single unstable periodic orbits with dotted curves, but now for this 3-D system double unstable periodic orbits appear which are denoted with squares, as well as complex unstable orbits which are denoted with crosses.

The families shown in Fig. 7 appear in pairs, stable-single unstable or single unstable-double unstable, with the I1A family changing to complex unstable in a short range of energies. The scheme is rather complex and new families of periodic orbits emerge from them which we have not followed. The region above the barrier of isomerization is unstable, but it has been shown that the rotating periodic orbits can be detected spectroscopically [46].

Fig. 8 shows some of the rotating type periodic orbits, and Fig. 9 a quantum eigenstate describing an isomerizing state, which we associate with the rotating periodic orbits.

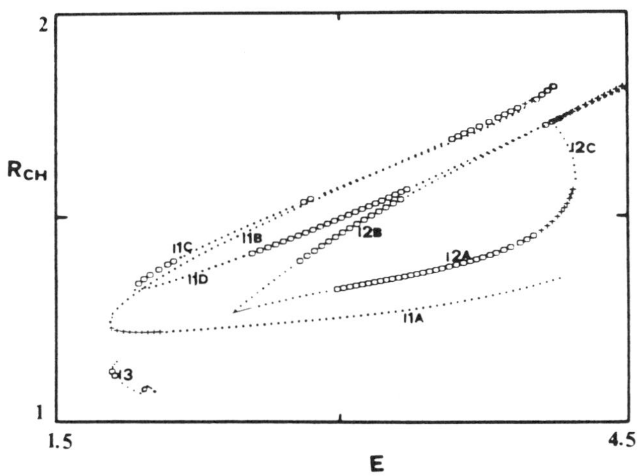

FIG. 7 Projection of the diagram of characteristics of *HCN* for the rotating type periodic orbits.

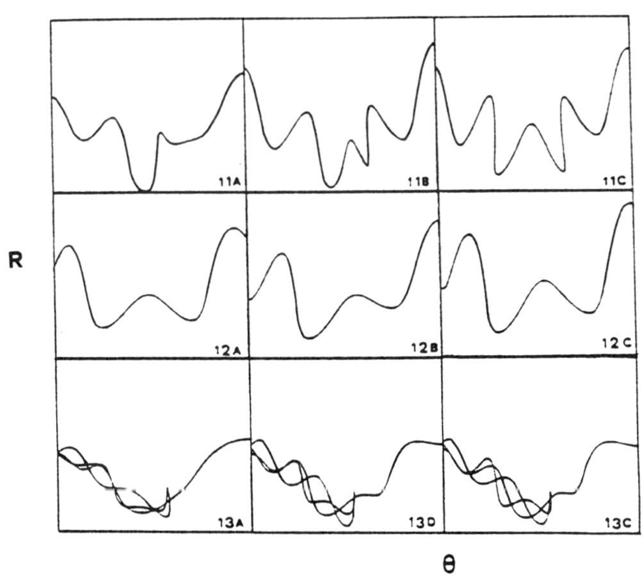

FIG. 8 Rotating type periodic orbits of *HCN*.

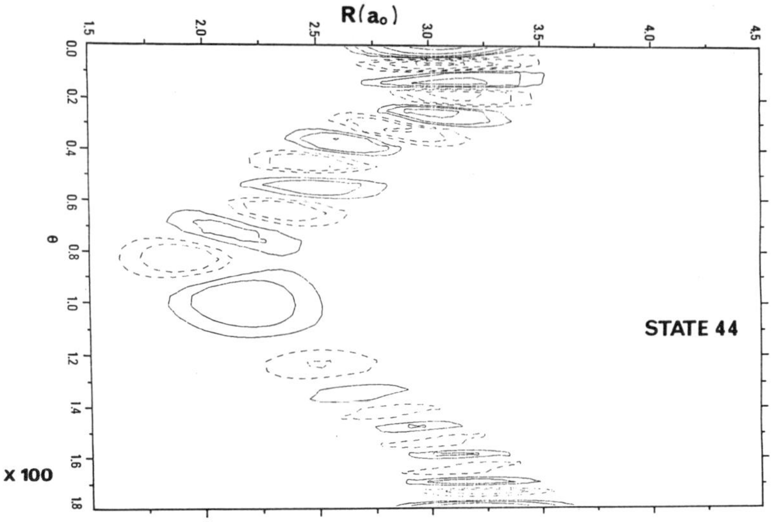

FIG. 9 A rotating type quantum state calculated with the bond CN frozen.

5. OTHER MOLECULES AND CONCLUSIONS

Periodic orbits of isomerizing (or rotating) type, i.e. orbits moving across a potential barrier, have been found for other systems as well. Pollak [49] studied the association of hyperspherical periodic orbits and resonances in atom-diatom reactions. Taylor and coworkers used similar type periodic orbits for the assignment of photodetachment spectra in H_2F [50], and HCl_2 [51]. Gomez and Pollak [52] used isomerizing T-shape periodic orbits to explain regularities in the low resolution photodissociation spectra of H_3^+.

Recently an interpretation of the structure of the photoabsorption spectrum of ozone was given by locating isomerizing periodic orbits, which involve simultaneous motion in the symmetric and antisymmetric modes [53]. Periodic orbits of the same type were found for CO_2 [54, 55] for an unbound electronic state, as well as for the doubly excited helium atom to mention a different system than triatomic molecules [56, 57].

Indications for the existence of rotating type periodic orbits in Ar_3 provide the calculation of Lyapunov exponents by Berry and coworkers [58]. They noticed that for energies above the barrier of linearization of argon trimer the Lyapunov exponents decrease something which could be anticipated from the existence of saddle-node families of periodic orbits.

All the above systems have not been studied systematically in order to confirm that the periodic orbits are of saddle-node type or bifurcations of a principal family. However, from these results we deduce the importance of the rotating (isomerizing) periodic orbits to find localized eigenfunctions (bound or resonances) at high excitation energies.

The location of saddle-node families of periodic orbits is not easy, nevertheless, it is important to know that such families exist above saddle points of the potential function.

References

[1] M. Quack. *Annu. Rev. Phys. Chem.*, 41:839, 1990.

[2] L. M. Raff, and D. L. Thompson. In M. Baer, editor, *Theory of Chemical Reactions*, page 1. CRC, 1985.

[3] M. C. Gutzwiller. *Chaos in Classical and Quantum Mechanics*, volume 1. Springer-Verlag, 1990.

[4] E. J. Heller . *Phys. Rev. Letters*, 53:1515, 1984.

[5] M. J. Davis. *J. Phys. Chem.*, 92:3124, 1988.

[6] A. Holle, G. Wiebusch, J. Main, B. Hager, H. Rottke, and K. H. Welge. *Phys. Rev. Letters*, 56:2594, 1986.

[7] D. Wintgen and H. Friedrich . *Phys. Rev.*, A36:131, 1987.

[8] J.-M. Mao, and J. B. Delos. *Phys. Rev.*, A45:1746, 1992.

[9] E. Abramson, R. W. Field, D. Imre, K. K. Innes, and J. L. Kinsey. *J. Chem. Phys.*, 83:453, 1985.

[10] J. P. Pique, Y. Chen, R.E. Field, and J. L. Kinsey. *Phys. Rev. Letters*, 58:4715, 1987.

[11] J. P. Pique, Y. M. Engel, R. D. Levine Y. Chen, R.E. Field, and J. L. Kinsey. *J. Chem. Phys.*, 88:5972, 1988.

[12] M. Broyer, G. Delacretaz, G. -Q Ni, R. L. Whetten, J. -P. Wolf and L. Wöste. *Phys. Rev. Letters*, 62:2100, 1989.

[13] M. Broyer, G. Delacretaz, G. -Q Ni, R. L. Whetten, J. -P. Wolf and L. Wöste. *J. Chem. Phys.*, 90:4620, 1989.

[14] J. M. Gomez Llorente, and H. S. Taylor . *J. Chem. Phys.*, 91:953, 1989.

[15] X. Yang, C. A. Ragashi, and A. M. Wodtke. *J. Opt. Soc. Am. B*, 7:1835, 1990.

[16] X. Yang, C. A. Rogaski, and A. M. Wodtke. *J. Chem. Phys.*, 92:2111, 1990.

[17] J. P. Pique, M. Joueux, J. Manners, and G. Sitja. *J. Chem. Phys.*, 95:8744, 1991.

[18] K. Yamanouchi, S. Takeuchi, and S. Tsuchiya. *J. Chem. Phys.*, 92:4044, 1990.

[19] D. E. Freeman, K. Yoshino, J. R. Esmond, and W. H. Parkinson. *Planet. Space Sci.*, 32:239, 1984.

[20] A. Carrington, and R. A. Kennedy . *J. Chem. Phys.*, 81:91, 1984.

[21] A. J. Lichtenberg, and M. A. Liberman. *Regular and Stochastic Motion*, volume 38. Applied Math. Sciences, Springer - Verlag, 1983.

[22] M. Founargiotakis, S. C. Farantos, G. Contopoulos, and C. Polymilis. *J. Chem. Phys.*, 91:1389, 1989.

[23] S. C. Farantos, M. Founargiotakis, and C. Polymilis. *Chem. Phys.*, 135:347, 1989.

[24] V. I. Arnold. *Mathematical Methods of Classical Mechanics*, volume 60, of *Graduate Text in Math.* Springer - Verlag, 1980.

[25] J. Milnor. *Morse Theory*, volume 51. Annals of Mathematics Studies, Princeton Univ. Press, 1963.

[26] V. I. Arnold, and A. Avez. *Ergodic Problems of Classical Mechanics.* Benjamin, New York, 1968.

[27] G. D. Birkhoff. *Acta Math.*, 47:297, 1925.

[28] A. Weinstein. *Inv. Math.*, 20:47, 1973.

[29] J. Moser. *Commun. Pure Appl. Math.*, 29:727, 1976.

[30] J. Moser. *Commun. Pure Appl. Math.*, 11:257, 1958.

[31] Ya. G. Sinai. *Amer. Math. Soc. Transl.*, 68:34, 1966.

[32] M. Kummer. *Commum. Math. Phys.*, 31:157, 1978.

[33] J. Roels. *J. Diff. Equat.*, 10:431, 1971.

[34] J. Guckenheimer, and P. Holmes. *Nonlinear Oscillations, Dynamical Systems, and Bifurcations of Vector Fields.* Springer-Verlag; Berlin, 1983.

[35] J. D. Hadjidemetriou. *Periodic Orbits and Stability.* University of Thessaloniki, 1988.

[36] V. A. Yakubovich, and V. M. Starzhinskii. *Linear Differential Equations with Periodic Coefficients*, volume 1. Halsted Press, 1975.

[37] M. A. Krasnosel'skii, and P. P. Zabreiko. *Geometrical Methods of Nonlinear Analysis*, volume 263. A series of comprehensive mathematics, Springer-Verlag, 1984.

[38] R. Seydel. *From Equilibrium to Chaos: Practical bifurcation and stability analysis.* Elsevier, 1988.

[39] W. H. Press, B. P. Flannery, S. A. Teukolsky, and W. T. Vetterling. *Numerical Recipes.* Cambridge Univ. Press, 1986.

[40] G. Contopoulos. *Astron. J.*, 75:96, 1970.

[41] L. Zachilas, and S. C. Farantos. *Chem. Phys.*, 154:55, 1991.

[42] S. C. Farantos, and J. Tennyson. In L. S. Cederbaum, A. Amann, and W. Gans, editors, *Fractals Quasicrystals, Chaos, Knots, and Algebraic Quantum Mechanics*, page 195. Kluwer Academic Publishers, 1988.

[43] S. C. Farantos, and M. Founargiotakis. *Chem. Phys.*, 142:345, 1990.

[44] K. K. Lehmann, G. J. Scherer, and W. Klemperer. *J. Chem. Phys.*, 77:2853, 1983.

[45] K. K. Lehmann, G. J. Scherer, and W. Klemperer. *J. Chem. Phys.*, 78:608, 1983.

[46] S. C. Farantos, J. M. Gomez Llorente, O. Hahn, and H. S. Taylor. *J. Chem. Phys.*, 93:76, 1990.

[47] M. Founargiotakis, S. C. Farantos, and J. Tennyson. *J. Chem. Phys.*, 88:1598, 1988.

[48] J.A. Bentley, J.-P. Brunet, R. E. Wyatt, R. A. Friesner, and C. Leforestier. *Chem. Phys. Letters*, 161:393, 1989.

[49] E. Pollak. In M. Baer, editor, *Theory of Chemical Reaction Dynamics*, volume 3, page 123. CRC, Boca Raton, FL., 1985.

[50] O. Hahn, and H. S. Taylor. *J. Chem. Phys.*, page in press.

[51] O. Hahn, J. M. Gomez Llorente, and H. S. Taylor. *J. Chem. Phys.*, 94:2608, 1991.

[52] J. M. Gomez Llorente, and E. Pollak. *Chem. Phys.*, 120:37, 1988.

[53] S. C. Farantos. *Chem. Phys.*, 159:329, 1992.

[54] K. C. Kulander, C. Cerjan, and A. E. Orel. *J. Chem. Phys.*, 94:2571, 1991.

[55] R. Schinke, and V. Engel. *J. Chem. Phys.*, 93:3252, 1990.

[56] G. Ezra, K. Richter, G. Tanner, and D. Wintgen. *J. Phys. B*, 24:L413, 1991.

[57] J. Müller, J. Burgdörfer, and D. Noid. *Phys. Rev.*, A45:1471, 1992.

[58] R. J. Hinde, R. S. Berry, and D. J. Wales. *J. Chem. Phys.*, 96:1376, 1992.

CLASSICAL AND QUANTUM MECHANICS OF THE DRIVEN DOUBLE WELL

David Farrelly* and John A. Milligan*†

*Department of Chemistry and Biochemistry
Utah State University, Logan, UT 84322-0300, USA

†Department of Chemistry and Biochemistry
University of California, Los Angeles, CA 90024-1569, USA

INTRODUCTION

Tunneling is one of the most alluring concepts in quantum mechanics and is often considered in introductory expositions of the subject with reference to the symmetric double well potential. The consequences of tunneling have long been recognized: In 1927 Hund[1] discussed tunneling in pyramidal molecules like ammonia where tunneling between two symmetric equilibria gives rise to splittings in vibrational spectra. Similarly other molecules (e.g., methanol) may contain barriers to internal rotation through which tunneling occurs, also giving rise to splittings in the spectra.[2] In addition to these relatively simple examples, tunneling is important in a wide variety of chemical processes[3] including biological situations. For example, electron transfer in proteins is often based on electron and nuclear tunneling in order to effect charge separation.[4,5] The strong distance dependence of tunneling has allowed the construction of the scanning tunneling microscope[6] and many devices in solid state physics are based on the tunnel effect.[7] Particularly fascinating is the prediction of mesoscopic tunneling of magnetic flux in superconducting quantum interference devices (SQUIDs).[8] In many of these situations a simple double well potential provides the most useful and generally applicable model.

In addition to tunneling through a physical potential energy barrier, a different type of tunneling can occur in which flux moves between two or more different regions in *phase space* that are separated by an effective barrier. This is usually referred to as dynamical tunneling and may, for example, connect classically disjoint regular volumes of phase space that are otherwise separated by chaotic regions.[9,10] The relationship between an essentially classical property (chaos) and a purely quantum phenomenon (tunneling) is clearly of much fundamental interest. From a more practical standpoint the relevance of tunneling and possibly chaos to devices like the Josephson junction also makes it worthwhile to investigate tunneling in chaotic systems.

A system that combines both tunneling and classical chaotic dynamics is the

Time-Dependent Quantum Molecular Dynamics, Edited by J. Broeckhove and L. Lathouwers, Plenum Press, New York, 1992

driven double well Hamiltonian or Duffing oscillator.[11,12] The one-dimensional double well problem itself is integrable, but the presence of a periodic driving term destroys integrability and gives rise to a chaotic separatrix layer. This system has been studied from a variety of angles, and recently, Lin and Ballentine[13] showed that tunneling rates are dramatically enhanced by the presence of the external field. In related work, Grossman and co-workers [14,15] using a Floquet analysis of the problem demonstrated that coherent suppression of tunneling is possible by a judicious choice of the frequency of the driving force. Suppression of tunneling by an exernal field is of considerable interest and could conceivably be observed experimentally. In this paper the problem of driving a double well at two different frequencies is studied and it is shown that tunneling can also be suppressed coherently by using two external fields whose frequencies are in a 2:1 ratio.

The paper is organized as follows; after a general discussion of the classical mechanics of the double well subjected to monochromatic and bichromatic driving forces, integration of the corresponding time dependent Schrödinger equation (TDSE) is described. The approach of Lin and Ballentine[13] is used in which a basis set expansion is made and the coupled equations of motion are integrated directly using an Adams-Moulton method. Rather than calculating the wavefunction itself a Husimi distribution is computed. The single frequency case is examined classically and quantum mechanically as is the effect of adding a second driving term having a different frequency. Finally, brief conclusions are presented.

CLASSICAL MECHANICS OF THE DRIVEN DOUBLE WELL

The Hamiltonian for a particle in a quartic double well potential subjected to two-frequency forcing may be written,

$$H = H_0 + H_1 \tag{1}$$

where,

$$H_0 = \tfrac{1}{2}p^2 - \tfrac{1}{2}ax^2 + \tfrac{1}{4}bx^4 \tag{2}$$

and,

$$H_1 = \lambda_1 x \cos\omega_1 t + \lambda_2 x \cos\omega_2 t \tag{3}$$

Throughout, $a = 20$, $b = 2$, $\lambda_1 = 10$, $\omega_1 = 6.07$, and ω_2 and λ_2 are varied. Atomic units are assumed. The frequency ω_1 is resonant with the energy difference between the ground and first excited states of the double well, neglecting tunneling splittings. H_0 describes the motion of a particle confined to a double well potential as shown in Fig. 1. The unperturbed eigenenergies below the barrier top are superimposed on the plot. In the absence of the second field, this is the same system as that studied by Lin and Ballentine[13]

The phase space plot of the unperturbed Hamiltonian H_0 is shown in Fig. 2. It is apparent that there are two different types of dynamics; motion below the top of the barrier in which the particle is confined (classically) to one well or the other and motion in the regime where the energy $E > 0$, i.e., above the top of the barrier. The two regions are separated by a classical separatrix. Phase space for the system contains two elliptic fixed points corresponding to the minima of the two wells, and a hyperbolic fixed point at the top of the barrier ($p = x = 0$). In the unperturbed case the problem is integrable and the classical equations of motion may be solved analytically in terms of elliptic functions.[11,12] It is also possible to obtain analytically sets of action-angle variables appropriate to the above barrier and below barrier cases.

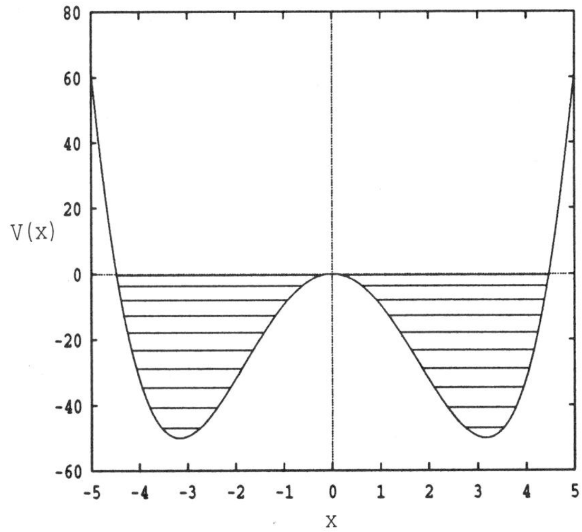

Figure 1. Double well potential energy curve with below barrier eigenenergies superimposed.

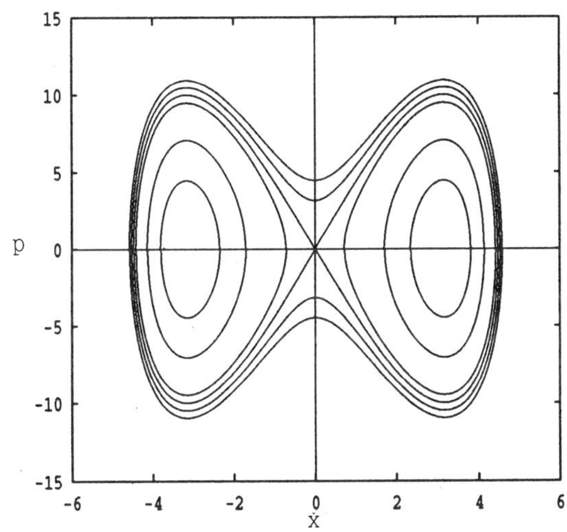

Figure 2. Phase space plot of the double well of eq. 1.

Addition of a single driving force to H_0 [i.e., $\lambda_2 = 0$ in eq. (3)] destroys the classical integrability of the system and gives rise to chaotic motion. This is evident in Fig. 3 which is a composite Poincaré surface of section (SOS) obtained by numerically integrating 10 classical trajectories with randomly chosen initial conditions. The surfaces of section were obtained by strobing the classical dynamics at times $t_n^m = \frac{2\pi}{\omega_1}(n + \frac{m}{4}), n = 0, 1, 2, ...,$ for each value of $m = 0, 1, 2, 3$. By doing this four different surfaces of section are generated corresponding to examining the dynamics at multiples of quarter periods. Note that while motion in the vicinity of the separatrix is quite strongly chaotic regular islands of regular dynamics persist in the two wells. The regular regions correspond to the existence of Kolmogorov-Arnold-Moser (KAM) surfaces or tori in an extended phase space.[12] These regions shrink as the maximum amplitude of the driving field is increased, but persist up to quite high fields.

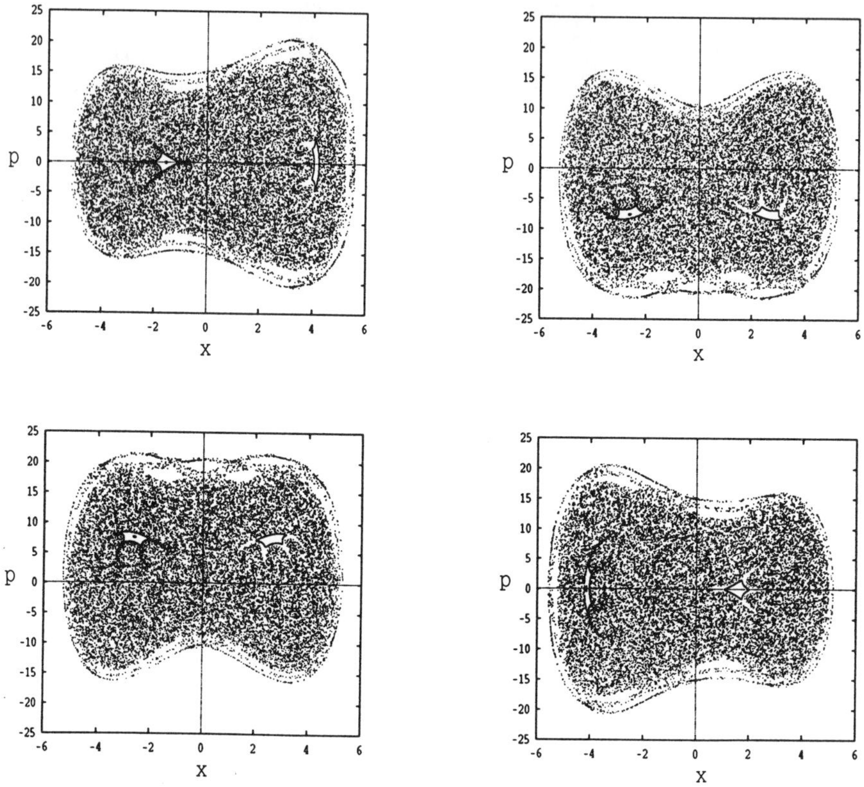

Figure 3. Poincaré surfaces of section strobed at quarter periods defined by $t_n^m = \frac{2\pi}{\omega_1}(n + \frac{m}{4}), n = 0, 1, 2, ...,$ for each value of $m = 0, 1, 2, 3$. Starting from top left and continuing clockwise, $m = 0, 1, 2, 3$. 10 classical trajectories were integrated using randomly selected initial conditions.

It is clear from an examination of Fig. 3 that the regular regions in the two-dimensional phase space rotate in each of the two wells as a function of time. At the separatrix an infinite number of resonances converge and the motion is always chaotic for any finite periodic driving force. In fact, motion in the vicinity of the separatrix for this system forms the basis of the well studied separatrix map.[12] Importantly, the regular regions (for the value of λ_1 used) are large in comparison to $\hbar$ and can thus support a number of quantum states. The symmetry of the problem is clearly apparent in that the motion in the right hand well mimics that in the left. In particular, this system is similar to the one-dimensional model of the hydrogen atom discussed by Charlie Cerjan in these proceedings[16] in that the Hamiltonian in the single frequency case possesses a discrete symmetry as pointed out by Peres,[17] namely,

$$H(p, x, t; \lambda_2 = 0) = H(-p, -x, t + \tfrac{\pi}{\omega_1}; \lambda_2 = 0) \tag{4}$$

Addition of a further driving term having a different frequency can have dramatic effects on the classical and quantum dynamics of Hamiltonian as well as non-Hamiltonian systems.[18] A second field can create new resonance zones that serve to connect the original resonances and thus enhance transport in phase space. However, for the double well potential, motion in the vicinity of the separatrix is not expected to be perturbed strongly by the existence of a second driving term because already an infinite number of classical resonances are starting to overlap in the separatrix

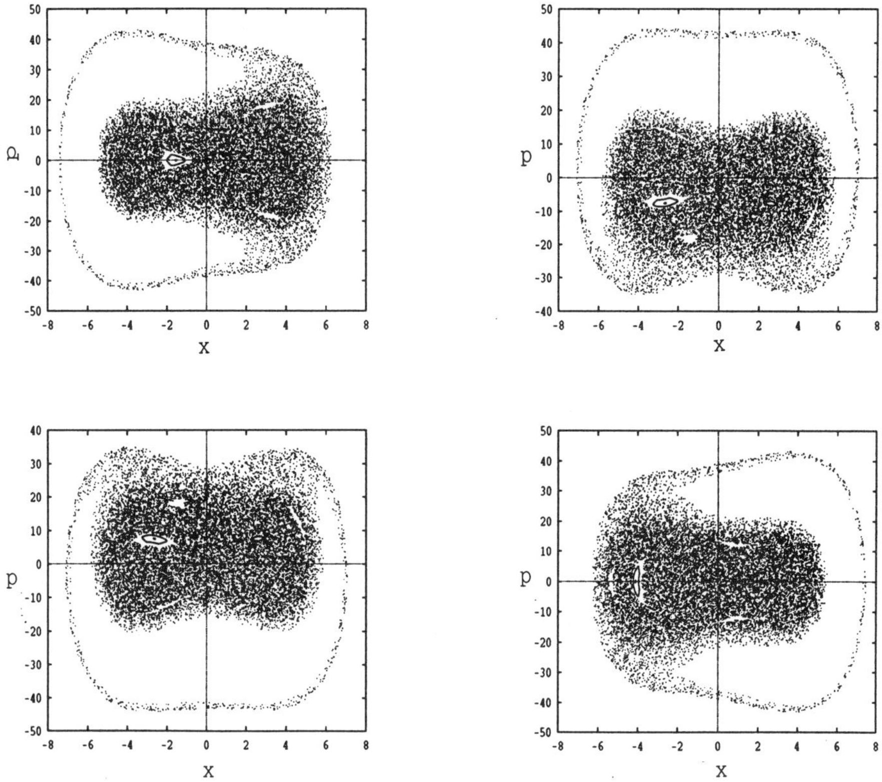

Figure 4. Poincaré surfaces of section for the case that $\omega_2 = 2\omega_1$ and $\lambda_2 = 2.5$.

region. This is borne out by SOS plots for the two-frequency case in Fig. 4 where it is seen that although new resonances are introduced outside of the separatrix region, the dynamics inside the separatrix layer is relatively similar to single frequency driving. The general picture revealed by Fig. 4 seems to be not especially sensitive to the choice of the second driving frequency. One important difference is, however, apparent in the case of a 2:1 ratio of driving frequencies; the symmetry between the two wells is broken quite markedly, indicated by the relative difference in size of the regular regions in the two wells (compare Fig. 3). This effect has significant consequences for the quantum behavior

QUANTUM MECHANICS AND HUSIMI DISTRIBUTIONS

The simplicity of the Hamiltonian (1) makes integration of the TDSE straightforward. Independent of how many driving terms are present the problem remains one dimensional in space. The TDSE was integrated in the single and double frequency cases by expanding the time dependent wavefunction $\Psi(t)$ in a harmonic oscillator basis and directly integrating the resulting set of 230 coupled first order differential equations. Rather than examining the wavefunction directly the Husimi function was computed,[13]

$$\rho = (2\pi\hbar)^{-1}| < \phi_{q,p}|\Psi(t) > |^2 \qquad (5)$$

where $\phi_{q,p}$ is a minimum uncertainty function that is Gaussian in position and momentum. The Husimi function provides a coarse grained phase space distribution and has several advantages over the Wigner distribution.[19,20] The basic idea of Lin and Ballentine was to start out a Gaussian wavepacket in one of the regular regions and follow its time evolution. They discovered that the tunneling rate between the two wells was significantly enhanced as compared to the undriven system, and secondly, tunneling between the two wells was coherent over many periods of the field. A packet started out in the chaotic region, by contrast, was found to spread or delocalize significantly on the same timescale. Fig. 5 shows computed Husimi distributions corresponding to starting out in the left hand well. After 115 periods the packet has tunneled coherently into the right hand well. Peres[17] has criticized the work of Lin and Ballentine on the grounds that the tunneling behavior can be explained simply by the existence of the symmetry of eq. (4). This in itself does not, however, explain why the rate of tunneling is enhanced so dramatically by the external field nor the localization of the Floquet states.[21]

Further insight is gained by examining not the Husimi function itself but its accumulated value (at each grid point) at the end of the total integration period. In Fig. 6 are shown the accumulated Husimi distributions corresponding to the quarter period surfaces of section of Fig. 3. Rotation of the Husimi distribution in the two wells is apparent and similar to the classical behavior (see Fig. 3). Additionally, the distributions are almost exactly symmetric upon reflection through the momentum axis. It is of interest to determine how flux exits the left well (where the packet was initially started out) and enters the right well. This can be seen by examining the accumulated Husimi distributions in Fig. 6. It is clear that while the transport appears to follow no particular channels, flux inexorably homes in on the target torus. In the classical system, integration for extremely long times reveals no penetration of the KAM tori.

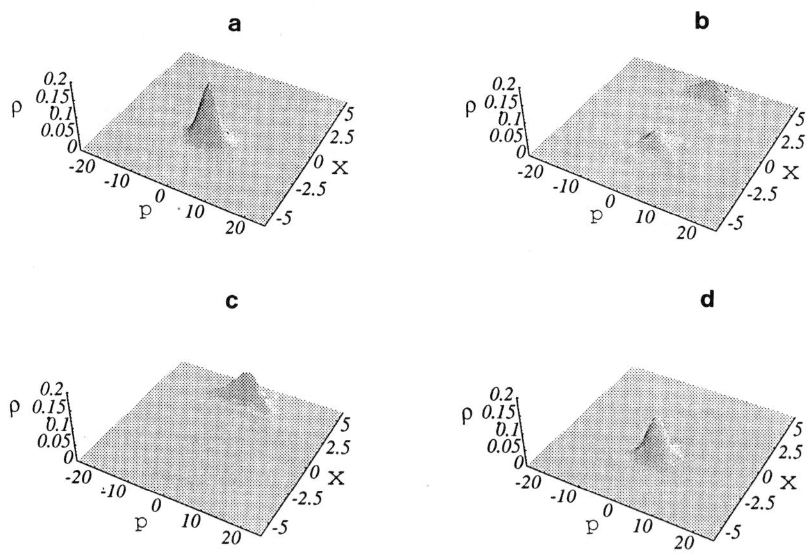

Figure 5. Husimi distributions for the single frequency case. The initial wavepacket had its maximum at $x = -1.5$ and $p = 0$. In (a) $t = 0$, in (b) $t = 60\tau$, in (c) $t = 115\tau$, and in (d) $t = 230\tau$.

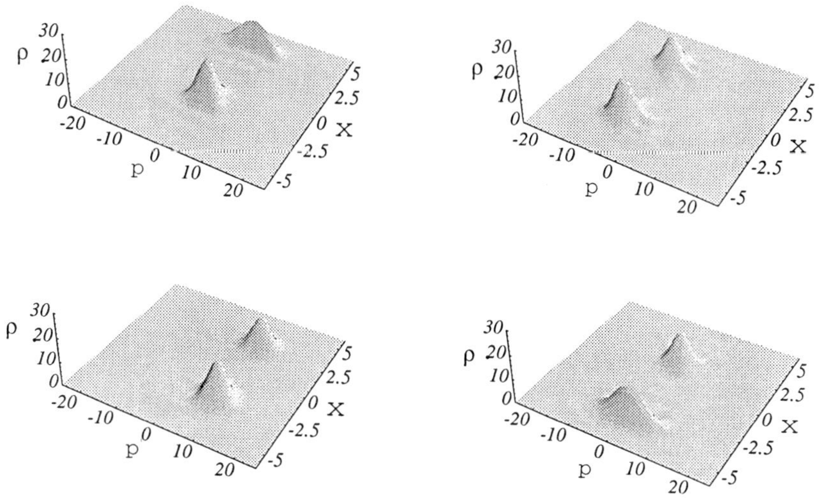

Figure 6. Accumulated Husimi distributions after 300 τ. At each grid point the value of the Husimi function is summed as a function of time. The initial state was the same as in Fig. 5. The four plots correspond to the quarter periods as defined in the caption to Fig. 3. As in Fig. 3 starting at the top left $m = 0$ and continuing clockwise, $m = 1, 2, 3$

The periodicity of the Hamiltonian allows a Floquet formulation of the problem which has also been useful in understanding the dynamics of a variety of "kicked" systems such as the one-dimensional kicked hydrogen atom. This approach has been used by Grossman and co-workers,[14,15] and involves introducing a unitary evolution operator that satisfies,[17]

$$U\Psi(x,0) = \Psi(x,\tau) \tag{6}$$

where $\tau = 2\pi/\omega_1$ is the period of the principal driving field. This essentially discretizes the problem and allows the wavefunction to be computed at any time $n\tau$ $(n = 0, 1, 2, ...)$ through the following,

$$\Psi(x, n\tau) = \sum c_k e^{in\phi_k} v_k(x) \tag{7}$$

where the v_k and c_k are defined through,

$$\Psi(x,0) = \sum c_k v_k(x) \tag{8}$$

and the v_k satisfy,

$$U v_k(x) = e^{i\phi_k} v_k(x) \tag{9}$$

For the double well driven at a single frequency the numerical results of Lin and Ballentine seem to indicate that some of the eigenfunctions v_k are localized in regular regions of phase; the fact that they belong to two symmetry classes (odd and even) means that the quasienergies will come as doublets. Grossman and co-workers[14,15] recognized that this could be exploited to completely supress tunneling by choosing an external field having the appropriate intensity and frequency so as to produce a crossing of the quasienergies and thus supression of tunneling. The timescale for tunneling is related to the splitting of the quasienergies and thus a crossing of the quasienergies shuts down the tunneling process completely. Their calculation achieved this by using frequencies comparable to the bare splitting of the levels in the unperturbed system.[14,15] Of course, because a finite basis set was used, mixing in of other levels can lift the degeneracy slightly. Nevertheless good localization of the packet was observed.[14,15] However, an alternative way of supressing tunneling exists; that is to destroy the discrete symmetry itself by use of a second driving term. This has the advantage that the very small frequencies used by Grossman and co-workers[14,15] are avoided.

In essence the idea is to use a single frequency field to dramatically enhance the tunneling rate and then to introduce a second field of lower intensity (a tickling field) to selectively turn on or off the coherent tunneling in the problem. This was tested by arranging that the second frequency be double the first, roughly coresponding to the energy difference between the ground and third excited states of the double well (neglecting splittings). The strength of the second field was 25% that of the principal field. Accumulated Husimi distributions for this case after 300τ are shown in Fig. 7 at each of the four quarter periods corresponding to the SOS plots in Fig. 4. The accumulated Husimi plots reveal that essentially no flux is transported from the left well over 300 periods of the field. The Husimi distribution merely rotates in the well (compare Fig. 4). It is apparent that the tunneling has been coherently destroyed and the packet remains localized in the left-hand well.

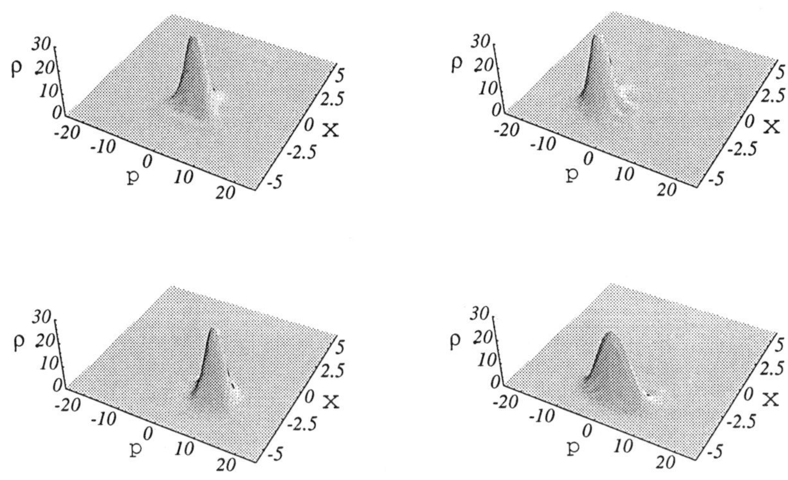

Figure 7. Husimi distributions accumulated after 300τ for the two frequency case. The initial state was the same as in Fig. 5. The four plots correspond to the quarter periods as defined in the caption to Fig. 3. As in Fig. 3 starting at the top left $m = 0$ and continuing clockwise, $m = 1, 2, 3$ The packet is localized coherently in the left hand well where it started. Compare Fig. 6.

In order to make a closer comparison with the classical mechanics, a swarm of 10,000 classical trajectories whose initial conditions satisfied the same probability distribution as the initial Husimi function was integrated for 10 periods of the principal field.

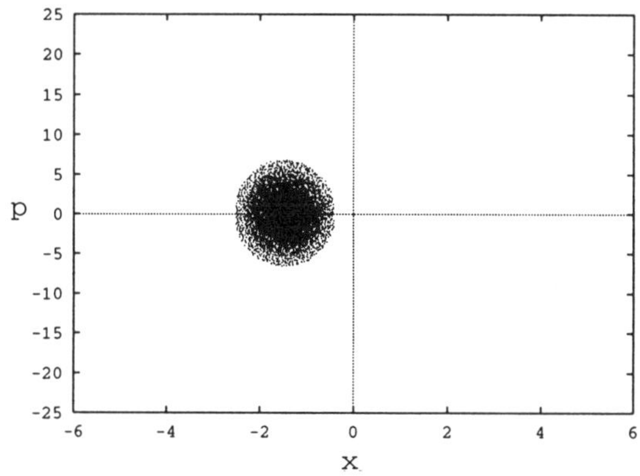

Figure 8. Swarm of 10,000 initial conditions satisfying the same probability distribution as the initial Husimi distribution.

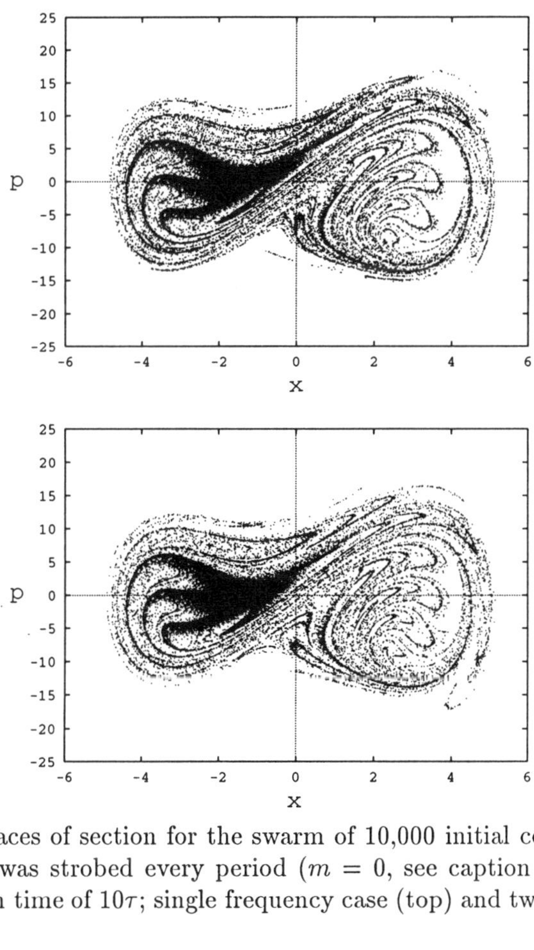

Figure 9. Surfaces of section for the swarm of 10,000 initial conditions in Fig. 8. The motion was strobed every period ($m = 0$, see caption to Fig. 3) for a total integration time of 10τ; single frequency case (top) and two frequency case (bottom).

Fig. 8 shows the initial swarm and Fig 9 shows composite SOS plots for the one and two frequency cases. The swarm and the Husimi distribution behave similarly in the single frequency case but in the two frequency case no tunneling occurs. If the packet is started in chaotic regions of phase space then rapidly delocalization occurs in both the one and two frequency cases. Examination of Fig. 9 reveals only subtle differences between the dynamics in the one and two frequency cases. The principal regular region of phase space in the right hand well in the two frequency case is smaller than that when only one frequency is present. However, it is still large enough to support quantum states. Most importantly the symmetry of eq. (4) has been broken (compare Figs. 3 and 4) and this leads to a dramatic decrease in tunneling in the same way that tunneling can be suppressed in a static symmetric double well by breaking the symmetry.[4,5,17]

CONCLUSIONS

By using two external driving forces tunneling in a driven double well system was supressed over hundreds of periods of the driving fields. The packet remained coherently localized in the regular region of phase space around which it was centered. Use of two driving fields avoids having to employ relatively small frequencies (comparable to the exponentially small splitting of the unperturbed eigenstates) in order to produce the desired effect. The main driving force has the effect of dramatically enhancing the tunneling rate while addition of a second tickling term suppresses tunneling coherently. It remains to investigate the quantal behavior for a variety of frequency ratios and to understand how and if the way the fields are turned on and off affects the rate and extent of tunneling. Future work will involve Floquet analysis of the one and two frequency cases to determine how localization of the Floquet states in regular regions of phase space occurs. Also to be studied classically and quantally is the effect of varying the frequencies, frequency ratios, relative phases and strengths of the two external driving terms. Conceivably a physical switch could be developed that relies on the ability to control tunneling in this system by use of two external fields.

ACKNOWLEDGEMENTS

We would like to thank the organizers for arranging the NATO workshop in Snowbird. DF thanks R.D. Levine for several motivating conversations and Julio Arce for early discussions of this problem.

REFERENCES

1. Z. Hund, Z. Phys., 43, 803 (1927).
2. E.B. Wilson, Jr., J.C. Decius, and P.C. Cross, "Molecular Vibrations: The Theory of Infrared and Vibrational Spectra," McGraw Hill, New York (1955).
3. R.D. Levine and R.B. Bernstein, "Molecular Reaction Dynamics and Chemical Reactivity," Oxford University Press, Oxford (1987).
4. R.A. Marcus and N. Sutin, BBA, 811, 265 (1985).
5. D.C. Rees and D. Farrelly, The Enzymes, *XIX*, 38 (1990).
6. S. Manne, H.J. Butte, S.A.C. Gold, and P.K. Hansma, Science, 56, 183 (1991).
7. F. Capasso, ed., "Physics of Quantum Electron Devices," Springer -Verlag, New York (1990).

8. C.D. Tesche, Ann. N.Y. Acad. Sci., 480, 36 (1986).

9. M.J. Davis and E.J. Heller, J. Chem. Phys., 75, 36 (1981).

10. O. Bohigas, S. Tomsovic, and D. Ullmo, Phys. Rev. Lett., 64, 1479 (1990).

11. P.J. Holmes, Philos. Trans. Roy. Soc. (London), 292, 41 (1979). L.E. Reichl and W.M. Zheng, Phys. Rev. A, 29, 2186 (1984).

12. A.J. Lichtenberg and M.A. Lieberman, "Regular and Stochastic Motion," Springer -Verlag, New York, (1983).

13. W.A. Lin and L.E. Ballentine, Phys. Rev. Lett., 65, 2927 (1990).

14. F. Grossman, P. Jung, T. Dittrich, and P. Hänggi, Z. Phys. B, 84, 315 (1991).

15. F. Grossman, T. Dittrich, P. Jung, and P. Hänggi, Phys. Rev. Lett. 67, 516 (1991).

16. C. Cerjan, these proceedings.

17. A. Peres, Phys. Rev. Lett., 67, 158 (1991).

18. J.E. Howard, Phys. Lett. A, 156, 286 (1991).

19. K. Husimi, Proc. Phys. Math. Soc. Jpn., 22, 264 (1940).

20. K. Takahashi and N. Saitô, Phys. Rev. Lett., 55, 645 (1985).

21. W.A. Lin and L.E. Ballentine, Phys. Rev. Lett., 67, 159 (1990).

22. T. Geisel, G. Radons, and J. Rubner, Phys. Rev. Lett., 57, 2883 (1986).

WAVE PACKET DYNAMICS OF RESONANCES IN UNIMOLECULAR DISSOCIATION

Stephen K. Gray

Theoretical Chemistry Group, Chemistry Division
Argonne National Laboratory
Argonne, IL 60439

1. INTRODUCTION

This paper discusses wave packet propagation and some recent three-dimensional applications to unimolecular dissociation. Emphasis is placed on direct generation of solutions to the time-dependent Schroedinger equation ($\hbar$ =1),

$$\frac{\partial|\psi(t)>}{\partial t} = -iH|\psi(t)> \quad , \tag{1}$$

and on extracting information out of such wave packets.

Wave packets are often associated with particular solutions to Eq. (1) that display, in some average manner, traces of classical trajectories. Heller's work[1] effectively exploits this picture and leads, for example, to intuitive descriptions of direct photodissociation processes. This semiclassical picture breaks down when resonances play a role in the dynamics. Wave packets then do not behave so simply, with the dissociation dynamics being generally slower and phenomena such as tunneling and interference playing important roles.

It is now possible, through the work of Kosloff and others, [2] to obtain accurate numerical solutions to Eq. (1) for systems with up to three degrees of freedom, and progress is expected in the study of some higher dimension systems, e.g., certain four-atom problems that require at least six degrees of freedom. Approximate wave packet methods, including time-dependent self-consistent field, [3] mixed classical-quantum dynamics,[4] and time-dependent golden rule[5] approaches may be of use in larger problems, since computer memory and speed limitations then make numerically accurate solutions of Eq. (1) quite difficult to obtain. The primary utility of wave packets in systems with a few degrees of freedom is in interpretation

Time-Dependent Quantum Molecular Dynamics, Edited by J. Broeckhove and L. Lathouwers, Plenum Press, New York, 1992

and visualization of dynamical processes. If all that is desired are certain energy-resolved properties, e.g., cross sections or resonance attributes, over a small energy range, or for a few specific energies, then time-independent scattering theory approaches[6,7] are usually much more efficient. This is easily seen. Time-independent methods construct stationary state solutions to Eq. (1), i.e., solutions to $H|\psi(E)\rangle = E|\psi(E)\rangle$. Since a wave packet is a large superposition of such stationary state solutions, it makes sense that the computation of a wave packet is often more difficult than generating just a few stationary state solutions. By the same reasoning, however, it is difficult to construct wave packets from time-independent scattering calculations. Thus a wave packet yields information that can complement time-independent scattering theory. In particular it will be shown here how mechanistic information can be garnered by inspection of the wave packet at various times.

Section 2 discusses aspects of methodology, including wave packet propagation techniques (Sec. 2.1), initial conditions appropriate for triatomic dissociation (Sec. 2.2), and the extraction of information out of wave packets (Sec. 2.3). Sec. 3 discusses two applications: vibrational predissociation of HeICl (Sec. 3.1) and the decay of resonances in the HCO radical (Sec. 3.2). Sec. 4 concludes.

2. METHODOLOGY

2.1 Propagation Techniques

The solution to Eq. (1) is

$$|\psi(t=\tau)\rangle = \exp(-i\tau H)|\psi(0)\rangle \quad , \tag{2}$$

where H has been assumed to be time-independent and $|\psi(0)\rangle$ is a given initial condition. Most wave packet calculations involve representing the state vector $|\psi(t)\rangle$ as a vector of complex numbers $\mathbf{c}(t)$,

$$|\psi(t)\rangle = \sum_k c_k(t)|k\rangle \quad , \tag{3}$$

with H represented by a corresponding matrix $\mathbf{H}$. This is true even when scattering is involved, since a scattering coordinate R is usually discretized, i.e. $|R_k\rangle$, $k = 1, \ldots, N_R$, replaces the continuous kets $|R\rangle$. The problem of evaluating Eq. (2) then amounts to repeated applications of the matrix $\mathbf{H}$ on a vector $\mathbf{c}$, whether one expands the exponential in a Taylor series, or whether a more efficient evaluation procedure, such as a Chebychev series,[8] or the Lanczos method,[9,10] is employed.

An efficient variation on the above approach is the following.[11] First, note that either equation

$$|\psi(t+\tau)\rangle = -|\psi(t-\tau)\rangle + 2\cos(\tau H)|\psi(t)\rangle \ , \quad (4)$$

or

$$|\psi(t+\tau)\rangle = \ |\psi(t-\tau)\rangle - 2i\sin(\tau H)|\psi(t)\rangle \ , \quad (5)$$

is formally equivalent to Eq. (2). One could, for example, use Eq. (5), once $|\psi(t=\tau)\rangle$ has been obtained (with Eq. (2)) to generate $|\psi(t=k\tau)\rangle$ over many time steps k =2,... Each new time step involves knowledge of the previous two steps, and the evaluation of an operator function acting on the current wave packet. (If $\sin(\tau H) \approx \tau H$, one has the familiar second order differencing scheme as used by Kosloff and Kosloff;[12] higher order variations, e.g., including the H^3 term, have also been used.[13]) If one writes

$$|\psi(t)\rangle = |\psi_R(t)\rangle + i\,|\psi_I(t)\rangle \ , \quad (6)$$

where $|\psi_R(t)\rangle$ is represented by Re[**c**] and $|\psi_I(t)\rangle$ is represented by Im [**c**], the following two equations are readily obtained from Eqs . (4) and (5):

$$|\psi_R(t+\tau)\rangle = -|\psi_R(t-\tau)\rangle + 2\cos(\tau H)|\psi_R(t)\rangle \quad (7)$$

and

$$|\psi_I(t+\tau)\rangle = \ |\psi_I(t-\tau)\rangle - 2\sin(\tau H)|\psi_R(t)\rangle . \quad (8)$$

H must be real for (7) and (8) to result; this is usually the case in many practical applications. The significant point is that Eq. (7) involves repeated applications of **H** on Re [**c**], i.e. real matrix-vector products, so that half the effort is involved in relation to acting **H** on **c** , as required to evaluate Eq. (2). Of course this just yields the real part of the wave packet. However to evaluate Eq. (8), thus obtaining the imaginary part as well, note that the sine operator in Eq. (8) acts on the *same* ket as the operator in Eq. (7). Usually, sufficient information to construct the act of the sine operator is generated by evaluating Eq. (7), so that Eq. (8) involves negligible additional work. Thus, up to a factor of two computational savings can be achieved with this approach. See Ref. [11] for more details.

In practice the particular matrix-vector products, either $\mathbf{H}\cdot\mathbf{c}$ or, with the alternative approach above, **H** · Re [**c**], may be divided into kinetic and potential parts. The Fourier method[11,12] then provides an accurate and efficient means of evaluating the kinetic contribution. In the application to HeICl to be discussed below, both the size of the problem and the relevant time scale are sufficiently small that the simple second order differencing approach[12,14] is used for propagation. However the HCO application is significantly larger, so that Eqs. (7) and (8) above, with a Lanczos procedure for evaluating $\cos(\tau H)|\psi_R(t)\rangle$ and $\sin(\tau H)|\psi_R(t)\rangle$, are used.[11]

2. 2 Initial Conditions for Laser Induced Dissociation of Triatomic Molecules

The applications that are of concern in Sec. 3 involve the dissociation of a triatomic molecule to form an atom and diatomic molecule, ABC $\rightarrow$ A + BC. The dynamics will be assumed to occur on a single Born-Oppenheimer potential surface, although the initial condition will be designed to reflect a non-adiabatic transition (due to laser induced absorption or emission) to it from another surface. Total angular momentum J = 0 will be assumed, which implies that with appropriate internal coordinates just three degrees of freedom suffice to describe the problem, and these are taken to be Jacobi coordinates (R, r, γ), with R = A to the BC center of mass distance, r = BC internuclear distance and γ = the angle between A, the BC center of mass and, say, the B atom. A coordinate space representation of the wave packet is then

$$\psi(R,r,\gamma,t) = \sum_{v,j} c_{v,j}(R,t)\chi_v(r)\varphi_j(\gamma) \ , \tag{9}$$

where $\{ \chi_v(r), \varphi_j(\gamma) \}$ represents a set of vibration-rotation states associated with the diatomic BC. In the applications below, asymptotic (both open and closed) BC vibration-rotation states, $\{ \chi_{v_F}(r), \varphi_{j_F}(\gamma) \}$, are used as basis set, although other choices are possible. The Hamiltonian operator and additional details may be found in Refs. [11] and [14].

The problem of determining $\psi(R,r,\gamma,t=0)$ depends on the specific experiments being simulated, or on the observable properties desired. For example, imagine there exists a lower potential surface X and that the experiments involve very short (e.g., femtosecond) laser pulses that excite ABC systems from their ground electronic and vibrational state up onto an excited potential surface B. If the electronic transition moment coupling the two surfaces can be assumed constant, then standard arguments lead to the initial state on the B surface being simply proportionate to the ground vibrational state of the X surface. (If the nuclear dependence of the transition moment is known, then the initial state on B would be the product of it and the ground vibrational state.) This state is not a stationary state of B, and hence nontrivial dynamics occurs. Furthermore, if the X and B surfaces are displaced such that the initial condition "lands" near the repulsive wall of B, then, assuming no resonances are important, the wave packet will exit the interaction region rather directly, as if following a classical trajectory. This represents the usual time-dependent picture of direct photodissociation.[1,6,7] Another advantage to this picture is that even if the experiments in question are energy-resolved ones -- and so technically the wave packet simulation bears little resemblance to the experimental situation because the laser is on for a long time -- it is still the case that a Fourier transform of the wave packet autocorrelation function leads to the energy-resolved photoabsorption cross section.[1]

Actually, there are situations that do not fall into the direct photodissociation category above, but can still be treated with wave packets. Consider vibrational predissociation in rare gas-halogen systems, e.g., HeI_2, HeICl, etc. Experiments often involve excitation from a lower state X to an upper state B, as discussed above. However, the excitation laser frequency is tuned to an $X \rightarrow B$ transition in the vicinity of a van der Waals resonance, such as a resonance corresponding to v' quanta of excitation in BC and zero-point excitation n' = b' = 0 in the van der Waals stretch and bend motions, all referred to the B state. Typically, decay times for such resonances are on the order of 10-1000 ps, so that time-resolved experiments can be based on picosecond lasers. In Ref. [15], for example, experiments were carried out with an initial excitation laser of duration 40 ps, and subsequent probing of the dynamics at time intervals on that order. This sort of initial excitation leads (via, e.g., time-dependent perturbation theory arguments) to the natural initial condition being some quasibound state on the B surface. A simple way of obtaining such states is to carry out a limited diagonalization of H within a bound state basis set. For example, one could use as basis a set of zero-order states $\psi_n(R)\eta_b(\gamma)\chi_v(r)$, where $\psi_n(R)$ represents A,BC stretching zero-order states and $\eta_b(\gamma)$ represents suitable bending zero-order states. As also stressed elsewhere,[11] it is *not necessary* to use a particularly large basis because the subsequent propagation of the state serves to pin down accurately the true resonance properties. The term metastable state is used for this sort of initial condition, which in essence is aimed at *targeting* particular (unimolecular) resonances. Such metastable states are appropriate both for simulation of real-time experiments, as in Ref. [15], or energy-resolved experiments, as discussed in Sec. IIC of Ref. [16]. In the latter case, of course, Fourier transforms of the wave packet must be used to extract energy-resolved information, as indicated in Sec. 2.3 below. Metastable states are employed as initial conditions in both applications discussed in Sec. 3.

2.3 Extraction of Information

Quantitative information includes resonance energies, rate constants (or decay widths) and product distributions. The wave packet autocorrelation function $C(t) = \langle\psi(0)|\psi(t)\rangle$, when Fourier transformed to yield a function of energy E, leads to the photoabsorption cross section (i.e. absorption spectrum). However, in the resonance decay situation, a practical difficulty occurs, namely the decay times can be so long that it is not practical to carry out a propagation to times such that C(t) is near zero. Unfortunately, this can lead to large uncertainties in decay widths inferred from the straightforward (discrete) Fourier transform of C(t). However, in the limit of one or more resonances dominating the dynamics, it is true that C(t) should be very accurately modeled by:

$$C(t) = \sum_s b_s \exp(-iE_s t)\exp(-i\Gamma_s t/2) \quad , \tag{10}$$

where b_s represents the probability of each resonance s being present in the wave packet, and the resonance positions and decay widths are given by $\{E_s, \Gamma_s\}$. Even if the duration of the wave packet calculation is such that C(t) has not completely decayed, the fact remains Eq. (10) should be accurate, so that one can attempt to fit Eq. (10) to the computed correlation function. This process can still be tricky.[11] However a reliable procedure -- as long as times longer than $2\pi/\Delta E_s$ are sampled, where ΔE_s represents the minimum energy splitting between resonances -- is given by Prony's method.[11,17] Prony's method reduces the difficult (nonlinear) fitting problem to a couple of linear least squares problems, and appears to be well suited to many quantum problems.

Energy-resolved product distributions $P_{vj}(E)$ may be obtained following Balint-Kurti, Dixon, and Marston[18] by Fourier transforming the amplitudes $c_{v,j}(R^*,t)$ of Eq. (9),where R^* represents a large value of R in the BC product channel region. Unlike the C(t) Fourier transform discussed above, good results for $P_{vj}(E)$ are often obtained without appeal to any sophisticated fitting procedures. (Again, a condition for this to be the case is that the times examined be longer than $2\pi/\Delta E_s$.)

The coordinate space wave packet is a complex function and involves three degrees of freedom and time. It is therefore difficult to properly visualize. The probability density

$$\rho(R,r,\gamma,t) = \psi^* \psi \sin\gamma \ , \tag{11}$$

is real and is such that ρ dR dr dγ is the probability the system is in (dR, dr, dγ) about (R, r, γ). Three-dimensional contour maps of ρ in coordinate space could be inspected at various snapshots in time. More simply, 2D contour plots of ρ could be inspected, with one of the coordinates either held fixed, or averaged over. What proves to be useful in the applications to be discussed is an average over r,

$$\bar{\rho}(R,\gamma,t) = \int dr\ \rho(R,r,\gamma,t) \ , \tag{12}$$

which can also be written as (using Eq. (9))

$$\bar{\rho}(R,\gamma,t) = \sum_v \rho_v(R,\gamma,t) \ , \tag{13}$$

with

$$\rho_v(R,\gamma,t) = \sin(\gamma) \left| \sum_j c_{v,j}(R,t)\ \varphi_j(\gamma) \right|^2 . \tag{14}$$

Eq. (12) represents the probability density for internal coordinates associated with the position of atom A, regardless of the BC internuclear distance. It is therefore particularly useful for developing a picture of the fragmentation dynamics. The associated vibrational channel probability densities given by Eq. (14) can also assist in developing a mechanistic picture, since sometimes vibrational transitions in the interaction region are relatively localized (e.g. occurring near the repulsive part of the potential wall), with dynamics away from such regions being essentially adiabatic.

3. APPLICATIONS

3.1 HeICl(n' = b' = 0, v') → He + ICl(v_F < v', j_F) Vibrational Predissociation

Vibrational predissociation involves transfer of vibrational energy from the ICl part of the molecule to the van der Waals modes. Calculations on v' = 3 excitations are outlined here, which can be compared with experimental results.[19] An appropriate initial condition,

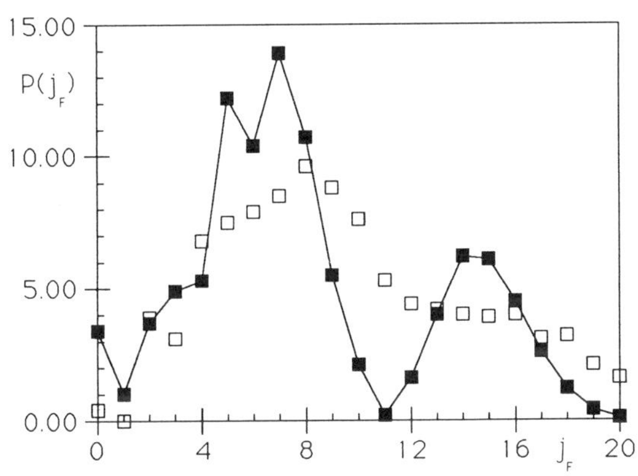

Figure 1. Energy-resolved rotational product distribution associated with ICl(v_F = 2, j_F) products arising from vibrational predissociation of HeICl(n' = b' = 0, v' = 3). The mean energy of the wave packet was used as energy. The solid line corresponds to the calculated results of Ref. [14], and the open squares correspond to the experimental results of Ref. [19].

corresponding to a metastable state with n' = b' = 0 and v' = 3 is calculated and then propagated in time, as discussed in Sec. 2 above. Additional details of the calculations (potential energy surface, etc.) may be found in Ref. 14. In this case it turns out just one resonance dominates the dynamics, and it correlates well with the nature of the initial metastable state. Figure 1 displays the wave packet based rotational product distribution for v_F = 2 and contrasts it with experimental results.[19] The v_F = 2 product channel is the dominant product channel in this case, consistent with a simple vibrational predissociation involving just a one quantum loss in ICl. Notice in particular the bi-modal character in the theoretical distribution. The experimental distribution tends to show a secondary structure too, but is not explicitly bi-modal. This is probably due to the experiments really being an average over several total angular momentum states. Experimental and time-independent theoretical work on the related $HeCl_2$ system have also shown bi-modality in rotational product distributions, and it has been suggested that interference effects may play some kind of role.[20] The calculations and analysis outlined here explicitly corroborate this speculation, and also explain why the analogous van der Waals systems, but with He replaced by Ne, do not show such strong bi-modality.[14]

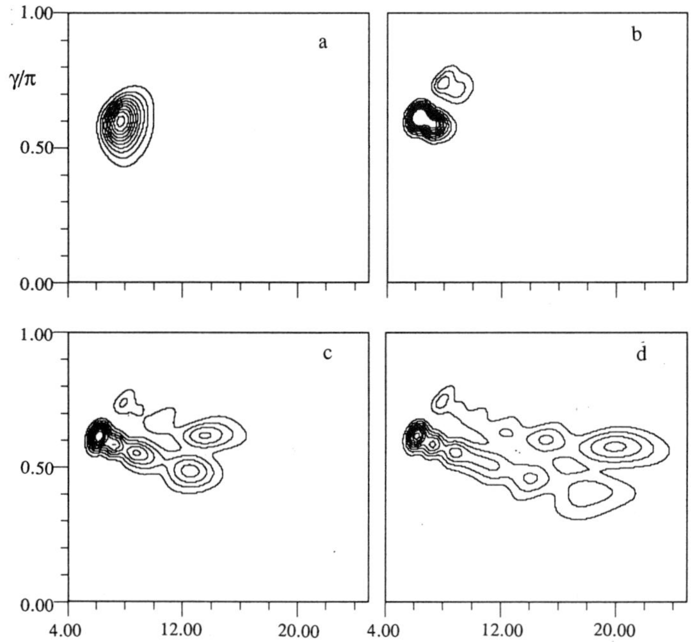

Figure 2. Reduced density plots for HeICl vibrational predissociation: (a) corresponds to the initial reduced density $\overline{\rho}(t=0)$; (b) - (d) correspond to the $v = 2$ contribution to the reduced density, $\rho_2(t)$, at times 0.2, 0.6 and 1.0 ps, respectively. The actual magnitudes of the contours in (b)-(d) are about 100 times smaller than in (a), reflecting the fact that only small amounts of density form in the reactive channels.

Fig. 2a depicts the reduced density $\overline{\rho}$, Eq. (12), at $t = 0$. This is the initial metastable state, so only $v = v' = 3$ contributes to the summation of Eq. (13). The maximum density occurs at $\gamma \approx 0.6\ \pi$ (or 110°), and $R = 7$ a.u., consistent with He being closer to Cl than I at equilibrium. Density in the $v = 3$ channel is essentially confined to the interaction region, because products with $v_F = 3$ cannot easily form on energetic grounds. (Technically, there could be high energy components to the wave packet that do decay along the $v = 3$ channel, but in practice the metastable state initial condition is sufficiently narrow energetically that this does not occur.) Dissociation to He + ICl therefore occurs via density dropping down from the v = 3 channel in the interaction region to the open channels, and the v = 2 channel makes up most of this reactive density. As time goes on, very small amounts of density begin to grow in the interaction region part of the v = 2 channel, and then move out to large R, corresponding to dissociation. Therefore, Figs. 2b-2d focus on the density $\rho_{v=2}$, Eq. (14). Fig. 2b displays ρ_2 at t = 0.2 ps. The fastest vibration present is the ICl vibration, which has period 0.2 ps, so that 0.2 ps may be thought of as a natural time unit for this problem. It is much shorter than the actual lifetime of the complex, which is estimated to be[14] $\hbar/\Gamma \approx 1000$ ps Interestingly, *two* lobes of density have formed at this relatively short time. Figs. 2c-2d display the reactive density at times 0.6 ps and 1.0 ps, respectively. It is clear that He + ICl products are associated with density flowing out of the two lobes that form at short times. The wave packet density always has an appearance similar to Fig. 2d, even for much longer times. It is natural to suspect that the bi-modality in the product distributions is related to the formation of the two lobes over short times.

Before demonstrating explicitly how the bi-modality arises out of the two lobes, it is worthwhile to explain why there are two lobes of reactive density (Fig. 2b) that appear shortly after beginning to propagate. By making a simple first order Taylor series approximation to the propagator, Eq. (2), it is easy to show[14] that reactive density after a short time τ is of the form

$$\rho_2(R,\gamma,\tau) \propto \left(\frac{\partial W}{\partial r}\right)^2_{r=\bar{r}} \bar{\rho}(R,\gamma,t=0) \ , \tag{15}$$

where W is the interaction potential (full potential minus the ICl diatomic interaction) and the partial derivative is evaluated at the ICl equilibrium distance associated with v'. Notice that $\partial W / \partial r$ represents a kind of force related to energy loss from ICl vibration due to the presence of He. When viewed as a function of R and γ it can be either negative or positive; it turns out to have a node, i.e. to go from negative to positive, almost exactly down the middle of the two lobes in Fig. 2b. If the initial density $\bar{\rho}(R,\gamma,t=0)$ is sufficiently broad that it is large on both sides of the node, Eq. (15) leads directly to the formation of two lobes, with one lobe being associated with positive force and the other with negative force. In the Ne cases (e.g. NeICl) it turns out that the initial density $\bar{\rho}(R,\gamma,t=0)$ is much more localized to one side of the nodal line. One then gets most of the reactive density in one of the lobes, and this is consistent with the lack of strong bi-modal product distributions in these cases.

A computational experiment serves to illustrate how the bi-modal product distribution arises.[14] Consider each of the short time lobes in Fig. 2b as an initial condition for a new wave packet propagation. In these calculations, one starts with only wave function in the v = 2 channel that corresponds to one or the other of the lobes in Fig. 2b. Rotational product distributions associated with the separate lobes are deduced and displayed in Fig. 3a. The solid curve in Fig. 3a corresponds to propagation of the lower lobe in Fig. 2b, and is seen to lead to a broad rotational distribution with a maximum at $j_F = 7$. The upper lobe in Fig. 2b

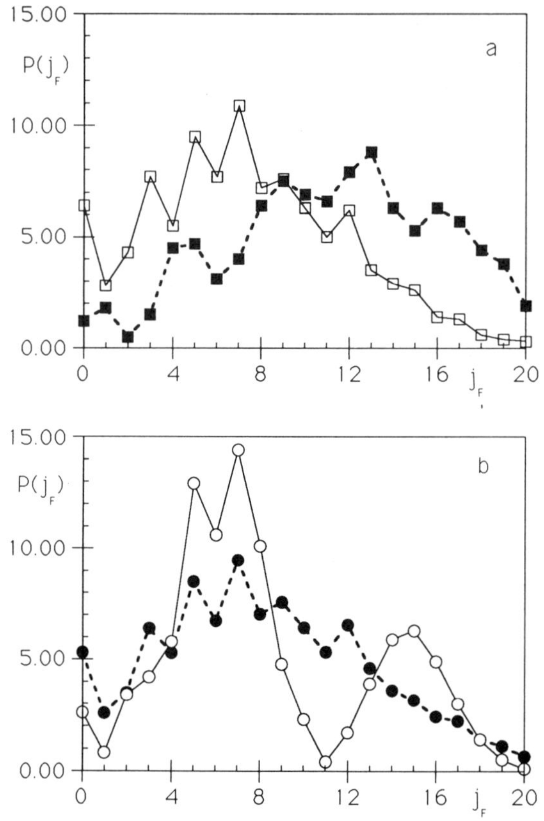

Figure 3. Rotational Product distributions inferred from propagation of the lobes in Fig. 2b. (a): The open squares connected by solid lines correspond to the result of propagating the lower lobe, i.e. the lobe centered on somewhat smaller γ values. The filled squares connected by dashed lines are the result of propagating the upper lobe. (b): The open circles connected by solid lines correspond to adding appropriately weighted wave packet amplitudes from the two propagations, and then squaring the result to obtain the probability distribution. The filled circles connected by dashed lines correspond to simply adding the (weighted) distributions in (a).

leads to the dashed curve in Fig. 3a, i.e. a broad distribution with maximum at a somewhat higher $j_F = 13$. Suppose the two distributions in Fig. 3a, each weighted by the integrated density of its associated lobe (the lower lobe has weight 0.79 and the upper lobe 0.21) are simply added. The result is the dashed curve in Fig. 3b. This does not look much like the full product distribution, Fig. 1. Suppose, however, one adds the appropriately weighted amplitudes (not probabilities) associated with products from the separate lobe propagations, and then squares the overall result to obtain a probability distribution. The result, shown as the solid line in Fig. 3b, is remarkably close to the true product distribution in Fig. 1. This shows clearly that it is an interference phenomenon that is leading to the bi-modality.

3.2 HCO(n', b',v') $\rightarrow$ H + CO($v_F = 0, j_F$) Resonance Decay

Ground electronic state HCO (formyl radical) has been shown both theoretically[21,22] and experimentally[23] to possess isolated resonances. The resonances of concern here are those that lie above the barrier to form products H + CO but below the energetic threshold to form vibrationally excited products. (Actually, HCO is an important system for combustion, and also exhibits a number of other interesting dynamical features, including a Renner-Teller coupling [24] of ground and first excited electronic states.)

A reasonable, *ab initio* potential energy surface is available for HCO, as well as time-independent estimates of the resonance properties.[21,22] Suppose the energy zero is taken to be separated H + CO, with CO at its equilibrium distance. Then the minimum energy corresponding to HCO is -6777.9 cm^{-1} and there is a barrier to form H + CO with energy 561.4 cm^{-1}. A recent wave packet study was able to reproduce the relevant resonance properties, and suggest a partial explanation of some features in the rotational product distributions.[11] It should be noted that such calculations are much more difficult than for, say, HeICl discussed above. This is because the forces involved are more chemical in nature, leading to the requirement of a larger basis set in the expansion of Eq. (9), as well as smaller grid spacings in the discretization of R. As illustration, the decay of a metastable state corresponding to n' = 0, b' = 4 and v' = 1 will be considered. The approximate mean energy of this state is 2160 cm^{-1}, which implies it is well above the barrier to form H + CO . (There are also 15 purely bound states, as well as 7 resonance states, that lie below this particular resonance.) Fig. 4a depicts the initial metastable state density, $\overline{\rho}(R,\gamma,t=0)$. The five islands are consistent with the b' = 4 bending quantum number assignment. Fig. 4b depicts the density $\overline{\rho}$ at t = 0.2 ps. The lifetime of this state has been estimated to be 6 ps, but examination of $\overline{\rho}$ at all times subsequent to t $\approx$ 0.2 ps yields a picture very similar to that in Fig. 4b: a protrusion of reactive density extends out from the interaction region. Notice that the protrusion is limited in angular extent, i.e. is contained mostly within the γ range ($\pi/4$, $\pi/2$). Actually, the transition state to form H + CO occurs at about $\gamma = \pi/4$ (and R = 4.2 a.u.), so that one has the result here that the density, when viewed along γ, is significantly localized to one side of the transition state.

Fig. 5 displays the rotational product distribution calculated for the resonance, and contrasts it with the result of time-independent scattering theory calculations.[21] The agreement is quite good. Notice that there is a $\Delta j_F = 3$ pattern in the maxima and minima, with, e.g., $j_F = 0$, 3, and 6 being maxima. Actually most of the resonances examined, despite having very different lifetimes and approximate quantum number assignments, lead to a similar $\Delta j_F = 3$ pattern.[11,21] It is also the case that these resonances lead to angular localization in the exiting reactive density as noted in the above paragraph. It is natural to suspect a corre-

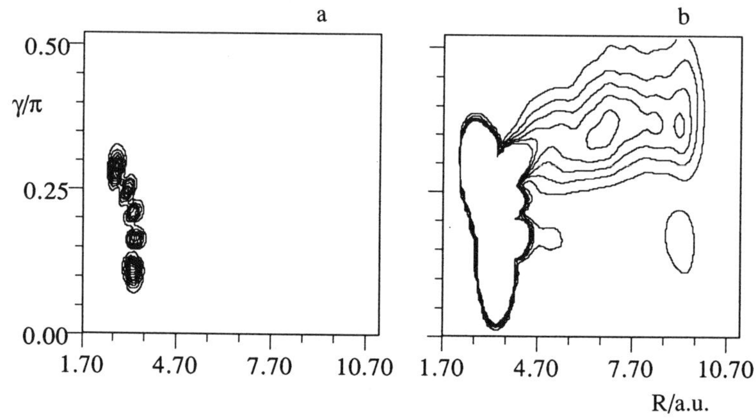

Figure 4. Reduced density plots for HCO resonance decay: (a) corresponds to the initial reduced density, $\bar{\rho}(t=0)$; (b) corresponds to $\bar{\rho}(t)$ at $t = 0.2$ ps. In this latter case, only low density contours (100 times smaller than the maximum density) are shown, since all reactive density is associated with such small values. (The R = 2.7-3.7 a.u. region still contains a structure similar to that in (a); it is just not evident when very low contours are displayed.)

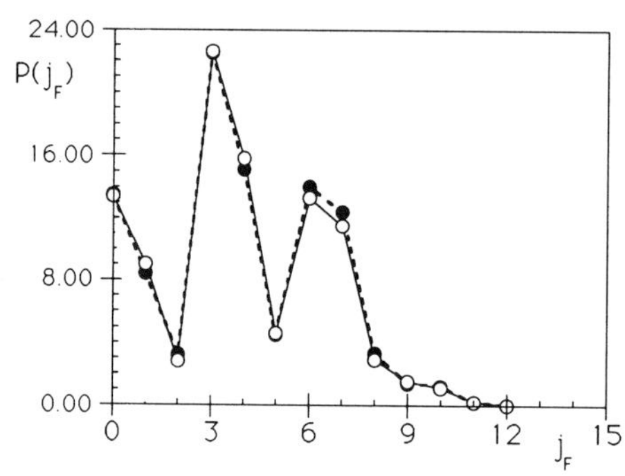

Figure 5. Rotational product distributions associated with the HCO (n' = 0, b' = 4, v' = 1) resonance. The filled circles correspond to the wave packet results, and the open circles are the time-independent scattering results of Ref. 21.

lation between such structured product distributions and the angular localization in the wave packet. A very simple analysis,[11] in fact, shows that the rotational states $\varphi_{j_F}(\gamma)$ which correspond to $j_F = 2$, 5, and 8 all have nodes right near the maximum of the exiting angular density. This in turn forces a pattern of minima at $j_F = 2$, 5, and 8, and, in turn, a $\Delta j_F = 3$ of maxima. (Of course the actual product distribution does not display a minimum for $j_F = 8$, but the probability is quite small for $j_F > 8$.) The analysis here is still incomplete in comparison with what has been outlined for HeICl above. In the HeICl case, it was also possible to also understand physically why the reactive density appeared as it did (cf. Eq. (15)), which then led to interesting consequences for the rotational product distributions. With HCO, the structured patterns in the product distributions can be correlated with angular localization of the exiting wave packets, but a full explanation of why the angular localization occurs remains to be given. (Part of the difficulty in obtaining a full explanation lies in the fact that while it would *appear* that the resonance decay here is very similar to van der Waals vibrational predissociation, i.e. v' = 1 describes CO in the complex, and the products are $v_F = 0$, there is really much more vibrational mixing present. To describe v' = 1 in the complex requires a superposition of up to 6 asymptotic CO vibrational states.)

4. CONCLUDING REMARKS

Wave packet calculations can offer insight into the mechanisms of chemical reactions, even when the dynamics is dominated by strong quantum effects. The two unimolecular dissociation reactions discussed here were dominated by resonances. These resonances led to lifetimes that were long in comparison to typical wave packet propagation times. However, it was still possible to extract out good estimates of the resonance attributes and product distributions. More importantly, inspection of the wave packet (or appropriate reduced densities) yielded in both cases useful insights into the nature of the dynamics. In the HeICl case, for example, an interference effect was identified as being the source of bi-modal product distributions. In HCO, angular localization of the exiting wave packet density was found to lead to very specific patterns in the product distributions.

Several future directions deserve to be mentioned. First, with respect to van der Waals vibrational predissociation, wave packet work[25] on the ArI_2 system shows that another interesting quantum effect, the accidental near degeneracy of van der Waals resonance states, plays an important role. This effect has also recently been seen in work[26] on $ArCl_2$. The NeI_2 system, for which there is extensive experimental lifetime data,[15] displays a simpler and more direct wave packet dynamics,[27] which has the advantage that theoretical calculations are also quite easy. There is therefore an opportunity to carry out a detailed optimization of the potential surface, along with a much more extensive experimental/theoretical comparison. Wave packet work on four-atom clusters such as He_2Cl_2 and Ne_2Cl_2, for which there is interesting experimental data regarding dissociation pathways,[28,29] has also begun.[30] Finally, as mentioned in Sec. 3.2, the HCO system displays an interesting non-adiabatic effect known as the Renner-Teller effect, due to its excited electronic state being a linear equilibrium geometry. In this effect, overall rotation can effectively mix electronic states and work has been initiated to accurately simulate it with wave packets.[31,32]

Acknowledgments: The author is grateful to F. Le Quéré for many helpful discussions. This research was supported by the Office of Basic Energy Services, Division of Chemical Science, U.S. Department of Energy, under Contract No. W-31-109-ENG-38.

REFERENCES

1. E. J. Heller, Acct. Chem. Res. 14:368 (1981).
2. R. Kosloff, J. Phys. Chem. 92:2087 (1988).
3. R.B. Gerber, V. Buch, and M. Ratner, J. Chem. Phys. 77:3022 (1982).
4. R. Kosloff and C. Cerjan, J. Chem. Phys. 81:3722 (1984).
5. (a) P. Villarreal, S. Miret-Artes, O. Roncero, G. Delgado-Barrio, J.A. Beswick, N. Halberstadt, and R.D. Coalson, J. Chem. Phys. 94:4230 (1991). (b) D.H. Zhang and J.Z.H. Zhang, J. Chem. Phys. 95:6449 (1991).
6. M. Shapiro and R. Bersohn, Annu. Rev. Phys. Chem. 33:409 (1982).
7. P. Brumer and M. Shapiro, Adv. Chem. Phys. 60:371 (1985).
8. H. Tal-Ezer and R. Kosloff, J. Chem. Phys. 81:3967 (1984).
9. T.J. Park and J.C. Light, J. Chem. Phys. 85:5870 (1986).
10. C. Leforestier, R. Bisseling, C. Cerjan, M.D. Feit, R. Freisner, A. Guldberg, A. Hammerich, G. Jolicard, W. Karrlein, H.D. Meyer, N. Lipkin, O. Roncero, and R. Kosloff, J. Comput. Phys. 94:59 (1991).
11. S. K. Gray, J. Chem. Phys. Vol. 96 (in press, 1992).
12. D. Kosloff and R. Kosloff, J. Comput. Phys. 52:35 (1983).
13. U. Manthe and H. Koppel, J. Chem. Phys. 93:345 (1990).
14. S.K. Gray and C. E. Wozny, J. Chem. Phys. 94:2817 (1991).
15. D.M. Willberg, M. Gutmann, J.J. Breen, and A.H. Zewail, J. Chem. Phys. 96:198 (1992).
16. S.K. Gray and C.E. Wozny, J. Chem. Phys. 91:7671 (1989).
17. S. Marple, Jr., "Digital Spectral Analysis with Applications," Prentice-Hall, Englewood Cliffs, N.J. (1987).
18. G.G. Balint-Kurti, R.N. Dixon, and C.C. Marston, J. Chem. Soc. Faraday Trans. 86:1741 (1990).
19. R.L. Waterland, J.M. Skene, and M.I. Lester, J. Chem. Phys. 89:7277 (1988).
20. J.I. Cline, B.P. Reid, D.D. Evard, N. Sivakumar, N. Halberstadt, and K.C. Janda, J. Chem. Phys. 89:3535 (1988).
21. S.-W. Cho, A.F. Wagner, B. Gazdy, and J.M. Bowman, J. Chem. Phys. 96:2799 (1992).
22. S.-W. Cho, A.F. Wagner, B. Gazdy, and J.M. Bowman, J. Chem. Phys. 96:2812 (1992).
23. A.D. Sappey and D.R. Crosley, J. Chem. Phys. 93:7601 (1990).
24. J.-C. Loison, S.H. Kable, P.L. Houston, and I. Burak, J. Chem. Phys. 94:1796 (1991).
25. S.K. Gray, to be submitted to Chem. Phys. Lett.
26. N. Halberstadt, J.A. Beswick, and K.C. Janda, J. Chem. Phys. 96:2404 (1992).
27. S.K. Gray, work in progress.
28. W.D. Sands, C.R. Bieler, and K.C. Janda, J. Chem. Phys. 95:729 (1991).
29. S.R. Hair, J.I. Cline, C.R. Bieler, and K.C. Janda, J. Chem. Phys. 90:2935 (1989).
30. F. Le Quéré and S.K. Gray, to be submitted to J. Chem. Phys.
31. R.N. Dixon, Chem. Phys. Lett. 190:430(1992).
32. E. M. Goldfield and S.K. Gray, work in progress.

TRANSITION REGION DYNAMICS OF THE REACTION

K + NaCl + hν → KCl + Na*

Bruce R. Johnson

Department of Chemistry and Rice Quantum Institute
Rice University
Houston, Texas, 77251

INTRODUCTION

The crossed-beam experiments of Maguire, et. al.,[1] have yielded what is believed to be a direct photoexcitation spectrum of the reaction complex in the collision of K and NaCl,

$$K + NaCl + h\nu \;\rightarrow\; [KClNa] \;\rightarrow\; KCl + Na^* \quad . \tag{1}$$

This excitation spectrum is measured by monitoring emission from Na* that is determined to arise only in the circumstances that K, NaCl, and the cw laser light all impinge on each other. More recent experiments by Barnes, et. al.,[2] have reproduced the earlier findings for the Cl reaction, and have produced new and striking results for the analogous reactions of Br and I analogs.

As shown in Fig. 1, all three K + NaX + hν (X = Cl, Br, I) excitation spectra possess a higher energy (blue) shoulder that increases in the direction of the Na D-line ($^2P_{3/2}$–$^2S_{1/2}$ transition) wavelength of 589.0 nm. The presence of scattered light in the experiment at wavelengths closer to the D-line makes it difficult to pursue this feature to shorter wavelengths. On the longer wavelength side, however, the K-NaBr and K-NaI excitation spectra each show an obvious lower-energy (red) feature that peaks around 610 nm. This leads to the suggestion that it is also present in the NaCl reaction, although much broader and peaked around 640 nm. The source of this (weak) structure is currently unknown, a situation which has sparked the present theoretical investigation into the quantum dynamics of Reaction (1).

Yamashita and Morokuma[3,4] have recently carried out ab initio calculations for the ground and first two excited states of the KClNa molecule, all of $^2A'$ symmetry. Dropping the doublet notation for convenience, these are labelled 1A', 2A', and 3A'. Global analytical potential energy surfaces for the

Time-Dependent Quantum Molecular Dynamics, Edited by J. Broeckhove
and L. Lathouwers, Plenum Press, New York, 1992

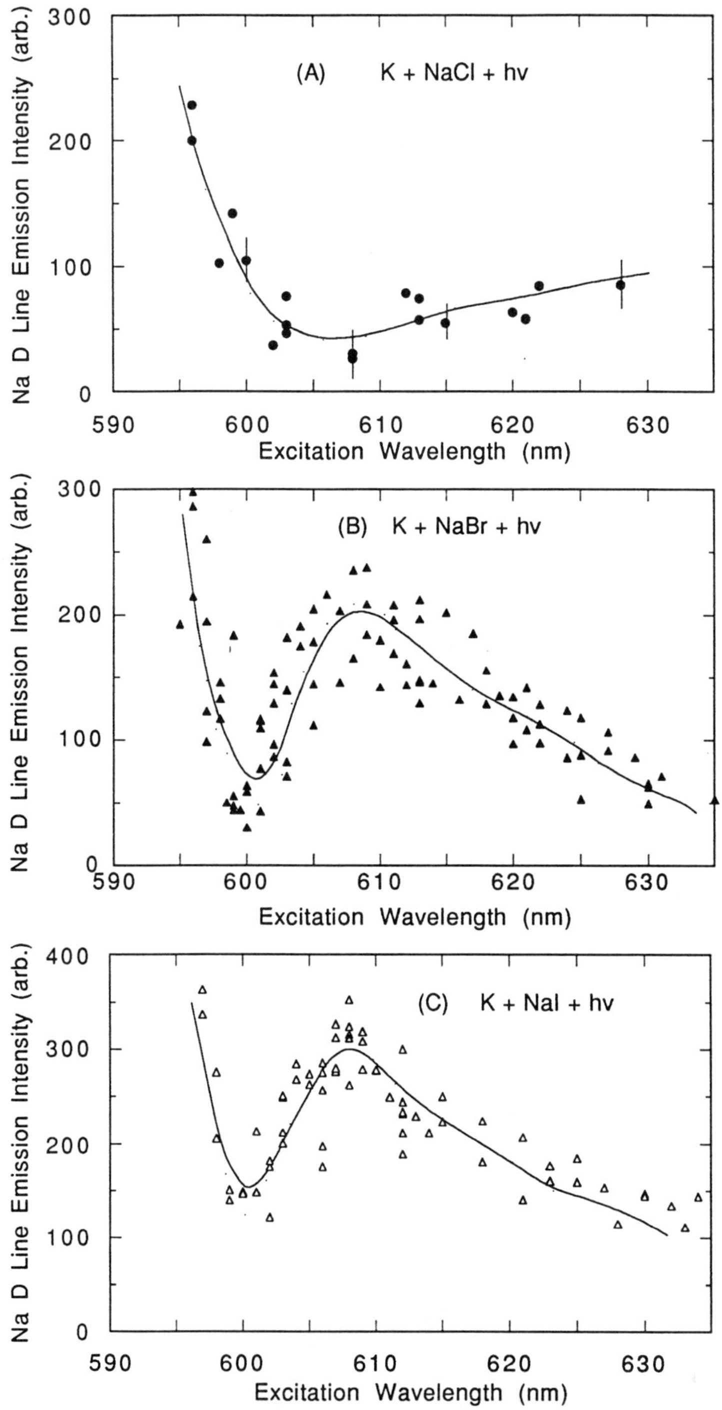

Fig. 1. Photoexcitation spectra as measured by Na* emission for (A) K + NaCl + hν, (B) K + NaBr + hν, (C) K + NaI + hν.

1A' and 3A' states were constructed, since absorption to the 3A' state was calculated to be the only process falling in the energy range of the experiment. Yamashita and Morokuma carried out classical trajectory calculations within the dressed-molecular-surface picture using Landau-Zener transition probabilities to approximate the rate of surface-hopping. Similar calculations were later performed in Ref. 2 to determine if a simple mass effect could explain the red feature seen in the NaBr reaction. No nonuniformity in the amount of Na* produced with excitation wavelength was found, leaving the origin of the red feature still in mystery. While not the initial focus in either case, it also turns out that neither set of classical calculations showed any sharp variation of Na* production in the wavelength regions where the blue features are observed. This is disturbing since the blue features in Fig. 1 can all be seen to vary in magnitude by factors of 2-3 within the range between 595 and 600 nm.

It is a critical point that the observed spectral structure is much narrower than the distribution of energies accessible under the conditions of the experiment, ca. 3350 cm^{-1} translational + internal initial energy.[1] This structure somehow survives the averaging over initial conditions, yet nothing has been seen in the classical calculations.

While the masses involved in the K + NaX systems are certainly large enough for quantum effects to be minimized, there is still the possibility that the disagreement between the classical calculations and the experimental findings derives from an interference effect most reliably calculated by performing fully quantum calculations. This is investigated here by numerical propagating the solution of the time-dependent wave equation for the laser-coupled electronic states through the interaction region. By examining the details of the potential surfaces and the positions of the spectral features (as far as we know them) relative to the Na D-line, we expect the most critical region to correspond to the Na being within 10 Å or so of the KCl molecule.

For this initial investigation, only collinear geometries are considered. While any quantitative analysis will require higher dimensionality (since the reaction is not direct), the reduced dimensionality of the restriction to collinear geometry facilitates the search for strong wavelength variations in the yield of Na*.

TIME-DEPENDENT CALCULATIONS

We need to solve the time-dependent dressed-molecule equation

$$\left[\begin{pmatrix} T + V_{1A'} + h\nu & \mu E \\ \mu E & T + V_{3A'} \end{pmatrix} - i\hbar\frac{\partial}{\partial t}\right]\begin{pmatrix}\psi_1\\ \psi_3\end{pmatrix} = \begin{pmatrix}0\\0\end{pmatrix}, \tag{2}$$

obtained by integration over the coordinates of the radiation field. Here ψ_1 and ψ_3 are the wave functions of the nuclear variables in electronic states 1A' and 3A'. Besides $V_{1A'}$ and $V_{3A'}$, Yamashita and Morokuma[3] also calculated the orientation-averaged electronic dipole moment μ for the transition. While it generally increases as the K-Cl-Na bond angle opens, it does not vary strongly with nuclear coordinates, and was therfore replaced approximately by an average value.[4] We also adopt the choice of field strength made by Yamashita and Morokuma, so that the coupling between states is constant, $\mu E = 0.1$ kcal/mole. This is actually a much greater coupling than is appropriate to the experiments, but the effects will primarily be found in the absolute amount of probability transferred between the states and the absolute yield is not measured in the experiments.

Slight changes have been made in the exit channels of the 1A' and 3A' surfaces used here in order to ensure that the asymptotic behavior as Na or Na* leaves the interaction region is reasonably correct over an extended range. This is accomplished[5] by including spin-orbit and leading electrostatic interactions (through quadrupole order) treating KCl as two point charges, i.e., K^+Cl^-. As it turns out, the differences between the original surfaces and those used here are relatively small over the region investigated in the present work.

Equation (2) is analogous to that for ordinary (field-free) non-adiabatic coupling between the nuclear-coordinate wave functions belonging to two different electronic states. The lack of derivative coupling in the present application implies that the precise analogy would be to a diabatic representation of the coupled states (although the latter is only well-defined in the diatomic case). Regardless of the extent of the analogy, we are free to use any of the methods that are used to solve non-adiabatic coupling problems for solution of Eq. (2).

The numerical method used is the Alvarellos-Metiu[6] generalization of the split-operator method. The split-operator method approximates the propagator for Hamiltonian $H = T + V$ and a small time step Δt by a product of exponential operators

$$\exp\left[-\frac{i\Delta t}{\hbar}H\right] \rightarrow \exp\left[-\frac{i\Delta t}{2\hbar}V\right]\exp\left[-\frac{i\Delta t}{\hbar}T\right]\exp\left[-\frac{i\Delta t}{2\hbar}V\right] , \tag{3}$$

where the error made is $O[(\Delta t)^3]$. Imposition of discrete grids in coordinate and momentum spaces then allows highly efficient FFT transformation between the two, where T and V are alternately diagonal. In the two-state generalization of the split-operator method,[6] one also transforms the two-state Hamiltonian back and forth between diabatic and adiabatic representations; this alternately eliminates derivative or coordinate coupling depending on whether one is evaluating T or V, respectively. This is a time-consuming step of the calculations. In the current applications, the transformation between representations is avoided at grid points where the two potential energies $V_{1A'}+h\nu$ and $V_{3A'}$ are both higher than a preset threshold. This threshold is chosen such that the results of the propagation are insensitive to its precise value.

The reaction is slightly exothermic in the ground state and slightly endothermic in the excited state. This causes some complications in that the probability amplitude tends to move off the numerical grid in the forward direction for the ground state (i.e., KCl + Na) and in the backward direction for the excited state (i.e., K* + NaCl). We have adopted the technique of Kosloff and Cerjan[7] which uses a window function to gently kill off the wave function at the ends of the grids. In our case we have found a simple $\cos^2 x$ window (0 <

$x \leq \pi/2$) spread over the last 12 grid points of the entrance and exit channels of both surfaces to suffice. The amount of probability density that is subtracted from the excited state exit channel (Na* + KCl) is accumulated over time as an approximate measure of the total flux leading to observation of Na* emission.

RESULTS

Figure 2 shows time slices of the absolute values of the wave functions on the appropriate potential surfaces for a laser wavelength of 600 nm. The initial wave function on the 1A' surface is the product of a v = 0 NaCl eigenstate and a complex gaussian in the distance between the K atom and the NaCl center of mass. The parameters of the latter are chosen to reproduce the $\approx 10^3$ m/sec relative velocity of the experiment and to be lower in energy spread than the initial translational distribution. The lone thicker contour around the wave function represents 2% of the next highest (solid) contour, and is intended to provide a picture of the fringes of the wave function.

Considerable dynamical complexity develops on the ground state surface during the next few hundred fs corresponding to the snarled trajectories found classically, although only a nominal amount of probability density is transferred to the excited state. After 400 fs, part of the 1A' wave function approaches the seam cutting across the exit channel (see Ref. 4) where the dressed surfaces cross and a considerable amount of probability density begins to transfer to the 3A' state.

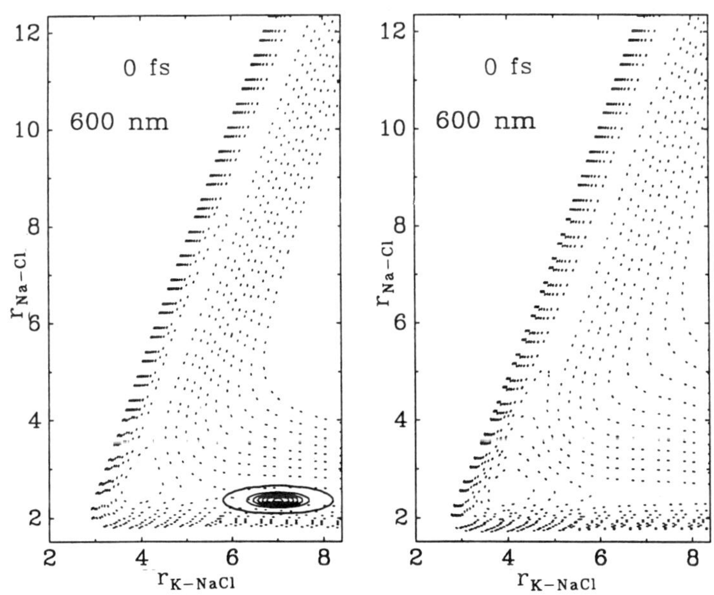

Fig. 2. Propagation of wave function through interaction region on the 1A' (left) and 3A' (right) potential surfaces. The thicker outer contour is only 2% of the value of the next highest contour. (continued)

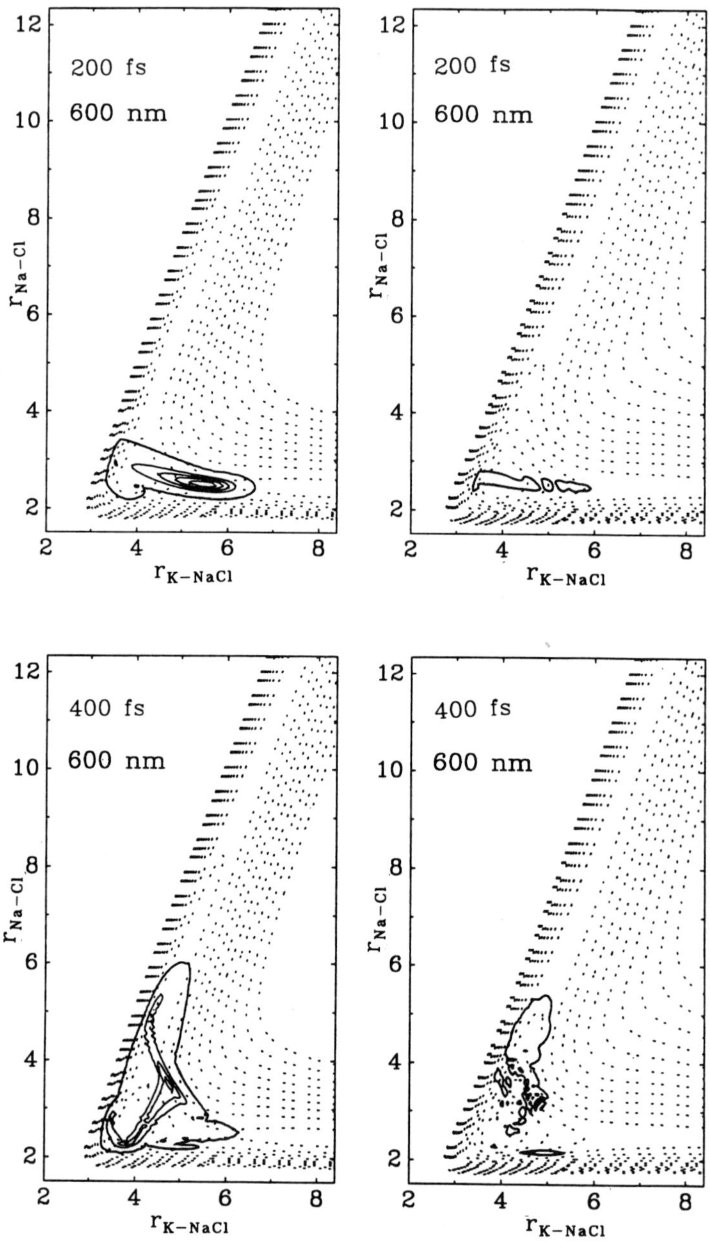

Fig. 2.(Continued)

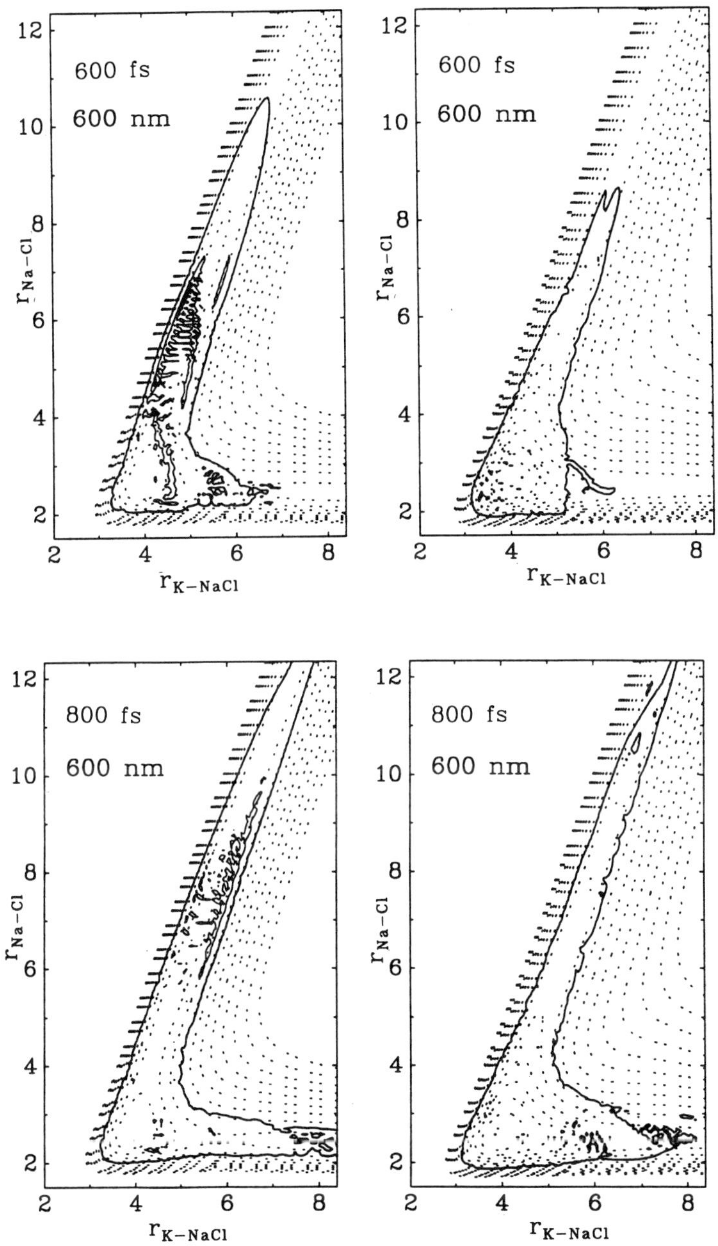
600 fs
600 nm
600 fs
600 nm
800 fs
600 nm
800 fs
600 nm
r_{Na-Cl}
r_{K-NaCl}

A part of the complexity gained by the wave function comes from traversal through a small attractive well in the near exit channel of the 1A' potential. If the collinear restriction is lifted, this small well connects to the absolute minimum of the three dimensional potential.

As the reaction proceeds further, extended probability transfer to the 3A' surface is found at all points in the exit channel reached by the 1A' wave function. Essentially the same behavior is found at shorter wavelengths except that the 3A' wave function reaches the boundary of the grid sooner. At longer wavelengths (e.g., 700 nm), the 3A' wave function never extends far out the exit channel, at least on the time scale (1 ps) of these calculations. This is clearly an effect of energy and momentum limitations.

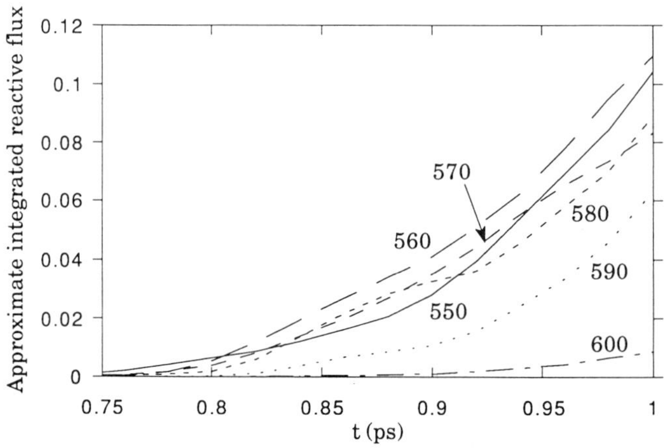

Fig. 3. Accumulated probability density that has left the grid on the 3A' surface in the direction of KCl + Na* as a function of time. Curves are labelled with wavelength (nm) of excitation.

In Fig. 3 is shown the time-integrated amount of probability density leaving the excited state surface in the direction of KCl + Na* for excitation wavelengths between 550 and 600 nm. While there is slight variation in the accumulations with wavelength, it is not dramatic. There is no indication of enhancement of the total flux at specific wavelengths. The yields for wavelengths longer than 600 nm are very small on the scale of Fig. 3, but similarly show no strong variations. Thus we draw essentially the same conclusion as reached in the classical trajectory calculations,[2,4] although the latter were only carried out for wavelengths above 590 nm.

The amount and prevalence of 3A' wave function produced for higher energies was somewhat surprising, given the original picture[4] of the potential surface intersection as a localized seam cutting across the exit channel. This has motivated a more extensive investigation in which it was found that the seam first crosses, then turns and parallels, the exit channel. This is reflected for collinear geometries in the very top of Fig. 4, which is the isosurface

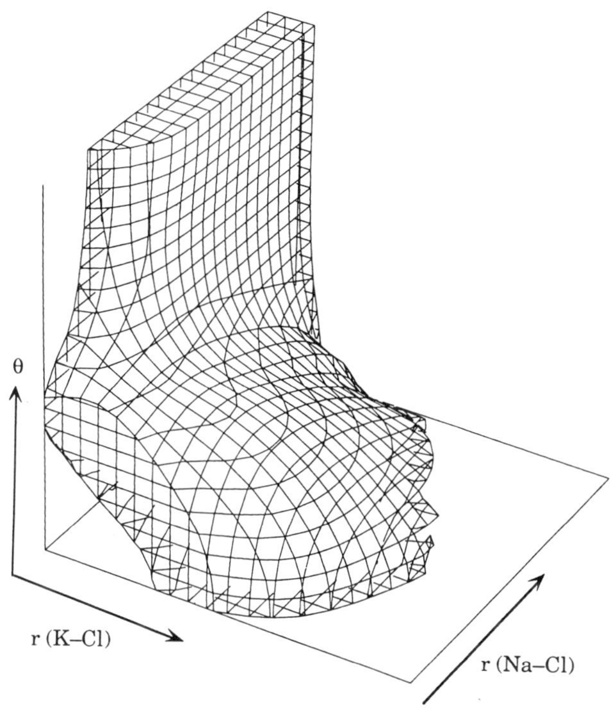

Fig. 4. Surface of degeneracy of $V_{1A'}$+hν and $V_{3A'}$ in the volume $r_{Na\text{-}Cl} \leq 10$ Å, $r_{K\text{-}Cl} \leq 10$ Å, $0 \leq \theta \leq \pi$ for excitation wavelength 600 nm. View is from entrance channel side. Top corresponds to collinear K-Cl-Na (θ = 180°).

describing exact degeneracy of the $V_{1A'}$ + hv and $V_{3A'}$ potentials in three dimensions at 600 nm. While the potentials are not superimposed onto this picture, the initial curving portion on the left is the seam which cuts across the exit channel. As one moves further right, the surface of degeneracy straightens out and has a nearly constant value of $r_{K\text{-}Cl}$. For all values of the bond angle $\theta > 90°$, one can see that the same qualitative pattern is followed, although the curving portion of the surface moves into the region of the molecular well (smaller bond lengths) as θ decreases in this interval. This agrees with the discussion given by Yamashita and Morokuma.[4]

The part of the surface which parallels the exit channel may also be identified as important in the production of probability density on the 3A' surface. In particular, close examination has shown that this parallel part of the seam in collinear geometries alligns significantly better with the ground state exit channel if the wavelength is decreased, and significantly worse if the wavelength is increased. Thus, the region involved in the increased transition probability to the 3A' state is delocalized throughout a very long stretch of the exit channel, and is not confined to the localized initial cut across. This aspect is largely insensitive to the bond angle for large angles.

It is evident that the observed signal should be much stronger for excitation on the short wavelength side of the Na D line. Unfortunately, this region is technically difficult because of the presence of interfering processes (e.g., NaK transitions).

DISCUSSION

Full confirmation that the collinear quantum calculations cannot produce the structure seen in the experiments must await more investigation with longer propagation times, denser sampling of wavelengths, and variation of the initial wavepacket width and vibrational state. Assuming that this view holds, it must still be investigated whether or not three-dimensional calculations can succeed where two-dimensional ones fail. On the basis of our investigations of the surfaces and their relative insensitivity to the K-Cl-Na angle (cf. Fig. 4) over important configuration ranges, it seems somewhat improbable.

Nevertheless, we are currently investigating a new variation of propagation of the time-dependent Schrödinger equation which will hopefully address the primary problem encountered here, the huge number of grid points required to converge the multidimensional problem. While tolerable in two dimensions, the simple uniform grid approach will be prohibitive in three. Instead, it is hoped that an adaptive grid method based on the relatively recent theory of wavelets[8,9] will provide an efficient means of placing grid points more selectively.

Other possible explanations exist for the observed structure in Figs. 1. A strong possibility is that a neglected electronic surface plays a role in the reaction dynamics. Yamashita and Morokuma[4] show that the 2A' will not contribute as the final state of photoabsorption, but Barnes, et. al.[2] have pointed out that a nonadiabatic transition from the 1A' to the 2A', followed by photon absorption to the 3A' state, could then give rise to a second excitation mechanism (cf., the red features in Figs.1). Also mentioned in Ref. 2 is the 1A" state, which correlates with the same $^2P_{3/2}$ Na* asymptote as does the 3A' state, thus providing plenty of opportunity for interference.

Finally, it is conceivable that the energy constraints of the reaction allow only a small higher-energy portion of the initial-energy distribution to contribute in a narrow window around the Na D line. If this is the case, the narrowness of the observed structure may be much more easily rationalized. This and the other points mentioned above are currently under active investigation.

ACKNOWLEDGMENTS

It is a pleasure to acknowledge scientific interactions with M. D. Barnes, P. R. Brooks, R. F. Curl, P. W. Harland, and J. L. Kinsey. This research was supported by NSF Grant No. CHE89-10975 and the Robert A. Welch Research Foundation (J. L. Kinsey) and NSF Instrumentation Grant No. CHE89-09777.

REFERENCES

1. T. C. Maguire, P. R. Brooks, R. F. Curl, J. H. Spence, and S. J. Ulvick, Photoexcitation of reaction complexes in the reaction K + NaCl → KCl + Na, J. Chem. Phys. 85:844 (1986).
2. M. D. Barnes, P. R. Brooks, R. F. Curl, P. W. Harland, and B. R. Johnson, Photoexcitation spectra of transition region species in reactions of potassium with sodium halides, J. Chem. Phys. 96:3559 (1992).
3. K. Yamashita and K. Morokuma, Theoretical Study of the Absorption Process during Chemical Reactions: Potential Surface and Classical Trajectory Study on K + NaCl, J. Phys. Chem. 92:3109 (1988).
4. K. Yamashita and K. Morokuma, A theoretical study of transition state spectroscopy: Laser dressed potential energy surface and surface hopping trajectory calculations on K + NaCl and Na + KCl, J. Chem. Phys. 91:7477 (1989).
5. R. F. Curl, unpublished.
6. J. Alvarellos and H. Metiu, The evolution of the wave function in a curve crossing problem computed by a fast Fourier transform method, J. Chem. Phys. 88:4957 (1988)
7. R. Kosloff and C. Cerjan, Atom-phonon interactions: A consistent quantum mechanical treatment, J. Chem. Phys. 81:3722 (1984).
8. I. Daubechies, Orthonormal bases of compactly supported wavelets, Comm. Pure Appl. Math. 41:909 (1988).
9. S. G. Mallat, A theory for multiresolution signal decomposition: The wavelet representation, IEEE Trans. Pattern Anal. and Mach. Intell. 11:674 (1989).

FEMTOSECOND WAVE–PACKET DYNAMICS ON STRONGLY COUPLED POTENTIAL ENERGY SURFACES

H. Köppel and U. Manthe

Theoretical Chemistry Group, University of Heidelberg
Im Neuenheimer Feld 253, 6900 Heidelberg, FRG

1. INTRODUCTION

The adiabatic or Born–Oppenheimer approximation constitutes one of the cornerstones in the theoretical description of molecular structure and dynamics[1 2]. It has proven very successful for ground electronic states, and also for many electronically excited diatomics[3]. The situation becomes more complex for polyatomic molecules where, due to the many nuclear degrees of freedom, the probability of degeneracies or near–degeneracies between the potential energy surfaces of electronically excited states increases enormously. Then the effects of interaction between these excited states, also called nonadiabatic or non–Born–Oppenheimer or vibronic coupling effects, come into play (see Refs. 4–8 for selected references on molecular spectroscopy and nonradiative decay).

In this contribution we hope to illustrate some salient features of nonadiabatic interactions in multi–mode systems and to underline their important role in molecular dynamics. As indicated by the title of this talk and the conference, the emphasis in on a time–dependent description of the phenomena. This offers on one hand a lucid picture and immediate intuitive understanding of time–dependent processes such as reactions or unimolecular decay. On the other hand also information on electronic spectra or resonance Raman excitation profiles can be conveniently extracted via Fourier transformations. We will first give a brief overview over the physical background of the problem and the numerical techniques used in the wave–packet dynamical calculations. A distinction will be made between systems with up to three

Time-Dependent Quantum Molecular Dynamics, Edited by J. Broeckhove and L. Lathouwers, Plenum Press, New York, 1992

vibrational degrees of freedom and those with three or more nuclear coordinates. While the former are treated with the general form of the kinetic and potential energies, for the latter we rely on a low–order power series expansion of the pertinent electron–vibrational coupling terms. The examples to be presented cover simultaneous dissociation and predissociation on a model systems mimicking HCN^+, a comparison between adiabatic and diabatic populations for systems related to $C_2H_4^+$ and a more complex example ($C_6H_6^+$) with three coupled electronic states and up to nine vibrational degrees of freedom.

2. PHYSICAL BACKGROUND

To describe nonadiabatic effects theoretically the usual product ansatz for the molecular wavefunctions Ψ in terms of an electronic factor ϕ_i and a nuclear factor χ_i (for the i–th electronic state) has to be abandoned. Fortunately it usually suffices to consider the strong coupling between a few (mostly two or three) close–lying electronic states. This means that Ψ has to be written as a sum of a few products only, referring to the strongly coupled subset S of states[8]

$$\Psi = \sum_{i \in S} \chi_i \phi_i \quad . \tag{1}$$

A key quantity responsible for the nonadiabatic effects is the matrix element of the nuclear momentum operator between different (adiabatic) electronic states[2]

$$g_{ij}^{(\alpha)} = \langle \phi_i | \frac{\partial}{\partial Q_\alpha} | \phi_j \rangle \tag{2a}$$

$$= \frac{\langle \phi_i | \frac{\partial H_{e\ell}}{\partial Q_\alpha} | \phi_j \rangle}{V_j(Q_\alpha) - V_i(Q_\alpha)} \quad . \tag{2b}$$

Here Q_α denotes a suitable intramolecular displacement and the integration is over the electronic coordinates only. $H_{e\ell}$ stands for the fixed–nuclear molecular Hamiltonian, $V_i(Q)$ for its i–th eigenvalue, the adiabatic potential surface corresponding to ϕ_i. The form (2b) represents the generalized Hellmann–Feynman theorem and shows explicitly the importance of degeneracies or near–degeneracies between V_i and V_j for the strength of the non–Born–Oppenheimer effects. Especially near conical intersections of these surfaces[9 10] the adiabatic wavefunctions are discontinuous and the coupling elements $g_{ij}^{(\alpha)}$ diverge[8]. Therefore they are expected to influence the nuclear motion severely. An example of such a conical intersection is displayed in Fig. 1. It represents the HCN^+ model system discussed in a later section.

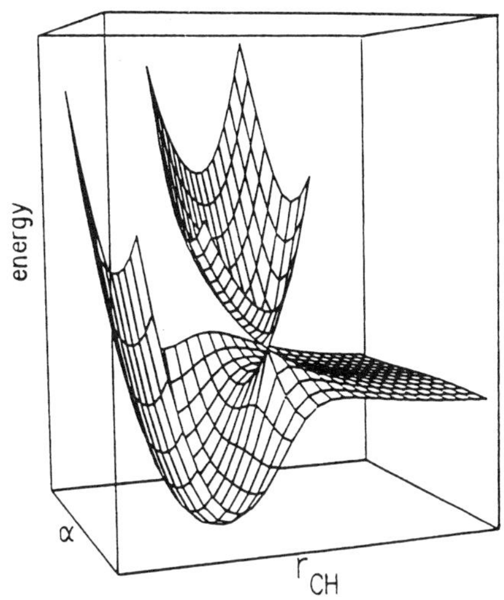

Fig. 1 Perspective drawing of the conical intersection representing the HCN^+ model system as a function of the bond angle α and the CH–distance r_{CH}.

In actual numerical calculations it is often convenient to switch to a diabatic representation of the coupled electronic states[11 12]. The diabatic basis functions pass smoothly through an avoided crossing and are well–behaved at conical intersections. Therefore the nuclear kinetic energy T_N can be taken to be diagonal to a sufficiently good approximation. It is rather the potential energy terms which cause the off–diagonal or coupling matrix elements in the diabatic electronic basis. Adopting a matrix notation, the Hamiltonian can be written as[6 8]

$$\mathcal{H} = T_N \underset{\sim}{1} + \begin{pmatrix} V_{11} & V_{12} \\ V_{12} & V_{22} \end{pmatrix} \tag{3}$$

for 2 interacting states. Here $\underset{\sim}{1}$ denotes the 2×2 unit matrix and the meaning of the potential matrix elements V_{ij} $(1 \le i, j \le 2)$ is self–evident. Such a form for $\mathcal{H}$ is chosen throughout this work.

The pointwise ab–initio determination of all potential matrix elements is tedious for general multi–mode problems. This holds also for the study of the ensuing nuclear dynamics. Up to three nuclear degrees of freedom can be treated nowadays in this way[13-15] and an example of such a study will be reported later on. One way to proceed to larger systems is to choose simplified equations governing the nuclear dynamics, such as harmonic diabatic potential curves V_{ii} $(1 \leq i \leq 2)$ and a corresponding form of T_N. Adopting in addition a first–order Taylor series for the coupling term V_{12} ("linear vibronic coupling"),

$$V_{12} = \lambda Q_c , \tag{4}$$

one obtains a structurally very simple form of the Hamiltonian matrix[8] (here Q_c denotes the coordinate of a suitable coupling mode and λ is called the vibronic coupling constant). The resulting class of matrices is, moreover, highly sparse and allows to treat three or four vibrational modes in a routine manner. Several millions of basis functions and up to ten nonseparable nuclear degrees of freedom have already been taken into account in this way (see below).

3. COMPUTATIONAL METHODS

The integration of the time–dependent Schrödinger equation involves two major computational steps, the evaluation of the matrix–vector product $\mathcal{H}\,\Psi(t)$ between the Hamiltonian (3) and the state vector at time t, and the solution of the coupled differential equations itself. For each we use a variety of different numerical techniques depending on the problem at hand.

For "linear vibronic coupling" problems the Hamiltonian and state vector are represented in a direct product basis of harmonic oscillator wavefunctions corresponding to the diabatic potential energy surfaces. This conventional basis set expansion is very efficient because of the above mentioned "structured sparsity" of the Hamiltonian matrix. The situation becomes more complicated for companion calculations performed occasionally in the adiabatic approximation[16]. This derives from the anharmonic form of the adiabatic potential surfaces which results even for harmonic diabatic potentials[8]. Then we use a discrete variable representation (DVR) as defined by the underlying harmonic oscillator basis[16 17]. The scheme becomes efficient for many modes with relatively small displacements by exploiting the tensor product form of the basis–to–grid transformation matrix.

A DVR scheme (together with a Fast–Fourier–Transformation (FFT)) is also our method of choice for triatomics with large–amplitude motion[13 18]. The three Jacobi coordinates are treated in different ways, the dissociative Jacobi coordinate by FFT methods, that corresponding to the diatomic fragment either by FFT or by (harmonic oscillator) DVR schemes. The latter holds also for the angular Jacobi coordinate θ except that the underlying basis functions are Legendre polynomials. The somewhat unusual FFT scheme for θ offers the advantage that the angular range to be covered by the grid can be chosen smaller than $0 \leq \theta < 180^0$ if desired. Moreover, by a suitable transformation of the Hamiltonian also interacting electronic states of different angular momenta can be treated with the same grid in both electronic states[18]. This reduces the computational effort considerably.

The coupled differential equations have been solved with a fourth–order differencing scheme (FOD)[16] and, more recently, with a short–time Lanczos (SIL) integrator[13 19]. The FOD scheme is a straightforward extension of the well–known second–order (SOD) scheme and faster than the latter (as well as predictor–corrector techniques) by a factor of ~ 3. It is, on the other hand, inferior to the SIL integrator by roughly the same amount. For our purposes the SIL integrator is even somewhat faster than the Chebyshev[20] method. Contrary to other work we employ it with a variable time step to adjust the accuracy of the integration to a desired pre–defined value[13].

4. APPLICATIONS

a) Dissociation and predissociation on coupled potential energy surfaces: HCN^+

We start with a three–dimensional calculation on general coupled surfaces mimicking important aspects of the photodissociation dynamics of the HCN^+ radical cation[13 18]. The initial state is the neutral ground state, the two coupled final states are represented by single–hole configurations with an electron ejected out of the CN–bonding π–orbital and out of the CH–bonding σ–orbital[21]. Fig. 1 displays a perspective drawing of the intersecting $^2\Sigma$ and $^2\Pi$ potential energy surfaces of these states for a fixed CN–distance. The CN–stretching motion is taken to be harmonic, with the same CN–distance (CH–distance) in the $^2\Sigma$ ($^2\Pi$) final state as in the neutral ground state. For the explicit form of the potentials and the values of the parameters used we refer to the original literature[13]. Here we only stress that in the dynamical calculation we have to deal with general potential energy surfaces because the potential ansatz is written in valence coordinates whereas the propagation is performed in Jacobi coordinates in order to keep the kinetic energy expressions simple. All calculations were donc for total angular momentum J−0.

What will be the fate of a wave–packet located at time t = 0 on the repulsive $^2\Sigma$ surface of Fig. 1, at a position defined by the minimum of the $^2\Pi$ surface? Depending on the strength of the vibronic coupling, a part of it will dissociate directly, while the rest will convert to the bound $^2\Pi$ state. In terms of diabatic electronic states the fragmentation for the latter, quasibound part of the wave–packet occurs via predissociation.

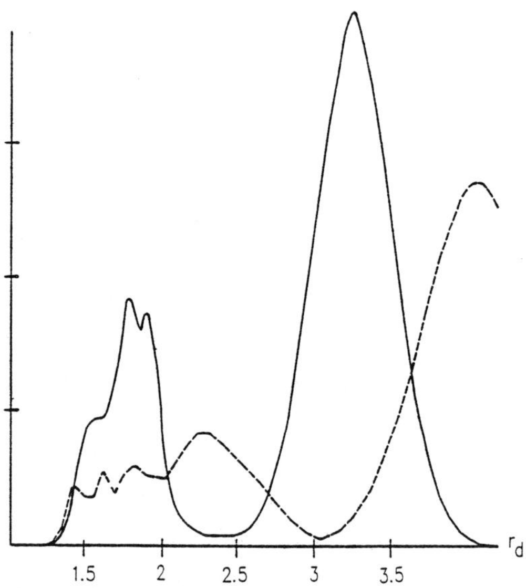

Fig. 2 Probability density to find the wave–packet at a defined value of the dissociative Jacobi coordinate r_d.
Full line: t=12 fsec; dashed line: t=17 fsec.

In Fig. 2 we assess the validity of this expectation more quantitatively. The probability density along the dissociative Jacobi coordinate r_d is shown for two different times, t = 12 fsec and t = 17 fsec. The vibronic coupling constant $\lambda = 0.9$ eV corresponds to three times the bending frequency in our energy units. Having reached the conical intersection at $r_d \approx 2$ after around 5 fsec the wave–packet splits into two parts. The directly dissociating part corresponds to the large Gaussian which steadily moves outwards. The quasibound part first moves towards smaller values of r_d and acquires a structured shape. At a later time (t = 12 fsec) it is being reflected at the inner turning point of the $^2\Pi$ potential surface. Upon approaching again the conical intersection (t = 17 fsec) a part of it dissociates by converting back to the $^2\Sigma$ surface. This is revealed impressively by that part of the packet which acquires a Gaussian shape (near $r_d \approx 2.4$) and moves further to the right.

This movement is accompanied by a substantial change in the (diabatic and adiabatic) electronic populations. These populations have also been studied and their analysis supports the interpretation given above. They show, in particular, that the quasibound part of the wave–packet moves indeed on the $^2\Pi$ potential energy surface. – Another important aspect is the intermode coupling between the motion in the dissociative coordinate and in the bending coordinate as well as that of the diatomic fragment. It leads to a considerable spreading of the wave–packet and slows down the predissociation of the quasibound part of the wave–packet although its energy is well above the dissociation asymptote[13]. The intermode coupling likewise affects the CN–vibration which becomes excited because of the different CN–distances in the initial and in the $^2\Pi$ electronic states.

To arrive at a more complete understanding of the dissociation dynamics we have also varied some system parameters such as the initial energy of the wave–packet and, in particular, the vibronic coupling strength[13]. The results reveal a transition from diabatic to adiabatic behavior which has also been observed for bound systems and appears to be a generic feature of the nuclear dynamics on conically intersecting potential energy surfaces. It is discussed in the following sub–section.

b) Comparison of adiabatic and diabatic electronic populations: $C_2H_4^+$

The nuclear dynamics in the coupled $\tilde{X}$ / $\tilde{A}$ electronic manifold of $C_2H_4^+$ was the earliest example of a wave–packet dynamical study in our group[22]. The "linear vibronic coupling" approach has been employed with three nonseparable nuclear degrees of freedom, the CH–stretching normal mode ν_1, the CC–stretching normal mode ν_2, and the torsional mode ν_4 as coupling mode[8, 22]. The system parameters have been obtained from ab initio Greens function calculations and subsequently been slightly readjusting so as to reproduce the experimental photoelectron spectrum of ethylene. In the original work only diabatic electronic populations have been studied for computational reasons[22]. More recently, we were able to calculate also adiabatic populations and compare them to the diabatic quantities[16 23].

In Fig. 3a we show the time–dependent populations of the $\tilde{A}\ ^2B_{3g}$ electronic state of $C_2H_4^+$. The full line refers to the diabatic electronic state. This is used as basis state in setting up the Hamiltonian matrix and is the state on which the wave–packet is placed at time t = 0. Due to the different bonding properties of the b_{3u} and the b_{3g} molecular orbitals there is a low–lying conical intersection occurring upon a shortening of the CC–bond[8]. It is reached by the wave–packet after half a period T_2 of the CC–normal mode ($T_2 \approx 20$ fsec) and leads to an "instantaneous" transfer of population to the lower surface. The implications of this behavior for the phaenomenon of ultrafast nonradiative decay have been pointed out before[22 24]. With instantaneous we mean here that the time–dependence for later times consists only in oscillations or fluctuations around the long–time limit $P_\infty \approx 0.28$. There is an interesting decay mechanism at later times which, however, refers to the amplitude of the oscillation, not to the (average) electronic population. The damping of the oscillations is the result of a purely quantum dynamical three–mode calculation and shows up also in the expectation values of nuclear position and momentum operators[25].

The adiabatic electronic populations are included as dashed lines in Fig. 3a and allow to understand this damping mechanism in some more detail. Due to the cusp occurring in intersecting adiabatic potential surfaces these calculations are considerably more cumbersome that those with diabatic states only[16]. Evidently, the adiabatic population drops to a much smaller long–time limit than the diabatic quantity and the characteristic oscillations are absent. This behavior can be traced to the different coordinate regimes where the off–diagonal or coupling element is effective[16 23]. In the adiabatic representation this regime is confined to a small neighborhood of the conical intersection. The latter is a subspace of dimension N–2 for N nuclear degrees of freedom and can be well avoided by the wave–packet once it moves on the lower adiabatic surface. In the diabatic representation the regime of effective coupling is of dimension N and less easy to avoid. In other words, the diabatic electronic states remain more strongly mixed for long times than the adiabatic states.

These findings help to understand the behavior of the diabatic population in Fig. 3a. As the wave–packet moves periodically on the lower adiabatic surface also the energy gap between the interacting states oscillates. This in turn induces a similar oscillation of the adiabatic–to–diabatic mixing angle[16] which leads to an analogous time–dependence of the diabatic electronic population. In companion calculations, the wave–packet has been propagated on the lower uncoupled adiabatic surface and the

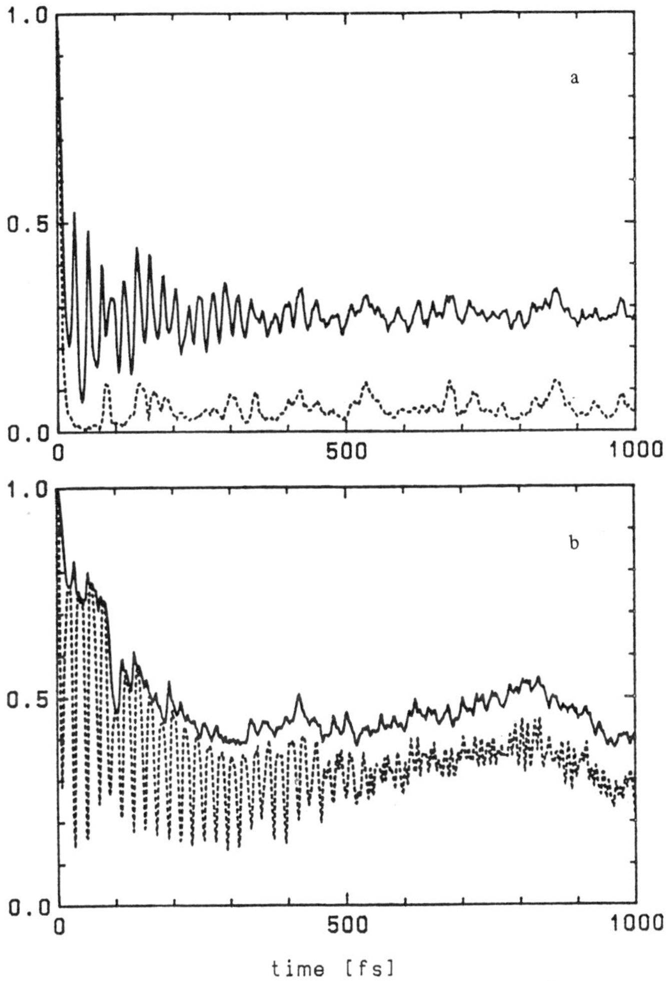

Fig. 3 Diabatic (full lines) and adiabatic (dashed lines) electronic populations for $C_2H_4^+$ – type systems with different values of the vibronic coupling constant λ.

a) Original ethylene, $\lambda \simeq 3.6\ \omega_c$; b) Reduced coupling, $\lambda \simeq 0.7\ \omega_c$.

same damped oscillations as in Fig. 3a have been obtained[16]. This proves that they are an essentially adiabatic effect and caused by the anharmonicity of the three–dimensional potential surface. It is only the noisy behavior at times beyond ~ 400 fsec for which the nonadiabatic effects are important[16].

How do these features of the nuclear dynamics depend on the system parameters? For vanishing vibronic interaction the systems becomes decoupled in the diabatic basis and the diabatic electronic population P_2^d will remain constant. Upon passing through a curve crossing the adiabatic populations will oscillate because the adiabatic–to–diabatic correlation differs on both sides of the crossing[11 12]. Such a situation prevails indeed in Fig. 3b where λ is smaller than in Fig. 3a by a factor of ~ 5. The remaining time–dependence of P_2^d is caused by the small finite value of λ but is considerably slower than typical vibrational periods T_{vib} ($T_{vib} < 40$ fsec).

Further calculations show that the transition from diabatic to adiabatic behavior occurs when λ roughly equals the frequency ω_c of the coupling mode, $\lambda \approx \omega_c$[23]. A similar transition is well–known from the theory for low–lying vibronic levels in Jahn–Teller systems[6]. Here we encounter the same phenomenon in the more complex high–energy dynamics when the conical intersection is well accessible for the wave–packet. It is caused by genuine multi–mode effects and exists also in other examples studied similarly, the coupled S_1–S_2 states of pyrazine and the A_1–B_2 system of NO_2 [23]. We emphasize that this behavior holds only after the wave–packet has passed one time through the conical intersection and undergone there an ultrafast nonradiative decay to the lower surface.

c) Dynamics in a three–state system with many vibrational modes: $C_6H_6^+$

We briefly report on a dynamical study in the coupled $\tilde{B}\ ^2E_{2g} - \tilde{C}\ ^2A_{2u}$ electronic manifold of $C_6H_6^+$ [26]. This three–state problem is characterized by a (nonadiabatic) Jahn–Teller (JT) effect in the $\tilde{B}$ state which affects even the low–energy vibronic states. In addition there is a high–dimensional conical intersection between the $\tilde{B}$ and the $\tilde{C}$ states, leading to a nonadiabatic dynamics also in the $\tilde{C}$ state. The occurrence of the latter intersection is driven, in particular, by the totally symmetric CC–stretching mode ν_2 and the JT active mode ν_{16} [26]. The JT active mode thus acts both through its intra–state and though its inter–state activity. The $\tilde{B}$–$\tilde{C}$ coupling is accomplished through two other degenerate modes, the out–of–plane modes ν_{19} and ν_{20}. Taking two out of the four JT active modes into account (even three may be important[27]) one arrives at nine nonseparable vibrational degrees of freedom which have been treated within the approximation of linear vibronic coupling[26]. CNDO–derived values of the coupling constants lead to a dimension of the secular matrix of ~ 6 Mio which could be treated with the aid of the

Lanczos algorithm. After adjusting some of these constants the experimental photoelectron spectrum could be well reproduced. This holds for the resolved structure in the energy range of the B̃ state ionization (up to 12 eV) as well as for the diffuse appearance in the energy range of the C̃ state ionization.

The ensuing wave–packet dynamical study has been performed with a reduced number of basis states (~ 1 Mio) and showed similar features of the C̃ state population decay as described above for ethylene [26]. These qualitative features (a fsec–population decay followed by damped oscillations) should therefore not depend on the structural details of the coupling mechanism. They are characteristic for the nuclear dynamics on conically intersecting potential energy surfaces. In the case of $C_6H_6^+$ the ultrafast C̃–B̃ nonradiative decay can explain that no emission from the C̃ state is observed (quantum yield $<10^{-4}$) although the C̃–X̃ transition is strongly dipole allowed [28]. The results illustrate that the linear coupling approach enables a full quantum dynamical study in systems with many vibrational modes and very large basis sets.

5. CONCLUSIONS

Conical intersections of potential energy surfaces provide a natural pathway for ultrafast (i.e. femtosecond) nonradiative relaxation in molecules. The importance of these processes is indicated by the lack of detectable emission in many radical cations[29] as well as higher excited singlet states of neutral species[30]. In this contribution we have reviewed a microscopic approach to treat the ultrafast population decay by fully quantum dynamical wave–packet techniques. The efficiency of the mechanism has been documented for a few examples and some salient features of the dynamics been explored by variation of selected parameters. Major characteristics appear to be independent of the details of the coupling mechanism and even of the type of nuclear motion (bound vs. dissociative). For example, as the vibronic coupling strength is increased also the dissociative HCN^+–type system reveals a transition from diabatic to adiabatic behavior as discussed above for the bound $C_2H_4^+$–type systems[13]. The adiabatic behavior has its origin in the lower dimensionality of the locus of degeneracy of the surfaces and in the fact that the wave packet fills the available phase space more or less uniformly[23]. This is another manifestation of the mode mixing effects which are typical for conical intersections[8].

Despite the enormous technical and computation progress in the past few years the exact quantum treatment on general coupled surfaces will remain extremely cumbersome for more than three degrees of freedom. This renders approximate treatments like the "linear vibronic coupling" approach utilized above very useful when dealing with larger molecules. Nevertheless, there is certainly a large class of problems where "exact" treatments[13-15] with three degrees of freedom will be required. In addition, a natural extension of this work as presented above is feasible, e.g. the study of total angular momentum $J \neq 0$. – Finally we mention that the electronic populations are amenable to a direct measurement via pump–probe experiments[31]. It would be highly desirable to assess in this way experimentally the time–dependent population dynamics of coupled electronic states.

ACKNOWLEDGEMENTS

It is a pleasure to acknowledge a fruitful collaboration with L.S. Cederbaum and W. Domcke on the vibronic coupling problem. This work has been financially supported by the Deutsche Forschungsgemeinschaft through SFB 91, Energietransfer bei atomaren und molekularen Stoßprozessen.

REFERENCES

1. M. Born and R. Oppenheimer, Ann. Phys. (Leipzig) 84, 457 (1927).
2. M. Born and K. Huang, Dynamical Theory of Crystal Lattices, Oxford Univ. Press, New York (1954).
3 G. Herzberg, Molecular Spectra and Molecular Structure, Vol. I, II, van Nostrand, New York (1950).
4. K.F. Freed, Topics Appl. Phys. 15, 23 (1976).
5. M. Desouter–Lecomte, D. Dehareng, B. Leyh–Nihant, M.Th. Praet, A.J. Lorquet and J.C. Lorquet, J. Phys. Chem. 89, 214 (1985).
6. I.B. Bersuker, The Jahn–Teller Effect and Vibronic Interactions in Modern Chemistry, Plenum, New York (1984).
7. R.L. Whetten, G.S. Ezra and E.R. Grant, Ann. Rev. Phys. Chem. 36, 277 (1985).
8. H. Köppel, W. Domcke and L.S. Cederbaum, Adv. Chem. Phys. 57, 59 (1984).
9. G. Herzberg and H.C. Longuet–Higgins, Disc. Faraday Soc. 35, 77 (1963).
10. T. Carrington, Disc. Faraday Soc. 53, 27 (1972); E.R. Davidson, J. Am. Chem. Soc. 99, 397 (1977).
11. W. Lichten, Phys. Rev. 131, 229 (1963); 164, 131 (1967).
12. F.T. Smith, Phys. Rev. 179, 111 (1969).
13. U. Manthe, H. Köppel and L.S. Cederbaum, J. Chem. Phys. 95, 1708 (1991).
14. B. Heumann, K. Weide and R. Schinke, J. Chem. Phys., to be published.
15. H. Guo, Three–dimensional Photodissociation of Methyl Iodide, submitted to J. Chem. Phys.

16. U. Manthe and H. Köppel, J. Chem. Phys. 93, 345 (1990).
17. D.O. Harris, G.G. Engerholm and W.D. Gwinn, J. Chem. Phys. 43, 1515 (1965); A.S. Dickinson and P.R. Certain, J. Chem. Phys. 49, 4209 (1968).
18. U. Manthe and H. Köppel, Chem. Phys. Lett. 178, 36 (1991).
19. C. Leforestier et al, J. Comp. Phys. 94, 59 (1991).
20. H. Tal–Ezer and R. Kosloff, J. Chem. Phys. 81, 3967 (1984).
21. J.P. Hansoul, C. Galloy and J.C. Lorquet, J. Chem. Phys. 68, 4105 (1978).
22. H. Köppel, Chem. Phys. 77, 359 (1983).
23. U. Manthe and H. Köppel, J. Chem. Phys. 93, 1658 (1990).
24. W. Domcke, H. Köppel and L.S. Cederbaum, in: "Stochasticity and Intramolecular Redistribution of Energy", R. Lefebvre and S. Mukamel, eds., Reidel, Dordrecht (1987) p. 217.
25. R. Schneider and W. Domcke, Chem. Phys. Lett. 159, 61 (1989), R. Schneider, W. Domcke and H. Köppel, J. Chem. Phys. 92, 1045 (1990).
26. H. Köppel, L.S. Cederbaum and W. Domcke, J. Chem. Phys. 89, 2023 (1988).
27. W. Domcke, private communication.
28. O. Braitbart, E. Castellucci, G. Dujardin and S. Leach, J. Phys. Chem. 87, 4799 (1983).
29. J.P. Maier, in: "Kinetics of Ion–Molecule Reactions", P. Ausloos, ed., Plenum, New York (1979) p. 437.
30. M. Kasha, Discuss. Faraday Soc. 9, 14 (1950).
31. W. Domcke and H. Köppel, Chem. Phys. Lett. 140, 133 (1987); G. Stock and W. Domcke, Phys. Rev. A45, 3032 (1992).

PSEUDO SPECTRAL METHODS FOR SOLVING

THE TIME DEPENDENT SCHRODINGER EQUATION

Ronnie Kosloff

Department of Physical Chemistry and
The Fritz Haber Research Center for Molecular Dynamics
The Hebrew University, Jerusalem 91904, Israel

I. Introduction

Just a decade ago time dependent quantum mechanical methods were an esoteric field shadowed by much more effective time independent methods. To mention early work McCullough and R. E. Wyatt [1] used a direct finite difference scheme to solve the $H + H_2$ reactive scattering problem. The numerical effort to do the calculation was formidable and although more effective finite difference schemes were developed [2] the use was limited to problems where direct time dependent methods have a significant advantage such as when the Hamiltonian has explicit time dependence [3] or when the process is coupled to more than one continuum [4].

In parallel to this early work Heller and coworkers [5], have developed a time dependent semiclassical scheme which had a great impact on the field of molecular dynamics. This work initiated new methods for obtaining observable results from time dependent quantum mechanical methods. Of particular importance were the formulas to obtain spectroscopic quantities such as the absorption or Raman spectra [6]. A more critical evaluation of the semiclassical method showed that they can supply insight into the molecular process but are very poor in obtaining quantitative results.

For many physical situations, time dependent methods have their own advantages. Because they are initial value problems, the methods are easy to implement. In particular the time dependent picture enables a simple treatment of continua and of the rearrangement problem in reactive scattering [7]. Besides these technical advantages, time dependent methods lead to a better interpretation of the physical mechanism under discussion. The time variable also enables a description of externally driven systems, by introducing time dependent Hamiltonians.

By now quantitative quantum mechanical time dependent methods have become established. The methods compete successfully with stationary methods in reactive scattering [7]. They have become the method of choice in simulations of gas surface encounters [8]. In simulations of photodissociation in particular for processes initiated by short strong pulses time dependent methods are the main working tool [9]. As a demonstration of the popularity of the field a whole volume of Computer physics communications edited by K. Kullander [10] deals with time dependent quantum mechanical methods.

An important step inthe establishment of time dependent quantum mechanical methods has been the introduction of pseudo-spectral methods [11]. The change is more than a technical improvement on algorithms. An elementary statment is that quantum mechanics is a non local theory. This characteristic has to be reflected in the methods we use to describe the quantum world. The foundation of the pseudo-spectral methods is a global functional base [12], therefore the nonlocal character of the representation is built

Time-Dependent Quantum Molecular Dynamics, Edited by J. Broeckhove
and L. Lathouwers, Plenum Press, New York, 1992

in the formulation. The basic attributes of quantum mechanics such as the commutation relations between conjugate operators are preserved in the pseudo-spectral description. The use of semilocal approximations such as the finite difference schemes must eventually lead to the violation of the uncertanty principles. Frequently the error does not appear immediately which is why many practical applications have succeeded, nevertheless the experimental ability to probe molecular dynamics as they evolve puts a new challenge on computational methods. In particular for processes which maintain their coherence for long periods of time the global approaches are superior to the semilocal approaches.

In this note a uniform global approach is suggested for the description of the position momentum phase-space as well as the time energy phase-space. This approach is based on global interpolation functions which are the key to the pseudo spectral methods.

II. Representation theory

Representation theory can be thought of as setting the stage on which dynamical events take place. In quantum mechanics this stage supplies the mathematical framework necessary to represent the state of a system and its evolution through time. In other words, an algorithm is needed to extract from the state of the system the values of measurements. Axiomatic quantum mechanics which originated in the work of Von Neumann, bases the description of the state in Hilbert space. Three elements constitute this Hilbert space: a) A functional domain with well defined boundary conditions. b) A set of vectors defined on this functional domain.(A crucial element is the definition of the scalar product of two vectors). c) Operators which map the vectors into new ones in the same functional domain.

In the construction of a calculation scheme a restriction of a discrete representation has to be imposed on the Hilbert space. Imagine viewing the functional space through a finite set of portholes which allow a local picture only. If this porthole view is dense enough, a complete picture can be reconstructed by using the continuous properties of the functional space to interpolate the observed data and obtain the full picture. The collocation technique which was originated by Gauss is just such a method [12].

2.1 General collocation method

Consider approximating an arbitrary function $\psi(x)$ by a set of functions $g_n(x)$:

$$\psi(x) \approx \sum_{n=0}^{N-1} a_n \, g_n(x) \tag{2.1.1}$$

where $g_n(x)$ are analytic in the domain of interest. The question is how to determine the expansion coefficients a_n. In the collocation method these expansion coefficients are determined by matching the solution on a set of N grid points:

$$\psi(x_j) \equiv \sum_{n=0}^{N-1} a_n \, g_n(x_j) \tag{2.1.2}$$

where x_j are the grid points (or collocation points) which are the sampling points discussed above. This equation is equivalent to a set of coupled linear equations. In matrix form they become:

$$\boldsymbol{\psi} = \tilde{\mathbf{G}}\, \mathbf{a} \tag{2.1.3}$$

where $\psi_j = \psi(x_j)$ and the matrix $\tilde{G}_{nj} = g_n(x_j)$. Provided $g_n(x_j)$ are linearly independent the solution becomes:

$$\mathbf{a} = \tilde{\mathbf{G}}^{-1}\, \boldsymbol{\psi} \tag{2.1.4}$$

The functional basis that is connected through the expansion coefficients a_n to the spatial grid, supplies the global picture. This provides the ability to define the scalar product of two functions. If:

$$\psi(x) = \sum_n a_n g_n(x) \qquad \text{and} \qquad \phi(x) = \sum_m b_m g_m(x) \tag{2.1.5}$$

then

$$< \psi|\phi > \equiv \sum_{nm} a_n b_m^* S_{nm}$$

where S_{nm} is the overlap matrix:

$$S_{nm} = \int_D g_n(x) g_m^*(x) dx \qquad (2.1.6)$$

where the integration is carried out over the domain D.

The operators in the discrete Hilbert space can be represented by the mapping of the expansion coefficients. Centeral to quantum mechanics is the calculation of nonlocal operators, such as the momentum operator $-i\hbar \frac{\partial}{\partial x}$

$$-i\hbar \frac{\partial}{\partial x} \psi(x) \approx -i\hbar \sum_{n=0}^{N-1} a_n \frac{\partial}{\partial x} g_n(x) \qquad (2.1.7)$$

The operation of these non local operators can be recast into the sampling points x_j:

$$-i\hbar \frac{\partial}{\partial x} f(x_j) \approx -i\hbar \sum_{n=0}^{N-1} a_n \frac{\partial}{\partial x} g_n(x_j) \qquad (2.1.8)$$

resulting in a discrete representation of the non local operation. The duality of the representation which on the one hand is represented by the values at sampling points and on the other hand, by a functional global representation, is responsible for the applicability of the method to represent processes in Hilbert space.

The choice of the functions $g_n(x)$ and of the sampling points x_j have great influence on the quality of the approximation as will be revealed in the next sections. Nevertheless the use of the primitive collocation method with arbitrary functions and loosely chosen sampling points, has the advantage of great flexibility. An example which has been well studied is the distributed Gaussian method [14] in which:

$$g_n(x) = A \exp\left(-(x - x_n)^2 / 2\sigma^2 \right) \qquad (2.1.9)$$

where usually the center of the Gaussians are located on the sampling points $x_n = x_j$.

2.2 Orthogonal collocation schemes

A great simplification in the collocation scheme is achieved if the set of expansion functions $g_n(x)$ obey the orthogonality relation:

$$\sum_n g_n(x_i) g_n(x_j) = \delta_{ij} \qquad (2.2.1)$$

allowing a direct inversion of equation (2.2.4):

$$a_n = \sum_j \psi(x_j) g_n(x_j) \qquad (2.2.2)$$

This means that the the expansion coefficients a_n are the discrete functional transform of the function f. On the other hand if

$$\int_D g_n(x) g_m(x) dx = \delta_{nm} \qquad (2.2.3)$$

then the scalar product in equation (2.2.5) is greatly simplified leading to:

$$< \psi|\phi > = \sum_n a_n b_n \qquad (2.2.4)$$

which using equation (2.2.9) is also:

$$< \psi|\phi > = \sum_j \psi(x_j) \phi(x_j) \qquad (2.2.5)$$

The result of the orthogonality relations is that the collocation functional expansion scheme becomes a discrete vector space with a unitary transformation between the discrete sampling points x_j and the discrete functional base a_n. The next sections will elaborate on functional sets which fulfill the orthogonality constraints equations (2.2.1) and (2.2.2).

2.3 The Fourier method in 1-D

The Fourier method is a special case of an orthogonal collocation representation. This method deserves special attention for two reasons: First, it can be shown that the limit of one point per unit volume in phase space can be approached. Secondly because of its great numerical advantages. Due to the FFT algorithm [13], the numerical effort scales semilinearly with the volume of phase space.

In the Fourier method, the orthogonal functions $g_n(x)$ are chosen as:

$$g_k(x) = e^{i2\pi kx/L} \quad , \quad k = -(N/2-1),\ldots,0,\ldots,N/2 \tag{2.3.1}$$

leading to N equally spaced sampling points where $x_j = (j-1)\Delta x$, and $\mathbf{L}$ is the length of the interval. using the relation $\mathbf{L} = N\Delta x$, the completeness relations of the Fourier expansion function become:

$$\sum_{k=-N/2-1}^{N/2} e^{i2\pi kx_i/L} e^{-i2\pi kx_j/L} = \sum_{k=-N/2-1}^{N/2} e^{i2\pi ki/N} e^{-i2\pi kj/N} = \frac{1-e^{i2\pi(i-j)}}{1-e^{i2\pi(i-j)/N}} = N\delta_{ij} \tag{2.3.2}$$

where the summation can be carried out because it is cast into a geometric series. The third equality in equation (2.3.2) follows because unless $i = j$, $|i-j|$ is always smaller than N.

The second orthogonality relation becomes:

$$\frac{1}{2\pi}\int_{-\pi}^{\pi} e^{imx'} e^{-inx'} dx' = \frac{1}{2\pi}\left[\frac{1}{i(m-n)} e^{i(m-n)x} \Big|_{-\pi}^{\pi}\right] = \delta_{nm} \tag{2.3.3}$$

where $x' = \pi(x/\mathbf{L} - 1)$, $0 \le x \le L$. These relationships imply that the domain D has periodic boundary conditions. The symmetry between i and k in equation (2.3.2) leads to the discrete version of equation (2.2.5):

$$\sum_{j=1}^{N} g_k(x_j)^* g_l(x_j) = \delta_{kl} N \qquad |k-j| < N \tag{2.3.4}$$

With the choice of Fourier expansion functions a wavefunction $\psi(x)$ is represented as:

$$\psi(x) \approx \sum_{k=-(N/2-1)}^{N/2} a_k e^{i2\pi kx/L} \tag{2.3.5}$$

The expansion coefficients a_k become the discrete Fourier expansion coefficients. One can use the orthogonality of the Fourier functions with equidistant sampling points to invert the relation giving:

$$a_k = \frac{1}{N}\sum_{j=1}^{N} \psi(x_j) e^{-i2\pi kx_j/L} \tag{2.3.6}$$

In quantum mechanics the coefficients a_k have an important interpretation since they represent the amplitude of the wavefunction in momentum space.

2.4 Representation of operators by the Fourier method

The main result of the previous section is the dual representation space: a discrete representation in coordinate as well as in momentum space. The discrete Fourier transform supplies the unitary transformation between the two spaces. This relation is in direct analogue to the continuous Fourier transformation changing a coordinate representation to a momentum representation. The special structure of the two spaces is used to convert the operator mapping of vectors to local operations thus reducing the number of operations from $O(N^2)$ to $O(N)$.

The partition of the Hamiltonian operator to the sum of local operators will now be presented as an example. The Hamiltonian operator is partitioned between kinetic and potential operators $\hat{\mathbf{H}} = \hat{\mathbf{T}} + \hat{\mathbf{V}}$. The strategy is to calculate each operator locally. The potential operator is already local in coordinate space, and therefore its operation is simply a multiplication of $V(x_j)$ by $\psi(x_j)$. A local operation of the kinetic energy operator is possible in momentum space where it becomes a multiplication by the kinetic energy discrete spectrum:

$$T(k) = \frac{\mathbf{P}^2}{2m} = \frac{\hbar^2\mathbf{k}^2}{2m} \tag{2.4.1}$$

and its operation on a wavefunction represented in momentum space becomes:

$$\frac{\hat{\mathbf{P}}^2}{2m}\bar{\psi}(k) = \hbar^2\frac{k^2}{2m}a_k \tag{2.4.2}$$

This result can also be obtained from equation (2.3.6) by differentiating the Fourier expansion functions twice. The result can be generalized for any local operator in momentum space. The algorithm for calculating the mapping of such operators is as follows: a) calculate the expansion coefficients a_k by the discrete Fourier transform. b) multiply each point in k space by the value of the operator at that point. c) transform the result back to the sampling space by an inverse Fourier transform. The total Hamiltonian mapping becomes the sum of potential and kinetic operators.

This spectral representation preserves the coordinate momentum commutation relations which are the heart of quantum mechanics [11].

2.5 Phase space representation of the Fourier method

At this point the phase space representation of the Fourier method should be considered. The minimum volume in phase space covered by the Fourier representation is calculated as follows: The length of the spatial dimension in phase space is $\mathbf{L}$, and the maximum momentum is $\mathbf{p}_{max}$. Therefore, the represented volume becomes $Volume = 2\mathbf{L}\cdot\mathbf{p}_{max}$, where the factor of two appears because the momentum range is from $-\mathbf{p}_{max}$ to $+\mathbf{p}_{max}$. Using the fact that $\mathbf{p} = \hbar\mathbf{k}$ the phase space volume can be expressed as:

$$Volume = 2\hbar\,\mathbf{L}\cdot\mathbf{k}_{max} = Nh \tag{2.5.1}$$

where N is the number of sampling points. Next, the sampling spacing Δx is related to the maximum wave vector via:

$$\Delta x = \frac{\pi}{|\mathbf{k}_{max}|} \tag{2.5.2}$$

Figure 2.1 expresses the relation between the volume in phase space, the unit volume, and the grid parameters Δx and N. The computational scaling properties of the Fourier method are a result of the scaling properties of the FFT algorithm which scales as $O(N\log N)$. As a result the Fourier method scales with phase the space volume as $O(\,Volume\,\log\,Volume\,)$.

A function that is bounded in momentum space is equivalent to the band limited Fourier transform of the function. Confinement of such a function to a finite volume in phase space is equivalent to a band limited function with finite support. The accuracy of a representation of this function is assured by the Whittaker-Kotel'nikov-Shannon sampling theorem. It states that a band limited function with finite support, is fully specified if the functional values are given by a discrete, sufficiently dense set of equally spaced sampling points. The number of points is determined by equation (2.5.1-2.5.2). This implies that a value of a point between sampling points can be interpolated with any desired accuracy. This theorem also implies a faithful representation of the n'th derivative of the function inside the supported interval.

As has been shown, a confined function in phase space can be represented with no loss of accuracy. In realistic problems, the wavefunction cannot be confined, which means that it is not strictly band limited with finite support. The idea of a wavepacket which is a wavefunction that is almost band limited is central to the use of the discrete representation. A wavepacket is a wavefunction that is semilocalized in phase space. The most well-known example is the Gaussian wavepacket. Although the wavefunction is not confined to a finite volume, the amplitude out of this volume in phase space converges exponentially to zero when the volume is increased. This exponential convergence is typical of a good representation of phase space.

2.6 Example: harmonic oscillator

The use of the Fourier expansion method to calculate the spectrum of the Harmonic oscillator can demonstrate the convergence of the method. For simplicity, consider an harmonic oscillator with mass 1 and frequency 1 ($\hbar = 1$). The problem is cast onto a finite grid with equally spaced sampling points.

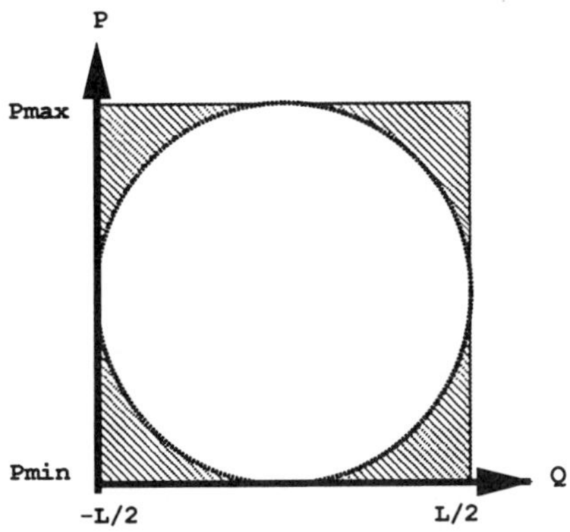

Figure 2.1. Schematic phase space of the harmonic oscillator

On this grid the maximum value of the potential energy is determined by the extent of the grid, leading to:

$$V_{\max} = \tfrac{1}{2}\left(\frac{L}{2}\right)^2 = \frac{\Delta x^2 N^2}{8} \tag{2.6.1}$$

where N is the number of sampling points and the grid is centered around zero. Also the maximum kinetic energy is limited by the maximum momentum which can be represented on the grid leading to:

$$T_{\max} = \frac{P_{\max}^2}{2m} = \tfrac{1}{2}\left(\frac{2\pi}{2\Delta x}\right)^2 = \frac{\pi^2}{2\Delta x^2} \tag{2.6.2}$$

A cutoff in energy in a discrete representation of Hilbert space is a common feature. For a balanced calculation the cutoff in the potential energy should match the cutoff in the kinetic energy leading to:

$$\Delta x_{opt} = \left(\frac{2\pi}{N}\right)^{\frac{1}{2}} \tag{2.6.3}$$

With the grid set, a finite expansion base is constructed by the functions $\psi_j(x) = \delta(x - x_j)$. The Hamiltonian matrix in this basis is calculated as follows: The potential energy matrix in this basis is diagonal leading to: $V_{jj} = V(x_j)$. To calculate the kinetic energy matrix elements, a discrete Fourier transform was applied to the expansion function ψ_j then it was multiplied by $\hbar^2 k^2 / 2m$ and back transformed. The resulting vector becomes the matrix element T_{ji}. At this stage the Hamiltonian matrix $\tilde{H} = \tilde{T} + \tilde{V}$ is diagonalized and the eigenvalues of the discrete representation are compared to the exact results. The next table shows the calculated eigenvalues using 8 , 16 and 32 sampling points with the optimal grid spacing Δx_{opt}:

Table 2.1. Convergence of the Fourier representation

n	Exact	8 points	16 points	32 points
1	0.5	0.4999760107111692	0.4999999998715793	0.4999999999999845
2	1.5	1.500539183698194	1.500000006153576	1.499999999999986
3	2.5	2.494397791604125	2.499999857637178	2.500000000000002
4	3.5	3.534291735531325	3.500002068787167	3.499999999999993
5	4.5	4.378294841394608	4.499978178076544	4.499999999999982
6	5.5	5.960743369677700	5.500169240572392	5.500000000000130
7	6.5	6.212358517586826	6.498928884935602	6.499999999997876
8	7.5	9.976917743504451	7.505122466888302	7.500000000025498
9	8.5		8.478529454896226	8.499999999731434
10	9.5		9.567379412999802	9.500000002414588
11	10.5		10.33263329166096	10.49999998073345
12	11.5		11.95570264326039	11.50000013337068
13	12.5		12.04789296489282	12.49999916184023
14	13.5		14.56012875187433	13.50000457741937
15	14.5		15.44949561677325	14.49997680742042
16	15.5		20.69252130835984	15.50010120025324
17				16.49957859092886
18				17.50146417275678

On comparing the 8 point with 16 point results, the exponential convergence of the expansion is demonstrated. For the ground state energy the error decreases six orders of magnitude by only doubling the number of grid points. Other states show similar behavior. When the number of points is doubled again to 32, the error in the ground state becomes saturated because the roundoff error of the double precision arithmetic used in the calculation overcomes the representation error. For unsaturated eigenvalues doubling the number of points from 16 to 32 reduces the error another five to six orders of magnitude.

Another issue is the number of converged eigenvalues that can be obtained in an n point representation. If the criteria is three significant digits it can be seen from the table that 3 eigenvalues are obtained for 8 points, 8 for 16 points, and 20 for 32 points. As a conclusion the fraction of significant eigenvalues increases with the number of points in the representation. Figure 2.2 shows the fraction of significant eigenvalues as a function of the logarithm of the number of points.

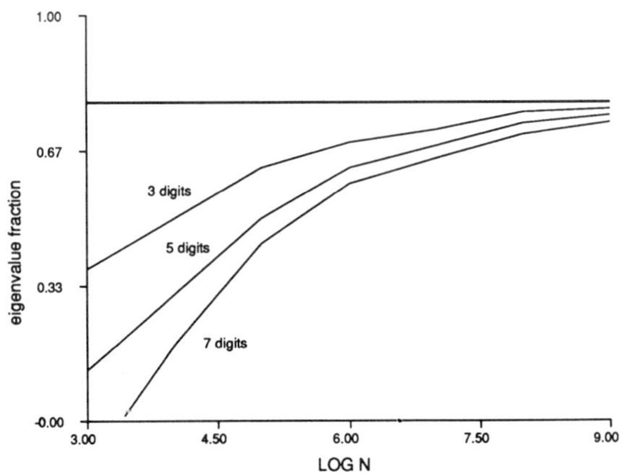

Figure 2.2. convergence of the spatial representation

A converged eigenvalue is defined when the number of significant digits is larger than a preset value. What can be seen in the figure is that the converged eigenvalue fraction increases with the number of points until it reaches saturation regardless of the number of significant digits used in the convergence criteria. The saturation phenomena occur because the Fourier method constructs a rectangular phase space. Using the balanced choice of Δx, the phase space becomes a square. On the other hand, due to energy conservation, the support for an eigenvalue in phase space has the shape of a disc, whereas the support for all eigenvalues up to a cutoff energy have the shape of a circle (or an ellipse in the general case). The area between the circumference of the circle and the square is waisted sampling space. Therefore the the maximum sampling efficiency is the ratio of the area of the circle to the area of the square leading to $\pi/4 \approx 79\%$. This is the asymptotic value represented by a bold line in figure 2.2. To illustrate further this point figure 2.3 shows the sampling ratio as a function of the grid spacing Δx.

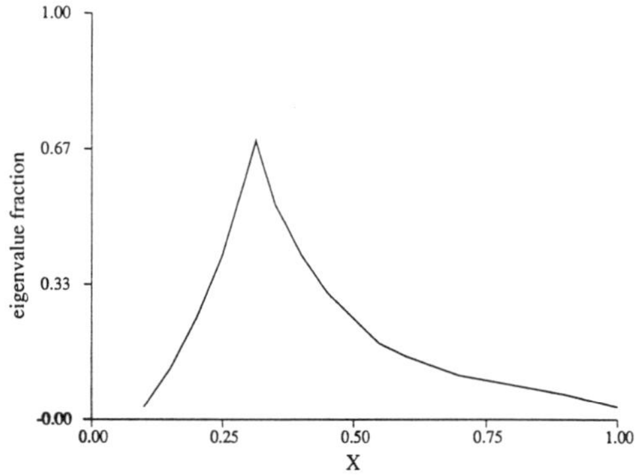

Figure 2.3. sampling efficiency as a function of Δx.

It is clear that a square in phase space offers the optimal choice.

The lesson from this example is that a careful choice of grid parameters can drastically change the amount of computation effort.

2.7 Fourier method in many dimensions [15]

The most simple generalization of the Fourier method for multidimensions is the use of an equally spaced Cartesian grid [13]. A more careful analysis leads to the observation that this grid is not isotropic in momentum space, so that different directions have different sampling intervals. A faithful representation of a multidimensional function by the Fourier method means that it has to be band limited. The symmetric construction of the coordinate and momentum representation mean that the description of the function is also periodic in momentum space. The representation therefore can be viewed as an infinite number of replicas of the original k space picture extending in all directions. The band limited property of the original function means that these replicas do not overlap. If *a priori* there is no preferable direction in space a cutoff in momentum can be represented as a sphere with radius p_{max}. A Cartesian grid in coordinate space is also Cartesian in momentum space therefore the non overlapping role can be visioned as packing spheres such that they touch their neighbours at $2d$ points for d dimensions. It is clear that the best sampled direction is in the diagonal one, but for an isotropic problem the volume between the spheres is waisted sampling volume. Although a completely isotropic grid is not possible, the sampling positions can be optimized to construct the optimum isotropic grid in momentum space. From the above picture it can be concluded that the optimal sampling points are equivalent to the centers of multidimensional densely packed hard spheres. (The problem of the optimal packing of hard spheres has been solved up to 23 dimensions). The free volume between the spheres is wasted sampling volume. The sampling efficiency can be defined as the ratio between the volume of the space filling spheres to the volume of the total space. Figure 2.4 demonstrates the situation in two dimensions. The Cartesian grid reaches a sampling efficiency of $\frac{\pi}{4} \approx 79\%$. S. Skewing the grid increases the sampling efficiency to $\frac{\pi}{2\sqrt{3}} \approx 91\%$. In multidimensions, the limit of one sampling point per unit volume is not obtainable even for optimal packing. Table 1 compares the sampling efficiency of a cubic grid with the optimal grid as a function of dimension.

Table 2. 2 Isotropic Sampling Efficiency

Dimension	Maximum sampling efficiency η_{max} (%)	Efficiency of cubic lattice η_{cub} (%)	Improvement factor η_{max}/η_{cub}
1	100.0	100.0	1.0
2	90.7	78.5	1.15
3	74.0	52.4	1.4
4	61.7	30.8	2.0
5	46.5	16.45	2.8
6	37.3	8.07	4.6
7	29.5	3.69	8.0
8	8.07	0.505	16.0

Source: Petersen and Middleton.

One can notice that the importance of optimal sampling is enhanced with dimensionality. Optimal sampling becomes extremely important for calculations beyond three dimensions. For example for six dimensions, 2.7 sampling points are needed per unit volume, compared to 12.4 points in the Cártesian cubic lattice. Nevertheless even the optimal sampling efficiency decreases with dimensionality. This fact poses an intrinsic limitation to the Fourier method for multidimensional calculations.

III. Time propagation and other propagators

3.1 General considerations

To model the evolution of a physical system, the vector describing the state of the system has to be followed through time. In quantum mechanics the evolution operator $\hat{U}(t', t)$ maps the state vector at time t to a new vector at time t'. The evolution operator is constructed by using our physical intuition that the forces that induce the dynamics are homogeneous, symmetric and continuous in time. These requirements imply a group property to the evolution operator with the form:

$$\hat{U}(t', t) = \hat{U}(t' - t) = e^{-\frac{i}{\hbar}\hat{H}(t'-t)} \tag{3.1.1}$$

the operator $\hat{H}$ is a continuous generator of the evolution, and to fulfill the symmetry requirement $\hat{H}$ has to be hermitian. Part from this requirement the operator $\hat{H}$ is unrestricted. It is customary to identify the operator $\hat{H}$ with the Hamiltonian of the system, which means that its expectation value becomes the energy. From this analysis one concludes that the stage on which the dynamical events are observed is the time energy phase space. Time is interpreted as a parameter and not an operator of a continuous dynamical group describing the evolution operator. From this description it seams that time and energy are not set on the same footing. Contrary to this situation, in the position momentum phase space a complete symmetry exists between the two variables. The different descriptions are manifested in the boundary conditions of the discrete descriptions. For the position momentum space the boundary values of phase space are imposed. In the time energy phase space only an initial boundary condition in time is imposed. The common strategy of both phase space descriptions is the use of global functional descriptions of the complete space. This nonlocal type of description assures the compliance with the uncertainty principles central to quantum mechanics.

The group property which is in the foundation of the time evolution operator $\hat{U}$ can be exploited by subdividing the time interval into segments:

$$\hat{U}(t + s) = \hat{U}(t)\,\hat{U}(s) \tag{3.1.2}$$

where for each segment there is an individual evolution operator. By repeating the segmentation process the individual evolution operator becomes a function of a very short time interval. Knowing the limit $\lim_{t \to 0} \hat{U}(t) = \hat{I}$ one can linearize $\hat{U}(t)$ for short time intervals. This is the base of a family of propagation techniques based on a short time expansion. Once a short time propagator is developed the global evolution operator can be reassembled. Similar techniques allow one to develop the evolution operator for problems with explicit time dependence.

To illustrate the use of this formal construction consider the following situation, a system is prepared in some initial state, for example by cooling the system one can reach the ground state. Then at time $t = 0$ an abrupt change is imposed on the system changing the potential. This can be realizable experimentally in the coloumb imaging technique. The abrupt change imposes the initial value of the state of the system. The purpose of the modeling is to follow the evolution of system in time under the influence of the new Hamiltonian. This situation is a generic example for which the solution of the time dependent Schrödinger equation

$$i\hbar \frac{\partial \psi(t)}{\partial t} = \hat{\mathbf{H}}\psi(t) \tag{3.1.3}$$

subject to the initial condition $\psi(0)$ describes the evolution in time. Examining the r.h.s of the Shrödinger equation, it can be identified as the mapping of the vector ψ caused by the hamiltonian $\hat{\mathbf{H}}$. Building on section II where a discrete version of the mapping has been worked out, it should be possible to use these discrete mapping techniques to solve for the evolution of the system. For a stationary Hamiltonian the solution of the time dependent Schrödinger equation becomes:

$$\psi(t) = e^{-\frac{i}{\hbar}\hat{\mathbf{H}}t}\,\psi(0) \tag{3.1.4}$$

This is the integral version of the time dependent Schrödinger equation and is identical to equation (3.1.3). Again comparing the r.h.s. of this integral form of the Schrödinger equation, a solution to the evolution operator is possible if a mapping procedure can be found which calculates the mapping caused by an exponential function of the Hamiltonian operator. The main topic of this section is finding effective methods which allow to calculate the mapping imposed on vectors in Hilbert space by mapping caused by a function of an operator. This generalization will allow the use of the same basic techniques for solving other equations.

3.2 Function of an operator

In this section the mapping operation is extended from the mapping caused by an operator to the mapping caused by a function f of the operator $\hat{\mathbf{O}}$:

$$\phi = f(\hat{\mathbf{O}})\psi \tag{3.2.1}$$

Equation (3.2.1) has only formal meaning and therefore has to be interpreted. The interpretation is based on the analytic properties of the function $f(z)$ which allow a convergent expansion in powers of the operator $\hat{\mathbf{O}}$. This formulation can directly take advantage of the ability which has been described in the previous chapter to perform the discrete mapping caused by linear operators in Hilbert space of an initial vector ψ into a new vector ϕ, $\phi = \hat{\mathbf{O}}\psi$. This means that equation (3.2.1) can be evaluated by a successive application of the basic mapping procedure. The next step is to utilize the spectral decomposition of the operator $\hat{\mathbf{O}}$

$$\hat{\mathbf{O}} = \sum_n \lambda_n \hat{\mathbf{P}}_n \tag{3.2.2}$$

where $\hat{\mathbf{O}}u_n = \lambda_n u_n$ and $\hat{\mathbf{P}}_n$ is the n'th projection operator $\mathbf{P}_n = |u_n><u_n|$. Using the property of the projection operators: $\hat{\mathbf{P}}_n\hat{\mathbf{P}}_m = \delta_{nm}\hat{\mathbf{P}}_n$, the spectral decomposition of the operator in equation (3.2.1) can be written as:

$$f(\hat{\mathbf{O}}_l) = \sum_n f(\lambda_n)\hat{\mathbf{P}}_n \tag{3.2.3}$$

This is a key step because now the problem of evaluating a function of an operator has been transformed to the evaluation of a function of a scalar. The power expansion and the spectral decomposition are the two basic methods of evaluating a function of an operator. Equation (3.2.3) has the nice feature of transforming the evaluation of the function f to be the evaluation of a scalar, but the formula requires the ability to perform the mapping of ψ caused by the projection operators $\phi_n = \hat{\mathbf{P}}_n\psi$. It will be found that these operators are closely analogous to the Lagrange interpolation polynomials. Because of the importance of these steps, a detailed description now follows. First consider the operator:

$$\hat{\mathbf{Q}} = \left(\hat{\mathbf{O}} - \lambda_l\hat{\mathbf{I}}\right) \tag{3.2.4}$$

This operator has the opposite effect than the projection operator, it eliminates from ψ the component belonging to the eigenvector l. This is now the basic building block. Now by successively applying these operators one can eliminate from the vector ψ all eigenvectors except the nth one with the result becoming a polynomial description of the nth projection operator:

$$\hat{\mathbf{P}}_n = (\hat{\mathbf{O}} - \lambda_N\hat{\mathbf{I}})(\hat{\mathbf{O}} - \lambda_{N-1}\hat{\mathbf{I}})\ldots(\hat{\mathbf{O}} - \lambda_{n-1}\hat{\mathbf{I}})(\hat{\mathbf{O}} - \lambda_{n+1}\hat{\mathbf{I}})\ldots(\hat{\mathbf{O}} - \lambda_1\hat{\mathbf{I}})(\hat{\mathbf{O}} - \lambda_0\hat{\mathbf{I}})\cdot 1/N \tag{3.2.5}$$

where N is a normalization term which becomes:

$$N = (\lambda_n - \lambda_N)(\lambda_n - \lambda_{N-1})\cdots(\lambda_n - \lambda_{n-1})(\lambda_n - \lambda_{n+1})\ldots(\lambda_n - \lambda_1)(\lambda_n - \lambda_0) \tag{3.2.6}$$

Inserting equation (3.2.5) into equation (3.2.3) constitutes a Lagrange interpolation polynomial procedure for evaluating the function of the operator. The polynomials can be evaluated recursively by successive application of the basic mapping of the operator in equation (3.2.4). The functional evaluation is at the interpolation points which match the eigenvalues of $\hat{\mathbf{O}}$. This reconstruction of equation (3.2.3) is exact because of the following property of interpolation polynomials:

$$P_{N+1}(x_n) \equiv f(x_n) \tag{3.2.7}$$

Numerically the Lagrange interpolation procedure has disadvantages mainly because N different polynomials have to be evaluated. This problem can be overcome by the use of Newtonian formulation of the interpolation polynomial:

$$\begin{aligned} f(z) &\approx a_0 + a_1(z - x_0) + a_2(z - x_1)(z - x_0) + a_3(z - x_2)(z - x_1)(z - x_0) + \ldots \\ &= \sum_{n=0}^{N-1} a_n \prod_{j=0}^{n-1} (z - x_j) \end{aligned} \tag{3.2.8}$$

where the coefficients a_k are determined by imposing the interpolation condition leading to :

$$\begin{aligned} a_0 &= f[x_0] = f(x_0) \\ a_1 &= f[x_0, x_1] = \frac{(f(x_1) - f(x_0))}{(x_1 - x_0)} \end{aligned} \tag{3.2.9}$$

And the rest of the coefficients are calculated by the recursive formula:

$$a_k = f[x_0, x_1, \ldots, x_k] = \frac{f(x_k) - a_0 - \sum_{l=1}^{k-1} a_l\,(x_k - x_0)\ldots(x_k - x_{l-1})}{(x_k - x_0)\ldots(x_k - x_{k-1})} \tag{3.2.10}$$

The coefficients a_k are called the divided difference coefficients and a common symbol for them is $a_k = f[x_0, x_1, \ldots, x_k]$. Using the results of equation (3.2.1) and by rewriting equation (3.2.2) and by replacing z by $\hat{\mathbf{O}}$ the function of the operator f is written as:

$$f(\hat{\mathbf{O}}) = a_0\hat{\mathbf{I}} + a_1(\hat{\mathbf{O}} - x_0\hat{\mathbf{I}}) + a_2(\hat{\mathbf{O}} - x_1\hat{\mathbf{I}})(\hat{\mathbf{O}} - x_0\hat{\mathbf{I}}) + \cdots \qquad (3.2.11)$$
$$= \sum_{n=0}^{N-1} a_n \prod_{j=0}^{n-1} (\hat{\mathbf{O}} - x_j\hat{\mathbf{I}})$$

where $\hat{\mathbf{I}}$ is the identity operator, $x_n = \lambda_n$, and the expansion coefficients a_n are calculated by equation (3.2.9) (3.2.10). This formulation leads to a recursive algorithm based on the elementary mapping, allowing one to construct the operation of the polynomial on an initial function ψ once an algorithm for the operation $\hat{\mathbf{O}}\psi$ exists. This recursive algorithm eliminates intermediate storage because unlike equation (3.2.3) the need for the eigenfunctions is eliminated. Summarizing the reformulation from equation (3.2.1) to equation (3.2.11) a finite recursive polynomial expansion of the function $f(\hat{\mathbf{O}})$ has been gained based on the mapping operation $\hat{\mathbf{O}}\psi$. However the algorithm is based on the solution of the more difficult problem of finding the set of eigenvalues of the operator $\hat{\mathbf{O}}$. As basic diagonalization procedures scale as $O(N^3)$ where N is the size of the vector ψ this approach is prohibitively expensive for realistic problems. Nevertheless an important lesson can be learned from the formulation, Choosing the interpolation points in regions where the eigenvalues of $\hat{\mathbf{O}}$ reside will effectively improve the convergence of the polynomial approximation.

3.3 Uniform approximation approach

As has been concluded from the previous section a detailed knowledge of the positions of the eigenvalues of $\hat{\mathbf{O}}$ exclusively determines the interpolation polynomial. As this detailed description is usually prohibitively expensive, a strategy of choosing the interpolation points has to be developed based on only a partial knowledge of the location of the eigenvalues. It will be shown that the number of interpolation points or the order of the polynomial can be reduced considerably from the number of eigenvalues of $\hat{\mathbf{O}}$.

First consider the situation when the operator $\hat{\mathbf{O}}$ is hermitian. As a result, all eigenvalues are located on the real axis and for most applications upper and lower bounds for the eigenvalues λ_n can be found. The strategy is then to use this partial knowledge to construct a uniform approximation of the function f on the interval on the real axis defined by the upper and lower bounds of the eigenvalues. For the Hamiltonian operator this wil be the energy axis. The original problem of approximating the operation of a function of an operator has been transformed to the problem of finding the best uniform approximation of a scalar function $f(z)$ on a closed interval $\lambda_{\min} < z < \lambda_{\max}$.

This problem is a classical problem in numerical analysis, that is finding a polynomial approximation so that the maximum error is minimum on this interval. The most general interpolation polynomial obeys the relation:

$$f(z) = P_N(z) + \frac{f^{N+1}(\xi(z))}{(n+1)!}(z - x_0)(z - x_1)(z - x_2)\cdots \qquad (3.3.1)$$

where x_n are sampling points and ξ which is a function of z is also in the interval. To minimize the error regardless of its position in the interval one can choose the sampling points x_n such that the product in the r. h. s of the equation

$$E(z) = \|(z - x_0)(z - x_1)(z - x_2)\cdots\| \qquad (3.3.2)$$

is minimum for arbitrary z in the interval. This product is a monic polynomial of degree $N+1$. The minimax criteria can now be traced to a condition on equation (3.3.2) that the maximum amplitude of $E(z)$ in the interval is minimum. The Chebychev polynomial has uniform amplitude in the interval therefore it obtains its maximum value N times with the value $1/2^{n-1}$. All other polynomials of the same degree become non uniform and therefore have larger maximum amplitudes (see figure 3.1). As a result they are inferior to the Chebychev polynomial in reducing the error term $E(z)$. Because the usual definition of the Chebychev polynomials is on the interval -1 to 1 , a shifting and scailling transformation is applied to the argument z:

$$z' = 2\frac{z - \lambda_{\min}}{\lambda_{\max} - \lambda_{\min}} - 1 \tag{3.3.3}$$

What has been found is that minimizing the error is equivalent to choosing the sampling points as the zeros of the $N+1$ Chebychev polynomial. One can proceed directly now by using the Newtonian formulation of the interpolation polynomial. This approach will be analyzed later, alternatively a direct expansion of the function of the operator in Chebychev polynomials is formulated [16]. Mathematically it will be shown that the two approaches are equivalent.

Explicitly for the function $f(z)$ an expansion by Chebychev polynomials becomes

$$f(z) \approx \sum_{n=0}^{N} b_n T_n(z') \tag{3.3.4}$$

where T_n is the Chebychev polynomial of degree n, and b_n are expansion coefficients defined by:

$$b_n = \frac{2}{\pi} \int_{-1}^{1} \frac{f(z)T_n(z')}{\sqrt{1 - z'^2}} dz' \tag{3.3.5}$$

Returning now to the problem of approximating $\phi = f(\hat{\mathbf{O}})\psi$, the approximation is obtained by replacing the argument z' by the operator $\hat{\mathbf{O}}$. This is done by the following steps:

a) Scale and shift the original operator $\hat{\mathbf{O}}$:

$$\hat{\mathbf{O}}' = 2\frac{\hat{\mathbf{O}} - \lambda_{\min}\hat{\mathbf{I}}}{\lambda_{\max} - \lambda_{\min}} - \hat{\mathbf{I}} \tag{3.3.6}$$

b) Calculate the expansion coefficients b_n using equation (3.3.5).

c) Using the recursion relation of the Chebychev polynomials, calculate ϕ_n where:

$$\phi_0 = \psi \tag{3.3.7}$$

$$\phi_1 = \hat{\mathbf{O}}'\psi$$

and

$$\phi_{n+1} = 2\hat{\mathbf{O}}'\phi_n - \phi_{n-1}$$

Again the calculation of the recursion is based on the ability to perform the discrete mapping caused by the operator $\hat{\mathbf{O}}$. Two basic operations are used for the recursion, multiplication (mapping) by an operator on a vector and an addition of another vector. d) Accumulate the result while calculating the recursion relation by multiplying b_n by ϕ_n and adding to the previous result:

$$\phi = \sum_{n-0}^{N} b_n \phi_n \tag{3.3.8}$$

e) Truncate the calculation when the desired accuracy has been obtained.

It should be noted that the Chebychev recursion relation which consumes most of the numerical effort is independent of the function f to be approximated. This means that many different functions f can be approximated simultaneously by repeating steps b d and e. The uniform nature of the approximation means the error is independent of the choice of the initial vector ψ.

Considering the time dependent Schrödinger equation: $i\hbar \frac{\partial \psi}{\partial t} = \hat{\mathbf{H}}\psi$

and its formal solution:

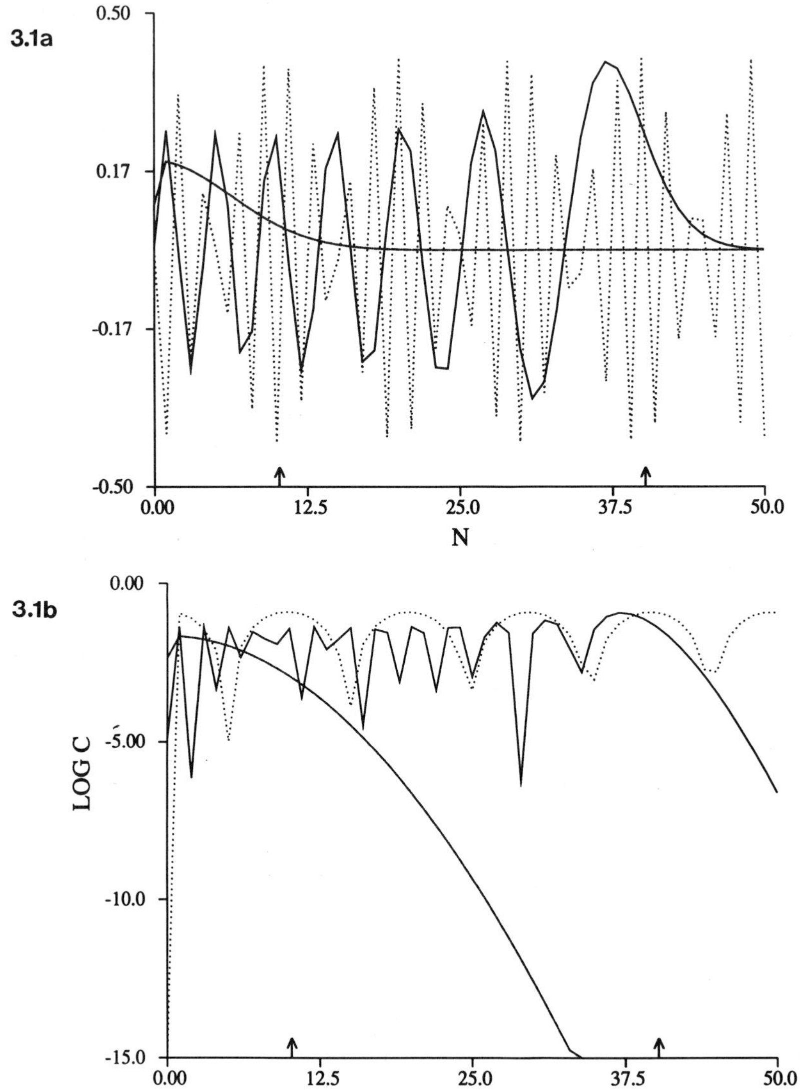

Figure 3.1a. Amplitude of the expansion coefficients of a Chebychev polynomial expansion for the functions e^{-izt} (solid line), e^{-zt} (broken line) and $\frac{1}{z-\omega}$ (dashed line) as a function of the order n. $\Delta Et/2$ is chosen to be 40, and the point is indicated by the right errow. Beyond this point the expansion coefficients of the real time propagator decay exponentially. The point $(\Delta E/2)^{\frac{1}{2}}$ is indicated by the left errow showing the point where the imaginary time propagation decays exponetially. Figure 3.1b shows the same picture where the coefficients are displayed in a logarithmic scale. The exponential convergence is of the propagators is clearly shown in comparison to the absence of convergence of the coefficients of the resolvant.

$$\psi(t) = e^{-\frac{i}{\hbar}\hat{\mathbf{H}}t}\,\psi(0) \tag{3.3.9}$$

the function f to be approximated is e^{-iz}. Following the steps of the algorithm, the range of energy $\Delta E = E_{\max} - E_{\min}$ represented by $\hat{\mathbf{H}}$ is estimated. The shift operation causes a phase shift in the final wavefunction which can be readjusted by multiplying ϕ by the phase factor:

$$\Phi = e^{\frac{i}{\hbar}(\frac{1}{2}\Delta E + E_{\min})t} \tag{3.3.10}$$

The expansion coefficients b_n are calculated from equation (3.3.5) to obtain:

$$b_0 = J_0\left(\frac{\Delta E \cdot t}{2\hbar}\right)\Phi \tag{3.3.11}$$

and

$$b_n = 2i^n J_n\left(\frac{\Delta E \cdot t}{2\hbar}\right)\Phi \qquad n \geq 1$$

where J_n are Bessel functions. The argument of the Bessel function $\frac{\Delta E \cdot t}{2\hbar}$ is related to the volume of the time energy phase space which is contained in the problem. The number of terms N needed to converge the expansion is determined by this volume. This is because of an asymptotic property of the Bessel function: when the order n becomes larger than the argument the Bessel functions J_n decreases exponentially fast (see figure 3.1).

The Chebychev vectors ϕ_n are functions of the normalized Hamiltonian and the initial wavefunction ψ. Therefore intermediate results can be obtained by calculating another set of coefficients b_n and repeating the resummation step. Figure 3.2 shows the relation between the number of terms needed to achieve convergence and the propagated time interval.

The linear relationship is clearly seen going up to very high orders of the polynomial. To check the limit of the expansion, a propagation for a time interval of 400 cycles of an oscillator has been tried. The resulting polynomial expansion was of the order of $N = 120000$. The norm energy and overlap with the initial state all achieved an accuracy of 14 digits. This behavior is typical of the uniform convergence of the method where all error indications are of the same order of magnitude.

Figure 3.3 shows the efficiency ratio defined as the time energy phase space volume divided by the number of terms in the expansion needed to obtain convergence: $\eta = (\Delta E \cdot t/2\hbar) / N$ as a function of time.

The efficiency approaches 1 for long time propagations. For short time intervals the efficiency decreases which means that the overhead of points needed to obtain a prespecified accuracy increases. This is the reason that the Chebychev propagation method is recommended for very long propagation times. As the error is uniformly distributed and can be reduced to the precision determined by the computer the Chebychev scheme is a fast accurate and stable method for propagating the time dependent Schrödinger equation. The method is not unitary and because of the uniform nature of the error distribution deviation from unitarity can be used as a check of accuracy. A vast number of applications in molecular dynamics stress this point. The same algorithm has been used for propagating the Liouville Von Neumann equation [15].

A small change in the function f leads to an effective method for obtaining the ground state and several low lying excited states. Considering the propagation of a vector ψ:

$$\phi = e^{-\hat{\mathbf{H}}\tau}\psi \tag{3.3.12}$$

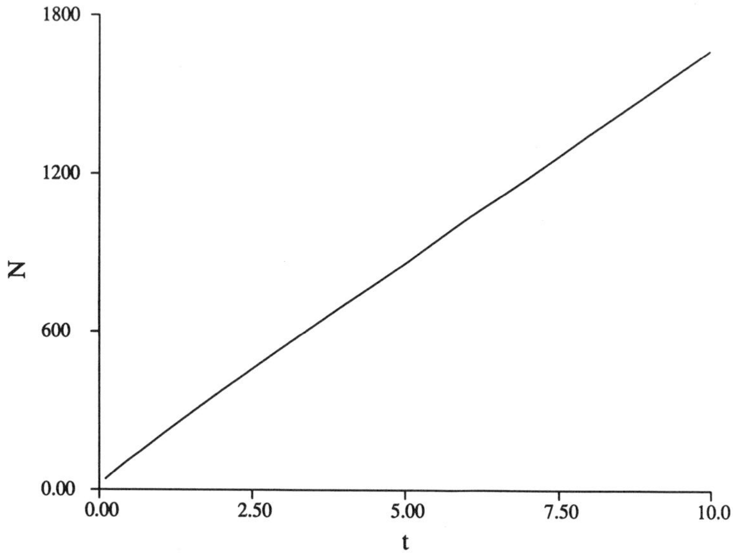

Figure 3.2. The Number of Chebychev terms in the expansion needed to obtain 12 digits accuracy as a function of time. The system used is an harmonic oscillator sampled by 64 spatial sampling points with grid spacing 0f $\Delta x = 0.295$, $m = 1$ and $\omega = 1$. Time is measured in periods of the oscillator. The energy range on the grid $\Delta E = 99.85$.

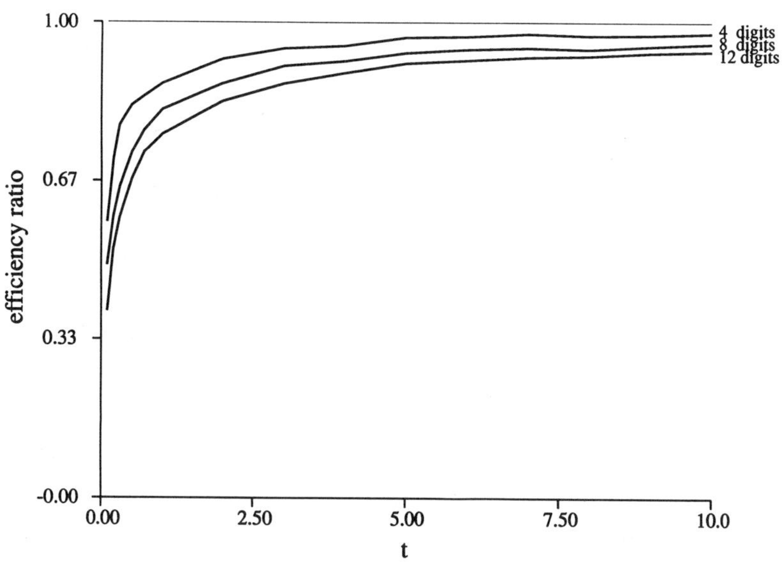

Figure 3.3. The efficiency ratio defined as $\Delta E \cdot t/2N$ where N is the number of expansion terms is plotted as a function of time. The three lines correspond to 12 digits of accuracy 8 digits of accuracy and 4 digits of accuracy. As time progresses the three lines reach their asymptotic value of 1. For short time the efficiency deteriorates reaching 50% for a time of 0.1 periods (40 terms).

when $\tau \to \infty$, ϕ approaches exponentially fast the ground state. A proof for this fact is obtained by expanding ψ in the eigenstates of $\hat{\mathbf{H}}$:

$$e^{-\hat{\mathbf{H}}\tau}\psi = \sum_n e^{-E_n\tau} u_n \tag{3.3.13}$$

which converges to the ground state at a rate of $e^{-(E_1 - E_0)\tau}$. By choosing $f(z) = e^{-z}$ the previous algorithm can be modified by performing analytic continuation to the expansion coefficients to obtain:

$$\begin{aligned} b_0 &= I_0(\alpha) \cdot N \\ b_n &= 2I_n(\alpha) \cdot N \end{aligned} \tag{3.3.14}$$

where $\alpha = -\dfrac{\Delta E \cdot \tau}{2}$ and I_n are modified Bessel functions and $N = e^{+(\frac{1}{2}\Delta E + E_{\min})\tau}$. These expansion coefficients converge exponentially because the modified Bessel functions I_n decay exponentially when $n > \sqrt{\Delta E \cdot \tau / 2}$. By comparison this convergence is faster than the coefficients b_n of equation (3.3.5). It should be noticed that equation (3.3.13) becomes the solution to the diffusion or heat equation when Hamiltonian operator is replaced by the appropriate diffusion operator. This means that the same algorithm can be used to solve these equations. The scaling of the numerical effort as a square root of time has physical significance in the diffusion equation, for which the higher eigenvalues loose their significance as time progresses until in equilibrium only the lowest eigenvalue is of importance. Consider for example obtaining the ground state of the Harmonic oscillator starting from a coherent initial state which means that the projections on the ground state and first excited state are approximately equal. The convergence is exponential with the relaxation constant of approximately as predicted from equation (3.3.13) for an harmonic oscillator of frequency 1. Considering that the numerical effort scales as $\sqrt{\tau}$ the numerical convergence is faster than exponential.

At this point it is appropriate to compare the Chebychev expansion of equation (3.3.20) to the Newtonian interpolation formula of equation (3.2.5). The connection can be worked out from equation (3.3.5) by replacing the analytic integration with a Gauss Chebychev quadrature of order N:

$$\bar{b}_n \approx \frac{2}{\pi} \sum_{l=1}^{N} \frac{f(x_l) T_n(x_l)}{(1 - x_l^2)^{\frac{1}{2}}} \tag{3.3.15}$$

The quadrature points x_n are the zeros of the Chebychev polynomial of degree $N + 1$. On inserting equation (3.3.15) into the Chebychev expansion (3.3.4) one finds that the Chebychev expansion has become a polynomial interpolation formula equivalent to (3.2.9) where the quadrature points are identical to the interpolation points:

$$\begin{aligned} \sum_{n=1}^{N} \bar{b}_n T_n(x_k) &= \frac{2}{\pi} \sum_{n=1}^{N} \sum_{l=1}^{N} \frac{f(x_l) T_n(x_l) T_n(x_k)}{(1 - x_l^2)^{\frac{1}{2}}} \\ &= \sum_{l=1}^{N} f(x_l) \frac{2}{\pi} \sum_{n=1}^{N} \frac{T_n(x_l) T_n(x_k)}{(1 - x_l^2)^{\frac{1}{2}}} = f(x_n) \end{aligned} \tag{3.3.16}$$

The last equation is due to changing the order of summation and the Christofel-Darboux formula. From equation (3.3.16) it can be concluded that the Chebychev expansion of equation (3.3.4) is equivalent to a polynomial interpolation when the sampling points are the zeros of the $N + 1$ Chebychev polynomial if the expansion coefficients are calculated using the Gauss-Chebychev quadrature scheme of order N. The sampling points become the quadrature points of the numerical integration. This means that an alternative uniform algorithm can be obtained using equation (3.2.9) with the zeros of the Chebychev polynomial as the sampling points x_n, together with the expansion coefficients a_n calculated by the divided difference procedure. To keep the dynamical range of the calculation in check it is desirable to use the scaled operator $\hat{\mathbf{O}}$ also for the Newtonian formulation. To keep the divided difference coefficient in check the stable interval is from -2 to 2. To obtain high orders the sampling points have to be staggered otherwise numerical instabilities occur. This point is quit amazing, noting that Newtons interpolation polynomials are invariant to a permutation of points. On the other hand numerically reordering the points completely changes the stability characteristics of the algorithm.

To connect to other work the derivation can be considered a pseudospectral scheme in time, similar to the DVR expansion method of Lill, Hamilton and Light [12] which is used for spatial coordinates.

3.4 Non uniform approximation.

The motto of this section is to find effective strategies to reduce the error in the polynomial interpolation procedure. Consider again the error in the polynomial expansion of equation (3.2.11)

$$R(z) = F(z) - P_N(z) = \frac{f^{N+1}(\xi)}{(N+1)!}(z - x_N)(z - x_{N-1})..(z - x_0) \qquad (3.4.1)$$

If one considers a particular point $\bar{z}$ then define it as a new interpolation point. Following the procedure that allows to add a point to the interpolation one obtains:

$$R(z) = f[x_0, x_1, \ldots, x_N, \bar{z}](\bar{z} - x_N)(\bar{z} - x_{N-1})..(\bar{z} - x_0) \qquad (3.4.2)$$

As has been described in section 3.2 the strategy of minimizing the error has been to minimize the product term irrespective of $\bar{z}$ confined to the interval -1 to 1 . As a result the quality of the interpolation is independent of the initial vector ψ. In many cases in particular when the vector ψ is spanned only by a small fraction of the spectral range of $\hat{\mathbf{O}}$ it is desirable to adopt the interpolation points to the actual vector ψ. This can be done by minimizing the norm of the error vector $< \chi | \chi >$:

$$\chi = R(\hat{\mathbf{O}})\psi = f[x_0, x_1, \ldots, x_n, \hat{\mathbf{O}}](\hat{\mathbf{O}} - x_N\hat{\mathbf{I}})(\hat{\mathbf{O}} - x_{N-1}\hat{\mathbf{I}}) \cdots (\hat{\mathbf{O}} - x_0\hat{\mathbf{I}})\psi \qquad (3.4.3)$$

where the minimization is in respect to the sampling points $x_0, x_1, \cdots, x_n$. For the general case evaluating the mapping of the operator $f[x_0, x_1, \ldots, x_n, \hat{\mathbf{O}}]$ is extremely difficult because the divided difference coefficients are rational fractions and not polynomials. To overcome this difficulty instead of the most general minimization procedure one can minimize the norm of the vector produced by the r.h.s of equation (3.4.3) only:

$$\bar{\chi} = \bar{R}(z)\psi = (\hat{\mathbf{O}} - x_n\hat{\mathbf{I}})(\hat{\mathbf{O}} - x_{n-1}\hat{\mathbf{I}}) \cdots (\hat{\mathbf{O}} - x_0\hat{\mathbf{I}})\psi \quad , \qquad (3.4.4)$$

with respect to the N sampling points. To simplify the minimization problem equation (3.4.3) can be rewritten as a power expansion:

$$\bar{\chi} = \bar{R}(z)\psi = \sum_{k=0}^{N+1} d_k \hat{\mathbf{O}}^k \psi \qquad (3.4.5)$$

where the minimalization procedure is now transformed to the d_k coefficients. Notice that $d_N = 1$ because the polynomial in equation (3.4.4) in monomic. The norm of χ with respect to the d_k coefficients is a quadratic expression:

$$< \chi | \chi > = \sum_{i=0}^{N} \sum_{j=0}^{N} d_i d_j < \phi_i | \phi_j > \qquad (3.4.6)$$

where $\phi_k = \mathbf{O}^k \psi$ is the Krylov space vector. The minimalization of the residuum (3.4.6) leads to the linear expression for the d_i vector:

$$\tilde{\mathbf{S}}\,\mathbf{d} = \mathbf{b} \qquad (3.4.7)$$

where the overlap matrix $\tilde{\mathbf{S}}$ is defined as:

$$S_{ij} = < \phi_i | \phi_j > \qquad (3.4.8)$$

and the **b** vector becomes:

$$b_i = -< \phi_i | \phi_{N+1} > \tag{3.4.9}$$

because the polynomial (3.4.4) is monomic $d_{N+1} = 1$. Once the coefficients d_k are known solving for the roots of $\bar{R}(z)$ gives the desired optimal interpolation points. A twist on this procedure is obtained if the primitive Krylov base is orhonormalized by a Grham Schmit procedure: $\xi_1 = \phi_1$ and $\xi_2 = \phi_2 - < \phi_2 | \xi_1 > \xi_1$ and etc. The result is a diagonal overlap matrix $\bar{\mathbf{S}}$. Then minimizing the residuum in this functional base $\bar{R}(z)\psi = \sum_{j=0}^{N+1} e_j \xi_j$ leads to: $e_k = 0$ for $k \neq N+1$ and $e_{N+1} = 1$. With the interpretation that the minimum residuum vector is orthogonal to all other vectors in the Krylov space. The consequences of this observation are that the use of an orthonormal Krylov subspace automatically minimizes the interpolation error. This means that an alternative procedure for obtaining the interpolation points is to find the eigenvalues of the Hamiltonian in the truncated orthogonal Krylov space. This can be done simultaneously with the creation of the Krylov space using the procedure:

$$\chi_0 = \psi \tag{3.4.10}$$

and

$$\hat{O} \chi_0 = \alpha_0 \chi_0 + \beta_0 \chi_1$$

and generally:

$$\hat{O} \chi_n = \beta_{n-1} \chi_{n-1} + \alpha_n \chi_n + \beta_n \chi_{n+1} \tag{3.4.11}$$

where:

$$\alpha_n = < \chi_n | \hat{O} | \chi_n > \tag{3.4.12}$$

and

$$\beta_{n-1} < \chi_{n-1} | \hat{O} | \chi_n > \tag{3.4.13}$$

This procedure due to Lanczos leads to a tridiagonal truncated representation of the operator $\hat{O}$ [16]. The eigenvalues of this truncated space can be now used as interpolation points in equations (3.2.9-3.2.11). The solution of the characteristic equation and the location of the zeros of the polynomial $\bar{R}(z)$ for the minimum residuum case coincide [17]. This is another case of a mathematically equivalent but algorithmically different procedure.

IV. Summary

The purpose of this note is to demonstrate a uniform approach based on a global functional base to describe quantum mechanical dynamical processes. The description of the position-momentum phase space and the time-energy phase space is based on a discrete sampling which through the collocation technique determines the global functional base. This note is far from exhausting all the possibilities. For the position momentum phase space only the Fourier method was described in detail because due to its simplicity it serves as a good illustration to the principles involved. In the time-energy phase space the Chebychev and the short iterative Lanczos (SIL) were put on equal footing. Other families of propagators in particular the split operator family [20] which have potential for a global description are left to future notes.

Acknowledgements

This research was partially supported by a grant from the Binational United States - Israel Foundation. The Fritz Haber Research Center for Molecular Dynamics is supported by the Minerva Gesellschaft für die Forschung, GmbH München, Federal Republic of Germany.

Reference

[1]E. A. McCullough and R. E. Wyatt *J. Chem. Phys.* **51**, 1253 (1969); **54** *,3578 (1971).*

[2]A. Askar and S. Cakmak, *J. Chem. Phys.* **68**, 2797 (1978).

[3]K. C. Kulander, *J. Chem. Phys.* **69**, 5064 (1978);

K. C. Kulander, K. R. Sandhya Devi and S. E. Koonin, *Phys. Rev A* **25**, 2982 (1982).

[4]C. LeForestier, G. Bergeron, and P. C. Hiberty, *Chem. Phys. Lett.* **84**, 385 (1981); C. LeForestier, *Chem. Phys.* **87**, 241 (1984)..

[5]E. J. Heller, *Acc. Chem. Res.* **14**, 368 (1981). E. J. Heller, J. Chem. Phys. **62**, 1544 (1975). E. J. Heller, *J. Chem. Phys.* **65**, 1289 (1976). E. J. Heller, *J. Chem. Phys.* **65**, 4979 (1976). E. J. Heller, *J. Chem. Phys.* **67**, 3339 (1977). R. C. Brown and E. J. Heller, *J. Chem. Phys.* **75**, 186 (1981). S. Y. Lee and E. J. Heller, *J. Chem. Phys.* **76**, 3035 (1982).

[6]D. Tannor and E. J. Heller, *J. Chem. Phys.* **77**, 202 (1982).

E. J. Heller, R. L. Sundberg and D. Tannor, *J. Chem. Phys.* **86**, 1882 (1982).

[7]V. Mohan and N. Sathyamurthy, *Comp. Phys. Rep.* **7**, 215 (1988). D. Neuhaser D. Kuri and M. Baer *Comp. Phys. Commun.* **63** 311 (1991).

[8]R.B. Gerber, R. Kosloff and M. Berman, *Comp. Phys. Rep.* **5**, 59 (1986); G. Billing *Comp. Phys. Rep.* **12** 383 (1990).

[9]S. O. Williams and D. G. Imre, J. Phys. Chem. **92**, 6636 (1988); M. V. Rama Krishna and R. D. Coalson, Chem. Phys. **120**, 327 (1988); W. T. Pollard, S.-Y. Lee, and R. A. Mathies, J. Chem. Phys. **92**, 4012 (1990).

[10]K. Kullander, *Comp. Phys. Commun.* **63** (1991).

[11]D. Kosloff and R. Kosloff, *J. Comp. Phys.* **52**, 35 (1983).

[12]D. Gottlieb and S. A. Orszag, " *Numerical Analysis of Spectral Methods: Theory and Applications*" (SIAM, Philadelphia 1977).

[13]J. W. Cooley and J. W. Tukey *Math. Comp.* **19**, 297 (1965);

C. Temperton , *J. Comp. Phys.* **52**, 1 (1983).

[14]J.C. Light I.P. Hamilton and J.V. Lill , *J. Chem. Phys.* **82**, 1400 (19855).

[15]R. H. Bisseling and R. Kosloff, *J. comp. Phys.* **76**, 243 (1988).

[16]H. Tal Ezer and R. Kosloff, *J. Chem. Phys.* **81**, 3967 (1984).

[17]M. Berman and R. Kosloff, *Comp. Phys. Commun.* **63**,1 (1991).

H. Tal Ezer M. Berman and R. Kosloff *J. Phys. A (1991).*

[18] C. LeForestier, R. Bisseling, C. Cerjan, M. Feit, R. Friesner, A. Guldberg, A Hammerich, G. Julicard, W. Karrlein, H. D. Meyer, N. Lipkin, O. Roncero and R. Kosloff, *J. Comp. Phys.* **94** 59 (1991).

[19]Hillel Tal Ezer, Ronnie Kosloff, and Charly Cerjan, *J. Comp. Phys.* **100** 179 (1992).

[20]M. D. Feit, J. A. Fleck Jr., and A. Steiger, *J. Comp. Phys.* **47**, 412 (1982).

M. D. Feit and J. A Fleck, *J. Chem. Phys.* **78**, 301 (1982).

DISTRIBUTED APPROXIMATING FUNCTIONS FOR REAL-TIME QUANTUM DYNAMICS

Donald J. Kouri

Department of Chemistry and Department of Physics
University of Houston
Houston, TX 77204-5641

David K. Hoffman

Department of Chemistry and Ames Laboratory
Iowa State University
Ames, IA 50011

INTRODUCTION

The application of path integrals to real-time quantum dynamics is of great interest because of the potential for circumventing severe difficulties which make standard scattering calculations for multiparticle systems unfeasible.[1-22] These difficulties are consequences of the extremely large matrix dimensions encountered in systems involving more than 3 atoms, or even 3 atom systems involving only heavy atoms. In addition to the rapid increase in computational effort with matrix dimension, the extremely large memory required as basis size becomes large is also a serious problem. However, the Feynman path integral utilizes the coordinate representation, short real-time propagator, so that the local potentials (which are normally encountered in chemical physics) are diagonal, and the kinetic energy portion of the propagator is of a relatively simple form. By compounding the short real-time propagator N-times (for the step τ, so that $t = N\tau$), the full propagator is obtained approximately as an NG-dimensional integral (the discretized Feynman path integral[1]). Here, G represents the number of atoms in the system (or if the center of mass motion has been separated, G equals the number of atoms less one). Because N is typically large, such a result suggests that one resort to Monte Carlo methods for evaluating the Feynman amplitude, but the extremely oscillatory nature of the exact short real-time propagator makes such a procedure totally impossible.[1-22] These interfering oscillations naturally suggest methods based on stationary phase ideas, and much effort has been expended developing Monte Carlo methods using the classical action to construct a

Time-Dependent Quantum Molecular Dynamics, Edited by J. Broeckhove and L. Lathouwers, Plenum Press, New York, 1992

bias sampling function.[9–10,12] These require the calculation of the classical action and become problematic when numerous classical trajectories contribute.

An alternative approach suggested by Makri[15] involves projecting out of the short, real-time free propagator contributions from momenta above a cutoff value, P_{max} (in magnitude). The justification for this is the idea that any particular, physically acceptable wavepacket has negligible momentum components above a sufficiently large P_{max} due to energy conservation. Makri's projection of the free propagator onto the subset of momenta less than or equal to P_{max} produces an "effective propagator" with greatly reduced oscillations. This Fourier-based effective propagator decays with the distance traveled as $(x - x')^{-1}$, as a direct consequence of the abrupt cutoff of momentum included in the plane wave spectral resolution of the free evolution operator.[15,23] Makri reported a Monte Carlo evaluation of the real-time Feynman path integral for the autocorrelation function of a Gaussian wavepacket in a Morse potential, but not for sufficiently long times to observe recurrance as the wave is reflected back by the potential.[15]

Yet another approach to the evaluation of real-time path integrals is that based on the coherent state representation.[24] Marchioro and co-workers[16,25] have shown that the coherent state representation of the short real-time evolution operator yields a Gaussian envelope in coordinate space after the momentum of the system is integrated (analytically). Compounding the coherent state representation of $exp(-iH\tau/\hbar)$ for N-time steps then yields a "coherent state discretized path integral," with Gaussian factors naturally occuring. Marchioro reports Monte Carlo evaluations for some simple model systems and has been able to follow the autocorrelation function through several recurrances. Unfortunately, the method still cannot be applied to sufficiently long times to enable general scattering problems to be done.

In addition to this work, the coherent state representation has also been utilized by Ohrn and co-workers[26] to treat atomic and molecular collisions in which the Born-Oppenheimer (approximate) separation of electronic and nuclear degrees of freedom has not been invoked. Nuclear motion has been treated classically leading to a problem involving electrons in the presence of time dependent forces. The theory of Lie groups is used to particularize the coherent state.

By contrast, we are focussing on path integral methods to treat the nuclear dynamics fully quantally, while taking account of the electrons by means of a single Born-Oppenheimer potential surface. The approach we have been exploring aims at a factorization of the real-time path integral into one whose integrand is the product of Gaussians times polynomials of degree M, which vary slowly in the region where the integrand is significant. The strategy we employ is based on the idea of distributed approximating functions (DAFs) which can be used to fit a given class of wavepackets to arbitrarily high accuracy and which are analytically and exactly freely propagated. The class of wavepackets (i.e., the "DAF-class") consists of all wavepackets which can be approximated sufficiently acccurately as an Mth-degree polynomial under the DAF envelope. The DAFs are constructed from Hermite functions[27] (the product of a Hermite polynomial with its Gaussian generator), so that they exactly reproduce any Mth degree polynomial under the DAF-envelope. It has been established that the Hermite functions are freely propagated exactly and analytically.[27] Because the DAFs involve the Gaussian generator of the Hermite polynomials, they are highly compact and lead to a path integral expression that automatically involves a Gaussian envelope (although due to the presence of an Mth-degree polynomial, where M is even, the overall envelope of the DAF is wider than the Gaussian generator of the Hermite polynomials). We are currently carrying out test calculations to see how well Monte Carlo integration of the DAF-form of the path integral works.[22,28]

In this paper, we focus on a discussion of the DAF approach to the calculation of path integrals. In so doing, we shall restrict our considerations to the basic stationary-DAF (SDAF)[22,27] form of the DAF-class propagators. We begin with a brief review of the basic ideas of real-time path integrals. We then review the DAF-class free propagator, pointing out the similarities and differences between the DAF approach and other approaches.

BASIC IDEAS OF PATH INTEGRALS

To facilitate the discussion of DAFs as an approach to real-time path integration and its similarities and differences with other approaches, we shall begin by giving a simple review of the basic ideas and properties of real-time path integrals. An excellent resource for the reader who desires more detail is the reference book by Schulman.[25] For time independent hamiltonian operators, $\hat{H}$. it is well known that the state of the system of interest at time t, $|\psi(t)\rangle$ is given in terms of the state at time $t = 0$ by the expression

$$|\psi(t)\rangle = exp(-i\hat{H}t/\hbar) \mid \psi(0)\rangle . \tag{1}$$

This reflects the most useful feature of time dependent quantum mechanics for such systems; namely, the initial value nature of the dynamics. The challenge is, of course, the construction of the operator $exp(-i\hat{H}t/\hbar)$, or at least the evaluation of its effect on the state or states of interest. The simplest procedure is to write the identity

$$exp(-i\hat{H}t/\hbar) = \left[exp(-i\hat{H}\tau/\hbar)\right]^N , \tag{2}$$

where

$$t = N\tau . \tag{3}$$

By making τ sufficiently small, one can, to any desired level of accuracy, express $exp(-i\hat{H}\tau/\hbar)$ in the so-called "symmetrized Trotter"[1,25] or "symmetric split operator"[29–30] form

$$exp(-i\hat{H}\tau/\hbar) \cong exp(-i\hat{V}\tau/2\hbar)\, exp(-i\hat{K}\tau/\hbar)\, exp(-i\hat{V}\tau/2\hbar) , \tag{4}$$

where $\hat{V}$ is the potential operator (assumed to be a function only of the position operator $\hat{x}$) and $\hat{K}$ is the kinetic energy operator (a function only of the momentum operator, $\hat{p}$). Obviously, the commutator $[\hat{K}, \hat{V}]$ is nonzero so that simultaneous eigenstates of the two operators cannot be found. Indeed, the difference between the left and right hand sides of Eq.(4) may be written explicitly in terms of such commutators, with the leading term varying as τ^4. Using Eq.(4) in Eq.(2) leads to the result

$$|\psi(t)\rangle = \left(e^{-i\hat{V}\tau/2\hbar}\, e^{-i\hat{K}\tau/\hbar}\, e^{-i\hat{V}\tau/2\hbar}\right)^N \mid \psi(0)\rangle . \tag{5}$$

Now the "standard" quantum mechanical prescription to compute the action of an operator on a state is to introduce a representation or basis, and the most natural bases are those which diagonalize the operators of interest. Thus, to proceed, we note that the eigenstates of the potenetial are those of the position

$$\hat{x} \mid x\rangle = x \mid x\rangle , \tag{6}$$

and since $\hat{V}$ depends only on $\hat{x}$,

$$e^{-i\hat{V}\tau/2\hbar} \mid x\rangle = e^{-iV(x)\tau/2\hbar} \mid x\rangle . \tag{7}$$

(Such interactions are said to be "local" and are diagonal in the coordinate representation.) Clearly, the commutator $[\hat{p}, \hat{x}]$ is nonzero, so that $exp(-i\hat{K}\tau/\hbar)$ will be nondiagonal (or as sometimes said, "nonlocal") in the coordinate representation. However, intuitively, it should not be hightly nonlocal since spatial derivatives of a function at a point are essentially determined by the values of the function in a small neighborhood about the point of interest. This suggests that the coordinate representation of $exp(-iK\tau/\hbar)$, for small τ, will be a highly banded matrix. These facts, along with the computational stratigem that multiplication of a vector by a highly banded matrix is suited perfectly to implementation on massively parallel processing supercomputers, makes it very attractive to try and develop computational schemes based on evaluating all quantities in the coordinate representation.

We therefore project the state vector after N-time steps, $|\psi(t)\rangle$, (Eq.(5)) onto the coordinate representative $\langle x_N|$,

$$\langle x_N \mid \psi(t)\rangle \;=\; \langle x_N \mid \left(e^{-i\hat{V}\tau/2\hbar}\, e^{-i\hat{K}\tau/\hbar}\, e^{-i\hat{V}\tau/2\hbar}\right)^N \mid \psi(0)\rangle \;. \tag{8}$$

We also insert resolutions of the identity after each of the N-factors $exp(-i\hat{V}\tau/2\hbar)\; exp(-i\hat{K}\tau/\hbar)\; exp(-i\hat{V}\tau/\hbar)$,

$$1 \;=\; \int_{-\infty}^{\infty} dx_j \mid x_j\rangle\langle x_j \mid \,, \tag{9}$$

where the subscript j denotes how many time steps have been taken before the $j-th$ identity resolution occurs. Noting that $\langle x_N|\psi(t)\rangle$ is the wavepacket $\psi(x_N|t)$ and $\langle x_0|\psi(0)\rangle$ is the wavefunction $\psi(x_0|0)$, and using Eq.(7), we obtain

$$\psi(x_N \mid t) \;=\; \int_{-\infty}^{\infty} dx_{N-1} \cdots \int_{-\infty}^{\infty} dx_0 \prod_{j=1}^{N} e^{\frac{-i\tau}{2\hbar}[V(x_j)+V(x_{j-1})]}$$
$$\langle x_j \mid e^{-i\hat{K}\tau/\hbar} \mid x_{j-1}\rangle\, \psi(x_0 \mid 0)\;. \tag{10}$$

The next step is to evaluate the matrix element $\langle x_j|exp(-i\hat{K}\tau/\hbar)|x_{j-1}\rangle$. However, we remark that Eq.(10) is already clearly in the form of a path integral (although discretized in time since $\psi(x_N|t)$ is the result of compounding N-discrete time steps, τ). To evaluate $\langle x_j|exp(-i\hat{K}\tau/\hbar)|x_{j-1}\rangle$, we do the standard quantum mechanical procedure of introducing a basis that diagonalizes $exp(-i\hat{K}\tau/\hbar)$, by inserting

$$1 \;=\; \frac{1}{\hbar}\int_{-\infty}^{\infty} dp \mid p\rangle\langle p \mid \,, \tag{11}$$

where the momentum eigenstates satisfy

$$\hat{p} \mid p\rangle \;=\; p \mid p\rangle \,, \tag{12}$$

$$\hat{K} \mid p\rangle \;=\; \frac{p^2}{2m} \mid p\rangle \;. \tag{13}$$

Then

$$\langle x_j \mid e^{-i\hat{K}\tau/\hbar} \mid x_{j-1}\rangle \;=\; \frac{1}{\hbar}\int_{-\infty}^{\infty} dp\, \langle x_j \mid p\rangle\, e^{-ip^2\tau/2m\hbar}\, \langle p \mid x_{j-1}\rangle \,, \tag{14}$$

and we note that the coodinate representation of the state of exact linear momentum (i.e., the wavefunction for a state of exactly known momentum) is a plane wave:

$$\langle x_j \mid p\rangle \;=\; \frac{1}{\sqrt{2\pi}}\, e^{ipx_j/\hbar} \,, \tag{15}$$

and $\langle p|x_{j-1}\rangle$ equals $\langle x_{j-1}|p\rangle^*$. Our coordinate representation of the free evolution operator is then

$$\langle x_j \mid e^{-i\hat{K}\tau/\hbar} \mid x_{j-1}\rangle = \frac{1}{2\pi\hbar}\int_{-\infty}^{\infty} dp\, e^{ip(x_j-x_{j-1})/\hbar}\, e^{-ip^2\tau/2m\hbar} . \tag{16}$$

This is easily evaluated analytically (by completing the square and performing a contour integration), and the result is

$$\langle x_j \mid e^{-i\hat{K}\tau/\hbar} \mid x_{j-1}\rangle = \left(\frac{m}{2\pi i\hbar\tau}\right)^{1/2} e^{im(x_j-x_{j-1})^2/2\hbar\tau} . \tag{17}$$

However, it is of interest to note that the integral in Eq.(16) is not convergent due to the integrand tending to zero sufficiently rapidly as the integration variable $p \to \pm\infty$. In fact, the modulus of the integrand in Eq.(16) is <u>constant</u>! The convergence of the integral results, in fact, as the consequence of ever increasingly rapid oscillations resulting in cancellation of the contributions as $p \to \pm\infty$; i.e., it converges due to extreme quantum interference. We shall see that this has fundamental consequences for the discretized (in time) Feynman path integral, which results from substituting Eq.(17) into Eq.(10). The path integral is thus

$$\psi(x_N \mid t) = \left(\frac{m}{2\pi i\hbar\tau}\right)^{N/2} \int_{-\infty}^{\infty} dx_{N-1} \cdots \int_{-\infty}^{\infty} dx_0$$
$$\times\, e^{\frac{i}{\hbar}\sum_{j=1}^{N}\left[\frac{m}{2\tau}(x_j-x_{j-1})^2 - \frac{\tau}{2}\{V(x_j)+V(x_{j-1})\}\right]}\, \psi(x_0 \mid 0) . \tag{18}$$

The connection to Feynman's notation is seen readily by noting that

$$\frac{m}{2\tau}(x_j - x_{j-1})^2 = \frac{m}{2}\frac{(x_j - x_{j-1})^2\tau}{\tau^2} \tag{19}$$

$$= \int_{(j-1)\tau}^{j\tau} dt\, K , \tag{20}$$

and

$$\frac{\tau}{2}\left[V(x_j) + V(x_{j-1})\right] \cong \int_{(j-1)\tau}^{j\tau} dt\, V , \tag{21}$$

where Eqs.(19)-(20) result from the fact that one has rectilinear motion due to the classical kinetic energy, and Eq.(21) represents a trapezoidal rule evaluation of the time integral of the potential. Combining Eqs.(18)-(21), we arrive at Feynman's result

$$\psi(x_N \mid t) = \left(\frac{m}{2\pi i\hbar\tau}\right)^{N/2} \sum_{\substack{\text{sum over}\\ \text{paths}}} e^{\frac{i}{\hbar}\sum_{j=1}^{N}\int_{(j-1)\tau}^{j\tau} dt\, L} , \tag{22}$$

where L is the usual classical Lagrangian,

$$L = K - V . \tag{23}$$

It is clear from the form Eq.(18) that most paths will produce extreme oscillations leading to destructive interference. This results from the facts that the integrand in Eq.(18) has constant modulus, and (a) as $|x_j - x_{j-1}|$ increases, the frequency of oscillation of the integrand increases proportional to $|x_j - x_{j-1}|$ (b) as the potential changes rapidly from x_j to x_{j-1}, there will be rapid oscillations. As a result, Eq.(18) or (22) cannot be evaluated by standard numerical integration methods.[1−22]

However, even looking at the coordinate representation matrix elements, $\langle x_j|exp(-i\hat{H}\tau/\hbar)|x_{j-1}\rangle$ and $\langle x_j|exp(-i\hat{K}\tau/\hbar)|x_{j-1}\rangle$, it is clear that totally counter-intuitive behavior is predicted, even though the expressions developed for these matrix elements are exact (for $exp(-i\hat{K}\tau/\hbar)$) or of arbitrarily high accuracy (for $exp(-i\hat{H}\tau/\hbar)$). We remind the reader that the standard interpretation of $\langle x_j|exp(-i\hat{K}\tau/\hbar)|x_{j-1}\rangle|^2$ is that it gives the "relative" probability of a transition from state $|x_{j-1}\rangle$ to state $|x_j\rangle$, under the action of the free evolution operator. That is, it gives the probability that a particle at the point x_{j-1} will be found at the point x_j, a time τ later. But clearly,

$$| \langle x_j \mid e^{-i\hat{K}\tau/\hbar} \mid x_{j-1}\rangle |^2 = \frac{m}{2\pi\hbar\tau} , \tag{24}$$

which
says the relative probability to travel the distance $|x_j - x_{j-1}|$ is constant, independent of the distance. Using

$$\langle x_j \mid e^{-i\hat{H}\tau/\hbar} \mid x_{j-1}\rangle = \left(\frac{m}{2\pi i\hbar\tau}\right)^{1/2} e^{\frac{i}{\hbar}\left[\frac{m}{2\tau}(x_j-x_{j-1})^2-\frac{\tau}{2}\{V(x_j)+V(x_{j-1})\}\right]} , \tag{25}$$

we also have

$$| \langle x_j \mid e^{-i\hat{H}\tau/\hbar} \mid x_{j-1}\rangle |^2 = \left(\frac{m}{2\pi\hbar\tau}\right) , \tag{26}$$

which says the relative probability to travel any distance $|x_j - x_{j-1}|$, in short time τ, is independent of the potential. (We remark that this probability is also independent of the potential, for any time τ, in the case of a harmonic oscillator, for which the exact coordinate representation full propagation can be obtained analytically.) Of course, we are seeing the manisfestation of the unusual nature of convergence of the integral, Eq.(16). The increasingly rapid oscillations which lead to convergence of Eq.(16) result from the infinite limits on the momentum; the infinitely high momenta will lead to correspondingly rapid oscillations in coordinate space due to the de Broglie relation between the spatial wavelength and momentum.[23] As was noted by Feynman, these rapid oscillations in Eqs.(18) or (22) lead to the fundamental connection between quantum and classical mechanics, through the stationary phase evaluation of the sum over all paths. This leads to the well known, elegant semiclassical treatment of scattering.[3–4]

However, the inclusion of infinitely large magnitude momenta in evaluating Eq.(16) can be viewed as the consequence of our having demanded exact or arbitrarily accurate matrix elements of the abstract operators $exp(-i\hat{K}\tau/\hbar)$ and $exp(-i\hat{H}\tau/\hbar)$, independent of the wavepacket to which they will be applied. Thus, the matrices are required to accurately propagate any physically allowed wavepackets. Even though the average momentum of any particular wavepacket is always finite, the requirement of being able to propagate any possible packet forces one to include infinite momenta, thus leading to the disasterous (from the standpoint of quadrature evaluations) rapid oscillations of the path integrand.

There is another aspect of the coordinate representations of the free and full evolution operators which should be mentioned. The use of position eigenstates intrinsically implies constant relative probabilities for traveling any distance $|x_j - x_{j-1}|$ in time τ. To see this, we consider the wavefunction for an eigenstate, at time $j\tau$, of the position operator, $|x_j\rangle$:

$$\psi_{x_j}(x) = \langle x \mid x_j\rangle \tag{27}$$

$$= \delta(x - x_j) . \tag{28}$$

Clearly, this is a state whose position uncertainty is $\Delta x \equiv 0$ (it describes a particle located exactly at x_j), and therefore whose momentum uncertainty is infinite. Obviously, such a state spreads infinitely rapidly, because it is a state in which all momenta, $-\infty < p_{x_j} < \infty$, have equal relative probability; viz.

$$\delta(x - x_j) \quad = \quad \frac{1}{2\pi\hbar} \int_{-\infty}^{\infty} dp\, e^{ip(x-x_j)/\hbar} \,. \tag{29}$$

Therefore, evolving such a state forward in time by <u>any</u> amount τ will yield a constant relative probability of being any distance $|x - x_j|$ from the starting point x_j.

The natural question is then, how can calculations (other than by semiclassical or stationary phase) be carried out? One way is to use what are known popularly among scattering theorists as "close coupling" truncations combined with discretization of the continuous spectra. Thus, one typically considers projection of the propagator onto a subspace. The standard procedure for truncations is based on consideration of (1) the range of the spectra contained in the initial wavepacket to be propagated. (2) the size of transitions which can be induced by the interaction during the collision. The truncated spectral resolution must include all values that can be significantly populated due to excitations of the values contained in the initial wavepacket. In the case of momentum, e.g., one considers the highest $|p|$ significantly contained in the initial packet, and the largest amount of momentum transfer likely to be caused by the potential, to estimate a maximum $|P_{max}|$. (In addition, if one intends to use Fast Fourier Transform methods (FFT), one must also discretize the momentum.) For the position quantum numbers one looks at the region where the initial packet is concentrated, and how far $exp(-i\hat{K}\tau/\hbar)$ will translate (and spread) the packet during the time required for the collision. This will fix the range of x over which the packet will be found. (Again, in FFT methods, x is discretized, and ΔX is fixed by $|P_{max}|$.) This leads to a sequential propagation involving matrix multiplication, which is typically done using discretized grids in p and x, and FFTs (see Feit and Fleck;[29] Kosloff and Kosloff[31]). The evolution due to the potential is carried out in the discretized coordinate representation, and that due to the kinetic energy is carried out in the discretized momentum representation, and the transformation between the two representations is carried out using FFTs.

Makri[15] was the first to point out that if one considers the effect on the propagator of such a truncation of the momentum to range from $-|P_{max}|$ to $+|P_{max}|$ is to eliminate the ever increasing oscillations. The effective free propagator that results is given by

$$\langle x_j \mid e^{-i\hat{K}\tau/\hbar} \mid x_{j-1}\rangle_{Makri} \quad = \quad \frac{1}{2\pi\hbar} \int_{-P_{max}}^{P_{max}} dp\, e^{-ip^2\tau/2\hbar m}\, e^{ip(x_j - x_{j-1})/\hbar} \,, \tag{30}$$

where the integral is evaluated by Makri using error functions (rather than quadrature and FFTs). For large $|x_j - x_{j-1}|$, Makri's "effective free propagator" behaves like

$$sin\left[P_{max}(x_j - x_{j-1})\right] \Big/ (x_j - x_{j-1}) \,, \tag{31}$$

so that the relative probability decays inversely with the distance traveled, and the oscillation frequency is determined by P_{max}. Using Eq.(30) in terms of error functions, it was possible to evaluate Eq.(10) by Monte Carlo integration, so long as N was not too large (i.e., for not too long a time).[15]

We now shall describe a different approach to wavepacket propagation, with the hope of being able to provide a useful framework both for real-time quantum Monte

Carlo calculations, and hightly banded discretized matrix propagation. The approach is based on the distributed approximating functions (DAFs)[22,27] for accurately fitting the wavepacket at each time step in its propagation. We shall note similarities to propagations based on coherent states (Ohrn and co-workers,[26] and Marchioro[16]), as well as to Makri's work.[15] In addition there are similarities to the orthogonal polynomial (quadrature) based discrete variable representations (DVRs) of Light and co-workers,[32] and to wavelets.[33] However, the DAFs are not the same as any of these, and the approach does not, in fact, correspond to the usual quantum mechanical basis set expansion method which yields a projection onto a subspace of the Hilbert space.

THE DAF-CLASS PROPAGATOR

As mentioned earlier, the DAF is constructed (1) to approximate to a specified accuracy any wavepacket, in the Hilbert space of L^2 functions, that can be approximated to an appropriate accuracy by a polynomial of degree M under the DAF-envelope (2) to be analytically and exactly freely propagated by $exp(-i\hat{K}\tau/\hbar)$. To achieve (1), the DAFs are constructed to exactly reproduce any polynomial of degree M, under the DAF envelope. To achieve (2), the DAFs are constructed from products of Hermite polynomials times their Gaussian generator (which we term "Hermite functions"). The Hermite functions differ from harmonic oscillator states in that the latter involve a factor of the square root of the Gaussian generator for the Hermite polynomials. As a result, the Hermite functions are not mutually orthogonal, but instead, they are the biorthogonal complement of the Hermite polynomials of the same argument. We note that the coherent state representation[16,24] is also nonorthogonal. Requirement (1) is expressed by the equation

$$\psi_{DAF}(x \mid 0) = \int_{-\infty}^{\infty} dx' \, F(x, x' \mid M, \sigma(0)) \, \psi(x' \mid 0) \, , \tag{32}$$

where $\sigma(0)$ is the variance of the Gaussian generator portion of the DAF. A discretized version is given by

$$\psi_{DAF}(x \mid 0) = \sum_{j=-\infty}^{\infty} F(x, x_j \mid M, \sigma(0), \Delta) \, \psi(x_j \mid 0) \, . \tag{33}$$

Here, $\psi_{DAF}(x|0)$ is the DAF-approximation to the wavepacket at time $t = 0$, $F(x, x'|M, \sigma(0))$ is the continuous DAF, $F(x, x_j|M, \sigma(0), \Delta)$ is the discretized DAF, and Δ is the grid spacing in x. In the discretized case, there is similarity to the discrete variable representation (DVR) of Light and co-workers[32] associated with Gauss-Hermite quadrature. However, the discrete DAF, $F(x_j, x_{j'}|M, \sigma(0), \Delta)$ does not satisfy the "discrete orthogonality" of equaling $\delta_{jj'}$. That is, it is not an interpolation which exactly reproduces the function at the grid points.[27] An additional requirement for the stationary DAFs is that, like one possible choice of coherent states[26] they be translationally invariant. Thus, one requires $F(x, x_j|M, \sigma(0), \Delta)$ and $F(x, x'|M, \sigma(0))$ to depend only on the difference $(x - x_j)$ and $(x - x')$, respectively. To construct the DAFs, we develop the Dirac delta function, $\delta(x - x')$, as an expansion in the Hermite functions:

$$\delta(x - x') = \sum_{n=0}^{\infty} a_n \, e^{-(x-x^2)/2\sigma^2(0)} \, H_n\left(\frac{x - x'}{\sqrt{2}\sigma(0)}\right) \, . \tag{34}$$

To exactly evaluate the a_n, one multiplies Eq.(34) by a Hermite polynomial, $H_{n'}(\frac{x-x'}{\sqrt{2}\sigma(0)})$, and integrates over x or x', obtaining

$$a_{n'} = \frac{H_{n'}(0)}{\sqrt{2}\sigma(0)} . \tag{35}$$

In obtaining this result for $a_{n'}$, we have explicitly used the biorthogonality of the Hermite polynomials and the Hermite functions. We also note that Eq.(35) vanishes unless n is even (the $\delta(x-x')$ is an even function). Truncation of the expansion at $n = M$ in Eq.(34) then gives an expression for the coordinate represenation identity with respect to polynomials of degree M:

$$\delta^{(M)}(x-x') \equiv F(x,x' \mid M, \sigma(0)) , \tag{36}$$

$$= \frac{e^{-(x-x')^2/2\sigma(0)^2}}{\sqrt{2\pi}\sigma(0)} \sum_{n=0}^{M/2} \left(-\frac{1}{4}\right)^n \frac{1}{n!} H_{2n}\left(\frac{x-x'}{\sqrt{2}\sigma(0)}\right) . \tag{37}$$

The DAF-approximation to a DAF-class wavepacket is obtained by substituting Eq.(37) into Eq.(33),

$$\psi_{DAF}(x \mid 0) = \int_{-\infty}^{\infty} dx' \frac{e^{-(x-x')^2/2\sigma(0)^2}}{\sqrt{2\pi}\sigma(0)} \sum_{n=0}^{M/2} \left(-\frac{1}{4}\right)^n \frac{1}{n!} H_{2n}\left(\frac{x-x'}{\sqrt{2}\sigma(0)}\right) , \tag{38}$$

for the continuous case, and for the discrete case, we evaluate Eq.(38) by trapezoidal rule quadrature for ψ_{DAF} on a grid:

$$\psi_{DAF}(x_l \mid 0) = \frac{\Delta x}{\sqrt{2\pi}\sigma(0)} \sum_{j=-\infty}^{\infty} e^{-(x_l-x_j)^2/2\sigma(0)^2} \times \sum_{n=0}^{M/2} \left(-\frac{1}{4}\right)^n \frac{1}{n!} H_{2n}\left(\frac{x_l-x_j}{\sqrt{2}\sigma(0)}\right) \psi(x_j \mid 0) . \tag{39}$$

Now the DAF exactly reproduces any polynomial of degree M. However, since no polynomial is L^2, the DAF cannot exactly fit any wavepacket in the Hilbert space. As a consequence, neither Eq.(38) or (39) can be exact, and the DAF-class of wavepackets depends on the desired accuracy of the DAF fit of the wavepacket. This implies that any given wavepacket in the DAF-class can only be fitted a finite number of times before the resulting approximate packet is no longer contained in the DAF-class. Closely related to this is the fact that the DAF is not a projection operator, but rather a "restricted identity" on the DAF-class. That is,

$$\int_{-\infty}^{\infty} dx'' \, F(x,x'' \mid M,\sigma(0)) \, F(x'',x' \mid M,\sigma(0)) \neq F(x,x' \mid M,\sigma(0)) . \tag{40}$$

The left hand side does yield, however, an identity for DAF-class wavepackets for which the error, induced by two successive DAF-fits, is less than the tolerance allowed for the polynomial fit under the DAF envelope. Note that this implies nonuniqueness of the restricted identity. By way of contrast, one may also view Makri's truncation of the range of momentum included in the effective propagator as a truncation of the Dirac delta function, $\delta(x-x')$, according to

$$\delta^{P_{max}}_{Makri}(x-x') = \frac{1}{2\pi\hbar} \int_{-P_{max}}^{P_{max}} dP \, e^{iP(x-x')} . \tag{41}$$

It is easy to show that

$$\int_{-\infty}^{\infty} dx''\, \delta_{Makri}^{(P_{max})}(x - x'')\, \delta_{Makri}^{(P_{max})}(x'' - x') \equiv \delta_{Makri}^{(P_{max})}(x - x') \,, \tag{42}$$

so that it is a unique identity on the projected subspace. This feature of the DAFs not being a projection is intimately connected with the ability to construct as a highly banded a representation of the free evolution operator as possible. To see this, we note that the more standard way of expressing the Dirac delta function in quantum mechanics is to use an orthonormal basis comprised of eigenstates of some Hermitian Hamiltonian, writing

$$\delta^{n_{max}}(x - x') = \sum_{n-0}^{n_{max}} \phi_n(x)\, \phi_n^*(x') \,, \tag{43}$$

where

$$\int_{-\infty}^{\infty} dx\, \phi_n^*(x)\, \phi_m(x) = \delta_{nm} \,, \tag{44}$$

or as in Makri's work,[15]

$$\delta_{Makri}^{(P_{max})}(x - x') = \frac{1}{2\pi\hbar} \int_{-P_{max}}^{P_{max}} dp\, (e^{ipx})^* \, (e^{ipx'}) \,, \tag{45}$$

$$\frac{1}{2\pi\hbar} \int_{-\infty}^{\infty} dx\, (e^{ipx})^* \, (e^{ip'x}) = \delta(p - p') \,. \tag{46}$$

If one takes the $\phi_n(x)$ to be harmonic oscillator eigenstates, the result is

$$\delta_{H.O.}^{(M)}(x - x') = \sum_{n=0}^{M} c_n\, e^{-x^2/4\sigma(0)^2}\, H_n\left(\frac{x}{2\sigma(0)}\right) e^{-x'^2/4\sigma(0)^2}\, H_n\left(\frac{x'}{2\sigma(0)}\right) . \tag{47}$$

In the above, each harmonic oscillator ($H.O.$) state contains the square root of a Gaussian orthogonality weight function, while the DAF associates the complete Gaussian orthogonality weight function with one Hermite polynomial. This should produce the most rapid decay of the truncated Dirac delta function possible, when Hermite polynomials are used. However, the appearance of the full Gaussian weight function leads to nonorthogonality of the Hermite functions, and to the DAF not being a projection. In addition, Eq.(46) does not explicitly contain basis states which reflect the translational invariance since they do not depend on the difference $(x - x')$. (However, Makri's plane wave basis does incorporate translational invariance due to the fact that the plane wave product $\phi_p(x)\phi_p^*(x')$ behaves like $exp(ip(x - x'))$.)

To propagate freely the wavepacket by time step τ, one applies $exp(-i\hat{K}\tau/\hbar)$ to the DAF, using

$$e^{-i\hat{K}\tau/\hbar}\, F(x, x' \mid M, \sigma(0)) = F(x, x' \mid M, \sigma(0), \tau) \,, \tag{48}$$

$$= \frac{e^{-(x-x')^2/2\sigma(\tau)^2}}{\sqrt{2\pi}\sigma(\tau)} \sum_{n=0}^{M/2} \left(-\frac{1}{4}\right)^n \times \frac{1}{n!}\, (\sigma(0)/\sigma(\tau))^{2n}\, H_{2n}\left(\frac{(x - x')}{\sqrt{2}\sigma(\tau)}\right) . \tag{49}$$

Here, $\sigma(\tau)^2$ is given by

$$\sigma(\tau)^2 = \sigma(0)^2 + i\hbar\tau/m . \tag{50}$$

The propagated wavepacket is given then by

$$\psi_{DAF}(x \mid \tau) = \int_{-\infty}^{\infty} dx' \, F(x, x' \mid M, \sigma(0), \tau) \, \psi_{DAF}(x' \mid 0) . \tag{51}$$

Although the DAFs are not eigenstates of the kinetic energy operator, they are still exactly freely propagated by $exp(-i\hat{K}\tau/\hbar)$ as a consequence of being constructed from the Hermite functions (as opposed, e.g., to $H.O.$ eigenstates). (This property of being exactly freely propagated is also shared by coherent states.[16,24]) The discretized version is obtained easily by introducing a trapezoidal rule approximation into Eq.(50), to obtain[22,27]

$$\psi_{DAF}(x_l \mid \tau) = \sum_{j=-\infty}^{\infty} F(x_l, x_j \mid M, \sigma(0), \Delta, \tau) \, \psi_{DAF}(x_j \mid 0) , \tag{52}$$

where

$$F(x_l, x_j \mid M, \sigma(0), \Delta x, \tau) = \frac{\Delta x \, e^{-(x_l - x_j)^2/2\sigma(\tau)^2}}{\sqrt{2\pi}\sigma(\tau)} \sum_{n=0}^{M/2} \left(-\frac{1}{4}\right)^n \times \frac{1}{n!} \left(\sigma(0)/\sigma(\tau)\right)^{2n} H_{2n}\left(\frac{x_l - x_j}{\sqrt{2}\sigma(\tau)}\right) , \tag{53}$$

and

$$\psi_{DAF}(x_j \mid 0) = \sum_{j'=-\infty}^{\infty} F(x_j, x_{j'} \mid M, \sigma(0), \Delta x, \tau = 0) \, \psi(x_{j'} \mid 0) . \tag{54}$$

We note that by Eqs.(50) and (53), one can interpret the DAF-procedure as leading to a restricted free propagator for DAF-class wavepackets.

In order to propagate the wavepacket in the presence of the potential, the complete DAF-class propagator, for short real-time step τ, is obtained as

$$G(x, x' \mid M, \sigma(0), \tau) = e^{-iV(x)\tau/2\hbar} \, F(x, x' \mid M, \sigma(0), \tau) \, e^{-iV(x')\tau/2\hbar} . \tag{55}$$

Clearly, the DAF-parameters must be chosen to ensure that application of $exp(-i\hat{V}\tau/\hbar)$ after each repeated free DAF-propagation, does not remove the wavepacket from the DAF-class. If we choose to interpret $F(x, x'|M, \sigma(0), \tau)$, $F(x_j, x_{j'}|M, \sigma(0), \tau, \Delta)$, or their corresponding $G(x, x'|M, \sigma(0), \tau)$ and $G(x_j, x_{j'}|M, \sigma(0), \tau, \Delta)$, as DAF-class propagators, then the consequence of the DAFs not being projection operators is that these propagators are <u>not</u> explicitly unitary. Indeed, they need only to be approximately unitary, to the same level as the accuracy defining the DAF-class of wavepackets. The construction of the DAFs should always take account of the number of time steps required for the propagation and the level of accuracy required for the dynamical process of interest. Within that number of time steps and accuracy level defining the DAF-class, the DAF-class propagator will display a similar level of accuracy of unitarity.

We are currently in the process of carrying out the initial computational tests of the DAF-approach. Easiest have been applications of the discretized DAFs. The evaluation of real-time path integrals using the continuous version of the DAFs is being studied, and it will likely be necessary to incorporate the potential (as well as the average packet momentum) into the DAF in order to achieve the best results. These modifications to the DAF-approach are also currently under study, and appear to hold promise both as computational tools as well as a new framework for probing the fundamental correspondence between classical and quantum mechanics. Although the DAF-approach shares some similarities to several popular approaches to quantum dynamics, such as the coherent state path integral approaches of Marchioro[16] and Ohrn and co-workers,[26] the Fourier-based effective propagator of Makri,[15] and the orthogonal polynomial-based DVR methods of Light and co-workers,[32] it is, in philosophy and in detail, different from each. There appears reason to hope that the DAF formulation can make some very useful contributions to the understanding of quantum dynamics.

ACKNOWLEDGEMENTS

DJK gratefully acknowledges partial support of this research under NSF grants CHE89-07429 and CHE92-00518. DKH gratefully acknowledges partial support of this research under NSF grant CHE92-01967. Ames Laboratory is operated for the Deparment of Energy by Iowa State University under Contract No. 2-7405-ENG-82.

REFERENCES

1. R. P. Feynman, Rev. Mod. Phys. 20, 367 (1948).
2. R. P. Feynman and A. R. Hibbs, Quantum Mechanics and Path Integrals (McGraw-Hill, New York, 1965).
3. P. Pechukas, Phys. Rev. 181, 166 and 174 (1969).
4. E. J. Miller, Adv. in Chem. Phys. 25, 69 (1974).
5. J. D. Doll, and L. E. Meyers, J. Chem. Phys. 71, 2880 (1979); J. D. Doll, J. Chem. Phys. 81, 3536 (1984).
6. D. Thirumalai, E. J. Bruskin, and B. R. Berne, J. Chem. Phys. 79, 5063 (1983); D. Thirumalai and B. J. Berne, Ibid. 79, 5029 (1983) and 81, 2512 (1984).
7. W. H. Miller, S. D. Schwartz, and J. W. Tromp, J. Chem. Phys. 79, 4889 (1983); K. Yamashita and W. H. Miller, Ibid. 82, 5474 (1985); J. W. Tromp and W. H. Miller, Faraday Discuss. Chem. Soc. 84, 441 (1987) and J. Phys. Chem. 90, 3482 (1986).
8. R. D. Coalson, J. Chem. Phys., 83, 688 (1985).
9. J. D. Doll and D. L. Freeman, Science 234, 1356 (1986); J. D. Doll, R. D. Coalson, and D. L. Freeman, J. Chem. Phys. 87, 1641 (1987).
10. V. S. Filinov, Nucl. Phys. B 271, 717 (1986).
11. B. J. Berne and D. Thirumalai, Ann. Rev. Phys. Chem. 37, 401 (1986).
12. N. Makri and W. H. Miller, Chem. Phys. Lett. 139, 10 (1987) and J. Chem. Phys. 90, 904 (1989).
13. N. E. Hendrikson and E. J. Heller, Chem. Phys. Lett. 148, 277 (1988).
14. J. R. Klauder and I. Daubechies, Phys. Rev. Lett. 52, 1161 (1984); I. Daubechies and J. R. Klander, J. Math. Phys. 26, 2239 (1985).
15. N. Makri, Chem. Phys. Lett. 159, 489 (1989) and Computer Phys. Commun. 63, 389 (1991).
16. T. L. Marchioro, J. Math. Phys. 31, 2935 (1990); T. L. Marchioro and T. L. Bech, J. Chem. Phys. 96, 2966 (1992).

17. L. I. Lolle, C G. Gray, J. P. Poll, and A. G. Basile, Chem. Phys. Lett., 177, 64 (1991).
18. J. D. Doll, D. L. Freeman, and T. L. Beck, Adv. Chem. Phys. 77, 61 (1990).
19. A. Sethia, S. Sanyal, and Y. Singh, J. Chem. Phys. 93, 7268 (1990).
20. D. Thirumalai and B. J. Berne, Computer Phys. Commun. 63, 415 (1991).
21. H. Kröger, Phys. Reports 210, 45 (1992); H. Kröger, K. J. M. Moriarty, and J. Potvin, Phys. Rev. A 42, 2661 (1990).
22. D. J. Kouri and D. K. Hoffman, J. Phys. Chem. (submitted); and in preparation.
23. See, e.g., J. L. Powell and B. Craseman, Quantum Mechanics (Addison-Wesley, Reading, MA, 1961).
24. R. Glauber, Phys. Rev. 131, 2766 (1963); J. R. Klauder, Ann. Phys. 11, 123 (1960); J. R. Klauder and B. Skagerstam, Coherent States (World Scientific, Singapore, 1985).
25. L. S. Schulman, Techniques and Applications of Path Integration (Wiley, New York, 1981).
26. Y. Ohrn, private communication.
27. D. K. Hoffman, N. Nayar, O. A. Sharafeddin, and D. J. Kouri, J. Phys. Chem. 95, 8299 (1991); D. K. Hoffman and D. J. Kouri, Ibid. 96, 1179 (1992).
28. B. M. Pettitt, X. Ma, W. Zhu, D. J. Kouri, and D. K. Hoffman, to be published.
29. M. D. Feit and J. A. Fleck, J. Chem. Phys. 78, 301 (1982); Ibid. 80, 2578 (1984).
30. P. DeVries, in NATO ASI Series, B171, 113 (1988).
31. D. Kosloff and R. Kosloff, J. Comput. Phys. 52, 35 (1983); R. Kosloff and D. Kosloff, J. Chem. Phys. 79, 1823 (1983).
32. J. V. Lill, G. A. Parker, and J. C. Light, Chem. Phys. Lett. 89, 483 (1982); J. C. Light, I. P. Hamilton, and J. V. Lill, J. Chem. Phys. 82, 1400 (1985).
33. See, e.g., Y. Meyer, Seminaire Bourbaki, 662 (1985-1986); C. K. Chui, Wavelets (Academic Press, San Diego, 1992); and references therein.

WAVE PACKET STUDIES OF THE PREDISSOCIATION OF H_3

Jeffrey L. Krause,* Ann E. Orel,** Byron H. Lengsfield* and
Kenneth C. Kulander*

*Theoretical Atomic and Molecular Physics Group
Lawrence Livermore National Laboratory
Livermore, California 94550

**Department of Applied Science
University of California, Davis
Livermore, California 94550

1. INTRODUCTION

The quantum dynamics of small molecules on multiple, coupled potential energy surfaces remains a challenging area of study. Even for the few systems for which accurate ground state potential surfaces are available, very little is typically known about the excited state surfaces, or the couplings among the ground and excited surfaces. However, recent advances in electronic structure theory have begun to make the characterization of these surfaces and couplings feasible, and improvements in dynamical methods allow them to be incorporated into a realistic description of the dynamics. In this paper we concentrate on the predissociation of H_3, the simplest polyatomic system. We use *ab initio* methods to obtain the necessary excited state surfaces and non-adiabatic couplings, and a time-dependent wave packet treatment of the dynamics.

The Rydberg states of H_3 are formed by attaching an electron to the stable H_3^+ core. The bound vibrational-rotational levels of these states are all unstable with respect to dissociation to the H_3 ground state via spontaneous radiative decay, predissociation, or both. One sheet of the unstable ground electronic state of H_3 correlates asymptotically to $H + H_2$ $X^1\Sigma_g$ and the other to $H + H_2$ $b^3\Sigma_u$, eventually leading to $H + H + H$. The two sheets of the ground state surface cross in a conical intersection, or seam, at D_{3h} (equilateral triangle) geometries. Both dissociation channels are energetically accessible from all of the Rydberg states of H_3. Since H_3^+ has D_{3h} symmetry,

Time-Dependent Quantum Molecular Dynamics, Edited by J. Broeckhove
and L. Lathouwers, Plenum Press, New York, 1992

the conical intersection occurs in the Franck-Condon region, and can have a significant influence on the dissociation dynamics.

Several experiments have been performed to study the H_3 system. Peterson et al.[1] mass select H_3^+ from a hydrogen discharge. The ions are neutralized in a cesium vapor, producing H_3 in a distribution of Rydberg states, predominantly $n = 2$ (where n refers to the principle quantum number of the Rydberg electron). Most of the Rydberg states produced by this method predissociate very rapidly. Peterson et al. measure the kinetic energy of the resulting hydrogen atoms, which allows them to infer a 3-body to 2-body branching ratio.

Using a similar source of H_3, Cosby and Helm[2] create a beam of H_3, which they allow to drift for $\sim 1\mu$s. By this time only the $2p\ ^2A_2''$ state remains bound. This state decays via rotational predissociation, so its rotationless, vibrationless ($N = K = 0$) level is very long-lived, decaying eventually by either spin-orbit predissociation, or by a very weak direct radiative transition.[2,3] Cosby and Helm excite this level with a laser to the $n = 3$ Rydberg states, which predissociate rapidly. They then measure the vibrational and rotational distributions of the resulting H_2.

The light masses involved in the H_3 system, and the availability of high-quality experimental data, as well as an accurate ground state potential energy surface,[4] make H_3 an attractive system to study theoretically. In addition, this system displays a variety of interesting dynamical processes. The excited state potential energy surfaces have their closest approach to the ground state surface far from the Franck-Condon region, allowing for a competition between radiative and predissociative pathways. Flux encountering the conical intersection can either continue along the "upper" adiabatic surface to 3-body products, or make a transition through the conical intersection to the "lower" adiabatic surface to give either 2-body or 3-body products. Interference between direct dissociation and indirect dissociation mediated by the conical intersection can alter such physical observables as the product state distributions.

In this paper, we present results of our first studies of the predissociation of the $3s$ Rydberg state of H_3. We describe briefly the method we used to calculate the non-adiabatic coupling matrix elements between the $3s$ state and the ground state, and then how we incorporate these couplings into a wave packet treatment of the dynamics. Finally, we show how the couplings affect the total dissociation rate, the product branching ratios and final state distributions.

2. METHOD

The $n = 3$ Rydberg states of H_3 all have energies in excess of 7.6 eV relative to $H(1s) + H_2\ X^1\Sigma_g(v = 0, J = 0)$, and as mentioned previously, can decay either to the 2-body or 3-body continua. For this reason, the dynamics calculations presented in this work were performed using a wave packet method. We solve the time-dependent

Schrödinger equation

$$i\frac{\partial}{\partial t}\psi(\mathbf{r},t) = H\psi(\mathbf{r},t), \tag{1}$$

in Jacobi coordinates r, R and θ, where r is the H_2 bond distance, R is the distance from the H atom to the center of mass of H_2, and θ is the angle between r and R. The time propagation is performed on the potential surface of Varandas et al.[4] with the Chebychev method,[5] and the kinetic energy operators are evaluated in the discrete variable representation (DVR).[6] We have used this method previously in 3-dimensional calculations of the spontaneous radiative decay of the $2p\ ^2A_2''$ state of H_3.[7]

The total dissociation rate can be calculated with the "short-time" method as the Fourier transform of the autocorrelation function of the initial state $\psi(\mathbf{r},0)$ with $\psi(\mathbf{r},t)$, the time evolved state on the dissociative surface[8]

$$R(E) \sim \int_{-\infty}^{\infty} dt\, e^{iEt} \langle \psi(\mathbf{r},0) \mid \psi(\mathbf{r},t)\rangle \,. \tag{2}$$

The total rate can also be obtained with the "long-time" method by projecting the time evolved wave function at asymptotic times onto final states[9]

$$R_{v,J}(E) \sim \lim_{t\to\infty} \left| \int\int\int dr_1 dr_2 d\theta\, \phi_t(r_2)\phi_J(\theta)\phi_{v,J}(r_1)\psi(r_1,r_2,\theta,t) \right|^2 , \tag{3}$$

where ϕ_t are translational functions, ϕ_J are rotational functions, $\phi_{v,J}$ are vibrational functions. The total 2-body rate is the sum of the partial rates,

$$R(E) = \sum_{v,J} R_{v,J}(E). \tag{4}$$

Note that in the short-time method, the propagation must be continued only as long as the initial wave packet remains in the Franck-Condon region, while the long-time method requires the wave packet to have reached the asymptotic regions of the potential surface. The H_3 potential is kinematically symmetric, so for a totally symmetric initial state, the projection onto final states can be performed in just a single asymptotic arrangement. Flux entering other arrangements (corresponding to alternative sets of Jacobi coordinates) is removed with absorbing boundary conditions.[7] At energies below the opening of the 3-body channel (4.75 eV) the short-time and long-time methods are related by a sum rule, and produce identical results if the propagation is performed accurately. At energies above 4.75 eV, the difference between the two methods gives an estimate of the 3-body dissociation probability.[10]

In the limit of first-order, time-dependent perturbation theory,[9] the initial state in the calculation, $\psi(\mathbf{r},0)$ is a product of $\phi_{v,J}(\mathbf{r})$, a ro-vibrational state of the initial bound potential energy surface, and the operator that couples the bound and dissociative surfaces. In the more familiar case of photodissociation, this operator is the dipole operator, and hence

$$\psi(\mathbf{r},0) = \mu(\mathbf{r})\phi_{v,J}(\mathbf{r}). \tag{5}$$

Predissociation is a radiationless transition resulting from a breakdown of the Born-Oppenheimer approximation. To first order, the operator coupling the surfaces is the derivative operator $d/d\mathbf{r}$, which describes the change in the total (electronic plus nuclear) wave function due to motion of the nuclei. Thus, calculating the predissociation rate requires non-adiabatic coupling matrix elements (NACMEs) of the form[11]

$$f_{\alpha}^{i,j}(\mathbf{r}) = \langle \psi_i(\mathbf{e};\mathbf{r})| \frac{\partial}{\partial r_\alpha} |\psi_j(\mathbf{e};\mathbf{r})\rangle_{\mathbf{e}}, \tag{6}$$

where $\psi_j(\mathbf{e};\mathbf{r})$ is defined by

$$H_e(\mathbf{e};\mathbf{r})\psi_i(\mathbf{e};\mathbf{r}) = V_i(\mathbf{r})\psi_i(\mathbf{e};\mathbf{r}), \tag{7}$$

and H_e is the electronic Hamiltonian. In analogy to the photodissociation problem, the initial state in the dynamics calculation is then

$$\psi(\mathbf{r},0) \sim \sum_{\alpha} f_{\alpha}^{i,j}(\mathbf{r})\frac{\partial}{\partial r_\alpha}\phi_{v,j}(\mathbf{r}). \tag{8}$$

Notice that in this approximation, the NACMEs must be calculated only at geometries where the initial state has significant density. Also note that in the case of non-adiabatic coupling, Eqs. 2-4 are defined only at the discrete energies of the Rydberg states. The energy E is the sum of the translational, vibrational and rotational energies of the dissociation fragments.

We calculate the $f_{\alpha}^{i,j}$ matrix elements analytically, using a method developed by Lengsfield and Yarkony.[11] Very large multi-reference configuration interaction wave functions are required to achieve converged results. We use atom-centered basis functions, as well as a Rydberg basis at the center of mass to accurately characterize both the ground-state surface and the Rydberg states. Table 1 shows our calculated excited-state energies and splittings for several $n = 3$ levels compared to experiment.[12] The agreement is quite good.

The NACMEs are transition amplitudes, which causes one additional complication. The phases must be calculated as well as the magnitudes (up to one overall phase factor, of course). We solve this problem by choosing one point on the surface as a reference, and then calculating the phase of all subsequent points relative to the reference point. This procedure works because the NACMEs are required over a limited enough region of configuration space that a single reference point is sufficient to describe the orbital character of the wave function. If the orbital composition of the wave function were changing rapidly, multiple reference points, and the transformations among them, would be necessary to define the phase.

3. RESULTS

As a first test of the method, we calculated the coupling between the two sheets of the ground state surface. Representative results are shown in Fig. 1, compared to

Table 1. Comparison between theory (this work) and experiment (Ref. 12) for H_3 vertical excitation energies, relative to the H_3 $2p\ ^2A_2''$ state.

State	Experiment (cm^{-1})	Theory (cm^{-1})
$3d\ ^2A_1'$	17 518	17 707
$3d\ ^2E''$	17 416	17 608
$3d\ ^2E'$	17 044	17 303
$3p\ ^2A_2''$	16 795	16 987
$3s\ ^2A_1'$	16 607	16 725
$3p\ ^2E'$	12 968	13 356

calculations using the method of Varandas et al.[4] The agreement is impressive, especially since the methods used to obtain the couplings are quite different. The surface of Varandas et al. is a fit to accurate electronic structure calculations for the lower adiabatic sheet of the H_3 potential surface. The upper sheet is obtained as an analytic continuation of the lower, and the couplings from a consideration of geometrical properties of the crossing.[13] This method is only expected to be accurate very near the intersection. Our results were derived from a large-scale configuration-interaction description of both sheets of the ground state surface, and direct calculation of the coupling. As can be seen in the figure, the method of Varandas et al. apparently works very well even at geometries relatively far from the crossing. This allows us to use the surface of Varandas et al. for the ground state dynamics.

Calculating NACMEs with proper phases when curve crossings occur is time-consuming, so our first calculations considered only 2-d geometries in which the H_3 is restricted to C_{2v} (isosceles triangle) geometries. In C_{2v}, the coupling between the A_1 and B_2 diabatic sheets of the ground state surface is zero. We chose to look first at the predissociation of the $3s$ state. This state couples to the A_1 sheet via motions in r and R, and to the B_2 sheet via out-of-plane motion in θ. Thus in 2-d, the predissociation of the $3s$ state reduces to a 2-surface problem.

Fig. 2 shows calculated NACMEs between the $3s$ and A_1 surfaces in the r and R directions at geometries near the equilibrium of the vibrationless $3s$ initial state. The couplings are quite smooth, and are fairly small in magnitude (compared, for example, to the magnitudes of the couplings in Fig. 1). Notice, however, that they are opposite in *sign*. This has an important effect in the dynamics, as shown in Fig. 3.

Fig. 3a shows a vibrationless initial state of H_3 superimposed on the A_1 surface. We compare this to the initial state resulting from the $d/d\mathbf{r}$ operator of Eq. 8 (Fig. 3b) and the state resulting from the dot product of $f_\alpha^{i,j}$ and $d/d\mathbf{r}$ (Fig. 3c). The initial density in Fig. 4a results from zero point motion in the symmetric stretch and bending vibrations on the $3s$ Rydberg state. Therefore, this state is a linear combination of

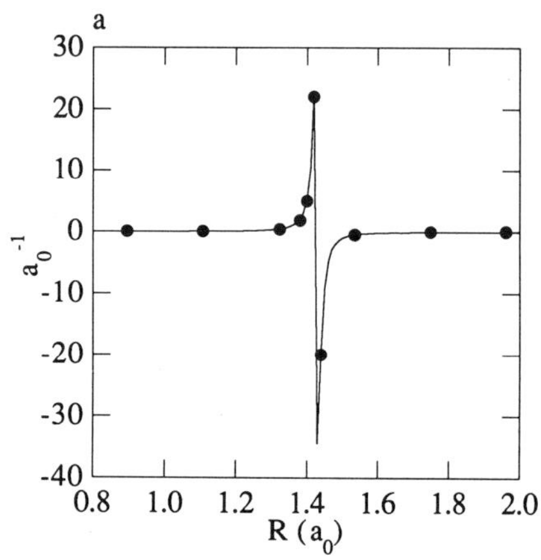

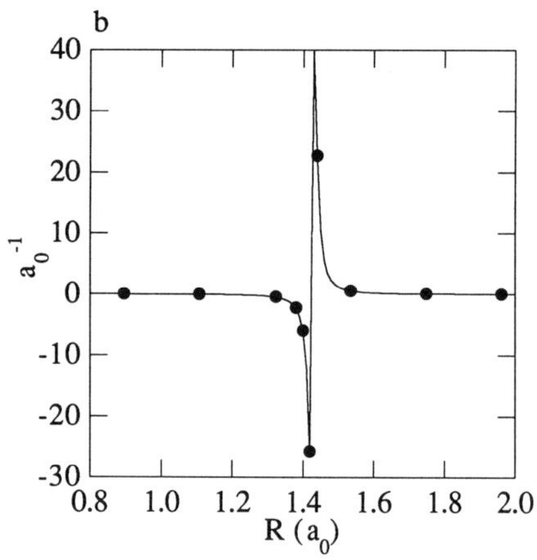

Figure 1. Non-adiabatic couplings between the two sheets of the ground state surface of H_3 as calculated using the methods of Ref. 4 (solid lines), and this work (filled circles) along a) the r, and b) the R directions at $\theta = 89.5°$.

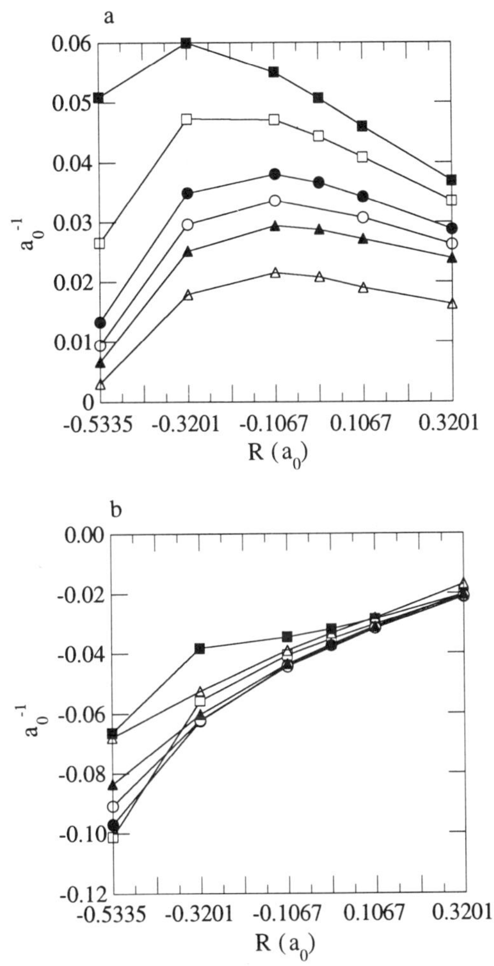

Figure 2. Non-adiabatic couplings between the $3s$ Rydberg state and the A_1 sheet of H_3, along a) the r and b) the R directions at r =-0.60 (filled squares), -0.36 (open squares), -0.12 (filled circles), 0.0 (open circles), 0.12 (filled triangles) and 0.36 a_0 (open triangles). In this figure, R and r refer to the distance from the equilibrium geometry ($r_{eq} = 1.65\ a_0$ and $R_{eq} = 1.4289\ a_0$.

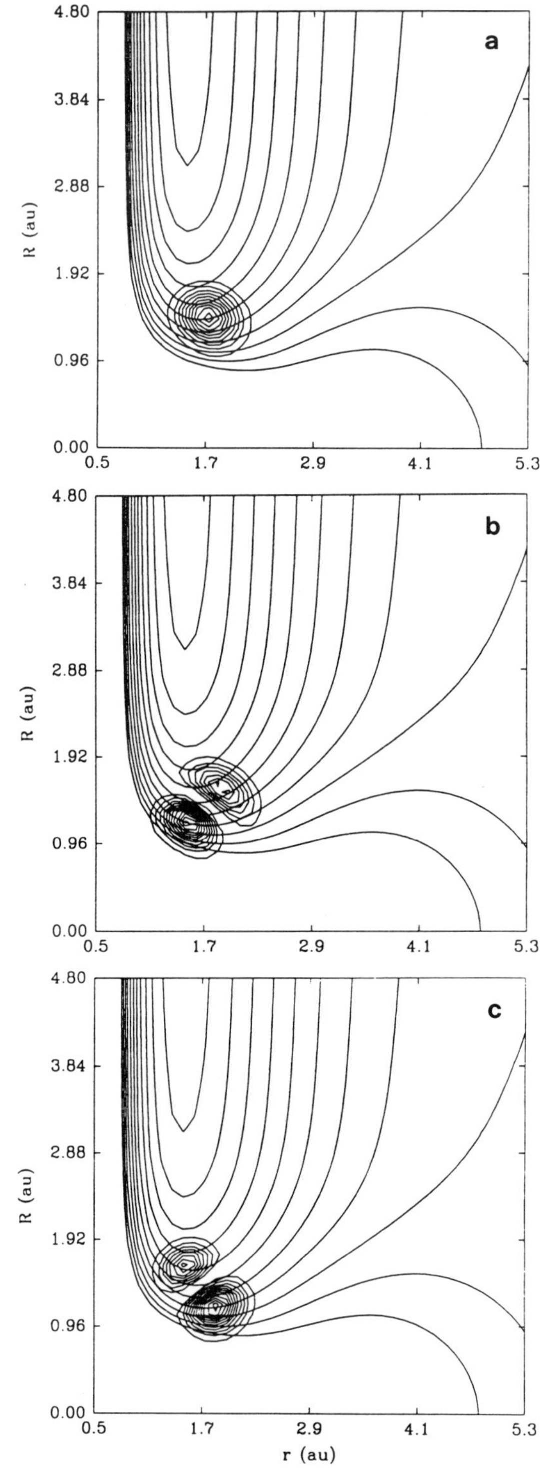

Figure 3. Contours of initial states representing a) the vibrationless initial state of the 3*s* Rydberg state, b) the gradient of the vibrationless initial state, and c) the non-adiabatic couplings dot the gradient of the initial state, superimposed on contours of the A_1 sheet of the ground state surface.

functions with a_1 and e_1 symmetry. The initial state in Fig. 3b, however, has a_1 symmetry and hence represents a symmetric stretch vibration, while the initial state in Fig. 3c has e_1 symmetry, and is a bending vibration. The 3s Rydberg state has A_1' symmetry in D_{3h}, and can couple to the E' ground state of H_3 only via an e_1 bending motion. Thus one result of including the NACMEs in the calculation is to ensure the proper symmetry of the initial state.

We compare the effects on the dissociation dynamics of including various levels of coupling by using the 3 different initial states of Fig. 3 in wave packet calculations. These initial states lead to rather different predissociation rates, and 3-body to 2-body branching ratios, as shown in Table 2. The branching ratios in the table were calculated from the difference between the "short-time" and "long-time" methods of Eqs. 2 and 3. The predissociation rates are relative, and correspond to the magnitude of the short-time rate. As shown in Fig. 4, the vibrational product state distributions also change somewhat depending on the initial state, though the effect is not dramatic. This is a reflection of the well-known vibrational adiabaticity of the H_3 ground state surface.[14] Using an excited initial state would lead to a much hotter vibrational distribution, and significantly more 3-body dissociation.[1,10]

4. CONCLUSIONS

In this paper we presented our first results on the predissociation of H_3 using accurate *ab initio* non-adiabatic couplings. We found that the couplings from the 3s state to the ground state are small and fairly smooth over the range of the vibrationless initial state, and so do not cause a dramatic change in the H_2 product state distribution. However, including the couplings did modify the dynamics in two important ways. First, as shown in Fig. 3, the coupling enforced the proper symmetry of the initial state, transforming it from the character of a symmetric stretch vibration to that of a bend. Secondly, the NACMEs alter the initial branching ratio to the coupled H_3 ground state. In 2-dimensions, the 3s state is coupled only to the A_1 diabatic surface. Away from 2-d, it is coupled to both sheets of the ground state surface, with an initial weight dictated by the magnitudes of the non-adiabatic couplings.

The importance of correctly characterizing the non-adiabatic couplings can be clearly seen in the experimental data.[10,15] For example, the 3s and 3d Rydberg states have similar energies, and similar potentials near the potential minima. In the absence of any information about the couplings to the ground state, the product state distributions from these two states would be exactly the same. However, the experimental product state distributions from the 3s and 3d states are in fact quite different.

The calculated vibrational distribution in Fig. 4 still does not agree well with experiment. The experimental distribution is much flatter, and peaks near $v = 7$. This is an indication that the 2-dimensional treatment of the dynamics that we have employed thus far is inadequate. Extending the calculations to 3-d will have several

Table 2. Effect of initial state on 2-body to 3-body branching ratio and relative predissociation rate

Initial State	2-body Yield (%)	3-body Yield (%)	Relative Rate[a]
Fig. 4a	84	16	9.1 (-4)
Fig. 4b	93	7	4.6 (-3)
Fig. 4c	82	18	1.4 (-2)

[a]Numbers in parenthesis indicate powers of ten

important effects. For one, in 3-d the 3*s* Rydberg state is coupled to both sheets of the ground state surface. The form of the coupling will be quite different away from $\theta = 90°$. In addition, away from C_{2v}, the two sheets of the ground state surface are coupled. One way to create additional vibrationally excited H_2 may be for flux landing initially on the upper adiabatic sheet to leak through the conical intersection to the lower surface.

We are continuing our studies on the predissociation of H_3 by calculating the non-adiabatic couplings at non-C_{2v} geometries. The NACMEs presented in this work were

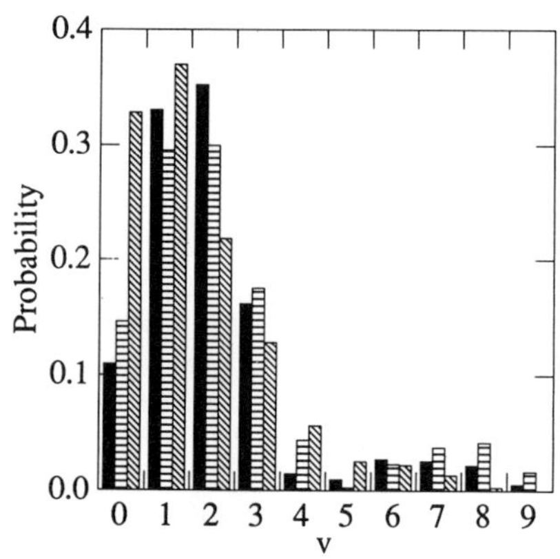

Figure 4. H_2 3*s* vibrational product state distributions deriving from the initial states shown in Fig. 4a (solid bars), Fig 4b (horizontal stripes) and Fig. 4c (diagonal stripes).

calculated from second-order configuration interaction wave functions. These wave functions were designed to obtain accurate excitation energies (as shown in Table 1), and to describe only that portion of the 3*s* potential surface necessary to obtain the $v = 0$ initial state. We are currently using larger configuration interaction wave functions to characterize the 3*s* and other Rydbergstates, so that initial states with higher vibrational excitation can be used in the dynamics calculations. We also plan to perform coupled 3-d calculations on the ground state surface. Such calculations are certainly possible, and should be quite informative about the role of the conical intersection in dissociation dynamics.[16] They should also allow us to make predictions for stimulated emission pumping experiments designed to explicitly probe the region of the intersection and the known scattering resonances on the lower surface.[17]

ACKNOWLEDGMENTS

The calculations presented here were performed at the San Diego Supercomputer Center. Work at Lawrence Livermore National Laboratory is performed under the auspices of the U.S. Dept. of Energy under contract number W-7405-ENG-48.

REFERENCES

1. J. R. Peterson, P. Devynck, Ch. Hertzler and W. G. Graham, J. Chem. Phys, in press.
2. P. C. Cosby and H. Helm, Phys. Rev. Lett. 61:298 (1988).
3. C. Bordas, P. C. Cosby and H. Helm, J. Chem. Phys. 93:6303 (1990); N. Bjerre, I. Hazell and D. C. Lorents, Chem. Phys. Lett. 181:301 (1991).
4. A. J. C. Varandas, F. B. Brown, C. A. Mead, D. G. Truhlar and N. C. Blais, J. Chem. Phys. 86:6258 (1987).
5. H. Tal-Ezer and R. Kosloff, J. Chem. Phys. 81:3967 (1984).
6. J. V. Lill, G. A. Parker and J. C. Light, Chem. Phys. Lett. 89:483 (1982); J. C. Light, I. P. Hamilton and J. V. Lill, J. Chem. Phys. 82:1400 (1985).
7. J. L. Krause, K. C. Kulander, J. C. Light and A. E. Orel, J. Chem. Phys. 96:4283 (1992).
8. E. J. Heller, J. Chem. Phys. 68:3891 (1978).
9. K. C. Kulander and E. J. Heller, J. Chem. Phys. 69:2439 (1978).
10. A. E. Orel and K. C. Kulander, J. Chem. Phys. 91:6086 (1989).
11. B. H. Lengsfield and D. R. Yarkony, Adv. Chem. Phys. 82:1 (1992).
12. G. Herzberg, J. T. Hougen and J. K. G. Watson, Can. J. Phys. 60:1261 (1982).
13. T. C. Thompson and C. A. Mead, J. Chem. Phys. 82:2408 (1985).
14. E. Pollak and M. S. Child, Chem. Phys. 60:23 (1981).
15. P. C. Cosby and H. Helm, unpublished.
16. See, for example, U. Manthe and H. Köppel, Chem. Phys. Lett. 178:36 (1991); G. Parlant and D. R. Yarkony, International Journal of Quantum Chemistry, in press; R. Schinke, this volume.
17. K. C. Kulander and J. C. Light, J. Chem. Phys. 85:1938 (1986).

TIME-DEPENDENT DYNAMICS OF AN ATOM IN AN INTENSE LASER FIELD

K. C. Kulander, J. L. Krause, K. J. Schafer, S. W. Allendorf,
J. K. Crane, K. S. Budil and M. D. Perry

Physics Department and Laser Program
Lawrence Livermore National Laboratory
Livermore, CA 94551

INTRODUCTION

In recent years we have seen very rapid advances in the technology of short pulse, very high intensity lasers. Lasers with pulse lengths of 0.1-1 ps and wavelengths from 0.2-1 μm can be focused to produce intensities from 10^{12} to above 10^{18} W/cm^2. One major use of these systems has been to study the response of atoms and molecules to such intense, well characterized electromagnetic fields. These pulses are short enough that neutral atoms survive to experience intensities where traditional perturbation expansions, based on the field-free states of the system, will fail completely to describe the dynamics of the system. Evidence for non-perturbative ionization can be found in the photoelectron energy spectra which contain numerous peaks separated by the photon energy. These peaks indicate a high probability of absorbing many more photons than the minimum required for ionization. This phenomenon, called above threshold ionization (ATI), has been observed in many experiments[1] and is reproduced theoretically from an energy analysis of the final state wave function obtained from a time-dependent calculation.[2] A second remarkable observation in high intensity experiments is the unexpectedly efficient production of *very* high-order harmonic radiation during the laser pulse. Harmonics up to the 109th order of a 140 fs Ti-sapphire (806nm) laser in neon have been reported.[3] Again, the time-dependent wave function contains the information that explains this highly nonlinear, non-perturbative behavior. The Fourier transform of the dipole induced in the electronic charge distribution gives the spectrum of emitted radiation. For many atomic systems, the calculated spectra agree quantitatively with the observed harmonic emission.[4] Finally, the excitation and ionization of an atom by two laser fields whose frequencies differ by a factor of two and whose relative phase is controlled, is an example of the "coherent control" of a quantum processes. Experimental[5] and theoretical[6] energy and angular distributions of the photoelectrons exhibit very strong phase-dependent effects.

Time-Dependent Quantum Molecular Dynamics, Edited by J. Broeckhove
and L. Lathouwers, Plenum Press, New York, 1992

In this paper we discuss recent calculations for krypton in single or dual laser fields which illustrate the multiphoton processes described above. In these calculations we have employed our single-active-electron (SAE) model which explicitly follows the time evolution of one of the valence electrons in the frozen, mean-field of the remaining electrons, the nucleus and the pulsed electromagnetic field. This model has been shown to be quite accurate for the rare gas atoms, at least partially because the neglected double or higher excitations involve states well above the ionization threshold.[7] In the next section we will give a brief description of the SAE model, emphasizing in particular the development of ion-core specific effective potentials. We present photoelectron and photoemission rates for krypton at 532nm and compare them to recent experimental results in the third section. Also we describe coherent control results for two-color ionization, where $\omega_2 = 2\omega_1$ and ω_1 corresponds to 1064 nm.

METHOD AND MODEL

A many electron atom in an intense, pulsed laser field is a formidable computational problem. This is because the Hamiltonian,

$$H(t) = H_o + V_I(t), \qquad (1)$$

is explicitly time dependent. H_0 is the (non-relativistic) atomic Hamiltonian and $V_I(t)$ is the interaction of the electron with the field. The interaction term is comparable in strength to the Coulombic terms in the atom. Therefore the method of solution must be able to treat these different interactions on an equal footing. Because the full multielectron problem is beyond present computational capabilities, we have developed a model based on the Hartree-Fock method in which the time-dependent wave function in approximated by a product of single particle orbitals. This approach has been shown to provide considerable insight into the excitation dynamics as well as quantitative predictions for the observed multiphoton processes.

If the laser is assumed to be linearly polarized along the z-axis and the field is strong enough to be treated semiclassically, then

$$V_I(t) = -e\sum_i z_i \varepsilon_o f(t)\sin(\omega t). \qquad (2)$$

Here ε_0 is the magnitude of the field and ω is the frequency. The pulse envelope $f(t)$ is typically chosen to be either a sine-squared or trapezoidal pulse. If we are interested in calculating quantities relevant to a particular intensity, $I_0 = c\varepsilon_0^2/8\pi$, we choose a pulse envelope that rises over several optical cycles to its maximum value (1.0) and then is held constant for 20-30 cycles. The pulse rise must involve at least a few cycles or the calculated results can be contaminated by unphysical transients.

The calculations reported here have been performed using the SAE approximation. The details of these calculations have been presented elsewhere[7] and will be repeated here only to the extent necessary to describe an improvement we have made in the effective potentials. In this model we hold the orbitals of all but one of the electrons in the atom fixed and allow the active electron to respond to the laser in the mean field of the remaining electrons and the nucleus. Because short pulse, high intensity experiments are generally performed in rare gases, some accommodation for the existence of a second, low-lying ionization limit in these species must be made. In the past we considered only ionization into the lowest continuum. In most short-pulse, high intensity experiments to date, ionization and, in particular, excitation into the manifold of states converging to the second ionization limit has been observed to be very important, especially in the lighter species. In our krypton studies, we have found this to be the case for the intensity range covered in these experiments.[8] Additionally, it is necessary to account correctly for the orbital occupancy of the active electron responsible for the emission.

The effective potentials for the active electron are generated according to the prescription given previously.[9] The potentials are constructed from the valence orbitals of the atom. We perform Hartree-Slater calculations on the ground and singly-excited states of the atom, adjusting the exchange-correlation parameter so that the orbital energy of interest agrees with the spectroscopic value. The valence orbital, which has a particular value of the orbital angular momentum, ℓ, is then used to construct an ℓ-dependent effective potential whose lowest eigenvalue is the ionization energy of that state. A separate calculation is carried out for each ℓ-value to generate the complete potential. We use the $\ell = 2$ potential for $\ell \geq 2$. We find this results in quite accurate excitation energies, much better than normally obtained in either Hartree-Fock or Hartree-Slater calculations. This procedure provides a model of the krypton atom which accounts only for the manifold of singly excited states based on the lowest ionic core, ${}^2P_{3/2}$. For all rare gases, a second manifold of states converges to the next spin-orbit component of the ion, the ${}^2P_{1/2}$ state. Therefore, we repeat the above procedure to obtain effective potentials for this second set of excited states. We assume that multiphoton excitations into these two manifolds are very weakly coupled so they can be treated separately. This assumption is reasonable because once one of the electrons is excited outside a particular core configuration, transitions into states with the other core requires two electrons to change state. Thus it is within the spirit of the SAE approximation to neglect such excitation pathways. The contributions to the induced dipole responsible for the harmonics from these separate calculations are added coherently.

We have constructed such core-specific effective potentials for all the rare gases. In xenon, for example, the splitting between the ion core limits is quite large, approximately 1.3 eV. Therefore the contributions from its second manifold are not too significant for the range of intensities and wavelengths studied here. However, for very short laser pulses, the atom can experience higher laser intensities and the contributions from these states will become relatively more important. We have found that for all the lighter rare gases, including krypton where the ionic cores are separated by 0.66 eV, the additional manifold of excited states is very important, and for some wavelengths and intensities, can be dominant.

Another important effect due to the spin-orbit coupling comes into play whether the upper ionic core is specifically involved or not. This is because the excitation dynamics is very sensitive not only to the ionization potential or binding energy of the active electron but also to m, the projection of the orbital angular momentum along the polarization axis. Since no spin-orbit terms are explicitly included in our Hamiltonian, we solve the Schrödinger equation in LS-coupling. This means that for a linearly polarized laser field, m is a good quantum number. Therefore, we need to assign weights to the valence orbitals according to their weights in the spin-orbit coupled states. In earlier calculations on xenon, the valence shell p-electrons with $m = 0$ were found to provide the dominant contribution to the ionization[10] and photoemission[11] rates. There are two $m = 0$ valence electrons in the p-shell, 4/3 associated with the $^2P_{3/2}$ core and 2/3 with $^2P_{1/2}$. Contributions to the ionization rates and harmonic emission strengths from the other four p-electrons in the valence shell of krypton which have $m = \pm 1$ are found to be unimportant for the intensities considered here. Thus we solve the time-dependent Schrödinger equation,

$$i\hbar\frac{\partial}{\partial t}\psi_j(t) = H_j(t)\psi_j(t) \tag{3}$$

for the 4/3 electrons with $m = 0$ that leave the ion in the $^2P_{3/2}$ state and the 2/3 electrons with $m = 0$ that leave the ion in the $^2P_{1/2}$ state. In Eq. (3), $\psi_j(t)$ is the active-electron orbital which at $t = 0$ is the valence pseudo-orbital with total angular momentum j, and H_j contains the appropriate effective potential. For the results discussed in this paper we solve Eq. (3) for each orbital, $\psi_j(t)$, for a laser which rises to its maximum intensity over five optical cycles and then has a constant intensity for the next 20-30 cycles. After the ramp, the transient excitations decay by ionization over the next few cycles. The ionization and photoemission rates are determined during the last part of the pulse. We obtain an ionization rate by monitoring the norm of the wave function. Our finite difference integration of Eq. (3) is carried out in a finite box with absorbing boundaries. As flux reaches the edges of the grid it is removed. The rate at which this occurs is defined to be the ionization rate.

The emission strengths at the harmonic frequencies are proportional to the square of the Fourier transform of the total induced dipole, which we typically calculate over the last five cycles of the pulse described above:

$$d_q = \frac{1}{T_f - T_i}\sum_j \int_{T_i}^{T_f} dt\, e^{-iq\omega t}\langle\psi_j(t)|z|\psi_j(t)\rangle. \tag{4}$$

If we are interested in the energy and angular distributions of the ejected electrons, we choose a pulse which also turns off over a few additional cycles (a trapezoidal pulse) and then perform an analysis of the final wave function using an energy window function.[2]

RESULTS

In this section we illustrate the methods we have described by considering the response of a krypton atom to an 532 nm laser field, presenting electron and photon emission rates for laser intensities well into the non-perturbative regime. We also consider the behavior of this atom in the combined fields of a 1064 nm laser and its second harmonic. As the relative phase

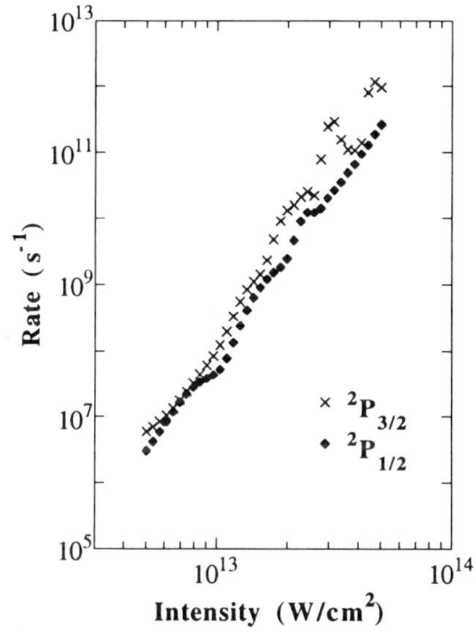

Fig. 1 Calculated partial ionization rates as functions of laser intensity for krypton at 532 nm.

between the fields is varied, the branching ratios for producing ions in the two lowest spin-orbit configurations and the angular distributions for the emitted electrons are shown to change dramatically.

Photoelectron and Harmonic Emission from Krypton

We have studied the excitation of krypton by 532 nm radiation for intensities from $5x10^{12}$ to $5x10^{13}$ W/cm^2. In Fig. 1 we show the partial rates for leaving the ion in each of its two spin-orbit states. Seven of these 2.33 eV photons are necessary to ionize the atom. There is enough excess energy to leave to ion in either the $J = 3/2$ or $J = 1/2$ states. The ionization rates exhibit structure as functions of intensity due to excited states which are Stark shifted into resonance. The ac Stark shifts of highly excited atomic states are generally found to be on the order of the ponderomotive energy of an unbound electron in the laser field given by $(e\mathcal{E}_o)^2/4m\omega^2$. At this wavelength the shift is 0.26 eV at 10^{13} W/cm^2. A second indication that the excited states are playing an important role in the ionization dynamics is the overall dependence of the rates on intensity which is very close to I^6 rather than I^7 as one would expect from perturbation theory for non-resonant ionization.

In Fig. 2 we show the measured[8] ionization rates to the two ion cores for approximately the same range of intensities shown in the previous figure. These results exhibit a strong resonance in the ionization rate in the $J = 1/2$-core near 7x10^{12} W/cm^2 which can be identified as a six-photon resonance with the $4f'$ state. The experimental results are obtained from the entire focal volume and therefore contain contributions from a range of intensities.

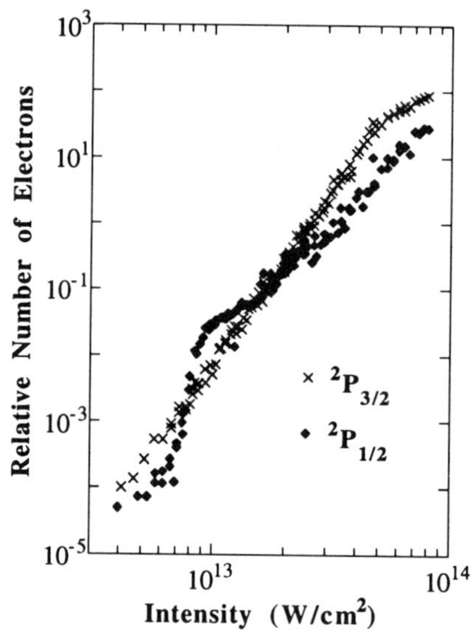

Fig. 2 Measured partial ionization rates as functions of laser intensity for krypton at 532 nm

The theoretical branching ratio near this resonance is smaller than the experiment by a factor of two to three. Although there are small uncertainties in the areas of the measured photoelectron peaks, the disagreement seems to lie outside the experimental error bars. We believe theoretical errors come from two sources. First, ionization rates below 10^8 s^{-1} are very difficult to obtain in a time-dependent, numerical calculation. The typical time step used in these calculations is on the order of 10^{-18} s, so the change in the norm of the wave function per time step for such low rates is less than one part in 10^{10}. Second, at these intensities the ionization rate for the $J = 3/2$-core manifold includes resonances with very high p-wave Rydberg states which are less accurately characterized in the large, but necessarily finite box used in our calculations. This could lead to an overestimate of the rate. The calculated total ionization rates, however, agree well with the measured rates, both showing the neutral medium begins to be depleted during the 75 ps pulse for peak intensities just above 3×10^{13} W/cm^2. At higher intensities also there is good agreement on the dominance of the $J = 3/2$-core rate.

Resonances can also enhance the photoemission rates. However direct comparison between our single atom emission rates and the macroscopically phase-matched harmonic intensities is inappropriate.[4] The observed harmonic strengths are sensitive to the focal parameters and the atomic density profile. Phase matching of the generated harmonic fields depends on the difference of the indices of refraction at the fundamental and harmonic frequencies which, due to their intensity dependence, vary throughout the medium. In Fig. 3 we display the experimental harmonic spectra as functions of the incident laser intensity. Clear resonant enhancements of the 5th and 7th harmonics are seen at 2 and 3×10^{13} W/cm^2. The structures in the harmonic signals are more pronounced than in the electron measurements.

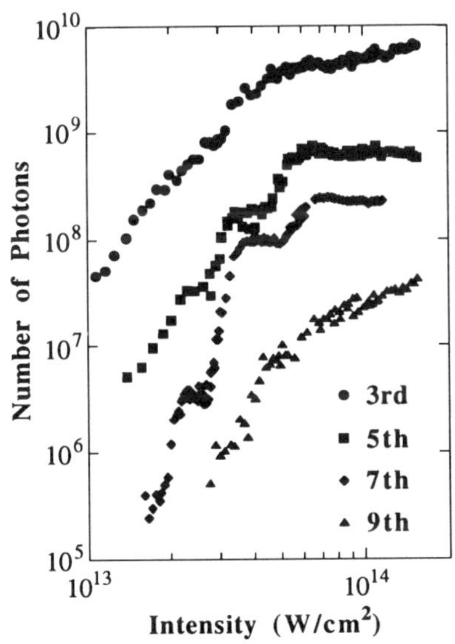

Fig. 3 Measured harmonic emission rates as functions of intensity for krypton at 532 nm.

This is due to phase matching effects during the propagation of the harmonic fields. The photoelectron signal experiences no equivalent interference effects. Above 4×10^{13} W/cm^2 all the signals saturate due to ionization of the medium and therefore depletion of the emitting source. The calculated single atom emission rates are shown in Fig. 4 for the 7th harmonic. For this case, most of the structure in the total emission rate comes from excitations of the electrons outside the $J = 3/2$ core. The three maxima can be associated, in order of increasing intensity, with 6-photon resonances with the $8p$, $7p$ and $4f$ states, respectively. The latter two

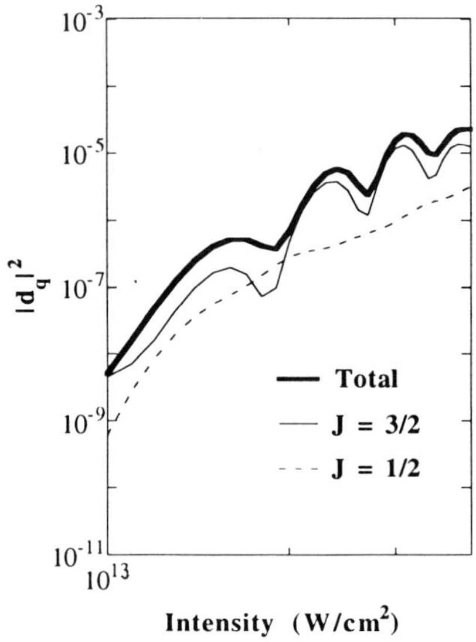

Fig. 4 Calculated pseudo-partial and total 7th harmonic emission rates as functions of intensity for krypton at 532 nm.

peaks seem to be reproduced in the experimental data. For comparison we also show the calculated 5th harmonic in Fig. 5. In this case the contributions from both manifolds are strong and show structure. As one can see, the effects of 6-photon resonances on a 5-photon process is relatively more complicated. The total single-atom rates do not follow the measured rates very closely, indicating the importance of phase matching effects. Phase effects can be expected to be more pronounced for the 5th harmonic which lies within the excited state manifolds as compared to the 7th which is above both ionization limits.

These calculations illustrate the application of the SAE method for atomic emission processes in the presence of a single laser field. In the next section, we consider the influence of a second laser field on the ionization rates and the angular distributions of the photoelectrons.

Two-Color Ionization

If a laser field is combined with its second harmonic, the reflection symmetry of the laser-atom interaction is broken and some very interesting ionization dynamics can be induced by varying the relative phase of the two fields. We assume that both lasers are linearly polarized along the same direction and have the same intensity. Then the interaction term given in Eq. (2) is modified to be

$$V_I(t) = -e\sum_i z_i E_o f(t)\{\sin(\omega t) + \sin(2\omega t + \phi)\}. \quad (5)$$

We have chosen to study the case with ω corresponding to a Nd/YAG laser whose wavelength is 1064 nm for which some experimental data have recently been obtained.[5]

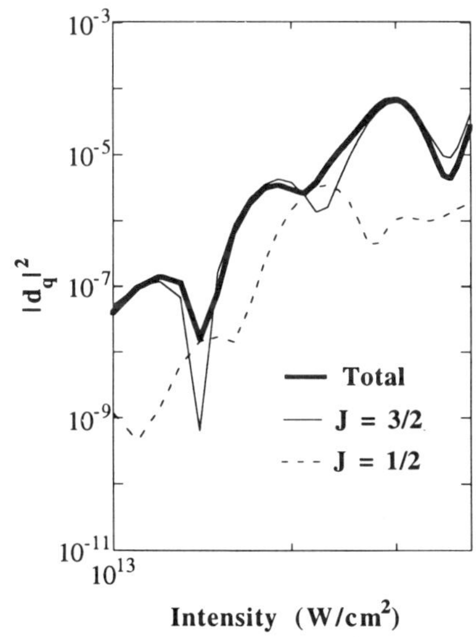

Fig. 5 Calculated pseudo-partial and total 5th harmonic emission rates as functions of intensity for krypton at 532 nm.

The combined field in Eq. (5) has quite different structure depending on ϕ. In Fig. 6 we show the field over three periods of the fundamental. At very high intensity, we expect ionization to occur at the peak of the combined fields due to tunneling through the suppressed Coulomb barrier. Therefore, the magnitudes of the field oscillations in this figure imply that in this regime ionization will be most rapid and the angular distribution of the photoelectron most asymmetric for $\phi = 90^o$.

Fig. 7 shows the partial and total ionization rates as functions of the phase where both fields have an intensity of $2x10^{13}$ W/cm^2. As can be seen in the figure the first of our expectations is fulfilled. The rates peak near 90°. However, the angular distributions, shown in Fig. 8, are much more complicated. In this figure, "forward" refers to the rate of emitting electrons with a positive component of velocity along the z-axis. At high intensity the electron distributions are strongly collimated along the polarization axis. A similar "backward"

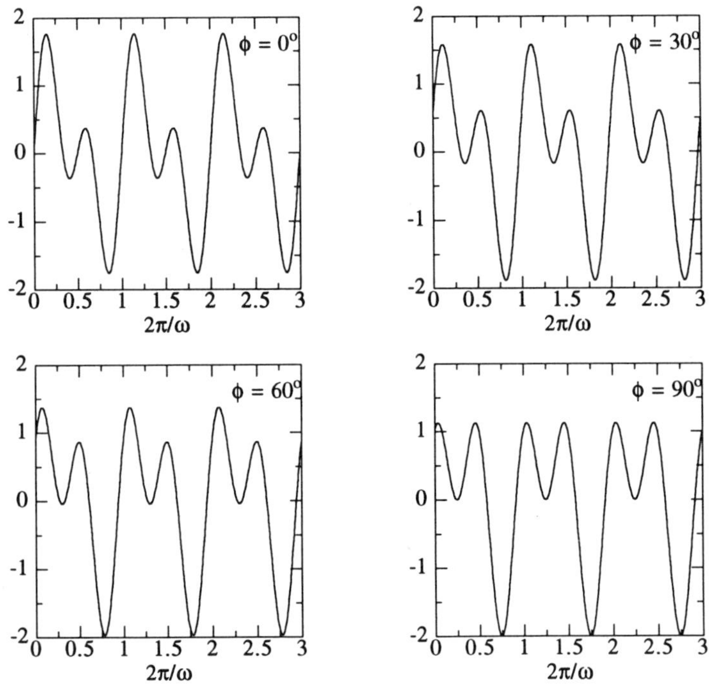

Fig. 6 Time-dependent electric field (in units of ε_o) for four choices of the phase angle ϕ.

distribution can be obtained by shifting the forward distribution $\pm 180^o$. This is illustrated in Fig. 9 which shows how the two separate angular distributions combine to produce a total rate which peaks at 90°. We see that even for the symmetric-looking pulse at $\phi = 0^o$, there is a strong asymmetry in the photoelectron emission. It is also interesting to note that although the ratio of the total rates for producing the different ion cores is relatively insensitive to the phase, the ratio changes dramatically as a function of the direction of electron emission.

Finally, we show in Fig. 10 that the distributions exhibit some intensity dependence even in the intense-field regime. The figure shows the total and partial distributions for the two

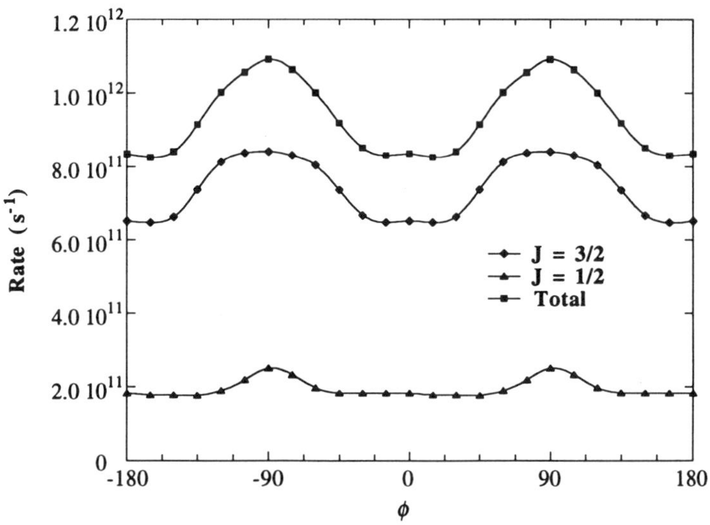

Fig. 7 Partial and total ionization rates as functions of the relative phase for krypton in a two-color (1064 - 532 nm) laser field. Laser intensities are both $2x10^{13}$ W/cm^2.

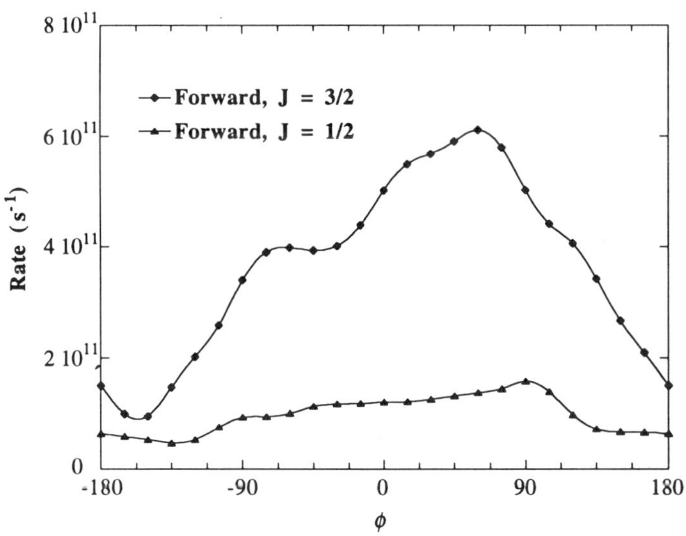

Fig. 8 Partial angular rates as functions of the relative phase for krypton in a two-color (1064 - 532 nm) laser field. Laser intensities are both $2x10^{13}$ W/cm^2.

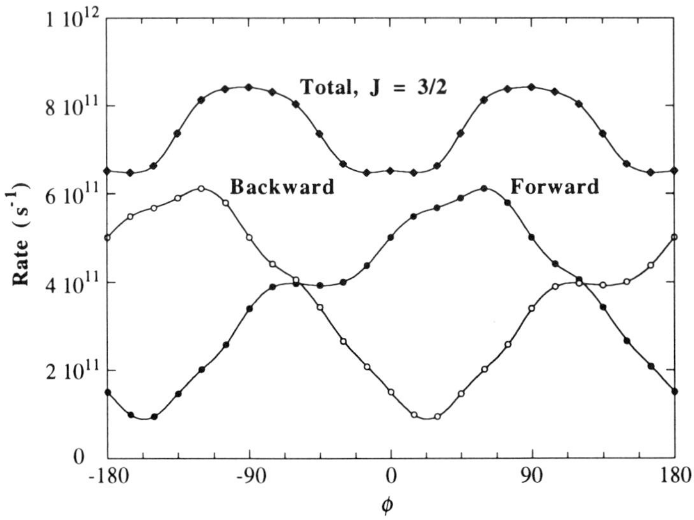

Fig. 9 $J = 3/2$ partial angular and total rates as functions of the relative phase for krypton in a two-color (1064 - 532 nm) laser field. Laser intensities are both $2x10^{13}$ W/cm^2.

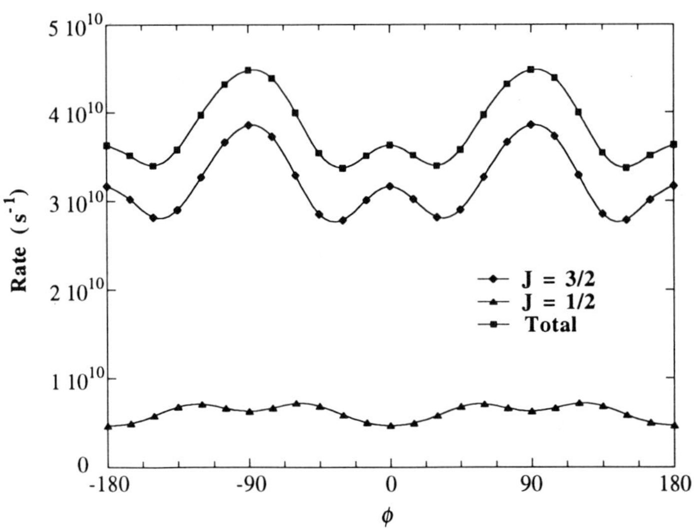

Fig. 10 Partial and total ionization rates as functions of the relative phase for krypton in a two-color (1064 - 532 nm) laser field. Laser intensities are both $1x10^{13}$ W/cm^2.

cores at laser intensities of 1×10^{13} W/cm^2. The partial rates are similar to those in the previous case, but they combine to give an additional peak in the rates near $\phi = 0^o$. Additionally, the ratio of the $J = 3/2$ to $J = 1/2$-core rates shows a stronger dependence on the phase. These results are consistent with the observations of Muller, et al.[5] They could not determine the absolute phase difference between their two fields, but did study the variations as a function of the phase. They found the production of the $^2P_{3/2}$ ion was much more likely than $^2P_{1/2}$, in agreement with the calculated branching ratio. These results show that quite appreciable asymmetries can be achieved through phase control of the ionization process.

The calculations we have presented in this section illustrate the application of our single-active-electron model of intense-field, multiphoton processes. We have found that it is very successful in modelling very high order, non-perturbative processes in atoms. Future directions of this research include applications to ions and the extension of these methods to treat multielectron excitations.

ACKNOWLEDGEMENT

This work has been carried out under the auspices of the U. S. Department of Energy at the Lawrence Livermore National Laboratory under contract number W-7405-ENG-48.

REFERENCES

1. G. Petit, P. Agostini and F. Yergeau, J. Opt. Soc. Am. B 4:765 (1987).
2. K. J. Schafer and K. C. Kulander, Phys. Rev. A 42:5794 (1990); K. J. Schafer, Comp. Phys. Comm. 63:427 (1991).
3. J. Macklin, J. D. Kmetec and C. L. Gordon III, to be published.
4. A. L'Huillier, K. J. Schafer and K. C. Kulander, J. Phys. B 24:3315 (1991), and references therein.
5. H. G. Muller, P. H. Bucksbaum, D. W. Schumacher and A. Zavriyev, J. Phys. B 23:2761 (1990).
6. K. J. Schafer and K. C. Kulander, Phys. Rev. A, in press.
7. K. C. Kulander, K. J. Schafer and J. L. Krause, Time-dependent studies of multiphoton processes, in: "Atoms in Intense Radiation Fields," M. Gavrila, ed. (Advances in Atomic, Molecular and Optical Physics, Supplement 1) Academic Press, New York (1992).
8. S. W. Allendorf, J. K. Crane, K. S. Budil and M. D. Perry, to be published.
9. K. C. Kulander and T. N. Rescigno, Comp. Phys. Comm. 63:523 (1991).
10. K. C. Kulander, Phys. Rev. A 38:778 (1988).
11. K. C. Kulander and B. W. Shore, J. Opt. Soc. B 7:502 (1990).

TIME-DEPENDENT TREATMENT OF REACTIVE AND DISSOCIATIVE COLLISIONS : A COUPLED-ARRANGEMENT-CHANNEL METHOD

Zeki C. Kuruoğlu

Department of Chemistry
Bilkent University
Bilkent, Ankara, Turkey

INTRODUCTION

This contribution will explore two topics within the time-dependent (TD) scattering theory: (i) formal and computational insights that can be gained from the momentum-space view of wave-packet evolution,[1] and (ii) a time-dependent coupled-arrangement-channel (TD-CAC) method[2,3] for reactive and dissociative collisons. Note that the TD-CAC method for rearrangement scattering is not necessarily tied to the momentum representation. However, time evolution of wave packets in momentum space have several novel features that deserve attention.

Computational implementations of the TD scattering theory have traditionally been based on the Schrödinger-picture coordinate-space propagation of wave packets, although interaction-picture and time-dependent integral-equation approaches have also been considered recently.[4,5] Wave packets move and spread in coordinate space. This gives rise to the so-called boundary-reflection problem, since one has to work in practice with finite approximation spaces which necessarily have finite support in coordinate space. Also, the moving and spreading of wave packets is basically through the action of kinetic energy operators. In contrast, momentum amplitudes of wave packets do not move or spread. A compact momentum-space wave packet remains compact. In other words, the effective momentum-space support of the wave packet (i.e., the momentum interval over which the probability density of the packet is greater than, say, 0.01) does not change. Scattering manifests itself in the form of phase modulation of the momentum amplitude of the wave packet.

Also, the momentum representation has a special place in the TD scattering theory in that the sharp-energy S-matrix elements are basically given by the ratio of the momentum amplitudes of the interacting and free wave packets at asymptotic times. For this ratio to give accurate and stable results, momentum amplitudes for both wave packets have to be determined to the same degree of accuracy. Problems that might arise from the use of a numerical wave packet for the full dynamics along with the exact wave packet for the free dynamics will be illustrated in the next section for two-particle collisions. A simple recipe to avoid this mismatch is to treat both time-evolution problems (with and without interaction) using identical numerical procedures. In this way, any errors that might arise from the approximate treatment of the kinetic energy operator will occur in both (numerical) wave packets and cancel out in the calculation of

Time-Dependent Quantum Molecular Dynamics, Edited by J. Broeckhove
and L. Lathouwers, Plenum Press, New York, 1992

the S-matrix. Note that, this problem is avoided in the interaction picture, since the exact solution of the free time-evolution is incorporated into the full time-dependent Schrödinger equation (TDSE). In addition, the interaction-picture wave packet does not move or spread even in coordinate space.[4] However, the interaction-picture Hamiltonian is time-dependent, a feature that is less than desirable from computational point of view. Furthermore, a meaningful interaction picture does not exist for reactive systems, due to the lack of a unique asymptotic Hamiltonian.

Note that a wave packet calculation can be expected to yield accurate S-matrix elements for only those momentum values that have appreciable amplitude in the momentum distribution of the initial wave packet, i.e, over the effective momentum-support of the wave packet. Therefore, accuracy requirement for the momentum amplitudes of the numerical wave packets is highest over this momentum interval. This requirement can be met with relative ease, e.g, via a nonuniform momentum-discretization involving a denser set of points over the effective momentum-support of the wave packet. The use of a nonuniform momentum-grid is feasible because the momentum-space wave packet does not move or spread. In this connection, we note that a coordinate-space discretization via the Fast Fourier Transform (FFT) can equivalently be viewed as momentum-space discretization. That is, FFT can be applied with the roles of coordinates and momenta reversed. However, FFT would involve a uniform discretization of the momentum variables. According to our above discussion, a uniform momentum-discretization is less than optimal.

Wave-packet propagation in momentum representation is not necessarily free of the boundary-reflection problem. The fineness of the momentum discretization determines the size of the corresponding finite region in the coordinate-space, and, hence, the maximum period of reflection-free propagation. Nevertheless, the momentum-space approach offers the possibility of extending the duration of the reflection-free propagation by employing a finer momentum-grid over only the effective momentum-support of the wave packet.

For reactive systems, the lack of a single set of variables capable of describing all types of asymptotic dynamics in separable form makes the space-discretization a nontrivial problem. Of course, this is the perennial problem of rearrangement scattering whether in time-independent (TI) or TD contexts. Among the attempts to develop a TD method for reactive systems, one line of approach[6] insists on keeping the simplicity of the single-arrangement (inelastic) wave-packet propagation by using the discretization of (or an expansion basis in) the Jacobi variables of just one arrangement. To account for the presence of other rearrangements, absorbing potentials are used.

Another approach is based on the simultaneous use of expansion bases (or discretization grids) in Jacobi variables of all important rearrangements[2,3]. This is basically a time-dependent adaptation of a coupled-arrangement-channel ansatz that has proved to be extremely efficient in handling reaction dynamics within the TI context.[7,8] In this type of approach, the approximation space for a three-particle system is constructed as the linear span of three rearrangement-channel subspaces. Each rearrangement subspace is spanned by (a subset of) the eigenstates of the asymptotic Hamiltonian for that rearrangement. We have shown in the TI context that such an approach also works above the dissociation threshold, provided the rearrangement subspaces are enlarged to include some pseudostates.[9,10] Note that such multi-arrangement bases are proving useful also in three-atom bound-state calculations.[11] In the present contribution, we discuss a multi-arrangement discretization, first briefly presented in [2], for time-dependent treatment of three particle scattering at energies above the dissociation (breakup) threshold The method has been tested on a three-particle model for which essentially exact results can be obtained by the Faddeev method.[12,13]

WAVE-PACKET PROPAGATION IN MOMENTUM SPACE

Time-Dependent Theory of Two-Particle Scattering

The two-particle Hamiltonian in the c.o.m frame is taken as $H = H_0 + V$, where H_0 is the relative motion kinetic energy operator and V is a rotationally invariant and short ranged interaction potential. The relative coordinate is denoted by $\mathbf{r}$ and its conjugate momentum by $\mathbf{p}$. Denoting the sharp-momentum states by $|\,\mathbf{p}\,\rangle$, we have $H_0\,|\,\mathbf{p}\,\rangle = E_p\,|\,\mathbf{p}\,\rangle$, where $E_p = p^2/2\mu$, with μ being the reduced mass. Denoting the coordinate states by $|\,\mathbf{r}\,\rangle$, we use the normalizations $\langle\,\mathbf{p}\,|\,\mathbf{p}'\,\rangle = \delta(\mathbf{p}-\mathbf{p}')$, and $\langle\,\mathbf{r}\,|\,\mathbf{r}'\,\rangle = \delta(\mathbf{r}-\mathbf{r}')$. Setting $\hbar = 1$, we have $\langle\,\mathbf{r}\,|\,\mathbf{p}\,\rangle = e^{i\mathbf{p}\cdot\mathbf{r}}/(2\pi)^{3/2}$.

Let the two-particle system be described at the distant past $t = t_P$ by the incoming wave-packet state $|\,\Phi(t_P)\,\rangle$ whose coordinate-support is outside the range of the potential.

$$|\,\Phi(t_P)\,\rangle \;=\; \int d\mathbf{p}\;|\,\mathbf{p}\,\rangle\; f(\mathbf{p})\,, \tag{1}$$

where $f(\mathbf{p})$ is a momentum-space wave-packet with an average momentum of $\mathbf{p}_0$. In the absence of the potential, the time evolution is given by

$$\langle\,\mathbf{p}\,|\,\Phi(t)\,\rangle \;=\; e^{-iE_p(t-t_P)}\,\langle\,\mathbf{p}\,|\,\Phi(t_P)\,\rangle \;=\; e^{-iE_p(t-t_P)}\,f(\mathbf{p})\,. \tag{2}$$

The state $|\Phi(t)\,\rangle$ will be referred to as the *in*-state. Note that the momentum probability density of the in-state is independent of time, viz., $|\langle\,\mathbf{p}\,|\,\Psi(t)\,\rangle|^2 = |f(p)|^2$. Hence, the momentum-support of the in-state does not change with time, i.e., a free wave packet does not move or spread in momentum space.

Let $|\Psi(t)\rangle$ denote the solution of the TD Schrödinger equation (TDSE),

$$i\,\frac{\partial\,|\,\Psi(t)\,\rangle}{\partial t} \;=\; H\;|\;\Psi(t)\;\rangle\,, \tag{3}$$

subject to the initial condition $|\,\Psi(t_P)\,\rangle = |\,\Phi(t_P)\,\rangle$. As the collision develops in time , $|\,\Psi(t)\,\rangle$ evolves eventually into an asymptotically free state, the *out*-state . Let us denote by t_F the asymptotic time after which the coordinate-space support of the wave packet $|\,\Psi\,\rangle$ is once again outside the range of the potential. That is, for $t \geq t_F$, the wave-packet is free and outgoing, with its further time-evolution being governed by H_0. The S-operator $\hat{S}$ is then defined as[14]

$$|\,\Psi(t)\,\rangle \;=\; \hat{S}\,|\,\Phi(t)\,\rangle\,, \qquad t \geq t_F\,. \tag{4}$$

Formally, this relation is justified only if the limits $t_P \to -\infty$ and $t_F \to +\infty$ are taken.

In computational implementation, the infinite-future and infinite-past limits are not only impracticable, but also not meaningful. This is because the TDSE is solved in practice on a finite dimensional approximation space. A finite dimensional approximation space implies a finite region in coordinate space. Depending on the nature of the approximation space, the boundary of this finite domain in coordinate space might be sharp or gradual. After some time t_{max}, which is a characteristic of the approximation space, the coordinate-space image of the wave packet will begin to be reflected from the boundary of the finite coordinate-space domain of the approximation space. Thus, in computational version of the TD scattering theory, the basic equation (4) will be valid only in the time interval $[t_F, t_{max}]$. Put another way, the approximation space should be large enough to allow a time period during which the wave packet is outgoing, outside the range of the potential, and free of boundary reflection.

To obtain the sharp-energy matrix elements of the S-operator, we project both sides of (4) onto the momentum state $\langle\, \mathbf{p} \,|$, and obtain

$$\langle \mathbf{p} \,|\, \Psi(t) \,\rangle \;=\; \int \mathrm{d}\mathbf{p}' \,\langle\, \mathbf{p} \,|\, \hat{S} \,|\, \mathbf{p}' \,\rangle \,\langle\, \mathbf{p}' \,|\, \Phi(t) \,\rangle \;. \tag{5}$$

The energy-conserving property of the S-operator reads[14]

$$\langle\, \mathbf{p} \,|\, \hat{S} \,|\, \mathbf{p}' \,\rangle \;=\; \langle\, p\hat{\mathbf{p}} \,|\, S \,|\, p\hat{\mathbf{p}}' \,\rangle \,\delta(E_p - E_{p'}) \;, \tag{6}$$

where S is the reduced scattering operator, and $\hat{\mathbf{p}}$ denotes the unit vector along $\mathbf{p}$. Hence, Eq.(4) becomes

$$\langle\, \mathbf{p} \,|\, \Psi(t) \,\rangle = \mu p \int d\hat{\mathbf{p}}' \,\langle\, p\hat{\mathbf{p}} \,|\, S \,|\, p\hat{\mathbf{p}}' \,\rangle \,\langle\, p\hat{\mathbf{p}}' \,|\, \Phi(t) \,\rangle \;. \tag{7}$$

with $t_{max} \geq t \geq t_F$.

To proceed further, we use the rotational invariance of the potential. For this purpose, let us introduce the partial-wave momentum states $|\, p\ell m \,\rangle$ via $|\, p\ell m \,\rangle = \int d\hat{\mathbf{p}} \,|\, \mathbf{p} \,\rangle\, Y^*_{\ell m}(\hat{\mathbf{p}})$, so that $|\, \mathbf{p} \,\rangle = \sum_{\ell m} |\, p\ell m \,\rangle\, Y_{\ell m}(\hat{\mathbf{p}})$. Defining the partial-wave position states $|\, r\ell m \,\rangle$ in a similar manner, our normalization convention implies $\langle\, p\ell m \,|\, p'\ell' m' \,\rangle = \delta_{\ell\ell'}\, \delta_{mm'}\, \delta(p - p')/p^2$, and $\langle\, \mathbf{r} \,|\, p\ell m \,\rangle = \sqrt{2/\pi}\; i^l\; Y_{\ell m}(\hat{\mathbf{r}})\, j_\ell(pr)$, with j_ℓ denoting the spherical-Bessel function. Any rotationally invariant operator A is expanded as

$$\langle\, \mathbf{p} \,|\, A \,|\, \mathbf{p}' \,\rangle \;=\; \sum_{\ell m} Y^*_{\ell m}(\hat{\mathbf{p}})\; Y_{\ell m}(\hat{\mathbf{p}}')\; A_\ell(p, p') \;. \tag{8}$$

where $A_\ell(p,p')$ $(\equiv \langle\, p\ell m \,|\, A \,|\, p'\ell m \,\rangle\,)$ is independent of m. For the scattering operator, we have $S_\ell(p,p') = \delta(E_p - E_{p'})\; S_\ell(p)$, where $S_\ell(p) = \mu p \,\langle\, p\ell m \,|\, S \,|\, p\ell m \,\rangle = e^{2i\delta_\ell(p)}$, with δ_ℓ being the phase shift.

Projection of Eq. (4) onto the partial-wave state $|\, p\ell m \,\rangle$ yields

$$S_\ell(p) \;=\; \frac{\langle\, p\ell m \,|\, \Psi(t) \,\rangle}{\langle\, p\ell m \,|\, \Phi(t) \,\rangle} \;. \tag{9}$$

This result presents us with two possibilities in numerical implementation: Either we solve the TDSE in full three-dimensional form using a general three-dimensional initial wave packet and subsequently take partial wave projections, or we work with one partial wave at a time by taking the initial wave packet in the form $\langle\, \mathbf{p} \,|\, \Phi(t_P) \,\rangle = \phi_\ell(p, t_P)\, Y_{\ell m}(\hat{\mathbf{p}})$. The second approach, which we adopt, involves the solution of the partial-wave time-dependent Schrödinger equation (PW-TDSE),

$$i\,\frac{\partial \psi_\ell(p,t)}{\partial t} \;=\; \frac{p^2}{2\mu}\,\psi_\ell\,(p,t) \;+\; \int p'^2\, dp'\; V_\ell(p,p')\; \psi_\ell\,(p',t) \;, \tag{10}$$

with the initial condition $\psi_\ell\,(p,\; t_P) \;=\; \phi_\ell\,(p,\; t_P)$. The partial-wave S-matrix is now given as

$$S_\ell(p) \;=\; \frac{\psi_\ell\,(p,t)}{\phi_\ell\,(p,t)} \;, \qquad \text{for } t_F \leq t \leq t_{max} \;. \tag{11}$$

From the unitarity of the S-matrix, viz., $|\; S_\ell \;|^2 \;=\; 1$, we find that $|\psi_\ell\,(p,t)|^2 = |\phi_\ell\,(p,t)|^2 \;=\; |\phi_l\,(p,t_P)|^2$ for $t \geq t_F$. That is, the momentum probability density of the out-state is exactly the same as that of the in-state. Scattering manifests itself in momentum space as the modification of the phase of the free wave packet.

We consider the s-wave scattering problem for the potential $V(r) = -e^{-r}$, with the reduced mass $\mu = 1$. The initial wave-packet is taken as an incoming gaussian :

$$\phi(r,0) = A\, e^{(r-r_0)^2/2d^2}\, \frac{e^{-ip_0 r}}{r} \,, \tag{12}$$

where $A^{-2} = d\sqrt{\pi}$, r_0 the average position, and d the width. Here we have suppressed the index $\ell = 0$, and set $t_P = 0$. The momentum-space representation of the initial wave packet reads

$$\phi(p,0) \equiv \langle p\,|\phi(0)\rangle = \frac{A'}{p}\left(e^{i(p-p_0)r_0}\, e^{-(p-p_0)^2d^2/2} + e^{i(p+p_0)r_0}\, e^{-(p+p_0)^2d^2/2}\right), \tag{13}$$

where A' is a normalization constant. Of course, the time-evolution under H_0 is simple: $\phi(p,t) = e^{ip^2t/2\mu}\,\phi(p,0)$. In the calculations, we used $p_0 = 5$, $r_0 = 10$ and $d = 0.60$, so that the probability density of the initial packet is greater than 0.01 in the p-interval $[2.0, 8.0]$.

To solve the PW-TDPSE numerically, an approximation space is constructed from a set of momentum-space local interpolation functions $u_n(p)$, $n = 1, 2, ..., N$. To this end, first a computational domain $[0, p_{max}]$ is introduced. Although the momentum amplitude of the initial packet is essentially negligible beyond $p = 12$, the cutoff p_{max} is determined by the requirement that the product $V_0(p,p')\phi(p',t)$ be negligible. The value $p_{max} = 24$ proved adequate for the present potential. The computational domain $[0, p_{max}]$ is divided into I subintervals and a set of N $(= 2I - 1)$ piecewise quadratic functions[15] are defined on this partition. The placement of the breakpoints for the partition is not uniform: a denser set of points is used over the momentum interval where the wave packet has appreciable amplitude. The calculations reported here used $I = 150$, employing the partition 0 (0.1) 10(0.2) 16 (0.4) 24 , where the notation $p\,(\Delta p)\,p'$ indicates that the interval $[p, p']$ is divided into equal subintervals of length Δp.

The wave packet is expanded on the approximation space via

$$\tilde{\psi}(p,t) = \sum_n c_n(t)\, \frac{u_n(p)}{p} \,. \tag{14}$$

Approximate wave packets are distinguished by a tilde. To obtain the approximate evolution equation we use the Galerkin method which requires $\langle u_n|(\,i\frac{\partial}{\partial t} - H\,)|\,\tilde{\psi}(t)\rangle = 0$. Another possibility would be to use the collocation criterion. Within the Galerkin approach, we obtain

$$\sum_{n'} \Delta_{nn'}\, \dot{c}_{n'}(t) = H_{nn'}\, c_{n'}(t) \,, \tag{15}$$

where $\Delta_{nn'} = \langle\, u_n \,,|\, u_{n'}\,\rangle = \int dp\, u_{n'}(p) u_n(p)$, and $H_{n'n} = \langle\, u_{n'}\,|\,H\,|\,u_n\,\rangle$. The overlap matrix Δ will be banded, but the Hamiltonian matrix will be essentially full, since the potential is nonlocal in the momentum representation. This feature represents perhaps the greatest disadvantage of the momentum-space approach. Computation of the potential matrix in a momentum-space basis, however, is not necessarily a computational bottleneck, since the coordinate-space representation of the basis functions, i.e., $\langle r|u_n\rangle$, can be computed analytically for simple piecewise polynomials.

We solved Eq.(15) using the central difference approximation for the time derivative. Denoting with δt the time step, the stability condition for this time-propagation scheme is roughly $p_{max}^2 \delta t/2\mu < 1$. In the calculations, we used $\delta t = 0.001$.

Table 1. Results of a two-particle wave packet calculation of s-wave scattering from the potential $V(r) = -e^{-r}$.

	Exact $S_0(p)$			Wave-packet calculation of $S_0(p)$			
				$\tilde{\psi}(p,T)\,/\,\tilde{\phi}(p,T)$		$\tilde{\psi}(p,T)\,/\,\phi(p,T)$	
p	Re	Im	T	Re	Im	Re	Im
2	0.614	0.789	10	0.611	0.771	0.606	0.775
			12	0.615	0.806	0.606	0.775
			14	0.620	0.787	0.628	0.823
			16	0.612	0.798	0.637	0.823
			18	0.622	0.796	0.672	0.903
			20	0.616	0.794	0.699	0.955
			22	0.621	0.798	0.748	1.030
			24	0.624	0.795	0.809	1.114
5	0.924	0.383	4	0.925	0.383	0.925	0.382
			6	0.924	0.384	0.969	0.403
			8	0.925	0.384	1.089	0.454
			10	0.925	0.384	1.315	0.549
			12	0.926	0.384	1.679	0.689
			14	0.926	0.383	2.032	0.779
			16	0.925	0.377	1.913	0.734
8	0.969	0.246	2	0.963	0.261	0.962	0.247
			4	0.969	0.246	1.042	0.229
			6	0.970	0.246	1.337	0.265
			8	0.970	0.246	1.955	0.350
			10	0.970	0.242	2.014	0.406
			12	0.971	0.241	1.381	0.285

Starting at about $t = 8$, the wave packet is outside the range of the potential, and Eq.(11) can be applied . The last two columns in Table 1 shows the S-matrix elements $S_0(p)$ determined from the ratio $\tilde{\psi}(p,t)/\phi(p,t)$. These are rather inaccurate, nonunitary, and unstable with respect to the sampling time. This is due to the fact that H_0 has been approximated in the numerical propagation of $\tilde{\psi}(t)$, whereas it is treated exactly in $\phi(t)$. To avoid this mismatch, the in-state is expanded as

$$\tilde{\phi}(p,t) \;=\; \sum_n b_n(t)\,\frac{u_n(p)}{p}\;. \tag{16}$$

The coefficients $b_n(t)$ satisfy (in matrix notation) $i\mathbf{\Delta}\dot{\mathbf{b}}(t) = \mathbf{H}_0\mathbf{b}(t)$ where $(\mathbf{H}_0)_{n'n} = \langle\, u_{n'} \mid H_0 \mid u_n \,\rangle$.

Although the momentum distribution of the exact in-state is smooth, $|\tilde{\phi}(p,t)|^2$ obtained with $I = 150$ and $\delta t = 0.001$ shows severe oscillations for $t > 5$, a clear indication of the basis-set deficiency in handling the H_0-dynamics. However, virtually the same oscillations occur in $|\tilde{\psi}(p,t)|^2$. As a result, when the S-matrix elements are calculated via

$$S(p) \;=\; \frac{\tilde{\psi}(p,t)}{\tilde{\phi}(p,t)}\,, \tag{17}$$

the errors arising from the finite-space approximation of the kinetic-energy operator cancel out.[1] The resulting S-matrix elements are quite accurate, as shown in Table 1.

The coordinate-space support of the present basis covers roughly the interval $0 \leq r \leq 60$. Of course, this interval does not have a sharp boundary, but rather a boundary zone. An incoming free particle with $p = 8$ starting at $r_0 = 10$ can be expected to reach the boundary at about $t = 9$, whereas one with $p = 2$ would leave the interval $[0, r_0]$ by about $t = 10$. Seemingly, this does not leave a time period for free propagation after the collision. In fact, the average position of the wave packet as a function of time also shows that a gradual reflection starts to occur after $t > 10$. However, Table 1 shows that the S-matrix can be extracted reliably over a much longer time period. For example, S_p with $p = 8$ is obtained quite accurately as late as $t = 12$. Although the free and fully interacting wave packets reach the boundary region at different times and, hence, experience different interference effects, the simple recipe of comparing numerical wave packets seems to be effective even in cancelling out such intereference effects, at least, at the early stages of the reflection phenomena.

THREE-PARTICLE SCATTERING

Kinematics and Arrangement Channels

Let $(\alpha\beta\gamma)$ stand for cyclic permutations of particle labels (123). A two-fragment partition $(\alpha)(\beta\gamma)$ of the three particles is referred to as the rearrangement channel α. The partition (1)(2)(3) is referred to as the breakup (dissociation) channel. The label α of the spectator particle in the partition $(\alpha)(\beta\gamma)$ is also used as a label for the pair $(\beta\gamma)$ The Jacobi coordinates for the arrangement $(\alpha)(\beta\gamma)$ are denoted by $\mathbf{r}_\alpha$ and $\mathbf{R}_\alpha$. Here $\mathbf{r}_\alpha$ is the internal relative coordinate of the pair $(\beta\gamma)$, and $\mathbf{R}_\alpha$ the relative position of the spectator particle α with respect to the center of mass of the pair $(\beta\gamma)$. The momenta conjugate to $\mathbf{r}_\alpha$ and $\mathbf{R}_\alpha$ are $\mathbf{p}_\alpha$ and $\mathbf{P}_\alpha$, respectively.

The total c.o.m Hamiltonian H of the system is given as $H = H_0 + V$. Here, H_0 is the kinetic energy operator, and V the total interaction. H_0 can be decomposed as $H_0 = k_\alpha + K_\alpha$, where $k_\alpha = p_\alpha^2/(2\mu_\alpha)$, and $K_\alpha = P_\alpha^2/(2\nu_\alpha)$, $\alpha = 1, 2, 3$. Its eigenstates are the plane-wave states $|\mathbf{p}_\alpha \mathbf{P}_\alpha\rangle$. It will prove convenient, especially in dealing with identical particles, to use the notation $|\,\mathbf{p}'\mathbf{P}'\,\rangle_\alpha$ for $|\mathbf{p}_\alpha = \mathbf{p}', \mathbf{P}_\alpha = \mathbf{P}'\rangle$.

The total-angular-momentum (J-representation) states in the coupling scheme of the αth arrangement are $|j\ell JM\rangle_\alpha$, where j is the rotational quantum number of the pair $(\beta\gamma)$, and ℓ the orbital angular momentum number of the particle α. In explicit notation, ${}_\alpha\langle\,\hat{\mathbf{p}}\,\hat{\mathbf{P}}\,|\,j\,\ell\,J\,M\,\rangle_\alpha = \mathcal{Y}_{j\ell}^{JM}(\hat{\mathbf{p}}, \hat{\mathbf{P}}) = \sum_{m_j m_\ell} \langle\, j\,\ell\, m_j\, m_\ell\,|\,J\,M\,\rangle\, Y_{j m_j}(\hat{\mathbf{p}})\, Y_{\ell m_\ell}(\hat{\mathbf{P}})$. The eigenstates of H_0 in the J-representation are

$$|\,p\,P\,j\,\ell\,J\,M\,\rangle_\alpha = \int d\hat{\mathbf{p}}\, d\hat{\mathbf{P}}\; \mathcal{Y}_{j\ell}^{JM*}(\hat{\mathbf{p}}, \hat{\mathbf{P}})\, |\,\mathbf{p}\,\mathbf{P}\,\rangle_\alpha\,. \tag{18}$$

The internal Hamiltonian for the pair $(\beta\gamma)$ is $h_\alpha = k_\alpha + V_\alpha$, where V_α is the potential between particles β and γ. The bound states of h_α are denoted $|\,\phi_{vj} j m_j\rangle_\alpha$ with energies $\epsilon_{\alpha vj}$. In momentum representation, we have ${}_\alpha\langle\,\mathbf{p}\,|\,\phi_{vj}\, j\, m_j\,\rangle_\alpha = \phi_{\alpha vj}(p)\, Y_{j m_j}(\hat{\mathbf{p}})$. Here, $\phi_{\alpha vj}(p) = c_j \int r^2\, dr\, j_j(pr)\, \phi_{\alpha vj}(r)$, with $c_j = i^j \sqrt{2/\pi}$, and $\phi_{\alpha vj}(r)$ being the radial wavefunction of the vibrational-rotational state. The time-dependence of the boundstates is simple: $|\phi_{\alpha vj}(t)\rangle = exp(-\epsilon_{\alpha vj}\, t)\,|\,\phi_{\alpha vj}\rangle$.

The asymptotic dynamics in the rearrangement-channel α is described by H_α $(\equiv K_\alpha + h_\alpha)$. Its J-representation eigenstates ("channel states") are $|\phi_{vj}\, P\, j\, \ell\, J\, M\rangle_\alpha$, with energies $E_{\alpha vjP} = \epsilon_{\alpha vj} + P^2/2\nu_\alpha$. The explicit momentum and coordinate representations of the channel states are

$${}_\alpha\langle\,\mathbf{p}'\,\mathbf{P}'\,|\phi_{vj} P j \ell J M\,\rangle_\alpha = \frac{\delta(P' - P)}{P^2}\, \phi_{\alpha vj}(p')\, \mathcal{Y}_{j\ell}^{JM}(\hat{\mathbf{p}}', \hat{\mathbf{P}}')\,, \tag{19}$$

$$ {}_{\alpha}\langle \mathbf{r}\mathbf{R} \,|\, \phi_{vj} P j \ell J M \rangle_{\alpha} = \sqrt{\frac{2}{\pi}}\; i^{\ell}\; j_{\ell}(PR)\; \phi_{\alpha v j}(r)\; \mathcal{Y}^{JM}_{j\ell}(\hat{\mathbf{r}}, \hat{\mathbf{R}}) \tag{20} $$

The space of all possible asymptotic channel states in a given rearrangement with fixed (JM) is characterized by the projector

$$ \Pi^{JM}_{\alpha} = \sum_{jv\ell} \int P^2\, \mathrm{d}P\, |\phi_{vj}(t) P j \ell J M\rangle_{\alpha}\; {}_{\alpha}\langle \phi_{vj}(t) P j \ell J M| \,. \tag{21} $$

Wave-Packet Description

Let us now consider an initial incoming wave packet, of total angular momentum J, representing a collision in which the particle α with orbital angular momentum ℓ_0 is incident on a bound-state of the pair $(\beta\gamma)$ with quantum numbers $(v_0 j_0)$:

$$ |\, \Phi^{JM}_{\alpha v_0 j_0 \ell_0}(0)\rangle = \int P^2\, \mathrm{d}P\, |\, \phi_{v_0 j_0} P j_0 \ell_0 J M\rangle_{\alpha}\; \xi_{\alpha \ell_0}(P) \,, \tag{22} $$

where $\xi_{\alpha\ell_0}(P)$ is an incoming wave packet for the relative motion of particle α with average momentum P_0, and average separation R_o. The average separation R_0 is taken well outside the the range of the channel interaction ($V - V_\alpha$). The time evolution of the initial state under H_α is given as $|\Phi_{\alpha v_0 j_0 \ell_0}(t) = |\phi_{v_0 j_0}(t)\, \xi_{\ell_0}(t)\, j_0 \ell_0 J M\rangle_{\alpha}$, where $\xi_{\ell_0}(P,t) = exp(-P^2 t/2\nu_\alpha)\, \xi_{\alpha\ell_0}(P)$.

Let $|\Psi^{JM}_{\alpha v_0 j_0 \ell_0}(t)\rangle$ denote the solution of the three-particle TDSE that coincide with $|\Phi^{JM}_{\alpha v_0 j_0 \ell_0}\rangle$ at $t = 0$. Then the fundamental result of the time-dependent scattering theory is that[14]

$$ \Pi^{JM}_{\beta}\, |\Psi^{JM}_{\alpha v_0 j_0 \ell_0}(t)\rangle = \hat{S}_{\beta\alpha}\, |\Phi^{JM}_{\alpha v_0 j_0 \ell_0}(t)\rangle \,, \tag{23} $$

as $t \to \infty$.

We can define the rearrangement components of the total wave packet via

$$ |\Psi^{\beta JM}_{\alpha v_0 j_0 \ell_0}(t)\rangle = \Pi^{JM}_{\alpha}\, |\Psi^{JM}_{\alpha v_0 j_0 \ell_0}(t)\rangle \,, \qquad \beta = 1,2,3 \,. \tag{24} $$

As $t \to \infty$, these components will be spatially separated, and the scattering into the rearrangement β will be solely contained in Ψ^{β}_{α}. The asymptotic part of the total wave-packet that corresponds to breakup can then be identified as

$$ |\Psi^{0JM}_{\alpha v_0 j_0 \ell_0}(t)\rangle = |\Psi^{JM}_{\alpha v_0 j_0 \ell_0}(t)\rangle - \sum_{\beta=1}^{3} |\Psi^{\beta JM}_{\alpha v_0 j_0 \ell_0}(t)\rangle \,. \tag{25} $$

We can thus write, as $t \to \infty$,

$$ |\Psi^{\beta JM}_{\alpha v_0 j_0 \ell_0}(t)\rangle = \hat{S}^{J}_{\beta\alpha}\, |\, \Phi^{JM}_{\alpha v_0 j_0 \ell_0}\rangle \,, \qquad \beta = 0,1,2,3, \quad \alpha = 1,2,3 \,. \tag{26} $$

To extract the sharp-energy matrix elements of the rearrangement S-operators, we use the energy conserving property of $S_{\beta\alpha}$, viz.,

$$ {}_{\beta}\langle \phi_{v'j'} P' j' \ell' J M |\, \hat{S}^{J}_{\beta\alpha}\, |\, \phi_{vj\ell} P j \ell J M\rangle_{\alpha} = S^{J}_{\beta v' j' \ell', \alpha v j \ell}(E)\; \delta(E_{\beta v' j' P'} - E_{\alpha v j P}) \,, \tag{27} $$

where $E = E_{\beta v' j' P'} = E_{\alpha v j P}$. Upon using (27) in (23), we find, as $t \to \infty$,

$$ S^{J}_{\beta v' j' \ell', a_0}(E_{\beta v' j' \ell' P'}) = \sqrt{\frac{\nu_\beta P'}{\nu_\alpha P}}\; \frac{{}_{\beta}\langle \phi_{v'j'}(t)\, P' j' \ell' J M \,|\, \Psi^{JM}_{\alpha}\rangle}{\xi_{\alpha l_0}(P,t)} \,, \tag{28} $$

where P is fixed by the requirement $E_{\beta v'j'P'} = E_{\alpha v_0 j_0 P}$.

The components $|\Psi^{\beta JM}_{\alpha v_0 j_0 \ell_0}\rangle$ can be further be decomposed as

$$_\beta\langle \mathbf{p}\,\mathbf{P}|\Psi^{\beta JM}_{\alpha v_0 j_0 \ell_0}(t)\rangle = \sum_{vj\ell} \phi_{\beta vj}(p,t) g^{\beta vj\ell}_{\alpha v_0 j_0 \ell_0}(P,t) \mathcal{Y}^{JM}_{j\ell}(\hat{\mathbf{p}},\hat{\mathbf{P}}) , \tag{29}$$

where

$$g^{\beta vj\ell}_{\alpha v_0 j_0 \ell_0}(P,t) = {}_\beta\langle \phi_{vj}(t) P j \ell J M \,|\, \Psi^{JM}_{\alpha v_0 j_0 \ell_0}(t)\rangle . \tag{30}$$

Note that $g^{\beta vj\ell}_{\alpha v_0 j_0 \ell_0}(P,t)$ asymptotically is the outgoing relative-motion packet in the $(\beta vj\ell)$-channel, while $g^{\beta vj\ell}_{\alpha v_0 j_0 \ell_0}(P,0) = \delta_{\beta\alpha}\,\delta_{vv_0}\,\delta_{jj_0}\,\delta_{\ell\ell_0}\,\xi_{\ell_0}(P)$ at $t=0$. Note also that the orthogonality property $\langle g^{\beta vj\ell}_{\alpha v_0 j_0 \ell_0}\,|\,g^{\gamma v'j'\ell'}_{\alpha v_0 j_0 \ell_0}\rangle = 0$ for $\beta \neq \gamma$ holds only as $t \to \infty$. Finally, we can rewrite Eq.(28) as

$$S^J_{\beta v'j'\ell',\alpha v_0 j_0 \ell_0}(E_{\beta v'j'P'}) = \sqrt{\frac{\nu_\beta P'}{\nu_\alpha P}}\,\frac{g^{\beta v'j'\ell'}_{\alpha v_0 j_0 \ell_0}(P',t)}{\xi_{\alpha \ell_0}(P,t)} . \tag{31}$$

Approximation Space

In principle, the approximation space can be built using functions in the Jacobi variables of any rearrangement. However, a basis or grid discretization in Jacobi variables of a single rearrangement will not be able to effectively describe the asymptotic configurations in other arrangements, unless an excessively large basis or fine discretization grid is employed. Of course, this difficulty is common to both TD or TI treatments. One method that proved succesful in the TI theory is based on the simultaneous use of expansion bases (or discretization grids) in natural variables of all important arrangements.[8,9] The present contribution explores a particular type of multi-arrangement basis for a TD treatment of reactive and dissociative collisions. A related approach in which breakup channel is treated on equal footing with the rearrangement channels has been discussed in [3].

In the present approach, we construct the full approximation space $\mathcal{S}$ as the linear span of three arrangement-channel subspaces $\mathcal{S}_\alpha$, $\alpha = 1,2,3$. Each subspace $\mathcal{S}_\alpha$ is spanned by a direct-product basis $\{\, u_{kj}(p_\alpha) w_{n\ell}(P_\alpha)\, \mathcal{Y}^{JM}_{j\ell}(\hat{\mathbf{p}}_\alpha,\hat{\mathbf{P}}_\alpha)\,\}$. The basis sets $\{\, u_{kj}(p)\,\}_{k=1}^{K}$ and $\{\, w_{n\ell}(P)\,\}_{n=1}^{N}$ are generated by a finite-element discretization of the variables p_α and P_α, just as in the two particle case. We assume that the set $\{u_{kj}(p)\}$ for a given j is large enough to allow an accurate expansion of the boundstates ϕ_{vj}.

Alternatively, we could use the vibrational states $\{\phi_{vj}\}$ as the p-basis in place of $\{u_{kj}\}$. In such a case, in order to be able to describe the breakup effects, we should augment the vibrational boundstates with pseudostates in the continuum. In fact, the use of the present finite-element basis is equivalent to such a pseudostate approach. Let $\{\tilde{\phi}_{\alpha vj}\}$ be the approximate eigenstates of h_α that results when h_α is diagonalized in $\mathcal{S}_\alpha$. The states $\{\tilde{\phi}_{\alpha vj}\}$ corresponding to negative eigenvalues will be approximations to the boundstates, while the positive-energy ones will be pseudostates in the continuum of pair α. Multi-arrangement pseudostate bases are quite effective in simulating breakup effects in the time-independent context.[9,10]

Note that the space $\mathcal{S}$ is not a direct sum of subspaces $\mathcal{S}_\alpha$ due to the well known nonorthogonality problem. The projector onto $\mathcal{S}$ can be constructed as

$$\Pi_{\mathcal{S}} = \sum_{\alpha jk\ell n} \sum_{\beta j'k'\ell' n'} |u_{kj} w_{n\ell} j \ell J M\rangle_\alpha \,(\mathbf{\Delta}^{-1})_{\alpha kjn\ell,\beta k'j'n'\ell'}\; {}_\beta\langle u_{k'j'} w_{n'\ell'} j'\ell' J M| , \tag{32}$$

where $\boldsymbol{\Delta}$ is the overlap matrix, viz.,

$$\Delta_{\alpha kjn\ell,\beta k'j'n'\ell'} = {}_\alpha\langle u_{kj}w_{n\ell}j\ell JM \,|\, u_{k'j'}w_{n'\ell'}j'\ell'JM\rangle_\beta . \qquad (33)$$

If the rearrangement subspaces are each pushed to completeness, overcompleteness problem would arise and the overlap matrix would become singular. If formal or numerical linear dependence arises, the inverse of the overlap matrix is to be taken in a pseudo-inverse sense. We note that the formulation of a scattering theory (TD or TI) on $\mathcal{S}$ can be done using two-Hilbert-space techniques[16] that can be useful when Δ is formally or numerically singular.

On the approximation space, the solution of the TDSE will be approximated by the expansion

$$|\tilde{\Psi}_{\alpha v_0 j_0 \ell_0}\rangle = \sum_{\beta kjn\ell} |u_{kj}w_{n\ell}j\ell JM\rangle_\beta \, C_{\beta kjn\ell}(t) , \qquad (34)$$

where initial state indices have been suppressed in the expansion coefficients. By requiring that the error $|\,\mathcal{E}\rangle \equiv (H - \frac{\partial}{\partial t}) \,|\, \tilde{\Psi}\rangle$ to be orthogonal to the approximation space, we obtain

$$i \frac{d\mathbf{C}(t)}{dt} = \boldsymbol{\Delta}^{-1}\mathbf{H}\ \mathbf{C}(t) , \qquad (35)$$

where $\mathbf{C}$ is the column vector of expansion coefficients $C_{\beta kjn\ell}$, and $\mathbf{H}$ the Hamiltonian matrix with elements ${}_\gamma\langle u'_{k'j}\, w_{n'\ell'}j'\ell'JM \,|\, H \,|u_{kj}w_{n\ell}\, j\ell JM\rangle_\beta$. For the case of three identical particles, this equation can be block diagonalized using the permutational symmetry, just as in the time-independent applications.

Application to a Three-Particle Model

Model. The three-particle problems with finite-rank separable potentials can be numerically solved within the Faddeev formalism using momentum-space techniques [12,13] that are by now well established. In fact, this momentum-space integral-equation approach is, currently, the only fool-proof method to treat three-particle problem above the breakup threshold. Therefore, to gauge the effectiveness of the TD-CAC method in handling the breakup channel at energies above the breakup threshold, we have chosen a simple model that is numerically solvable within the Faddeev method. The model consists of three identical particles (of unit mass) whose total interaction is pairwise additive, with the pair potentials being rank-1 separable and restricted to act only on the s-wave states of the pairs. The pair potentials are taken as $V_\alpha = |\chi_\alpha\rangle\lambda_\alpha\langle\chi_\alpha|$, with $\langle \mathbf{p}_\alpha|\chi_\alpha\rangle = (\beta^2 + p_\alpha^2)^{-1}$. Note that each pair potential supports at most a single boundstate. We took $\beta = 1.444$, and λ was chosen to give the boundstate energy $\epsilon = -0.0537$. With these parameters, this model is used in Nuclear Physics as a prototype of the n+d scattering.[13]

We further restrict our attention to zero total-angular-momentum state. Since $j = \ell = J = 0$ and there is just one vibrational boundstate, the labels $(vj\ell J)$ will be suppressed in the subsequent discussion. In particular, the rearrangement S-matrix elements will carry only the rearrangement indices, viz, $S_{\alpha\beta}(E_P)$ with $E_P = \epsilon + 3P^2/4$.

For three identical particles , Eq. (35) can be block-diagonalized according to the irreducible representations of the permutation group S_3.[17] For the present model with all angular momenta restricted to zero, the totally symmetric (S) and mixed-symmetry (E) matrix elements can be defined via

$$O_{\mathrm{S}}^{k'n',kn} = O_{1k'n',1kn} + 2\ O_{2k'n',1kn} , \qquad (36)$$

$$O_{\mathrm{E}}^{k'n',kn} = O_{1k'n',1kn} - O_{2k'n',1kn} , \qquad (37)$$

where $O = H,\ \Delta$. Similarly, the irreducible-representation S-matrices for rearrangement are defined as $S_{\mathrm{S}} = S_{11} + 2S_{21}$ and $S_{\mathrm{E}} = S_{11} - S_{21}$.

Table 2. Comparison of Faddeev and Wave-Packet Calculations of $S_E(E_P)$.

P		Faddeev	Wave-packet calculation T = 8	T = 12	T = 16	T = 20
1.2	Re	0.442	0.472	0.454	0.476	0.471
	Im	0.761	0.740	0.797	0.789	0.789
1.6	Re	0.769	0.765	0.777	0.775	0.781
	Im	0.501	0.491	0.483	0.490	0.494
2.0	Re	0.902	0.900	0.902	0.905	0.905
	Im	0.307	0.317	0.321	0.323	0.322
2.4	Re	0.957	0.961	0.962	0.963	0.959
	Im	0.186	0.184	0.184	0.183	0.183
2.8	Re	0.980	0.978	0.983	0.977	0.976
	Im	0.114	0.117	0.113	0.124	0.115

Computational Details. The initial relative-motion wave-packet $\xi(P)$ has the same form as in Eq.(13), with $P_0 = 2.0$, $R_0 = 9.0$, and $d = 2.0$. The effective momentum-support of this packet is in the interval $1.0 \leq P \leq 3.0$. The computational momentum-domain was restricted by the cut-off values $P_{max} = 6.0$, and $p_{max} = 7.2$. For each variable, the interval $[\,0, x_{max}\,]$ was divided into I_x subintervals, where $x = p$ or P, and $(2I_x - 1)$ quadratic interpolation functions[15] were defined on each partition. The grid points in P and p were unevenly distributed, with most of the points covering the effective momentum support of the initial packet. We used $I_p = 12$ and $I_P = 33$ for the E-symmetry case, and $I_p = 13$ and $I_P = 49$ for the S-symmetry calculation. The corresponding matrix dimensions for the S and E symmetry cases were, respectively, 2465 and 1495. The time-propagation was done using the central difference scheme with a time step of 0.005.

Results and Discussion. The irreducible-representation S-matrices S_τ, τ=E, or S, extracted from the wave-packet calculations are shown in Tables 2 and 3 for a number of collision energies within the energy distribution of the initial wave packet. Although the discretizations used for the variables p and P are quite coarse, the wavepacket results agree with the reference solutions to better than 5% in most cases. As in the two-particle case, the success of the method hinged strongly upon the use of the numerical free state $\tilde{\phi}(p,t)\tilde{\xi}(P,t)$ in Eqs.(30) and (31). Although the $\tilde{\phi}(p,t)$ was in good agreement with $\phi(p,t)$, the free relative-motion packet $\tilde{\xi}(P,t)$ was plagued by unphysical oscillations P for $t > 8$. However, virtually the same oscillations occured in $\tilde{g}^\tau(P,t)$.

Note that the wave packet results for the E-symmetry case are in general more accurate than those for the S-symmetry case, despite the fact that a smaller basis has been used for the former. This is due to the fact that the E-symmetry states of the system are weakly interacting compared to the S-symmetry states. For example, the three-particle system posseses two bound states of S-symmetry, while it is nonbinding in the E-symmetry case. Also, the total breakup probabilities range from $\sim$ 23% at $P = 1.2$ to $\sim$ 2% at $P = 2.8$ for the E-symmetry. In contrast, the breakup probabilities for the S-symmetry range from $\sim$ 92% to $\sim$ 27% over the same momentum interval. Clearly, the breakup channel plays a dominant role in the S-symmetry state. It is remarkable that the coupled-arrangement-channel method can take account of such strong breakup effects with a relatively small multi-rearrangement basis.

Table 3. Comparison of Faddeev and Wave-Packet Calculations of $S_s(E_P)$.

P	% Breakup		Faddeev	Wave-packet Calculation T = 8	T = 12	T = 16
1.2	92.3	Re	-0.275	-0.315	-0.271	-0.283
		Im	-0.040	-0.018	-0.017	-0.042
1.6	86.1	Re	0.003	0.036	-0.008	0.008
		Im	0.373	0.363	0.402	0.392
2.0	64.5	Re	0.361	0.361	0.340	0.353
		Im	0.474	0.445	0.450	0.455
2.4	43.1	Re	0.622	0.642	0.627	0.634
		Im	0.427	0.401	0.384	0.412
2.8	27.2	Re	0.783	0.753	0.758	0.743
		Im	0.339	0.329	0.312	0.367

CONCLUSIONS

The principle advantage of performing the wavepacket calculation directly on the momentum space is the possibility of concentrating the numerical effort on the momentum interval where the packet has appreciable amplitude and we need the most accuracy. One can use a very fine discretization over this significant interval, while the rest of the computational domain can be covered with a coarser discretization grid. This is also beneficial in connection with the boundary-reflection problem. In our experience, the use of a fine grid over a portion of the total computational interval has been sufficient to guarantee relatively long periods of reflection-free propagation. In addition, the insight gained from the behavior of the wave packets in momentum space during time evolution has lead us to a simple method of cancelling out any phase errors that may accumulate in numerical wave packets, especially at large times and for high momentum components, due to an inadequate treatment of kinetic energy operators. That rather accurate S-matrices can be obtained via Eq.(17) from wave packets infected with numerical inaccuracies has been demonstrated.

Our wave packet calculations on a model three-particle problem with a dominant breakup channel show that the coupled-arrangement-channel ansatz is capable of describing the wave packet dynamics in reactive and dissociative collisions. Compared with the single-arrangement-plus-absorbing-potential approach,[6] the principle difficulty of the TD-CAC method is in the calculation of overlap and Hamiltonian matrices and in the formation of $\Delta^{-1}H$. However, efficient algorithms have been developed for calculating such matrices within the TI-CAC context[8,11]. Moreover, the TI approach also requires similar matrix inversions, but at every collision energy. It is plausible that the basis sets used in the TI treatments of reactive scattering will also prove efficient in the TD version.

ACKNOWLEDGEMENTS

Research reported in this work has partially been supported by National Research Council of Turkey (TUBITAK) through Grant No. TBAG-1088. Computing support from Bilkent University and Brown University is gratefully acknowledged.

REFERENCES

1. Z. C. Kuruoglu and F. S. Levin, to appear in Physical Review A (1992).
2. Z. C. Kuruoglu and F. S. Levin, Phys. Rev. Lett. 64 : 1701 (1990).
3. Z. C. Kuruoglu, Phys. Rev. A 42 : 6314 (1990).
4. J. Z. H. Zhang, J. Chem. Phys. 92 : 324 (1990).
5. D. K. Hoffman, O. Sharafeddin, R. S. Judson and D. J. Kouri, J. Chem. Phys. 92 : 4167 (1990);
   O. A. Sharafeddin, R. S. Judson, D. J. Kouri, and D. K. Hoffman, J. Chem. Phys. 93 : 5580 (1990).
6. D. Neuhauser, R. Judson, and A. Perski, Chem. Phys. Lett. 181 : 151 (1991);
   R. S. Judson, D. J. Kouri, D. Neuhauser, and M. Baer, Phys. Rev. A 42 : 351 (1990);
   D. Neuhauser and M. Baer, J. Chem. Phys. 91 : 4651 (1989).
7. W .H. Miller, J. Chem. Phys 50 : 407 (1969);
   D. W.Schwenke, D. G. Truhlar and D. J. Kouri, J. Chem. Phys. 86 : 27 (1987).
8. J. Z. H. Zhang and W. H. Miller, J. Chem. Phys. 88 : 4454 (1988);
   J. Z. H. Zhang, D. J. Kouri, K. Haug, D. W. Schwenke, Y. Shima, and D. G. Truhlar, J. Chem. Phys. 88 : 2492 (1988).
9. Z. C. Kuruoglu and F. S. Levin, Phys. Rev. Lett. 48 : 899 (1982).
10. Z. C. Kuruoglu, Phys. Rev. C 44 : 1354 (1991); Phys. Rev. A 44 : 7307 (1991).
11. P. N. Day and D. G. Truhlar, preprint.
12. L.D. Faddeev, Sov. Phys. JETP 12 : 1014 (1961);
    E. O. Alt, W. Grassberger and W. Sandhas, Nucl. Phys. B 2 : 167 (1967).
13. W. Glockle, "The Quantum Mechanical Few Body Problem," Springer, Berlin (1983).
14. J. R. Taylor, "Scattering Theory," Wiley, New York (1972);
    R. G. Newton, "Scattering Theory of Waves and Particles," Springer, New York (1982).
15. P. M. Prenter, "Splines and Variational Methods," Wiley, New York (1975).
16. C. Chandler and A. G. Gibson, J. Math. Phys. 14 : 2336(1977);
    Gy. Bencze, C. Chandler and A. G. Gibson, Nucl. Phys. A 390 : 461 (1982).
17. J. Z. H. Zhang and W. H. Miller, J. Chem. Phys. 91 : 1528 (1989).

INTRAMOLECULAR DYNAMICS

F. Remacle and R.D. Levine

The Fritz Haber Research Center for Molecular Dynamics

The Hebrew University of Jerusalem

Jerusalem 91904, Israel

INTRODUCTION

The temporal evolution of an initially prepared nonstationary state of a polyatomic molecule is discussed. Particular attention is given to the short time evolution which can be probed by spectroscopic experiments. It is suggested that the time evolution is often sequential, with separation of time (or, equivalently, of frequency scales). The relevance to mode selective chemistry (in both the gas phase and in solution) is emphasized. In particular it is argued that the very shortest time scale identified from the spectrum is typically shorter than the time scale for mode mixing which is of interest in statistical theories of unimolecular dissociation.

BACKGROUND

On the one extreme are direct bimolecular reactions[1]. A well studied prototype is the abstraction of an H atom by Fluorine[2–4] $RH+F \rightarrow R+HF$. Even for a large polyatomic, molecule RH, much of the reaction exoergicity is carried away as internal excitation of HF. The mechanical picture is quite clear: The newly formed hot HF bond is coupled to R for one or, at most, a few of its (very rapid) vibrations. There is not enough time for the energy in the 'hot spot' to be dissipated over all the available degrees of freedom of RHF. Our question is 'how rapidly does such a hot spot delocalize' and, even more important for us, 'is the dissipation a simple process or does it have fairly well defined stages'. That there is an opposite extreme where, given enough time, the initial hot spot is very delocalized is clearly borne out by the chemical activation experiments[5–10] especially, those of

Time-Dependent Quantum Molecular Dynamics, Edited by J. Broeckhove and L. Lathouwers, Plenum Press, New York, 1992

Rabinovitch and coworkers. Chemical clocking[5,11,12] in particular provides striking evidence for exceedingly fast delocalization. Some of the reported rates are just an order of magnitude slower than typical vibrational frequencies.

Our question is whether the loss of the selectivity of the initial excitation, in highly excited molecules, is indeed so rapid. We shall bring forth arguments in support of a much more gradual and sequential loss. In particular we shall argue that the rapid sub-ps initial stage will have the same temporal character in a purely harmonic molecule. At the end of this first stage the molecule has by far not sampled the majority of its energetically accessible phase space. Hence we shall conclude that the fast initial dephasing is definitely not sufficient to establish the representative coverage of phase space which is assumed in the RRKM approach.

The reason for reexamining this question are the recent high resolution spectroscopic experiments of higher overtones of polyatomic molecules[12–20]. Adopting a time dependent point of view[21,22] such high resolution experiments provide insight on the temporal evolution of the initially localized excitation[22–30]. These, and the corresponding studies in the frequency domain[31–36] furnish what, from the standpoint of chemical kinetics, is complementary the short time information.

The information provided by both collisional clocking and the inversion of spectroscopic data is essentially indirect information. Ultimately, real time monitoring[37–40] will allow a direct probe of the dissociation process. For the time being, overtone spectroscopy provides useful insights on the first dozen ps or so following the $t = 0$ initiation of a non stationary molecular state. It would be of considerable interest to extend the time domain into the sub-ns range. This however is unlikely to be realized in the immediate future. Such time intervals are long on the spectroscopic scale as they require a frequency resolution which is typically unrealistic at the high excitation energies of interest. Computations are equally better at shorter times, for both numerical reasons and for the need to know finer and finer details of the Hamiltonian in order to obtain a higher resolution spectrum or, equivalently, an accurate longer time propagation.

THE EXPLORATION OF PHASE SPACE

Of the many possible spectroscopic techniques we here select for discussion the two that most readily correspond to the information of interest. The normalized absorption spectrum $S(\omega)$ [22] and the normalized resonant Raman excitation spectrum $\sigma(\omega)$[41,42]. (Note that there are other Raman cross sections, most notably the Raman emission cross section, which is not here discussed). The two types of spectra are related to two time correlation functions. The absorption spectrum is the Fourier transform of the time autocorrelation function $C(t)$ of an initial non-stationary state $\phi(0)$

$$C(t) = \int \exp(-i\omega t)S(\omega)d\omega$$

$$C(t) = < \phi(0)|\phi(t) > \ .$$

The spectral intensity p_n, at the energy of the n'th eigenstate ψ_n,

$$S(\omega) = \sum_n p_n \delta(\omega - \omega_n) \ ,$$

is related to $\phi(0)$ by

$$p_n = | < \psi_n | \phi(0) > |^2 \ .$$

The time autocorrelation function determines the 'survival probability', $|C(t)|^2$ of the non stationary state, $|C(t)|^2 = | < \phi(t)|\phi(0) > |^2$.

It is equally of interest to determine not only where the initial state is not, but where it is. This information is provided by the time crosscorrelation function $< \phi_f(0)|\phi_i(t) >$ which is related to the resonant Raman excitation cross section

$$\sigma_{fi}(\omega) = | \int_0^\infty < \phi_f | \phi_i(t) > \exp(i\omega t) dt |^2 \ .$$

Note however that unlike $S(\omega)$ and $C(t)$ which are a Fourier transform pair, the time cross correlation function determines the Raman excitation cross section but the inversion of this relation is less immediate. The direct route is thus to compute the time crosscorrelation function from the dynamics and to obtain the excitation profile as the one-sided Fourier transform. Recently we have however developed a method of analysis for extracting the time crosscorrelation function from the Raman excitation profile. $C(t)$ on the other hand can either be computed or determined directly from the experimental data, figure 1.

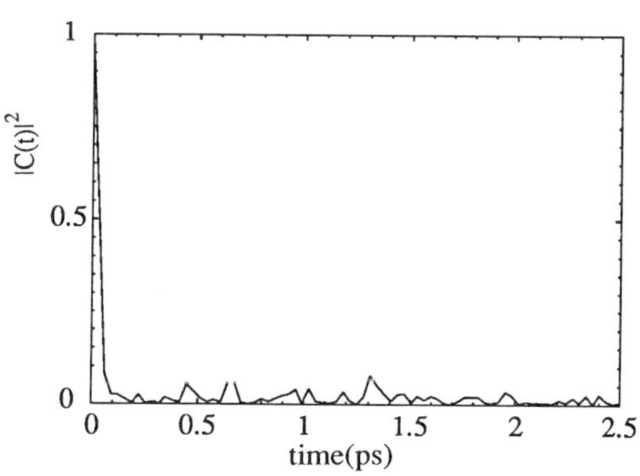

Figure 1. The survival probability $|C(t)|^2$ determined by a Fourier transform of the excitation spectrum of acetone in the visible and near UV. High resolution experimental data of H. Zuckermann, Y. Haas, M. Drabbels, J. Heinze, L. Meerts, J. Reuss, and J. van Bladel.

The short time evolution of $C(t)$ has been extensively discussed[22–30]. In particular, considerable attention was given to the notion of the correlation hole[15,19,28,29,43]. Here, however, we draw attention to a much more prominent feature, namely the very rapid drop of $|C(t)|^2$ from an initial value of unity to an (average) value smaller by one or more orders of magnitude, figure 1. The point is that this drop occurs on a time scale of vibrational periods and does not provide evidence for intramolecular vibrational energy redistribution. In other words, consider two Hamiltonians H_0 (whose spectrum is regular) and $H = H_0 + V$ where V includes the diverse anharmonic couplings. While V will cause local level repulsions, splittings, etc., the envelope of the spectrum of H_0 and H will be essentially identical. Since the gross structure in $S(\omega)$ is the same, the short time dependence of $C(t)$ will be the same.

That the initial time evolution does not discern the fine spectral details (which indicate the presence or absence of chaos) is not a uniquely quantal phenomena. It can just as well be understood on purely classical considerations. For historical reasons we phrase the discussion in terms of Slater's model of unimolecular dissociation[44]. Recall that in the Slater model the molecule is perfectly harmonic. The extension of the bond to be broken is written as a linear superposition of normal modes. Each mode with its own phase, which varies with time in accord with its harmonic frequency. As the phases vary, so does the extension of the bond to be broken. Dissociation is said to take place when, due to a phase matching, the bond is extended to a critical value. Now let us run this mechanical system backwards in time: An initially extended bond rapidly returns to a more average value due to the dephasing of the normal modes, each beating with its own frequency. The analogy with the optical excitation is established by thinking of the initial bond extension as resulting from a vertical Condon transition. The initial drop of $|C(t)|^2$ corresponds to the Slater-like dephasing process of the non-interacting modes.

In a purely harmonic Slater-type model, the initial dephasing leading to the contraction of the extended bond will eventually lead to a rephasing and an elongation of the bond. There is however an important proviso since real molecules are anharmonic. $|C(t)|^2$ does not recur back to its initial value of unity but to a lower value. It is this, much more gradual, relaxation of $|C(t)|^2$ which reflects the sampling of phase space. The same is true for a quantal time evolution.

ANALYSIS OF SEQUENTIAL EVOLUTION IN PHASE SPACE

The molecular eigenstates provide one suitable basis for discussing the time evolution of $C(t)$. In that basis the time history is entirely expressed in terms of dephasing since the eigenstates themselves are stationary

$$C(t) = \sum_n p_n \exp(-iE_n t/\hbar) \quad .$$

It is the distribution of the intensities, p_m, and the spacings $E_n - E_m$, as reflected in the spectral autocorrelation function $G(\omega)$, the so called 'power spectrum'[28]

$$|C(t)|^2 = \int d\omega \exp(-i\omega t)G(\omega)$$

$$G(\omega) = \sum_n \sum_m p_n p_m \delta(\omega_n - \omega_m - \omega)$$

that determines the different time regimes in $|C(t)|^2$ if such regimes exist.

$G(\omega)$ is the weighted distribution (by $p_n p_m$) of spacings in the spectrum. Note that $G(\omega)$ describes not only the distribution of spacings[45] but also the fluctuation of intensities.[46] Our understanding of molecular spectra is sufficient for us to recognize that $G(\omega)$ may well contain several peaks. The one at highest spacings (or shortest times) corresponds to spacings of different vibrational states. Next are contributions from the dominant anharmonicities (the largest resonant couplings), etc. One sees very clearly the existence of several scales in $G(\omega)$ for spectra taken at different resolutions (sometimes referred to as 'clumps', see, e.g. Refs. 15,18,47). At each increasing scale there is a new set of lines, which merge into a clump when the resolution is reduced, figure 2.

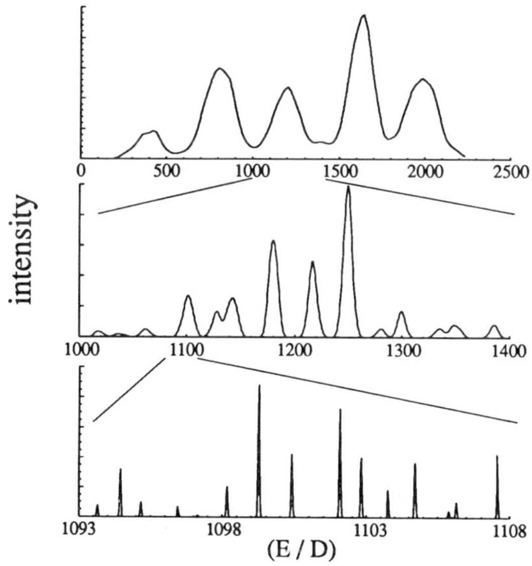

Figure 2. A (computed) spectrum at three levels of resolution, plotted vs. the frequency measured in units of the mean spacings between eigenstates. A group of resolved lines at the highest resolution (bottom panel) appears as a single feature (a 'clump') at a lower resolution. A group of clumps appears as a single clump at the lowest resolution (uppermost panel).

The nested structure of the observed spectrum suggests that it may be useful to consider an alternative basis set, corresponding to a zeroth order Hamiltonian H_0. The initially excited optically 'bright' states correspond to a small subset of states of H_0. Elsewhere[48] we have discussed the nested structure of the perturbation, V, $V = V_1 + V_2 + \cdots$, which gives rise to a sequence of Hamiltonians H_i, $H_1 = H_0 + V_1$, $H_2 = H_1 + V_2, \ldots$, such that it is the eigenstates of H_i which span the phase space sampled by the i'th stage. In terms of the eigenstates of H_0, each stage corresponds to a larger subset of states which are coupled to the initially prepared state. One can therefore define projection operators M_i, which project onto the subspace of states of H_0 relevant at the i'th stage, such that

$$M_i = M_{i-1} + \tilde{M}_i = \sum_{r=0}^{i} \tilde{M}_r \quad .$$

Here, $\tilde{M}_r$ is the projection on the new set of states brought into coupling at the r'th stage and $M_0 = \tilde{M}_0$.

The experimental observation of a clump structure in the spectrum means that one can compute a survival probability for a given clump. Therefore one can determine from the observed spectrum not only $|C(t)|^2$,

$$|C(t)|^2 = Tr(M_0 \rho(t)) \quad ,$$

where $\rho(t) = |\phi(t) >< \phi(t)|$, but also

$$\begin{aligned} |C_i(t)|^2 &= Tr(M_i \rho(t)) \\ &= \sum_i Tr(\tilde{M}_i \rho(t)) \quad . \end{aligned}$$

In the maximum entropy approach,[27,48] we determine $\rho(t)$ as the most spread out density subject to the constraints imposed by the spectrum, i.e. the values of the survival probabilities $|C_i(t)|^2$, $i = 0, 1, \ldots$. Taking into consideration that $Tr(\rho(t)) = 1$ and $\tilde{M}_{i-1}\, \tilde{M}_i = 0$, one obtains for $\rho(t)$

$$\rho(t) = \exp(\sum_{r=0}^{R} \mu_r(t) \tilde{M}_r)/Z$$

where Z insures the normalization of $\rho(t)$

$$Z = Tr(\exp(\sum_{r=0}^{R} \mu_r(t) \tilde{M}_r))$$

and R is the highest level of nesting evident in the spectrum, $\tilde{M}_{R+1} = I - \sum_{r=0}^{R} \tilde{M}_r$ and $\mu_{R+1}(t) \equiv 0$. Two very suggestive forms of $\rho(t)$, obtained using the orthogonality of the $\tilde{M}_i$'s are

$$\rho(t) = Z^{-1} \sum_{r=0}^{R+1} \exp[\mu_r(t)] \tilde{M}_r$$

showing that $\exp(\mu_r(t))Tr[\tilde{M}_r]/Z$ is the fractional occupancy of the r'th subspace and the nested form

$$\rho(t) = Z^{-1}\{\tilde{M}_{R+1} + \exp(\lambda_R(t))\{\tilde{M}_R + \exp(\lambda_{R-1}(t))\{\tilde{M}_{R-1} + \cdots$$
$$+ \exp(\lambda_1(t))\{\tilde{M}_1 + \exp(\lambda_0(t))\tilde{M}_0\}\cdots\} \ .$$

Here the $\lambda_r(t)$'s are the Lagrange multipliers conjugate to the survival probabilities $|C_r(t)|^2$ and are related to the $\mu_r(t)$'s by

$$\begin{aligned} \lambda_R &= \mu_R \\ \lambda_{R-1} &= \mu_{R-1} - \mu_R \\ &\cdot \\ &\cdot \\ \lambda_0 &= \mu_0 - \mu_1 \end{aligned}$$

A plot of the entropy of $\rho(t)$ for a system manifesting a separation of time scales is shown in figure 3. (The corresponding frequency spectrum is shown in figure 2). A detailed discussion of sequential time evolution is available in Ref. 48. Earlier closely relevant work includes Refs. 26, 27 and 30.

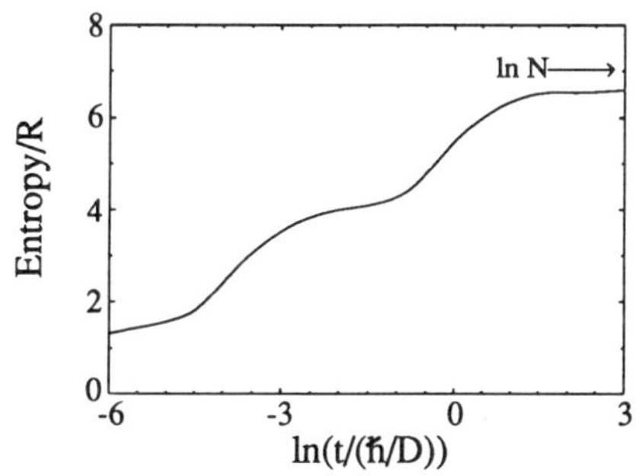

Figure 3. The entropy (in units of R) vs. time (in units of $\hbar$/mean spacings, logarithmic scale) computed for the density in phase space of maximal entropy. The survival probabilities vs. time are those determined by the spectrum shown in figure 2. Note the clear separation into three time regimes (the logarithmic abscissa scale makes it possible to show the different regimes on the same plot). N is the number of eigenstates so that for a strictly uniform covering of phase space the long time value of the entropy should be $R \ln N$. As can be seen the actual asymptotic value is somewhat lower. This is due to the fluctuation of the spectral intensities.[46]

In a wider context, the separation of time scales, as reflected in the sequential evolution stages in phase space, is a particular case of impulsive coupling. Impulsive coupling, manifested in a variety of intramolecular dynamics systems[49] e.g., van der Waals clusters, energy rich molecules, reactions in liquids[50] and on surfaces,[51] is the rather sudden switch from a motion confined to a smaller region of phase space to a less confined mode. It results from terms in the Hamiltonian which are 'non-resonant' on a shorter time scale and thus not included for shorter time dynamics.

DYNAMICS

An alternative to extracting the time correlation functions from the observed spectra is to compute them. This is particularly attractive because it is the earlier time dynamics which determine the more gross features of the spectrum. The considerable interest in time-dependent methods as reflected in these proceedings is no doubt related to this inherent advantage. Our own interest is however in the dynamics as such with special relevance to unbound systems. The possibility of mode selective chemistry,[52] as mentioned in the introduction, is an important motivation. Here, we review two recent studies, one in the gas phase[53] and one in solution,[54] which were based on the use of classical dynamics: An ensemble of initial conditions which represents the initially prepared non stationary state, is generated. Trajectories are then propagated and the time correlation function is computed.

Four molecules at energies both below and above their dissociation threshold were studied in the gas phase[53]: CH_3ONO, C_2H_4, SiF_4 and the large (by our standards) trinitro *s*-triazine (RDX). Initial excitations included both mode selective and random deposition of the energy. The computations were run for 10 ps, which is short for thermal unimolecular reactions where the excess energy is low. We have however examined all four molecules also at an energy where their lifetimes for dissociation are comparable or shorter than 10 ps.

The central conclusion of this work, parts of which are still in progress, is that on the timescale of 10 ps mode mixing is very incomplete. This is reflected both in the behavior of the time correlation functions and in the resulting spectra. Figure 4 is a typical outcome for SiF_4 at an excess energy of 75.8 kcal/mole. The time autocorrelation of the SiF stretch modes even at longer times is still distinct from that of the FSiF bends. Many more results are available in the full report.[53]

For all molecules there is a very clear initial drop of $|C(t)|^2$. As expected from the earlier discussion of the Slater model, this drop is most extreme for the many mode RDX molecule. However, this drop (quite moderate for SiF_4, figure 4), does not represent any loss of selectivity of the initial excitation. Taking figure 4a as an example, the initial drop of $|C(t)|$ means that a single SiF stretch is not a good zero order mode. Rather, it should be better thought of as a linear superposition of the stretch normal modes. These rapidly dephase with respect to one another, but the energy remains in the stretches. Plucking an SiF stretch shows no spectral activity in the bend region. The same incomplete energy scrambling is seen even well above the dissociation threshold, figure 5. (Note that the energy of SiF_4 is high enough for dissociation to take place in less than 10 ps.).

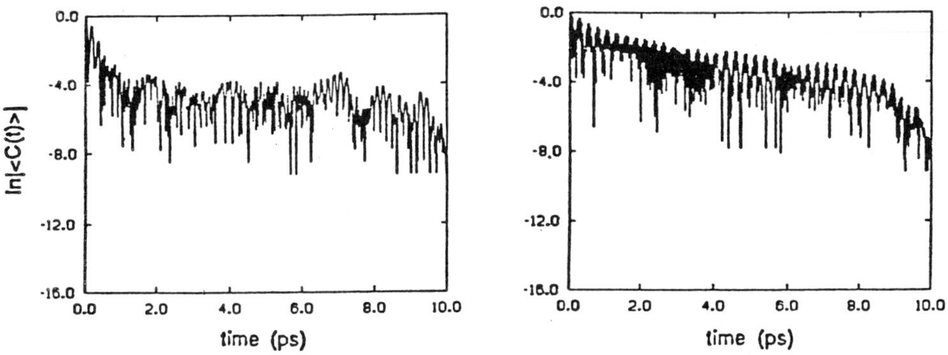

Figure 4. Ln $|C(t)|$ vs. time (in ps) for SiF_4 excited to the $v = 4$ level of all normal modes (excess energy of 75.8 kcalmol^{-1}). Left panel: For the SiF stretch mode. Right panel: for the FSiF bend mode. (Adapted from ref. 53).

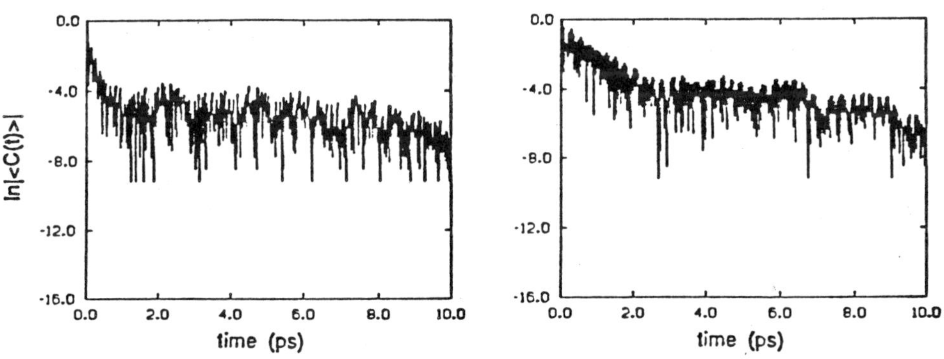

Figure 5. Same as figure 4 but for SiF_4 excited to well above the dissociation threshold. The lifetime is shorter than 10 ps yet the density in phase space is not uniform for times well over 2 ps. (Adapted from ref. 53).

The onset of mode mixing occurs later and precludes the full recurrence of $|C(t)|$. It is important to note that in these polyatomics, the recurrence time can be quite long. A simple estimate is as follows. Say the initially prepared state is a stretched bond. Recurrence means that this bond is again stretched to the same extent. The mean time between two such events is just what is computed in the Slater theory: In that model, the dissociation lifetime of a molecule equated to this mean recurrence time.

SiF_4 is a prototype of molecules which have been excited by infrared multiphoton absorption. Such experiments have typically tended to support extensive energy scrambling in the pumped molecule.[1,55] Our work suggests that this conclusion may well reflect the long duration of the typical CO_2 lasers used in these experiments. (At least in the ns scale if not longer). It should therefore be of interest to return to such experiments but with pulse lengths in the ps range.

The computations in the liquid[54] studied the time evolution of HOCl pumped to the $v = 6$ overtone of OH. This provides sufficient energy to dissociate the weak OCl bond. Yet there are good reasons to expect the very non-resonant energy flow from the OH to the OCl stretch to be quite hindered[56] and to proceed primarily via the bend. The novel feature, very evident in the simulations,[54] is the significant enhancement of this intramolecular energy transfer by collisions with the surrounding Ar liquid. This was particularly evident when the spectrum $S(\omega)$ is computed not over the entire time history (as it should be) but over a narrower time window. This results, of course, in a broad spectrum, devoid of fine structure. However, the three vibrational modes of HOCl are so different in their frequency that for a time window of, say, 1 ps, one can readily discern the three (two stretches and the bend) spectral regions. Such studies clearly demonstrate that the intramolecular vibrational energy redistribution is slow enough that the collisions with the atoms of the liquid can exercise a beneficial effect. From an overall kinetic point of view, the density of the liquid has additional roles.[54] It can cage the dissociation products. It can also remove energy by collisions with the activated HOCl molecule. Even so, at low densities, the primary role of the surrounding Ar atoms is to enhance the intramolecular energy transfer.

CONCLUDING REMARKS

As can be seen from the results presented here, the selectivity of the initial excitation may well persist on time scales comparable to the lifetime of an unbound molecule. This means that the system has not enough time to sample all the energetically accessible phase space before dissociation occurs.

In highly excited polyatomic molecules, the recurrence time $t_r = \hbar\rho(E)$ ($\rho(E)$ is the density of states), for which the sampling of phase space is by definition completed, is very long. From the standard expression[57] of the RRKM rate constant $k(E)$

$$k(E) = \frac{N^{\neq}(E - E_0)}{\hbar\rho(E)} \quad ,$$

($N^{\neq}(E-E_0)$ is the number of states of the transition state and E_0 the threshold energy), one can see that the fraction of molecules $\exp(-N^{\neq}(E-E_0))$ that have not dissociated by t_r is very small since $N^{\neq}(E-E_0)$ is typically very large. It is therefore not complete sampling of phase space which needs to be assumed for the validity of statistical theories of unimolecular reactions. In other words, the probing of phase space by the dissociation process is a coarse grained one.

We have recently argued[30] that more than the completeness of the sampling of phase space, it is its representativity which matters to insure the validity of statistical theories. It is well known[46,58,59] that the individual rate constants of the quasibound states in an energy interval $(E, E+dE)$ fluctuate around an average value $k(E)$ which can be estimated by the RRKM theory. Therefore, by representative sampling, we mean that the average rate calculated from the fraction of the quasibound space which has been sampled agrees with the value obtained from RRKM theory. In the case where the localized initial excitation sequentially explores regions of phase space coupled to different product channels, selectivity might be expected.[60]

We have studied[30] the competition between the two processes: mode mixing and dissociation. Here, the two relevant observables are the autocorrelation function $|C(t)|^2$ which probes the persistence of the memory of the initial excitation and the survival probability of the bound subspace B,

$$P(t) = Tr[B\rho(t)] \ .$$

The entropy $S_B(t)$ of the ME density matrix normalized in bound space $B\rho(t)/P(t)$ is the measure of the volume sampled.

Particular attention has been devoted to the amount of mixing of the initial state and to the role of overlapping resonances. The fact that resonances overlap can be easily understood from the RRKM equation of the rate constant: the ratio of the width $\hbar k$ and the mean spacing $1/\rho(E)$ is large, $(\hbar\rho(E)k = N^{\neq}(E-E_0) \gg 1)$. This means that the rate matrix[61] Γ is off-diagonal or, in classical terms, that there are recrossings at the barrier. Or, in a more pictorial language, there is a healing[62] of the broken bond. Under such circumstances, the RRKM rate constant is only an upper bound and the system might have enough time to sample phase space before dissociating. Our numerical results[30] show that due to recrossings the rate of exploration of phase space is indeed increased but not enough to lead to a representative sampling before the dissociative process occurs.

Incomplete mixing and sequential exploration of phase space depend in a very sensitive fashion on the sharpness of the localization on the initial state. We find that even a small amount of spreading in the initial state (less than 10% of the available phase space is initially accessed) is sufficient to provide a substantial sampling of phase space before dissociation and thus to insure the validity of statistical theories.

ACKNOWLEDGMENT

F. Remacle is a 'chargé de recherches' of FNRS at the Université de Liège, Belgium. We thank Prof. J.C. Lorquet for his interest and support of this work. This work was supported by the Air Force Office of Scientific Research and by the Commissariat Général de la Communauté Française de Belgique. The Fritz Haber Research Center is supported by the Minerva Gesellschaft für die Forschung, mbH, Munich, Germany.

REFERENCES

1. R. D. Levine and R. B. Bernstein, "Molecular Reaction Dynamics and Chemical Reactivity", Oxford University Press, N.Y. (1989).
2. J. D. McDonald, *Ann. Rev. Phys. Chem.*, 30:29 (1979).
3. B. S. Agrawalla and D. W. Setser, p. 157 *in*: "Gas-Phase Chemiluminescence and Chemi-Ionization", A. Fontijn, ed., North Holland, Amsterdam (1985); E. Arunan, S. J. Wategaonkar and D. W. Setser, *J. Phys. Chem.*, 95:1539 (1991).
4. J. J. Sloan, *J. Phys. Chem.*, 92:18 (1988).
5. I. Oref, D. Schuetzle and B. S. Rabinovitch, *J. Chem. Phys.*, 54:575 (1971).
6. J. D. Rynbrandt and B. S. Rabinovitch, *J. Phys. Chem.*, 75:2164 (1971).
7. A. N. Ko, B. S. Rabinovitch and K. J. Chao, *J. Chem. Phys.*, 66:1374 (1977). A. N. Ko and B. S. Rabinovitch, *Chem. Phys.*, 30:361 (1978).
8. S. Chervinsky and I. Oref, *J. Phys. Chem.*, 79:1050 (1975); *ibid*, 81:1967 (1977).
9. R.A. Coveleskie, D. A. Dolson and C. S. Parmenter, *J. Phys. Chem.*, 89:645 (1985); *ibid*, 89:655 (1985); K. W. Holtzclaw and C. S. Parmenter, *J. Chem. Phys.*, 84:1099 (1986); D. A. Dolson, K. W. Holtzclaw, D. B. Moss and C. S. Parmenter, *J. Chem. Phys.*, 84:1119 (1986).
10. I. Oref and B. S. Rabinovitch, *Acc. Chem. Res.*, 12:166 (1979).
11. C. S. Parmenter, *J. Phys. Chem.,* 86:1735 (1982).
12. M. Quack, *Ann. Rev. Phys. Chem.,* 41:839 (1990).
13. C. E. Hamilton, J. L. Kinsey and R. W. Field, *Ann. Rev. Phys. Chem.,* 37:493 (1986).
14. P. M. Felker and A. H. Zewail, *Adv. Chem. Phys.,* 70:265 (1988).
15. J. P. Pique, Y. Chen, R. W. Field and J. L. Kinsey, *Phys. Rev. Lett.* 58:475 (1987).
16. K. Yamanouchi, S. Takuchi and S. Tsuchiya, *J. Chem. Phys.,* 92:4044 (1990).
17. X. Luo and T. R. Rizzo, *J. Chem. Phys.,* 94:889 (1991).
18. K. Yamanouchi, N. Ikeda, S. Tsuchiya, D. M. Jonas, J. K. Lundberg, G. W. Anderson and R. W. Field, *J. Chem. Phys.*, 95:6330 (1991).
19. J. P. Pique, M. Joyeux, J. Manners and G. Sitja, *J. Chem. Phys.,* 95:8744 (1991).
20. K. K. Lehmann, B. H. Pate and G. Scoles, p. 17 *in*: "Mode Selective Chemistry", J. Jortner, R. D. Levine and B. Pullman, eds., Kluwer, Dordrecht (1991).

21. a) E. J. Heller, *J. Chem. Phys.,* 72:1337 (1980); b) E. J. Heller, *Accs. Chem. Res.*, 14:368 (1981).

22. E. J. Heller, *Phys. Rev. A*, 35:1360 (1987).

23. P. Pechukas, *J. Phys. Chem.,* 86:2239 (1982).

24. M. Lombardi, P. Labastie, M. C. Bordas and M. Broyer, *J. Chem. Phys.,* 89:3479 (1988).

25. J. C. Lorquet and V. B. Pavlov-Verevkin, *J. Chem. Phys.,* 93:520 (1990).

26. J. C. Lorquet, Y. M. Engel and R. D. Levine, *Chem. Phys. Letts.*, 175:461 (1990).

27. F. Remacle and R. D. Levine, *Chem. Phys. Letts.,* 181:307 (1991).

28. R. D. Levine and J. L. Kinsey, *Proc. Natl. Acad. Sci USA*, 88:11133 (1991).

29. J. Wilkie and P. Brumer, *Phys. Rev. Letts.,* 67:1185 (1991).

30. F. Remacle and R. D. Levine, *J. Phys. Chem.*, 95:7124 (1991).

31. S. C. Farantos, J. M. Gomez Llorente, O. Hahn and H. S. Taylor, *Int'l J. Quant. Chem. Quant. Chem. Symp.,* 24:429 (1990).

32. S. H. Tersigni, P. Gaspard and S. A. Rice, *J. Chem. Phys.*, 92:1775 (1990).

33. H. S. Taylor, *Accs. Chem. Res.*, 22:263 (1989).

34. E. Pollack and C. Schlier, *Accs. Chem. Res.*, 22:223 (1989).

35. J. S. Hutchinson, *in*: "Lasers, Molecules ad Methods", J. O. Hirschfelder, R. E. Wyatt and R. D. Coalson, eds., Wiley, New York (1989).

36. C. C. Martens and G. S. Ezra, *J. Chem. Phys.*, 87:284 (1987); C. C. Martens, M. J. Davis and G. S. Ezra, *Chem. Phys. Lett.*, 142:519 (1987); L. E. Field and G. S. Ezra, *J. Chem. Phys.*, 90:6378 (1989).

37. L. R. Khundkar and A. H. Zewail, *Ann. Rev. Phys. Chem.,* 41:15 (1990).

38. A. H. Zewail and R. B. Bernstein, *C & W News*, 66:24 (1988); D. W. Willberg, M. Gutmann, J. J. Breen and A. H. Zewail, *J. Chem. Phys.*, 96:198 (1992).

39. U. Banin, A. Waldman and S. Ruhman, *J. Chem. Phys.*, 96:2416 (1992).

40. D. L. Phillips, A. B. Myers and J. J. Valentini, *J. Phys. Chem.*, 96:2039 (1992).

41. E. J. Heller, R. L. Sundberg and D. Tannor, *J. Phys. Chem.,* 86:1822 (1982).

42. S. C. O'Brien, C. Kittrell, J. L. Kinsey and B. R. Johnson, *J. Chem. Phys.,* 96:67 (1992).

43. L. Leviandier, M. Lombardi, R. Jost and J. P. Pique, *Phys. Rev. Lett.*, 56:2449 (1986).

44. N. B. Slater, "Theory of Unimolecular Reactions", Methuen, London (1959).

45. Th. Zimmermann, L. S. Cederbaum, H. D. Meyer and H. Köppel, *J. Phys. Chem.*, 91:4446 (1987).

46. R. D. Levine, *Adv. Chem. Phys.*, 70:53 (1987).

47. Y. Chen, R. W. Field and J. L. Kinsey, *J. Chem. Phys.,* 88:5972 (1988).

48. F. Remacle and R. D. Levine, to be published.

49. R. D. Levine, "Intramolecular Dynamics", Invited Lectures, *in* "Perspectives in Quantum Mechanics, pp.41-55, 6'th International Congress of Quantum Chemistry, J. Jortner and B. Pullman, eds., Kluwer, Dordrecht (1989).

50. K. R. Wilson and R. D. Levine, *Chem. Phys. Letts.*, 152:435 (1988).

51. T. A. Holme and R. D. Levine, *Surf. Sci.*, 216:587 (1989).

52. R. D. Levine and J. Jortner, *in*: Mode Selective Chemistry, J. Jortner, R. D. Levine and B. Pullman, eds., Kluwer, Dordrecht (1991).

53. T. Sewell, D. L. Thompson and R. D. Levine, to be published.

54. Y. S. Li, R. M. Whitnell, K. R. Wilson and R. D. Levine, to be published.

55. D. W. Lupo and M. Quack, *Chem. Rev.,* 87:181 (1987).

56. D. Weeks and R.D. Levine, to be published.

57. P. J. Robinson and K. A. Holbrook, "Unimolecular Reactions", Wiley, New York (1972).

58. M. Shapiro and M. S. Child, *J. Chem. Phys.*, 76:6176 (1982).

59. R. D. Levine, *Ber. Bunsenges. Phys. Chem.*, 92:222 (1988).

60. F. F. Crim, *Science*, 249:1387 (1990).

61. R. D. Levine, "Quantum Mechanics of Molecular Rate Processes", Clarendon Press, Oxford (1969).

62. G. G. Hall and R. D. Levine, *J. Chem. Phys.*, 44:1567 (1966).

DISCRETE VARIABLE REPRESENTATIONS IN QUANTUM DYNAMICS

John C. Light

Department of Chemistry and James Franck Institute
University of Chicago
Chicago, IL 60637

ABSTRACT

A derivation of discrete variable representations (DVR's) is given, with emphasis on formal properties and relations to the classical orthogonal polynomials. These "Gaussian quadrature" DVR's are compared with several different DVR's proposed recently. Finally, some cautions and brief comparisons of accuracy are noted for quantum solutions for large amplitude motions of triatomic molecules (H_2O in this case).

INTRODUCTION

As evidenced by the many recent publications in chemical physics, numerical techniques based on discrete coordinate or "pointwise" representations of the Schrodinger equation have been increasingly successful for the study of quantum dynamics of relatively small systems[1-3]. This has been partially due to the availability and economy of large scale computers, notably in the form of high performance work stations, but perhaps even more is due to the development of improved representations and algorithms for the solution of the Schrodinger equation for physically interesting systems. Because computers now allow representations of wave functions in terms of long vectors, or components of solutions, the numbers of elements of the relevant vector spaces may now be much larger, provided we can evaluate, manipulate, and operate on them with conceptual and numerical ease.

Over the past several years, discrete "pointwise" representations have been developed[4-7] which provide this "ease", and they have been very effective in a number of static and dynamical quantum problems. These include the vibration-rotation states of molecules[8-14], the electronic structure of molecules[7], inelastic and reactive scattering of atoms and molecules[15-17], atom/molecule collisions with surfaces[18], interactions of small molecular systems with

Time-Dependent Quantum Molecular Dynamics, Edited by J. Broeckhove and L. Lathouwers, Plenum Press, New York, 1992

radiation fields, photodissociation[19-21], etc. In this article we shall focus on the characteristics of these discrete representations, compare the formal bases of several recently proposed discrete representations, briefly discuss boundary conditions; and finally look at some simple examples of accuracy.

Until recently, "pointwise" representations in quantum mechanics, which usually refered to representations of wave packets on Cartesian coordinate grids, were most often used to underlie the represention of time evolution of quantum systems, i.e. wave packet dynamics[1]. Although grid representations have been used for many years in hydrodynamics and other areas[22], their use in quantum dynamics was sporadic until recently. I believe the primary reason for this was that quantum problems, particularly electronic structure calculations, required a level of accuracy which was not available by these methods, given the computers of the time, and the methods then in use. These grid methods were based primarily on finite difference approximations of differential operators or on quadratures relying on "finite element" methods. Although both of these approaches yield sparse representations of the Hamiltonian operator, they often require a dense set of points (or "nodes") for high accuracy. Thus the early work of McCullough and Wyatt[23], Askar et al[24], Frost et al[25], and Gilbert et al[26] could not be effectively extended to more than two dimensions or to very few solutions of the Schrodinger equation for higher dimensions.

Some 25 years ago, Harris, Engerholm, and Gwinn (HEG)[27] proposed a simple means of evaluating potential matrix elements by generating a quadrature appropriate to the basis, i.e. by diagonalizing the coordinate operator in the function basis. This was subsequently shown by Dickinson and Certain (DC)[28] to be equivalent to a Gaussian quadrature (for bases based on the classical orthogonal polynomials under consideration, such as harmonic oscillator functions). Although HEG and DC proposed the use of this transformation for evaluation of matrix elements in the basis representation, the explicit use of the mathematically equivalent "pointwise" and "function" representations as representations, has only recently become widespread with common use of discrete Fourier Transforms (DFT's or FFT's)[1,6,18] and Finite Basis Representation - Discrete Variable Representation pairs (FBR-DVR's)[4,9,12-17]. The FFT's were introduced for quantum dynamical problems by Kosloff et al [6], in conjunction with powerful algorithms for time evolution of wave packets such as Feit and Fleck's split exponential propagation[29], Tal Ezer and Kosloff's Chebychev propagation[30], and the iterative Lanczos propagation of Park et al[31]. The advantages of the grid and FFT approaches over normal basis representations are twofold. First, it permits the action of the Hamiltonian on a vector to be carried out <u>without</u> explicit prior evaluation of the potential matrix elements in a basis, i.e. by simply evaluating the potential at a set of equally spaced points. Second, the transformations to and from the conjugate momentum representation, which permit simple evaluation of the action of the kinetic energy operators on the wave packet, scale very efficiently, requiring O(N lnN) operations versus N^2 for direct full matrix vector multiplication. Recently, the advantages of the discrete FT approach

(DFT) were extended to the fast evaluation of the action of angle type differential operators on wave packets[32].

The discrete variable representations (DVR's), *qua representations*, were introduced some ten years ago by Lill et al [4] and independently by Blackmore and Shizgal (under the name discrete ordinate method)[5]. The DVR's, the main topic of this paper, are now recognized to be very flexible and efficient in the mathematical representation of quantum dynamical problems. Because DVR's can be generated which are appropriate for a variety of generic potential types, they may permit a representation for a variety of problems which is sufficiently simple, compact, and efficient to be the representation of choice. The most important advantage, however, is for multidimensional problems using orthogonal coordinate descriptions, in which the direct product DVR's are simple to evaluate and the Hamiltonian matrix is extremely sparse and simple to generate.

However, even though discrete variable representations are now used extensively and by many groups, the lure of "novel" DVR's or "analytic" DVR's, or even "Generalized DVR's" have lead to a number of papers in which formally novel approaches are proposed. However, after the "new" DVR's are formally introduced, the numerical examples given often use only DVR's fitting the original Gaussian quadrature definition of DVR's (or perhaps the DFT). One major exception to this is the Lobatto quadrature approach of Manolopolous and Wyatt[33], which, although not a DVR based on classical orthogonal polynomials, is DVR based on a well defined quadrature of high accuracy. A very recent proposal by Clary[34] may also be useful. Thus it seems appropriate to examine, relate, and compare these generic approaches.

In this brief paper, therefore, we will look first at the "standard" definition of DVR's in terms of the classical orthogonal polynomial function bases (FBR's) and their related Gaussian quadratures[4,28]. We will then examine properties of a DVR defined for a general quadrature rule as proposed by Muckerman et al[35,36]. These are in theory more general than the Lobatto quadrature DVR proposed by Manolopolous et al or the DFT-DVR proposed by Balint-Kurti et al[37]. The recent "novel" DVR, proposed by Colbert and Miller[38] as "divorced" from a function basis will also be noted. After these comments, we will move toward a more constructive note and look at two rather general problems, the generation of DVR's appropriate for functions of specific symmetry; and the important role of boundary conditions in choosing a FBR-DVR which will provide good convergence.

GAUSSIAN QUADRATURE DVR'S

Although the use of spectral or pseudo-spectral methods in numerical simulation has a relatively long history, often being associated with Orszag[22], they were applied primarily to hydrodynamic problems and used equally spaced Cartesian grids with Chebychev polynomials or DFT's. Because quantum dynamical systems are often bounded with smoothly varying potentials and are not periodic, we often wish to use basis sets better adapted to the problem than particle-in-a-box functions would be. We note, however, that $\{\sin(n\theta)\}$ are Chebyshev polynomials of the second kind (in $\cos(\theta)$), and thus may be used to generate one simple DVR corresponding to equally spaced points with DVR functions satisfying zero boundary conditions. The full set of classical orthogonal polynomials[39] may be used to generate a full set of DVR's of Gaussian quadrature accuracy, and of surprising flexibility. These include most of the standard basis sets of quantum mechanics (except for the Slater and Gaussian functions of electronic structure calculations) such as $\{\sin(n\theta)\}$, $\{\cos(n\theta)\}$, (associated) Legendre polynomials, Jacobi and Laguerre polynomials, harmonic oscillator functions (Hermite polynomials), Sturmian functions, etc. Even the solutions for the one dimensional Schrodinger equation for the Morse potential are of this form for a suitable (exponential) coordinate variable.

The classical orthogonal polynomials come in three basic types: Jacobi polynomials:

$$P_n^{(\alpha,\beta)}(x) \; ; \qquad x \in [-1,1]; \qquad w(x) = (1-x)^{\alpha}\,(1+x)^{\beta}$$

Laguerre polynomials:

$$L_n^{(\alpha)}(x) \; ; \qquad x \in [0,\infty) \; ; \qquad w(x) = x^{\alpha} \exp\,(-x)$$

Hermite polynomials:

$$H_n(x) \; ; \qquad x \in (-\infty,\, \infty); \qquad w(x) = \exp\,(-x^2)$$

All are related to solutions of a second order differential (eigenvalue) equation:

$$P(x)\, y_n'' + Q(x)\, y_n' + R(x)\, y_n = \lambda_n\, y_n$$

and the polynomials satisfy a three term recursion relation:

$$p_{n+1}(x) = (a_n + b_n x)\, p_n(x) - c_n\, p_{n-1}(x)$$

and are orthogonal:

$$\int w(x)\, p_n(x)\, p_m(x)\, dx = \gamma_n^2\, \delta_{nm}$$

Each is associated with a Gaussian quadrature defined by points $\{x_n\}_N$, and weights, $\{w_n\}_N$, such that the Nth order quadrature of a function, $f(x)$,

$$I_N = \int w(x)\, f(x)\, dx = \Sigma_k^N\, w_k\, f(x_k)$$

is exact provided $f(x)$ is a polynomial of order $2N - 1$ or less. This means, of course, that the polynomials of order n,m are themselves orthogonal over the quadrature:

$$\gamma_n^2\, \delta_{nm} = \Sigma_k^N\, w_k\, p_n(x_k)\, p_m(x_k) \qquad\qquad n,m < N$$

The orthogonality of the polynomials and the three term recurrence relation provide an analytic tri-diagonal matrix representation of the coordinate operator. The essential properties of the polynomials are, of course, unchanged by a linear transformation which shifts and scales the coordinate. In Table I we give examples of quadrature (DVR) points. Each basis consists of the lowest twelve polynomials times the appropriate weight function. The points are the zeros of the thirteenth polynomial, and indicate some of the variety of nodal structure available.

TABLE I. DVR Points for Orthogonal Polynomials (Ndvr = 12)

Chebyshev polynomials of the second kind, range x=arccos (theta) scaled to (8, 0) (note x's are equally spaced):

7.3846	6.7692	6.1538	5.5383	4.9231	4.3077
3.6923	3.0769	2.4615	1.8462	1.2308	0.6154

Legendre polynomials, range x=cos (theta) (-1,1):

-0.9816	-0.9041	-0.7699	-0.5873	-0.3678	-0.1252
0.1091	0.3678	0.5873	0.7699	0.9041	0.9816

Associated Legendre Polynomials (order 2), range (-1,1):

-0.9376	-0.8352	-0.6939	-0.5202	-0.3222	-0.1091
0.1091	0.3222	0.5202	0.6939	0.8352	0.9376

Jacobi polynomials (alpha=.5, beta =5), range (-1,1):

-0.8377	-0.6875	-0.5112	-0.3150	-0.1069	-0.1044
0.3101	0.5015	0.6705	0.8100	0.9140	0.9783

Gauss-Hermite points (harmonic oscillator), range fixed to (-2,2):

-2.0000	-1.5531	-1.1721	-0.8215	-0.4873	-0.1616
0.1616	0.4873	0.8215	1.1721	1.5531	2.0000

Laguerre polynomials, range (0,oo):

0.1157	0.6118	1.5126	2.8338	4.5992	6.8445
9.6213	13.0061	17.1169	22.1511	28.4880	37.0991

Associated Laguerre polynomials (order 2), range (0,oo):

0.4902	1.3238	2.5421	4.1645	6.2180	8.7408
11.7872	15.4366	19.8105	25.1111	31.7262	40.6489

Sturmian functions (Laguerre Polynomials), range scaled to (.5,8)

0.5000	0.6301	0.8368	1.1235	1.4952	1.9589
2.5252	3.2089	4.0336	5.0379	6.2964	8.0000

We now examine how DVR's are related to these sets of functions. The standard variational approach in quantum mechanics is based on a truncated matrix representation of the Hamiltonian in a Hilbert space. Given a set of N orthonormal functions on $x \varepsilon (a,b)$, $\{\varphi_n\}$, the variational equations for the eigenvalues, E_i, and the eigenvectors, $\mathbf{c}_i$, are

$$(H - EI)\mathbf{c} = (K + V - EI)\mathbf{c} = 0 \tag{1}$$

where the matrix elements are defined by <u>exact</u> integrals such as

$$K_{nm} = -\hbar^2/2m \int \varphi_n \nabla^2 \varphi_m \, dx \quad \text{and}$$
$$V_{nm} = \int \varphi_n V(x) \varphi_m \, dx \tag{2}$$

The DVR and other collocation schemes usually approximate the exact evaluation of these integrals and change the representation. In the "standard" DVR, using as a basis one of the classes of classical orthogonal polynomials above, the approximation is equivalent mathematically to evaluating the integral over the potential by an N point Gaussian quadrature. We examine this first.

Given an N function orthonormal basis (the FBR), $\{\varphi_n(x)\}_N$, on $x \varepsilon (a,b)$, made up one of the classical orthogonal polynomials, $p_n(x)$, times the square root of the weight function, $w^{1/2}(x)$, one can normally evaluate quite simply the exact matrix representation of the kinetic energy operator in x, K^{FBR}, and the tri-diagonal (exact) matrix representation of the polynomial coordinate, x, X^{FBR}. Following HEG[27], the coordinate matrix is diagonalized to obtain the DVR "points" as eigenvalues, the orthogonal FBR-DVR transformation matrix, T, and the orthonormal DVR basis functions, $\{\theta(x)\}_N$:

$$X^{DVR}(\text{diag}) = T^T X^{FBR} T \tag{3}$$

$$\theta_n(x) = \Sigma_l T_{ln} \varphi_l(x) \tag{4}$$

As shown by Dickinson and Certain[28], the eigenvalues of the coordinate operator, x_n, are the N Gaussian quadrature points for the Nth order quadrature appropriate to the functions, and the transformation matrices have elements

$$T_{ln} = w_n^{1/2} p_l(x_n) = [w_n/w(x_n)]^{1/2} \varphi_l(x_n) \tag{5}$$

For real basis functions, T is an orthogonal matrix, and thus we have the "discrete orthonormality" relations among the DVR functions on the quadrature:

$$\Sigma_m (w_m/w(x_m))\, \theta_n(x_m)\, \theta_l(x_m) = \Sigma_m T_{nm} T_{lm} = \delta_{nl} \tag{6}$$

This also leads to the "Kronecker delta" property of the DVR basis:

$$(w_l/w(x_l))^{1/2}\ \theta_n(x_l) = \Sigma_m T_{mn}(w_l/w(x_l))^{1/2}\varphi_m(x_l) = \Sigma_m\ T_{mn}\ T_{ml} = \delta_{nl} \quad (7)$$

This "Kronecker delta function" property of the DVR basis is quite interesting -- each DVR basis function is a constant at "it's own" DVR point, and is zero at all others.

The <u>approximation</u> made in the DVR, which accounts for much of its simplicity, is that of HEG; replace the exact potential matrix in the DVR by the <u>diagonal</u> matrix of the potential evaluated at the DVR points:

$$V^{exact,DVR}{}_{nm} = (T^T\ V^{FBR}\ T)_{nm} \dashrightarrow V^{DVR}{}_{nm} = V(x_n)\ \delta_{nm} \quad (8)$$

Using Eq. 7, this follows from the evaluation of the potential matrix in the DVR (or FBR) by the N point Gaussian quadrature, and is therefore exact to polynomial terms of order (2N - 1) in the product of the φ's and the potential:

$$V^{FBR} \approx T\ V^{DVR}\ T^T \quad (9)$$

Because the quadrature is <u>exact for the basis functions</u>, the orthonormality of the FBR basis functions is preserved in the DVR basis. Thus when the full Hamiltonian is written in the DVR,

$$H^{DVR} = T^T\ K^{FBR}\ T + V^{DVR} \quad (10)$$

the secular equation

$$(\ H^{DVR} - E\ I\ )c' = 0 \quad (11)$$

is exactly equivalent to that for H^{FBR} with the potential matrix evaluated by the N point Gaussian quadrature, and there is <u>no approximation</u> of the overlap matrix.

Finally, we note that the operations required to generate the one dimensional DVR's are simple both mathematically and computationally. The classical orthogonal polynomials have three term recurrance relations which permit analytic evaluation of the coordinate matrix, and the kinetic energy matrix is usually also simple to evaluate in the FBR using the second order differential equations for the polynomials. The physical coordinates used usually must be scaled and shifted to the appropriate range. Examples of the utility of Sturmian (Laguerre) functions and Jacobi polynomials[9] have been published, as well as the more usual Lagrange and Chebyshev polynomial bases.

However, the major advantage of DVR's is the simplicity afforded for multidimensional systems[3]. For multidimensional problems using orthogonal coordinate systems, the Schrodinger equation has no mixed derivatives and the kinetic energy operators are therefore separable.

$$H = \sum_i K_i + V \qquad (12)$$

Thus a direct product DVR requires only the <u>separate</u> evaluation of the DVR and the kinetic energy matrices for each dimension. Since the potential is <u>diagonal</u> in the DVR, and the kinetic energy terms will not "mix" dimensions for orthogonal coordinates, the Hamiltonian matrices in the DVR are very sparse for higher dimensions. Since the "standard" DVR's maintain orthonormality of the basis, the overlap matrices are unit matrices. Thus large ($\approx 10^4$ to 10^5) basis sets can be used and handled easily. This approach has been used by our group and a number of others[8-17].

COLLOCATION

The above equations also represent a collocation method using specific functions and points for which one obtains maximum accuracy. The more general collocation procedure is to require that the Schrodinger equation be solved exactly at a set of N points, $\{x\}_N$, given a set of N basis functions, $\{\chi\}$[40]. Thus the normal collocation equations are a "mixed representation":

$$\sum_m \langle x_n | H - E | \chi_m \rangle c_m = 0 \qquad (13)$$

For H = K + V, and operating explicitly with the kinetic energy operator, we have the matrix form

$$(\chi'' + V^{diag}\chi - E\chi) c = 0 \qquad (14)$$

where

$$\chi_{mn} = \chi_n(x_m) \text{ and } \chi''_{mn} = (K\,\chi_n)\big|_{x=x_m} \qquad (15)$$

This is equivalent to the standard DVR if three conditions are fullfilled:

a) χ is an orthogonal matrix ;

b) $\chi'' = K'\,\chi$, i.e. the action of the kinetic energy operator on any basis function produces a function entirely representable in terms of the basis; and,

c) the basis is orthonormal.

Under these conditions, multiplying Eq. (14) on the left by χ^T yields the quadrature approximation to the basis representation,

$$(\chi^T K'\,\chi + \chi^T V^{diag}\,\chi - EI\,)\,c = 0. \qquad (16)$$

The collocation equations are usually <u>not</u> of this form, since one wants the freedom to choose the points rather arbitrarily, without regard to (a). When this is the case, the inverse of the matrix χ must exist in order that N solutions of Eq (14) exist. Even in this case, however, there is no guarantee that the N eigenvalues (from Eq. (14)) will be real. A measure of the stability or accuracy of the procedure may be obtained from the eigenvalues of the "quadrature overlap matrix", $\mathbf{S}' \equiv \chi^T \chi$. Small eigenvalues, λ, will cause amplification of errors as $O(1/\lambda)$[41]. For the Gaussian quadrature (standard) DVR's, the eigenvalues are all unity, and no amplification occurs.

The advantages of general collocation approaches are twofold. First, one can choose a more or less arbitrary set of points for the collocation, and very easily construct Eq (14). The accuracy may be determined by convergence as N increases. Second, and by far the most important advantage is that the functions, $\{\chi\}$, and points, $\{x\}$, can both refer to a d-dimensional space. Thus there is no requirement that the basis be of the direct product variety. This is very useful when the obvious basis is not of the direct product form (e.g. spherical harmonics). A major disadvantage when compared with the direct product DVR, however, is that the representation in Eq (14) is full, not sparse as in the DVR. This tends to limit the collocation to two dimensions.

ALTERNATIVE DVR'S

At this point I would like to analyse briefly some "alternative" DVR's, namely those proposed by Muckerman[35], Manolopolous and Wyatt[33], Kanfer and Shapiro[42] (and recently Clary[34]), and by Miller[38].

Muckerman[35] provides a succinct approach to the formation of a DVR, apparently for a general quadrature rule consisting of a set of points $\{x_l\}_N$ and weights $\{w_l\}_N$, and a set of orthonormal functions, $\{\varphi_l\}_N$. The goal is to transform to a set of DVR basis functions, $\{\theta_l\}_N$, satisfying the Kronecker delta condition of Eq (7). We show that this is possible <u>only</u> if the products of all functions in $\{\varphi\}_N$ are integrated exactly by the N point quadrature over the positive weight function, $w(x)$.

The construction proceeds by

a) formally expanding the $\{\theta_l\}$'s in terms of the $\{\varphi_l\}$'s:

$$\theta_l(x) = \sum_n \varphi_n(x) \langle\varphi_n | \theta_l\rangle \tag{17}$$

b) using the quadrature to evaluate the overlap;

$$\langle\varphi_n | \theta_l\rangle = \sum_k (w_k/w(x_k))\, \varphi_n(x_k)\, \theta_l(x_k) \quad \text{and,} \tag{18}$$

c) assuming that the desired "Kronecker delta" relation (Eq. (7)) holds for the θ's, which yields the DVR basis from Eqns (18) and (17),

$$\langle\varphi_n \mid \theta_l\rangle = (w_l/w(x_l))^{1/2}\ \varphi_n(x_l) \qquad \text{or,} \qquad (19)$$

$$\theta_l(x) = \Sigma_n\ \varphi_n(x)\ (w_l/w(x_l))^{1/2}\ \varphi_n(x_l) \qquad (20)$$

We see that Eq. (20) is identical to Eq. (5) with

$$T_{ln} \equiv (w_n/w(x_n))^{1/2}\ \varphi_l(x_n) \qquad (21)$$

The desired "Kronecker delta" property, Eq. (7), is satisfied for this set of DVR functions <u>only</u> if the transformation in Eq.(21) is orthonormal. This is exactly true only if the original functions, {φ}, are orthogonal and are integrated <u>exactly</u> by the quadrature, i.e.

$$\delta_{nm} = \Sigma_k(w_k/w(x_k))\ \varphi_n(x_k)\ \varphi_m(x_k) \qquad \text{or} \qquad (22)$$

$$I = T^T\ T \qquad (23)$$

In this case, of course, the DVR-FBR relationship is similar to that for the standard DVR's. The discrete Fourier transform (DFT) appears to be of this type since the transformation of Eq (21) is orthogonal (actually unitary), but is not generated by diagonalization of a coordinate as in Eq (3). There do not appear to be examples of this DVR to date which are not either standard DVR's or DFT's[37].

Manolopolous and Wyatt[33], on the other hand, started with a Lobatto quadrature and Lobatto shape functions, which are the appropriate Lagrangian interpolating polynomials for this quadrature. Thus they defined the (N + 2) DVR basis functions directly:

$$\theta_i(x) \equiv \Pi_{j=0}^{N+1}\ (x-x_j)/\ (x_i-x_j) \qquad (j{=}i \text{ omitted; } i,j{=}0,N{+}1) \qquad (24)$$

These functions automatically satisfy relations similar the "Kronecker delta" relation of Eq.(7) and the discrete orthonormality condition of Eq (6). The Lobatto quadrature integrates polynomials exactly to order (2N+1) for the N+2 point quadrature. However, the overlap matrix, evaluated to be a unit matrix by the Lobatto quadrature is not exact (the contribution of the highest order polynomial term to each element is not correct). In practice, however, Manolopolous et al found the use of a unit matrix for the overlap matrix yielded accurate results. The advantage of this DVR is that the end points of the interval, x=a,b, are included explicitly and thus one can impose appropiate boundary conditions at a and b separately. It is interesting to note that in this case the DVR basis, the Lobatto shape functions, are the <u>primary</u> basis, and they determine the non-uniform nodal pattern.

Finally, Kanfer and Shapiro[42] and recently Clary[34], have proposed a DVR which is not a collocation, but which is obtained directly by diagonalization of a coordinate. Since the basis functions are not polynomials in this coordinate, this does not generate a Gaussian quadrature DVR; the transformation matrices are orthogonal but are not related to the basis by Eq (5); and the DVR basis does not satisfy the Kronecker delta relation of Eq (7). Thus the accuracy of the DVR quadrature approximation, replacing the exact transformed potential matrix by the potential at the DVR points as in Eq (8), is not known *a priori* . The advantage of this procedure is that one may be able to reduce the number of DVR points required since, presumably, the basis can be chosen to "fit" the potential.

Colbert and Miller[38] recently proposed a "novel" "basis free" DVR with an infinite number of equally spaced points, from which one abstracts the finite subset desired. The kinetic energy operator on this set can be defined either as the $N \to \infty$ limit of an infinite order finite difference approximation, or as the $N \to \infty$ (N/L finite) limit of the particle in a box functions (Chebyshev polynomials). For infinite ranges, $(-\infty,\infty)$ or $(0,\infty)$ this is equivalent to a normal Chebyshev DVR. However, on a finite range, $(0,\pi)$ for example, the procedure suffers an error of $O(1/N)$ due to a normalization problem. Although this is an interesting way to "derive" a DVR, it does not differ on an infinite range, and on the finite range it is preferable to use a standard Chebyshev DVR for equally spaced points.

Finally I would like to point out an equivalence which is well known to numerical analysts between real symmetric tri-diagonal matrices, polynomials, and quadrature schemes. This is pointed out in books by Parlett[43] and others[44] Given a symmetric tri-diagonal matrix, T, the eigenvalues are unchanged if all the off-diagonal elements, β_{i+1}, are made positive, defining T_+ (we assume, for simplicity, that all β's are non-zero). There then exists a three term recursion relation which defines a set of orthogonal polynomials in terms of the elements of T_+, for which T_+ is the coordinate matrix, i.e. $T_+ = X^{FBR}$. Thus diagonalization of T_+ yields the related quadrature points as eigenvalues, and the weights as the first elements of the eigenvectors. In theory, at least, the process can be reversed. That is, starting with DVR points and positive weights, $\{x_i\}_N$ and $\{w_i\}_N$, there exists a unique set of orthogonal polynomials for which this quadrature is exact, i.e. for which Eqs (6) and (7) hold. It is not clear, however, that this provides a practical route to an "infinitely flexible" DVR, and this is probably worth looking into further. However, the large set of known classical orthogonal polynomials which have well defined "standard" DVR's provides at this time a rather wide choice of distributions of points, weights, and boundary conditions.

ACCURACY AND CONVERGENCE

I will close with a few remarks (cautions) on the accuracy and convergence of DVR's. First, as is well known, since a quadrature approximation is used to define a DVR (Eq. (8) and (9)), a DVR is not variational in the same sense as Eq (1) and (2) - the eigenvalues of Eq. (10) may be below the true values. The typical behaviour is that as the DVR basis is increased, some eigenvalues may increase slightly, then approach the true values from above as the quadrature error becomes negligible compared to the basis set truncation error. While this is not often a serious problem, it is worth keeping in mind since a balance of these errors may simulate convergence.

Second, it is important that the basis and therefore the related DVR be constrained to satisfy the appropriate boundary conditions. If this is not done, then the quadrature errors may be severe, and convergence may be so slow that one believes convergence has been attained when, in fact, it has not been. This has been a problem with angular DVR's where, for example, effective potential terms like $(\sin\theta)^{-2}$ have singular matrix elements in a Legendre polynomial basis, but appear to be finite in the related DVR. Since these terms are usually repulsive, forcing the wave function to zero at $\theta = 0$ and π, a DVR in a Legendre polynomial basis will yield the correct result, but only very slowly. A major controversy[13,45-47] over the vibrational states of H_3^+ has recently been traced to a lack of convergence due to a basis (and resulting DVR) which did not satisfy the appropriate boundary conditions.

The last brief example concerns truncation of a DVR basis[48]. It is tempting (and has been recommended) to exclude DVR basis functions when the associated DVR points are in regions of very high potential energy since, presumably, the desired wave functions will be zero in these regions of space. However, the DVR basis is not entirely localized, and some penetration of the true wave functions into regions of high potential energy should occur. Truncation of the basis prevents this and often raises the calculated eigenvalues above their true values. This problem can be easily avoided by permitting some amplitude at all DVR points. In the sequential diagonalization/truncation procedure we use for multi-dimensional problems, this means, for example, keeping some 2-dimensional basis functions at every DVR point of the 3rd dimension. An example of the difference this makes is given below in Table I. In our recent calculations of the J=0 vibrational states of H_2O, we compare the number of levels converged to a given accuracy when a strict energy cut-off is used in the 2-D basis at each angle (the "last" dimension), versus the accuracy obtained when the same energy cut-off was used but with a minimum of eight 2-D basis functions kept at each DVR angle. We note that the cut-off was much higher in energy (42000 cm^{-1}) than the eigenvalues in question (22000 cm^{-1}).

It is interesting to note the differences between the convergence of the symmetric and asymmetric vibrational manifolds. It is due to the fact that for

TABLE II. Energy and number of highest vibrational level of H_2O converged

Vibrational Manifold	N2-D_{min} (at each angle)	Energy (number) of level converged to: 0.05 cm^{-1}	0.20 cm^{-1}
Symm	0	19800 (80)	21300 (96)
	8	22500 (111)	25000 (148)
Asym	0	22300 (73)	24600 (98)
	8	22500 (75)	25000 (104)

small HOH angles, the asymmetric stretch vibrational states automatically keep the H-atoms away from each other, but the symmetric states require a large correlation of bend and stretch states to do this.

CONCLUSIONS

The use of DVR's has proven very useful for large amplitude motion of multi-dimensional systems. The standard DVR's, used to generate a multi-dimensional direct product basis, provide a simple and sparse representation for the Hamiltonian which permits the use of very large bases to obtain very accurate results for many states. For the standard DVR's the implicit quadratures are of Gaussian quadrature accuracy, and convergence is rapid if the basis is chosen to satisfy the boundary conditions. Alternative DVR's have been discussed. The Lobatto quadrature DVR of Manolopolous and Wyatt is useful where the boundary conditions at the end points must be included explicitly (although the Jacobi polynomials also provide this flexibility). The general DVR proposed by Muckerman appears to be accurate only if the quadrature is exact for the basis functions. An example is the discrete Fourier transform.

It is expected that because of their simplicity, DVR's and associated solution algorithms will be developed for more complex problems in the future.

ACKNOWLEDGEMENT

This research was supported in part by grants NSF-CHE-8806514 and DE-FG02-87ER13679.

REFERENCES

1. R. Kosloff, J. Phys. Chem. 92, 2087 (1988) (Feature Article).
2. K.C. Kulander, Ed., "Time Dependent Methods for Quantum Dynamics", Comput. Phys. Commun. 63 (1991).

3. J.C. Light, R.M. Whitnell, T.J. Park, and S.E. Choi, in "Supercomputer Algorithms for Reactivity, Dynamics and Kinetics of Small Molecules, A. Lagana Ed., NATO ASI Series C. Vol 277 (Kluwer, Dortdrecht, 1989) pp. 187-214.
4. a) J.V. Lill, G.A. Parker, and J.C. Light, Chem. Phys. Lett. 89, 483 (1982); b) J.C. Light, I.P. Hamilton and J.V. Lill, J. Chem. Phys. 82, 1400 (1985).
5. B. Shizgal and R. Blackmore, J. Comput. Phys. 55, 313 (1984); R. Blackmore and B. Shizgal, Phys. Rev. A 31, 1855 (1985).
6. D. Kosloff and R. Kosloff, J. Comput. Phys. 52, 35 (1983); R. Kosloff and D. Kosloff, J. Chem. Phys. 79, 1823 (1983).
7. R.A. Friesner, Chem. Phys. Lett. 116, 39 (1985); J. Chem. Phys. 85, 1462 (1986); ibid, 86, 3522 (1987).
8. Z. Bacic and J.C. Light, J. Chem. Phys. 85, 4594 (1986); Z. Bacic and J.C. Light J. Chem. Phys. 86, 3065 (1987).
9. R.M. Whitnell and J.C. Light, J. Chem. Phys. 89, 3674 (1988); ibid, 90, 1774 (1989).
10. J.C. Light and Z. Bacic, J. Chem. Phys. 87, 4008 (1987).
11. Z. Bacic, R.M. Whitnell, D. Brown and J.C. Light, Computer Phys. Commun. 51, 35 (1988).
12. S.E. Choi and J.C. Light, J. Chem. Phys. 90, 2593 (1989); ibid, 92, 2129 (1990); J.C. Light and S.E. Choi, in "Mode Selective Chemistry", J. Jortner and B. Pullman, Eds. (Kluwer, Dortdrecht, 1991) pp 67-86.
13. J. Tennyson and J.R. Henderson, J. Chem. Phys. 91, 3815 (1989); Chem. Phys. Lett. 173, 133 (1990).
14. S. Carter and W. Meyer, J. Chem. Phys. 96, 2424 (1992); J.R. Henderson, J. Tennyson and B.T. Sutchiffe, J. Chem. Phys. 96, 2426 (1992).
15. R.M. Whitnell and J.C. Light, J. Chem. Phys. 86, 2007 (1987).
16. T.J. Park and J.C. Light, J. Chem. Phys. 91, 974 (1989); ibid, 94, 2946 (1991).
17. T. Seideman and W.H. Miller, J. Chem. Phys. 96, 4412 (1992).
18. R. Kosloff and C. Cerjan, J. Chem. Phys. 81, 3722 (1984).
19. R.W. Heather and J.C. Light, J. Chem. Phys. 79, 147 (1983).
20. J. Krause, A. Orel, K.C. Kulander and J.C. Light, J. Chem. Phys. 96, 4283 (1992).
21. U. Manthe and H. Koppel, Chem. Phys. Lett. 178, 36 (1991).
22. S.A. Orszag, Phys. Rev. Lett. 26, 1100 (1971); D. Gottlieb and S.A. Orszag, Numerical Analysis of Spectral Methods (Arrowsmith) SIAM, Philadelphia, 1977.
23. E.A. McCullough and R.E. Wyatt, J. Chem. Phys. 51, 1253 (1969); ibid 54, 3578 (1971).
24. A. Askar, A. Cakmak and H. Rabitz, Chem. Phys. 33, 267 (1978); ibid, 72, 5287 (1980).
25.- A.A. Frost, R.E. Kellogg, B.M. Gumarc and J.D. Scargle, J. Chem. Phys. 35, 827 (1961).
26. T.L. Gilbert and P.J. Bertoncini, J. Chem. Phys. 61, 3026 (1974); A. Altenberger-Siczek and T.L. Gilbert, J. Chem. Phys. 64, 432 (1976).
27. D.O. Harris, G.G. Engerholm, and W.D. Gwinn, J. Chem. Phys. 43, 151 (1965).

28. A.S. Dickinson and P.R. Certain, J. Chem. Phys. 49, 4209 (1968).
29. M.D. Feit, J.A. Fleck, Jr. and A. Steiger, J. Comput. Phys. 47, 412 (1982); M.D. Feit and J.A. Fleck, Jr., J. Chem. Phys. 78, 301 (1983); ibid, 80, 2578 (1984).
30. H. Tal-Ezer and R. Kosloff, J. Chem. Phys. 81, 3967 (1984).
31. T.J. Park and J.C. Light, J. Chem. Phys. 85, 5870 (1986).
32. C.E. Dateo, V. Engel, R. Almeida and H. Metiu, Computer Phys. Commun. 63, 435 (1991).
33. D.E. Manolopolous and R.E. Wyatt, Chem. Phys. Lett. 152, 23 (1988).
34. J. Echave and D.C. Clary, Chem. Phys. Lett. in press.
35. J.T. Muckerman, Chem. Phys. Lett. 173, 200 (1990); F.J. Lin and J.T. Muckerman, Computer Phys. Commun. 63, 538 (1991).
36. R.V. Weaver, J.T. Muckerman and T. Uzer, this volume.
37. C.C. Marston and G.G. Balint-Kurti, J. Chem. Phys. 91, 357 (1989).
38. D.T. Colbert and W.H. Miller, J. Chem. Phys. 96, 1982 (1992).
39. Good references are: M. Abramowrtz and I.E. Stegum, "Handbook of Mathematical Functions", Nat. Bur. Std., Applied Mathematics Series 55 (U.S. GPO, Washington, D.C. 1964); P. Dennery and A. Krzywicki, "Mathematics for Physicists" (Harper and Row, New York, NY 1967).
40. A.C. Peet and W. Yang, J. Chem. Phys. 90, 1746 (1989); ibid 92, 522 (1990); Chem. Phys. Lett. 153, 98 (1988).
41. R. Friesner, Personal Commun.
42. a) S. Kanfer and M. Shapiro, J. Phys. Chem. 88, 3964 (1980); b) See also J . M. Hutson, J. Comput. Phys. 56, 165 (1984).
43. B.N. Parlett, "The Symmetric Eigenvalue Problem" (Prentice-Hall, Engelwood Cliffs, N.J. 1980).
44. A. Ghizzetti and A. Orsicini, "Quadrature Formulae" (Academic, NY 1970).
45. S. Carter and W. Meyer, J. Chem. Phys. 93, 8902 (1990).
46. Z. Bacic and J.Z.H. Zhang, J. Chem. Phys. 96, 3707 (1992); Chem. Phys. Lett. 184, 513 (1991).
47. P.N. Day and D.G. Truhlar, J. Chem. Phys. 95, 6615 (1991).
48. S.E. Choi and J.C. Light, Private Commun.

COMMENT ON TIME-DEPENDENT FORMULATIONS OF QUANTUM MECHANICS

Jan Linderberg

Department of Chemistry
Aarhus University
DK-8000 Aarhus C, Denmark

ABSTRACT

An attempt is made here to formulate a procedure for the calculation of the evolution operator in quantum mechanics from a variational functional defined as a finite time integral of the appropriate Lagrangian density. It is demonstrated that the errors in the computed operator matrix elements are of the second order in the trial wave function error.

INTRODUCTION

Lecturing on the time-dependent variational principles in quantum mechanics should involve substantial background on the historical evolution of the topic and a critical evaluation of the merits of alternative realizations. This will not be so here. Numerous review papers are available and there is little to add at this time. Emphasis will instead be placed on the inclusion of notions from the finite element method in the well established action integral formulation.

A single state example is used in the next section to illustrate the workings of the finite element method. The case of a single parameter path in a Hilbert space is developed in the third part and an evolution operator integrator is

Time-Dependent Quantum Molecular Dynamics, Edited by J. Broeckhove and L. Lathouwers, Plenum Press, New York, 1992

suggested in the fourth. Comments on the more general forms are offered in the concluding section.

A SIMPLE CASE

Variational formulations have a long and important history in all branches of theoretical physics, both as a formal, comprehensive tool in expounding the theory and as an effective means of doing detailed calculations. The particular case of a pure state time evolution in quantum mechanics is chosen here to introduce a few notions related to the use and interpretation of variational results.

A single pure state is characterized by an amplitude, $\Psi(t)$, which is a function of the time t. The Lagrangian density

$$L(t) = (i/2)\left[\Psi^*(t)\frac{d\Psi(t)}{dt} - \frac{d\Psi^*(t)}{dt}\Psi(t)\right] - E|\Psi(t)|^2 , \qquad (1)$$

enters an action integral in conjunction with terms that are designed to give correct boundary value conditions. Thus we write

$$S = \int_0^T dt L(t) + (i/2)\left[\Psi^*(T)\lambda_T - \lambda_T^*\Psi(T) - \Psi^*(0)\lambda_0 + \lambda_0^*\Psi(0)\right]$$
$$- (i/2)\left[\lambda_T^* U^* \lambda_0 + \lambda_0^* U \lambda_T\right] \qquad (2)$$

and obtain from the condition that the functional is stationary that

$$i\frac{d\Psi(t)}{dt} - E\Psi(t) = 0; \qquad (3)$$
$$\Psi(0) = \lambda_0; \qquad \Psi(T) = \lambda_T; \qquad (4)$$
$$\Psi(0) = U^*\lambda_T; \quad \Psi(T) = U\lambda_0. \qquad (5)$$

These are, together with their complex conjugate counterparts, the Euler equations for this problem. The desired solution,

$$U = \exp(-iET), \qquad (6)$$

is readily found.

The single state case does not call for approximate forms, but in order to exemplify the workings of the variational form, it is illustrative to work out the result of a linear *ansatz* for the amplitude. We find, for

$$\Psi(t) = \Psi(0)(1 - {}^{t}/_{T}) + \Psi(T)({}^{t}/_{T}), \tag{7}$$

that the functional becomes a sesquilinear form,

$$S = ({}^{i}/_{2}) \begin{vmatrix} \Psi^*(0) & \Psi^*(T) & \lambda_0^* & \lambda_T^* \end{vmatrix} \begin{vmatrix} 2iET/_3 & iET/_3+1 & -1 & 0 \\ iET/_3-1 & 2iET/_3 & 0 & 1 \\ 1 & 0 & 0 & -U^* \\ 0 & -1 & U & 0 \end{vmatrix} \begin{vmatrix} \Psi(0) \\ \Psi(T) \\ \lambda_0 \\ \lambda_T \end{vmatrix} \tag{8}$$

Proper solutions are found when the four by four matrix is singular, that is when

$$0 = U + U^* + ({}^{iET}/_3)(U - U^*) + UU^*({}^{E^2T^2}/_3 - 1) - 1. \tag{9}$$

A unitary solution obtains when

$$U = \frac{1-ir}{1+ir} \; ; \; r = \frac{ET}{\sqrt{16-E^2T^2} - 2} \tag{10}$$

and the positive square root is chosen. The linear form is clearly inadequate for long time intervals but gives a Taylor series for the evolution matrix element which is correct through second order. Wave function amplitudes have a ratio which is inaccurate already at first order in the time parameter,

$$\Psi(T) = \Psi(0)[1 - {}^{3iET}/_4 + \ldots] \tag{11}$$

and so do the boundary values,

$$\lambda_T = \lambda_0 \; [1 - {}^{5iET}/_4 + \ldots]. \tag{12}$$

It is clear that a better approximation to the trial function will result in a phase error which is proportional to the value of the first term, the action integral, in the functional.

The linear approximation gives an alternative solution when the square root is chosen negative. It was discarded on account of not giving the smallest value for the action integral. A general approximation, $\Phi(t)$, to the wave function leads to a form

$$S = (i/2) \begin{vmatrix} 1 & \lambda_0^* & \lambda_T^* \end{vmatrix} \begin{vmatrix} 2iA & -\Phi^*(0) & \Phi^*(T) \\ \Phi(0) & 0 & -U^* \\ \Phi(T) & U & 0 \end{vmatrix} \begin{vmatrix} 1 \\ \lambda_0 \\ \lambda_T \end{vmatrix} \tag{13}$$

which gives the result that the phase of the evolution operator is

$$arg(U) = arg[\Phi^*(0)\Phi(T)] + arcsin\left[\frac{A}{|\Phi^*(0)\Phi(T)|}\right]$$

which confirms the assertion that the phase error is of second order in the wave function error.

GENERAL TIME-INDEPENDENT HAMILTONIANS

Treatment of the general case of time-independent Hamiltonian systems in a finite dimensional Hilbert space can be directly transcribed from the single state case by means of the introduction of appropriate scalar products. The functional becomes

$$S = \int_0^T dt\ [(i/2)\langle\Psi(t)|\frac{\partial\Psi(t)}{\partial t}\rangle - (i/2)\langle\frac{\partial\Psi(t)}{\partial t}|\Psi(t)\rangle - \langle\Psi(t)|H|\Psi(t)\rangle]$$
$$+ (i/2)[\langle\Psi(T)|\lambda_T\rangle - \langle\lambda_T|\Psi(T)\rangle - \langle\Psi(0)|\lambda_0\rangle + \langle\lambda_0|\Psi(0)\rangle]$$
$$+ (i/2)[\langle\lambda_T|U|\lambda_0\rangle - \langle\lambda_0|U^\dagger|\lambda_T\rangle], \tag{14}$$

and the wave function as well as the boundary values are now elements of a Hilbert space where the Hamiltonian H is Hermitian and the evolution operator U is unitary.

There is no difficulty in realizing that the case where there exists a readily available diagonal representation of the Hamiltonian reduces to a set of simple, single state problems. It is more interesting to consider spaces generated by non-linear parameters. The case of a single, complex parameter space is exemplified by a state vector which depends on time through a variable:

$$\Phi = \Phi[z(t)], \tag{15}$$

and it is assumed that the derivatives exist

$$\frac{\partial\Phi}{\partial z} \equiv \Phi_z[z(t)]; \qquad \frac{\partial^2\Phi}{\partial z^2} \equiv \Phi_{zz}[z(t)]. \tag{16}$$

The action integral is then expressed as

$$A = \int_0^T dt\ [(i/2)\langle\Phi|\Phi_z\rangle_t(\frac{dz}{dt})-(i/2)(\frac{dz^*}{dt})\langle\Phi_z|\Phi\rangle_t-\langle\Phi|H|\Phi\rangle_t] \tag{17}$$

and by optimizing this expression one may determine the path from an initial $z(0)$ and obtain the final $z(T)$. Thus one may map out the evolution operator elements such as was indicated for the single state problem. It should be realized that a slight generalization of the state space to include a phase factor $exp(-i\varepsilon t)$ permits us to assume that the optimized value of the action is zero.

EVOLUTION OPERATOR INTEGRATOR

Small time intervals are generally necessary in order to integrate the equations of motion in quantum mechanics of composite systems and the procedures are universally developed from the differential equations. The variational formulation described above can be used to derive an integrator based on the finite element method. The simplest form of a path in the parameter space is the straight line and with no loss of generality it is assumed that the initial point is given by $z = 0$ and that the interval is the unit of time so that $T = 1$. The variational parameter is $z(T) = z(1) = Z$. Matrix elements in the action integral may be complicated functions of time and their integrals are not always conveniently available so we develop the expression for the action integral based on the end point formula, for instance

$$\int_0^T dt\ \langle\Phi|\Phi_z\rangle_t = (i/2)[\langle\Phi|\Phi_z\rangle_0 + \langle\Phi|\Phi_z\rangle_1]. \tag{18}$$

The action integral depends on the parameter values at the initial and final point and the stationary value of the functional requires that small displacements at either point should leave the action integral invariant. Thus we investigate the effect of the substitutions

$$\Phi \rightarrow \Phi + \delta z \Phi_z \ ; \ \Phi_z \rightarrow \Phi_z + \delta z \Phi_{zz} \tag{19}$$

at both points. A first order change in the action integral is induced and the Euler equations are

$$0 = \langle\Phi_z|\Phi\rangle_0 + \langle\Phi_z|\Phi\rangle_1 + \langle\Phi_z|\Phi_z\rangle_0 Z - \langle\Phi_{zz}|\Phi\rangle_0 Z^* + 2i\langle\Phi_z|H|\Phi\rangle_0 \tag{20}$$

$$0 = \langle\Phi_z|\Phi\rangle_0 + \langle\Phi_z|\Phi\rangle_1 - \langle\Phi_z|\Phi_z\rangle_1 Z + \langle\Phi_{zz}|\Phi\rangle_0 Z^* - 2i\langle\Phi_z|H|\Phi\rangle_0 \tag{21}$$

Two additional equations are obtained as the complex conjugates of these and by some rearrangement the conditions can be formulated as a matrix equation:

$$0 = \begin{vmatrix} \langle\Phi_z|\Phi_z\rangle_0 & \langle\Phi_{zz}|\Phi\rangle_0 \\ \langle\Phi|\Phi_{zz}\rangle_1 & \langle\Phi_z|\Phi_z\rangle_1 \end{vmatrix} \begin{vmatrix} Z \\ -Z^* \end{vmatrix} + \begin{vmatrix} \langle\Phi_z|\Phi\rangle_0 + \langle\Phi_z|\Phi\rangle_1 + 2i\langle\Phi_z|H|\Phi\rangle_0 \\ \langle\Phi|\Phi_z\rangle_0 + \langle\Phi|\Phi_z\rangle_1 - 2i\langle\Phi|H|\Phi_z\rangle_1 \end{vmatrix} \tag{22}$$

There is an implicit dependence on the variable in the matrix elements with subscript one and the equation system cannot be solved in terms of formal expressions. Iterative solutions are then a possibility and they are conveniently defined from the optimization of the action integral.

Generalization of the preceding analysis to spaces generated through multidimensional parameter domains are immediate and lead to equations such as those above where the matrix and vector elements are to be interpreted as matrices and vectors themselves.

OUTLOOK

There are two new elements presented in this exposition. Admission of the evolution operator into the variational functional allows the direct determination of results which are correct through second order in the wave function error.

Application of the finite element method to the optimization procedure for the functional offers a means of including final state information, based on the variational principle, in the integrator. The method is not yet implemented and no judgment will be given as to its potential effectiveness to generate accurate time integrations at a manageable cost.

ACKNOWLEDGMENTS

Support for part of this work has been given by the Carlsberg Foundation and Julie von Müllens Fund. Seminal inspiration has been derived from my collaboration with Yngve Öhrn. It is pleasure to acknowledge the economic support from NATO, through the organizers, which enabled my participation in the workshop.

BIBLIOGRAPHY

Early applications of the variational method are due to Dirac and Frenkel. The field was admirably reviewed by P. W. Langhoff, S. T. Epstein, and M. Karplus in *Rev. Mod. Phys.* **44**, 602 (1972) and later developments are covered by P. Kramer and M. Saraceno in "*Geometry of the Time-Dependent Variational Principle in Quantum Mechanics*" (Springer Verlag, Berlin 1981).A recent review is provided by R. Kosloff in *J. Phys. Chem.* **92**, 2087 (1988). Much of the inspiration for the present effort has been derived from Öhrn and collaborators, see E. Deumens, Y. Öhrn, and B. Weiner, *J. Math. Phys.* **32**, 1166 (1991) and B. Weiner,E. Deumens, and Y. Öhrn, *J. Math. Phys.* **32**, 2413 (1991).

REAL TIME PATH INTEGRALS WITH QUASI-ADIABATIC PROPAGATORS: QUANTUM DYNAMICS OF A SYSTEM COUPLED TO A HARMONIC BATH

Nancy Makri

School of Chemical Sciences, University of Illinois
505 S. Mathews Avenue, Urbana, Illinois 61801

I. INTRODUCTION

The model of a "system" coupled to a harmonic "bath" is useful in many areas of chemistry and physics. Even though modeling the environment as a collection of harmonic oscillators involves in many cases a strong (and sometimes hard to justify) assumption, the model has been found to offer a surprisingly successful – and perhaps the only possible – theoretical description of many complex phenomena. Despite its simplicity, the system-bath model can adequately describe phenomena such as friction and incoherent relaxation encountered in the condensed phase. The idea of expressing a complex problem in terms of a system coupled to a bath has also found considerable utility in the area of gas-phase dynamics, where it leads to reaction path descriptions of reactive processes.

The quantum mechanics of system-bath problems is particularly rich and remains poorly understood. Nearly all known analytic and numerical quantum mechanical methods have been applied to system-bath Hamiltonians, in an attempt to elucidate some intriguing features that those systems possess. Even though the quadratic bath assumption constitutes an enormous simplification, the problem is highly nonlinear due to the coupling with the system degree of freedom. As a result, analytic solutions are possible only in special cases. Nevertheless, simple approximations have been able to provide valuable physical insight in certain regimes (e.g. weak coupling, adiabatic limit, etc. [1]). However, many realistic problems cannot be described by any of the above limits, as they involve intermediate coupling strengths and time scales of the same order. Understanding the behavior of these systems depends on one's ability to perform reliable numerical calculations.

The dynamical description of the problem requires knowledge of the (time-dependent) density matrix. Feynman's path integral approach[2,3] offers a unique advantage in the case of a system coupled to a harmonic bath. Because quadratic Lagrangians lead to Gaussian integrals in the path integral formalism, the harmonic bath can be integrated out *analytically* in this case, reducing the problem to a path integral that involves only the system degree of freedom with the effect of the bath included *via* an influence functional.[2,4] The latter involves interactions nonlocal in time, and therefore the path integral over the system does not correspond to an ordinary (local) Schrödinger equation; thus, it cannot be evaluated by iterative techniques. Since in the conventional formulation the path integral is a multidimensional integral, Monte Carlo techniques[5] offer the only conceivable way of evaluating it. Monte Carlo methods work best if the integrand is positive, which

Time-Dependent Quantum Molecular Dynamics, Edited by J. Broeckhove
and L. Lathouwers, Plenum Press, New York, 1992

is the case if one is interested in *equilibrium* properties (i.e., averages over the Boltzmann operator). Indeed, Monte Carlo path integral methods have managed to provide extremely useful information regarding the equilibrium statistical mechanical properties of many-body quantum systems. Recent work in this area[6–13] has attempted to extend Monte Carlo path integral techniques to deal with real time, where the integrand is oscillatory; the developed methodology[6–13] has been fairly successful in calculating short time dynamical properties. The *long real time* dynamics of multidimensional problems, though, is extremely hard to extract in most interesting cases, because Monte Carlo techniques fail to converge due to the phase interference nature of quantum mechanics. For these reasons the nonlocal path integral that arises in the case of a system coupled to a harmonic bath is extremely hard to evaluate.

In this article we describe a different approach[14] to the dynamics of a system coupled to a harmonic bath. The path integral is formulated in terms of an *adiabatic* propagator for the system, with an influence functional that incorporates the *non-adiabatic corrections*. With parameters typical of problems in chemical physics, converged results are achieved with a very small number of time slices in this formulation, such that the nonlocal path integral for the system can be evaluated by *quadrature*. The idea and the resulting expressions for the influence functional are presented in section II. Section III discusses some preliminary applications of the quasi-adiabatic propagator formulation to the tunneling dynamics of a symmetric double well potential coupled to a bath of ten harmonic oscillators. Finally, some concluding remarks appear in section IV.

II. QUASI-ADIABATIC PROPAGATOR AND INFLUENCE FUNCTIONAL

We consider the standard system-bath Hamiltonian, which has the form

$$\hat{H} = \hat{H}_s + \hat{H}_Q - \sum_{i=1}^{F} f_i(\hat{s})\hat{Q}_i . \tag{2.1a}$$

Here $\hat{H}_s$ describes the system,

$$\hat{H}_s = \frac{\hat{p}_s^2}{2m} + V(\hat{s}), \tag{2.1b}$$

and

$$\hat{H}_Q = \sum_{i=1}^{F} \hat{H}_i = \sum_{i=1}^{F} \frac{\hat{P}_i^2}{2m} + \frac{1}{2} m\,\omega_i^2 \hat{Q}_i^2 \tag{2.1c}$$

is the Hamiltonian for the F harmonic oscillators $\{Q_i\}$ which are linearly coupled to the system and which constitute the bath.

The adiabatic approximation is known to provide a reasonable zeroth order description of many problems in chemistry. However, correction terms to the adiabatic approximation are usually very hard to calculate. This is so because the size of the problem to be solved in order to correct the adiabatic approximation to order n grows typically (e.g., if adiabatic basis functions are used) as n^F.

Our purpose is to exploit the adiabatic approximation as a good zeroth order representation in the path integral context. The advantage that we shall gain this way is that the harmonic bath can be eliminated, and thus the numerical effort associated with the nth order correction term will now scale *linearly* with the number of bath degrees of freedom. On the other hand, if the adiabatic approximation is a reasonable zeroth order approximation to the dynamics, we expect that the path integral (which will include non-adiabatic correction terms) will converge very rapidly as the number of time slices is increased, such that we may be able to evaluate it by quadrature, thus avoiding Monte Carlo procedures (which would have to deal with the "sign problem").

We begin by partitioning the Hamiltonian into a form suggested by the adiabatic approximation. I.e., Eq. (2.1a) is split into an adiabatically displaced harmonic oscillator Hamiltonian,

$$\hat{H}_Q' \equiv \sum_{i=1}^{F} \hat{H}_i' = \sum_{i=1}^{F} \frac{\hat{P}_i^2}{2m} + \frac{1}{2} m\,\omega_i^2 \left[\hat{Q}_i - \frac{f_i(\hat{s})}{m\,\omega_i^2} \right]^2, \tag{2.2}$$

and a rescaled Hamiltonian for the system, in which $V(s)$ is replaced by the potential along the adiabatic path $Q_i(s) = f_i(s)/m\,\omega_i^2$:

$$\hat{H}_s' = \frac{\hat{p}_s^2}{2m} + V(\hat{s}) - \sum_{i=1}^{F} \frac{f_i(\hat{s})^2}{2m\,\omega_i^2}. \tag{2.3}$$

Using this partitioning, we split the time evolution operator as follows:

$$\exp(-i\hat{H}\Delta t/\hbar) \approx \exp(-i\hat{H}_Q'\Delta t/2\hbar)\,\exp(-i\hat{H}_s'\Delta t/\hbar)\,\exp(-i\hat{H}_Q'\Delta t/2\hbar). \tag{2.4}$$

Since the shifted oscillator Hamiltonians depend only parametrically on the system coordinate, the coordinate representation of the propagator that results by using Eq. (2.4) is

$$\begin{aligned} &\langle s_k \mathbf{Q}_k | e^{-i\hat{H}\Delta t/\hbar} | s_{k-1} \mathbf{Q}_{k-1} \rangle \\ &\quad\approx \langle s_k | e^{-i\hat{H}_s'\Delta t/\hbar} | s_{k-1} \rangle \prod_{i=1}^{F} \langle Q_{i,k} | e^{-i\hat{H}_i'(s_k)\Delta t/2\hbar} e^{-i\hat{H}_i'(s_{k-1})\Delta t/2\hbar} | Q_{i,k-1} \rangle. \end{aligned} \tag{2.5}$$

(We use the indices $k-1$ and k to label the coordinate points because we plan to use Eq. (2.5) as the short time propagator in a path integral.) The propagator that appears in Eq. (2.5), with $\hat{H}_i'$ given by Eq. (2.2), is easily shown to be equal to

$$\begin{aligned} &\langle Q_{i,k} | e^{-i\hat{H}_i'(s_k)\Delta t/2\hbar} e^{-i\hat{H}_i'(s_{k-1})\Delta t/2\hbar} | Q_{i,k-1} \rangle \\ &\quad= \left[\frac{m\,\omega_i}{2\pi i\hbar\sin\omega_i\Delta t} \right]^{1/2} \exp \frac{im\,\omega_i}{2\hbar\sin\omega_i\Delta t} \{ (q_{i,k}^2 + q_{i,k-1}^2)\cos\omega_i\Delta t \\ &\quad - 2q_{i,k}q_{i,k-1} + 2\Delta\lambda_{i,k}\Delta q_{i,k}\cos\omega_i\Delta t/2 + \Delta\lambda_{i,k}^2\cos^2\omega_i\Delta t/2 \}, \end{aligned} \tag{2.6}$$

where

$$\Delta\lambda_{i,k} \equiv \lambda_{i,k} - \lambda_{i,k-1}, \qquad \lambda_{i,k} \equiv \frac{f_i(s_k)}{m\,\omega_i^2},$$

$$q_{i,k} \equiv Q_{i,k} - \lambda_{i,k}, \quad \text{and} \quad \Delta q_{i,k} \equiv q_{i,k} - q_{i,k-1}.$$

With this particular splitting of the Hamiltonian the error due to the factorization of the time evolution operator, Eq. (2.4), is proportional to

$$\Delta t^3 \sum_{i=1}^{F} [\hat{p}_s^{\,2}, [\hat{p}_s^{\,2}, (\hat{Q}_i - \frac{f_i(\hat{s})}{m\,\omega_i^2})^2]].$$

The inner commutator in the last equation is a measure of the non-adiabaticity of the problem; i.e., the derived propagator is *exact* (for *any* value of Δt) if only points on the adiabatic path, $Q_i = f_i(s)/m\,\omega_i^2$, are important.

The above expression for the propagator leads to the following simple picture of the dynamics. The first (bath-independent) factor in Eq. (2.5) is the *exact* quantum propagator for motion along the adiabatic path. and therefore this part of the system-bath propagator corresponds to adiabatic dynamics; non-adiabatic effects are accounted for in Eq. (2.5) by Franck-Condon factors in the displaced bath propagator [cf. the $\Delta\lambda_{i,k}$ terms in Eq. (2.6)]. For moderately non-adiabatic problems this correction factor is accurate for fairly large Δt.

Our strategy is to use the path integral formulation of time-dependent quantum mechanics in order to improve systematically the accuracy of the non-adiabatic correction factor. We can accomplish this by incorporating the derived quasi-adiabatic propagator in the path integral. Splitting the total time t into N time slices $\Delta t = t/N$ yields the discretized path integral

$$<s_f \mathbf{Q}_f \,|\, e^{-i\hat{H}t/\hbar} \,|\, s_0 \mathbf{Q}_0> = \int ds_1 \cdots \int ds_{N-1} \int dQ_{1,1} \cdots \int dQ_{1,N-1} \cdots \int dQ_{F,1} \cdots \int dQ_{F,N-1}$$
$$\times \prod_{k=1}^{N} <s_k \mathbf{Q}_k \,|\, e^{-i\hat{H}\Delta t/\hbar} \,|\, s_{k-1} \mathbf{Q}_{k-1}>, \qquad (2.7)$$

and use of equations (2.5) and (2.6) leads to the following path integral expression for the system-bath propagator:

$$<s_f \mathbf{Q}_f \,|\, e^{-i\hat{H}t/\hbar} \,|\, s_0 \mathbf{Q}_0> = \int ds_1 \cdots \int ds_{N-1} \int dQ_{1,1} \cdots \int dQ_{1,N-1} \cdots \int dQ_{F,1} \cdots \int dQ_{F,N-1}$$
$$\times \prod_{k=1}^{N} <s_k \,|\, e^{-i\hat{H}_s^{'}\Delta t/\hbar} \,|\, s_{k-1}> \prod_{i=1}^{F} \left[\frac{m\,\omega_i}{2\pi i\,\hbar \sin\omega_i \Delta t} \right]^{1/2} \exp \frac{i m\,\omega_i}{2\hbar \sin\omega_i \Delta t} \{ (q_{i,k}^2 + q_{i,k-1}^2) \cos\omega_i \Delta t$$
$$- 2 q_{i,k} q_{i,k-1} + 2\Delta\lambda_{i,k} \Delta q_{i,k} \cos\omega_i \Delta t/2 + \Delta\lambda_{i,k}^2 \cos^2\omega_i \Delta t/2 \}, \qquad (2.8)$$

where $s_N \equiv s_f$ and $Q_{i,N} \equiv Q_{i,f}$.

The part involving the bath variables appears in the last equation as a Gaussian integral and therefore can be performed analytically to give [14]

$$<s_f \mathbf{Q}_f \,|\, e^{-i\hat{H}t/\hbar} \,|\, s_0 \mathbf{Q}_0> = \int ds_1 \cdots \int ds_{N-1} \prod_{k=1}^{N} <s_k \,|\, e^{-i\hat{H}_s^{'}\Delta t/\hbar} \,|\, s_{k-1}> I(s_0, s_1, \cdots, s_N) \quad (2.9a)$$

In the last equation I is an influence functional that describes the effect of the bath on the dynamics of the system and is given by the following expression:

$$I(s_0,s_1,\cdots,s_N) = \prod_{i=1}^{F}\left[\frac{m\omega_i}{2\pi i\hbar\sin\omega_i\Delta t}\right]^{1/2}\prod_{k=1}^{N-1}(2\cos\omega_i\Delta t-2\cos k\pi/N)^{-1/2}$$

$$\times\exp\frac{im\omega_i}{2\hbar\sin\omega_i\Delta t}\{(q_{i,N}^2+q_{i,0}^2+2\sum_{k=1}^{N-1}\lambda_{i,k}^2)\cos\omega_i\Delta t + 2(\lambda_{i,N-1}Q_{i,N}+\lambda_{i,1}Q_{i,0}) - 2\sum_{k=1}^{N}\lambda_{i,k}\lambda_{i,k-1}$$

$$+ (\cos^2\omega_i\Delta t/2-2\cos\omega_i\Delta t/2)\sum_{k=1}^{N}\Delta\lambda_{i,k}^2 + 2\cos\omega_i\Delta t/2(\Delta\lambda_{i,N}Q_{i,N}-\Delta\lambda_{i,1}Q_{i,0})$$

$$-\frac{1}{N}\sum_{k=1}^{N-1}[\sigma_{i,k}-\sin\frac{k\pi}{N}Q_{i,0}-\sin\frac{(N-1)k\pi}{N}Q_{i,N}]^2(\cos\omega_i\Delta t-\cos\frac{k\pi}{N})^{-1}\}, \tag{2.9b}$$

where

$$\sigma_{i,k} = \sum_{j=1}^{N-1}\sin\frac{jk\pi}{N}[-2\lambda_{i,j}(\cos\omega_i\Delta t-\cos\omega_i\Delta t/2)+(\lambda_{i,j-1}+\lambda_{i,j+1})(1-\cos\omega_i\Delta t/2)]. \tag{2.9c}$$

Since the influence functional involves a simple product over the bath degrees of freedom,[2] the numerical effort associated with the evaluation of Eq. (2.9) grows *linearly* with the size of the bath.

Equation (2.9) is the main result of this section. The path integral is expressed in terms of a system propagator which corresponds to motion on the potential energy surface along the adiabatic path, times an influence functional which accounts for non-adiabatic effects. Thus, the present idea provides a natural way to calculate non-adiabatic corrections to the adiabatic approximation within the path integral framework.

As we emphasized in the introduction, the presence of the non-local influence functional in Eq. (2.9a) prohibits use of iterative procedures (e.g. matrix multiplication) for the evaluation of the path integral. The practical advantage offered by our formulation is that accurate results are obtained with very small values of N even for fairly long time t. Specifically, if the dynamics is mostly adiabatic, no time discretization of the path integral is required. For problems relevant in chemistry, though, low frequency vibrations give rise to significant deviations from the adiabatic approximation; in such cases the path integral accounts for the non-adiabatic effects, approaching rapidly the exact dynamical result as N is increased. The numerical calculations presented in the next section indicate that (with parameters relevant to rate processes in chemistry) virtually exact results are achieved with $N<6$. This number is perhaps small enough to permit use of Monte Carlo methods for the evaluation of the $N-1$ dimensional path integral in complex or purely real time. A much more accurate procedure, though, would be to evaluate the path integral of Eq. (2.9) on a grid by *numerical quadrature*; this is possible with the present formulation because the number of integrations required to produce accurate results is small. Finally, in strongly non-adiabatic problems, where the proposed scheme may not converge to satisfactory accuracy even with the maximum realistic value of N (which is probably equal to 6 with modest values of F on present day computers), Richardson extrapolation[15] can be employed to extrapolate the noise-free path integral results obtained for $1/N$=1, 1/2, 1/3,... to the desired $1/N$=0 limit.

Finally, we show here how the derived path integral expression with the quasi-adiabatic propagator can be used to propagate a wavefunction. Let $\Psi_0(s,Q_1,\ldots,Q_F)$ be the wavefunction that describes the initial state. Clearly, one is interested in time correlation functions, or in probability densities averaged over the bath, and *not* in the propagated wavefunction in multidimensional space. Therefore we define a time-dependent *survival amplitude function* by integrating the product of the propagated and initial wavefunctions over the bath coordinates:

$$\chi(s;t) \equiv \int dQ_{1,N} \cdots \int dQ_{F,N}\ \Psi(s, Q_{1,N}, \ldots, Q_{F,N}; t)\, \Psi_0^*(s, Q_{1,N}, \ldots, Q_{F,N}). \tag{2.10}$$

Expressing the evolved wavefunction in terms of the path integral representation of the propagator given by Eq. (2.9) allows the reduced wavefunction to be written in the following form:

$$\begin{aligned}\chi(s;t) = &\int ds_0 \int ds_1 \cdots \int ds_{N-1} \int dQ_{1,0} \cdots \int dQ_{F,0} \int dQ_{1,N} \cdots \int dQ_{F,N} \\ &\times \Psi_0^*(s_N, Q_{1,N}, \ldots, Q_{F,N})\ \Psi_0(s_0, Q_{1,0}, \ldots, Q_{F,0}) \\ &\times \prod_{k=1}^{N} \langle s_k | e^{-i\hat{H}_s' \Delta t/\hbar} | s_{k-1} \rangle\ I(s_0, s_1, \cdots, s_N),\end{aligned} \tag{2.11}$$

where $s_N \equiv s$. If the initial state is a product of an arbitrary function of s times a Gaussian (or, more generally, a harmonic oscillator eigenfunction) in the bath coordinates the integrals over the endpoints $Q_{1,0}, \ldots, Q_{F,0}$ and $Q_{1,N}, \ldots, Q_{F,N}$ can be performed analytically, resulting in an N-dimensional integral to be performed by quadrature.

III. NUMERICAL APPLICATIONS: TUNNELING DYNAMICS OF A SYSTEM COUPLED TO A HARMONIC BATH

In this section we apply the idea described in section II to study the tunneling dynamics of a symmetric double well potential linearly coupled to a harmonic bath. The Hamiltonian is that of Eq. (2.1) with $f_i(\hat{s}) = c_i \hat{s}$; the system potential $V(\hat{s})$ is given by a quartic minus a quadratic term, and the parameters (mass, barrier height, etc.) are typical of proton transfer processes.

In order to assess the accuracy and the convergence characteristics of the proposed scheme, we originally[14] performed calculations with a bath of a single harmonic oscillator ($F=1$), for which we can generate exact results for comparison. We computed the off-diagonal element of the density matrix,

$$\rho = \langle -s_0 - Q_0 | e^{-\beta\hat{H}} | s_0 Q_0 \rangle \tag{3.1}$$

where $\pm(s_0 Q_0)$ are the coordinates of the two potential minima. This quantity is closely related to the tunneling rate[16] and is therefore very sensitive to the coupling terms in the Hamiltonian. Table 1 shows the results of calculating the off-diagonal density matrix element at 300 K from the path integral expression given in Eq. (2.9) with $\omega = 300\ \text{cm}^{-1}$ (typical of low frequency molecular vibrations). The path integral (which includes the non-local influence functional) was evaluated by quadrature (i.e., trapezoid rule); the calculation required storage of just one 80×80 matrix and (depending on the value of N) CPU time ranging from a few seconds to a few hours on a SPARCstation 2.

The convergence rate of the path integral is seen to be very fast in this formulation. Even though the single step quasi-adiabatic propagator is off by a factor of 3 for the largest value of the coupling, use of the former in a path integral with $N=5$ reduces the error to 5% or smaller for all values of the coupling constant considered. Also shown in Table 1 are the results of applying Richardson extrapolation[20] (a rational polynomial fit) to the data obtained for various values of N, in order to extrapolate to the $N \to \infty$ limit. This procedure is applicable here and indeed very successful, because the extrapolation is based on data free of statistical error; Richardson extrapolation would not have been very reliable if the path integral had been evaluated by Monte Carlo. It is seen that the extrapolated results are accurate to better than 1% for all values of the coupling constant considered, even when the coupling of the bath to the system is strong enough to decrease the value of the density matrix (i.e., diminish tunneling) by four orders of magnitude.

Table 1. Off-diagonal density matrix, Eq. (3.1), for the double well at $T \approx 300$K

c (a.u.)	$N=1^{a)}$	$N=2^{a)}$	$N=3^{a)}$	$N=4^{a)}$	$N=5^{a)}$	$N \to \infty^{b)}$	Exact$^{c)}$
0.000	8.18	8.18	8.18	8.18	8.18	8.18	8.18
0.001	6.07	6.19	6.23	6.24	6.25	6.26	6.27
0.002	2.48	2.69	2.76	2.78	2.79	2.81	2.81
0.003	5.63E-1	6.77E-1	7.14E-1	7.26E-1	7.32E-1	7.44E-1	7.43E-1
0.004	7.16E-2	9.97E-2	1.09E-1	1.12E-1	1.13E-1	1.16E-1	1.16E-1
0.005	5.20E-3	8.72E-3	9.85E-3	1.03E-2	1.05E-2	1.09E-2	1.09E-2
0.006	2.20E-4	4.63E-4	5.39E-4	5.70E-4	5.85E-4	6.03E-4	6.13E-4

$^{a)}$ Quadrature evaluation of the path integral with the influence functional, Eq. (2.9)
$^{b)}$ Richardson extrapolation
$^{c)}$ Exact results, computed by using the split propagator algorithm [17]

Next, we present some preliminary results that involve the *real* time zero-temperature dynamics of the double well system, in the case where the bath consists of ten harmonic oscillators and is described by the spectral density

$$J(\omega) = \frac{\pi}{2} \sum_{i=1}^{F} \frac{c_i^2}{m\,\omega_i} \delta(\omega - \omega_i) = \eta\omega \tag{3.2}$$

($F=10$).

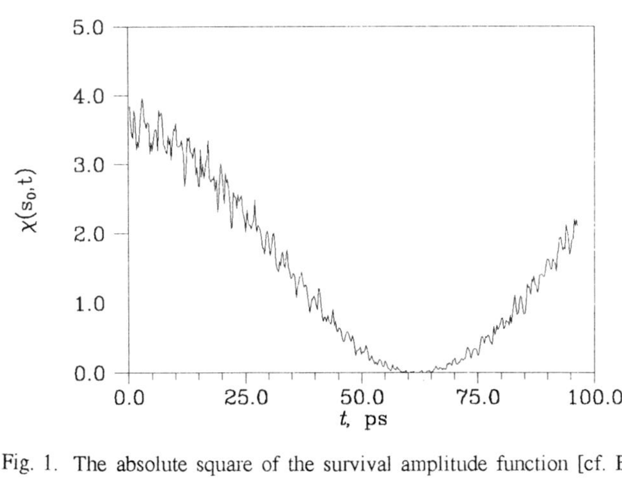

Fig. 1. The absolute square of the survival amplitude function [cf. Eq. (2.10)] at $s=-s_0$, the coordinate of the left minimum of the double well potential, as a function of time, as obtained by using the single-step quasi-adiabatic propagator in the case where the bath consists of ten harmonic oscillators.

We have calculated the survival amplitude function $\chi(s;t)$ defined by Eq. (2.10) at various times for an initial state that is a Gaussian-centered at the left potential minimum. The survival amplitude function looks roughly like a Gaussian in s, whose height oscillates on the time scale of the tunneling period. The single-peak shape of χ is due to Franck-Condon factors that arise when we integrate over the bath. I.e., at time equal to half the tunneling period, $\chi(s;t)$ is equal to zero at all points, whereas the full system-bath wavefunction has equal amplitudes in both potential wells.

In addition to the tunneling time scale, the wavefunction (and thus χ) evolves on a much shorter time scale, which corresponds to the vibrational frequencies in the individual potential wells. This short time scale gives rise to high frequency fluctuations in the height and width of $\chi(s;t)$ of about 10%. Fig. 1 shows the value of $|\chi(s;t)|^2$ as a function of time at $s=-s_0$, the

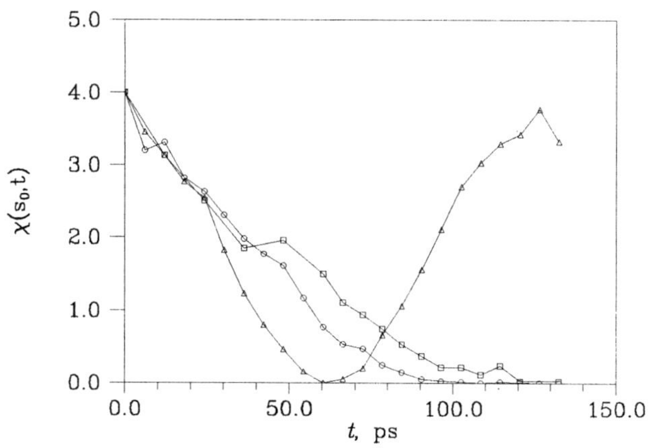

Fig. 2. The same quantity as in Fig. 1, but obtained by using the path integral representation of the propagator [cf. Eq. (2.11)] for N=1,2, and 3. The high frequency fluctuations of $|\chi(s;t)|^2$ are not seen in this picture because fewer points are shown. Triangle, N=1. Circles, N=2. Squares, N=3.

coordinate of the left minimum, as obtained by using the single-step quasi-adiabatic propagator [N=1 in Eq. (2.11)]. This result is only qualitatively correct, because there are significant non-adiabatic effects.

Fig. 2 shows the same quantity as obtained from the path integral calculation with N=1,2, and 3. The results have not fully converged with these values of N, but the convergence rate (as well as the calculations in the case of a one-oscillator bath shown in Table 1) indicates that they are converged to within a few percent. Calculations with N=4 and 5 and extrapolation to the $N \rightarrow \infty$ limit are certainly realistic as they should require just ten times more computer time than in the case of the one-oscillator bath; such results will be reported in a future publication.

It is worth noting that the time steps Δt employed in these calculations are larger than the shortest time scale in the problem (i.e., the vibrational period in the potential wells) by three orders of magnitude, and yet the method is numerically stable and accurate. Path integral calculations based on conventional short time propagators, on the other hand, would be completely meaningless if the time step exceeded the vibrational period; i.e., thousands of time steps would have been necessary to study the long time dynamics with conventional methods, and thus such calculations would not have been possible in practice.

IV. CONCLUDING REMARKS

Perhaps the most appealing feature in path integral calculations is that the result is *exact* once the procedure has converged. The feature that distinguishes the present formulation from more conventional path integral approaches is that converged path integral results can now be obtained (with parameters typical of rate processes in chemistry) for long time, much longer than the shortest time scale present. This is possible with the present approach because the path integral is now constructed to utilize a good zeroth order representation, which permits use of large time steps. The fact that the resulting path integral can often be evaluated by quadrature means that Monte Carlo techniques (which lead to sizable statistical errors when used in real time) can be avoided, yielding accurate results which can be extrapolated to the desirable zero time step limit.

The preliminary results presented here are very encouraging. We believe that the quasi-adiabatic propagator formulation of the path integral will find considerable utility in quantitative studies of tunneling dynamics in the condensed phase and in fully quantum calculations for intramolecular dynamics. Such calculations are in progress.

REFERENCES

1. See, for example, A. J. Leggett, S. Chadravarty, A. T. Dorsey, M. P. A. Fisher, A. Garg, and W. Zwerger, *Rev. Mod. Phys.* **59**, 1 (1987).
2. R. P. Feynman, *Rev. Mod. Phys.* **20**, 367 (1948); R. P. Feynman and A. R. Hibbs, *Quantum Mechanics and Path Integrals,* McGraw-Hill, New York, 1965.
3. L. S. Schulman, *Techniques and Applications of Path Integration*, Wiley, N.Y., 1981.
4. R. P. Feynman and F. L. Vernon, *Ann. Phys.* **24**, 118 (1963).
5. N. Metropolis, A. W. Rosenbluth, M. N. Rosenbluth, H. Teller, and E. Teller, *J. Chem. Phys.* **21**, 1087 (1953); J. P. Valleau and S. G. Whittington in *Modern Theoretical Chemistry,* B.J. Berne, Ed., Phenum, New York, 1977, vol. **5**, pp. 137-168.
5. D. Thirumalai and B. J. Berne, *Chem. Phys. Lett.* **116**, 471 (1985); D. Thirumalai and B. J. Berne, *Ann. Rev. Phys. Chem.* **37**, 401 (1986).
6. E. C. Behrman, G. A. Jongeward, and P. G. Wolynes, *J. Chem. Phys.* **79**, 6277 (1983); E. C. Behrman and P. G. Wolynes, *J. Chem. Phys.* **83**, 5863 (1985); P. G. Wolynes, *J. Chem. Phys.* **87**, 6559 (1987); R. E. Cline, Jr., and P. G. Wolynes, *J.Chem. Phys.* **88**, 4334 (1988); B. A. Mason, K. Hess, R. E. Cline, and P. G. Wolynes, *Superlattices and Microstructures* **3**, *421 (1987).*
7. J. D. Doll, *J. Chem. Phys.* **81**, 3536 (1984); J. D. Doll and D. L. Freeman, *Science* **234**, 1356 (1986); J. D. Doll, R. D. Coalson, and D. L. Freeman, *J. Chem. Phys.* **87**, 1641 (1987).
8. J. Chang and W. H. Miller, *J. Chem. Phys.* **87**, 1648 (1987).

9. V. S. Filinov, *Nucl. Phys. B* **271**, 717 (1986).

10. N. Makri and W. H. Miller, *Chem. Phys. Lett.* 139, 10 (1987); N. Makri and W. H. Miller, *J. Chem. Phys.* 89, 2170 (1988).

11. J. D. Doll and D. L. Freeman, *Adv. Chem. Phys.* 73, 120 (1988); J. D. Doll, D. L. Freeman, and M. J. Gillan, *Chem. Phys. Lett.* **143**, 277 (1988); J. D. Doll, T. L. Beck, and D. L. Freeman, *J. Chem. Phys.* **89**, 5753 (1988); T. L. Beck, J. D. Doll, and D. L. Freeman, *J. Chem. Phys.* 90, 3181 (1989).

12. N. Makri, *Chem. Phys. Lett.* **159**, 489 (1989); N. Makri, *Comp. Phys. Comm.* **63**, 389 (1991).

13. C. Mak and D. Chandler, *Phys. Rev. A* **41**, 5709 (1990); C. H. Mak and D. Chandler, *Phys. Rev. A* **44**, 2352 (1991).

14. N. Makri, *Chem. Phys. Lett.* (in press).

15. W. H. Press, B. P. Flannery, S. A. Teukolsky and W. T. Vetterling, *Numerical Recipes*, Cambridge University Press, 1989.

16. D. Chandler and P. G. Wolynes, *J. Chem. Phys.* **74**, 4078 (1981); D. M. Ceperley and G. Jacucci, *Phys. Rev. Lett.* **58**, 1648 (1987).

17. M. D. Feit, J. A. Fleck, Jr., and A. Steiger, *J. Comp. Phys.* **47**, 412 (1982).

FEMTOSECOND SPECTROSCOPY OF RETINAL PROTEINS

L. A. Peteanu, W. T. Pollard, and R. A. Mathies

Department of Chemistry
University of California
Berkeley, CA 94720

R. W. Schoenlein, S. L. Dexheimer, Q. Wang, and C. V. Shank

Lawrence Berkeley Laboratories
University of California
Berkeley, CA 94720

INTRODUCTION

Recent advances in the generation of ultrashort light pulses[1,2] have made possible the study of chemical and biological reactions with unprecedented time resolution. Two of the most rapid reactions that have been studied thus far occur in the biological pigments rhodopsin and bacteriorhodopsin. Both are intrinsic membrane proteins which contain visible absorbing, retinal prosthetic groups. Rhodopsin contains 11-*cis* retinal which is converted to the all-*trans* isomer following absorption of a photon (Fig. 1).[3,4] Bacteriorhodopsin contains all-*trans* retinal which photoisomerizes to the 13-*cis* form (Fig. 2). The protein stores a portion of the energy of the absorbed photon and uses it to drive subsequent ground state conformational changes.[5,6] In rhodopsin, this process leads ultimately to visual excitation,[7-10] while in bacteriorhodopsin, a proton is pumped across the membrane forming an electrochemical potential gradient that is used to synthesize ATP.[11] Both isomerization reactions are very rapid, occurring on timescales equal to or more rapid than the isomerization rates of small organic molecules in solution.[12,13]

Femtosecond dynamic absorption spectroscopy has proven to be a useful tool for understanding the mechanism of these important biological reactions. The sections which follow contain an outline of the methods used for measuring femtosecond transient absorption spectra, a qualitative presentation of the theoretical framework on which our understanding of the spectra is based, as well as several of the results which have been obtained in studies of bacteriorhodopsin and rhodopsin over the last few years.

Time-Dependent Quantum Molecular Dynamics, Edited by J. Broeckhove and L. Lathouwers, Plenum Press, New York, 1992

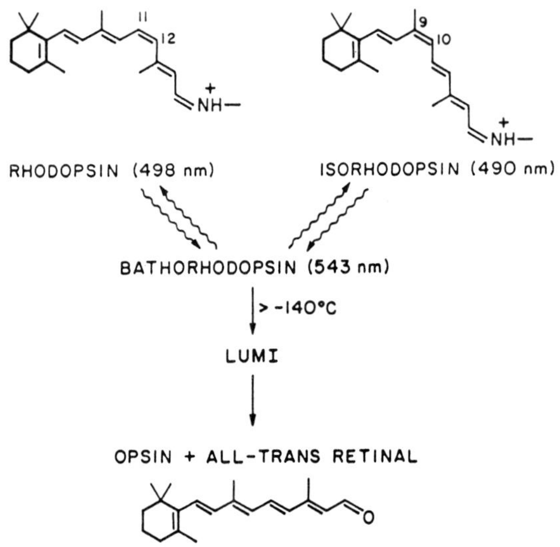

Fig. 1. Bleaching scheme of the 11-*cis* rhodopsin and 9-*cis* isorhodopsin pigments. Numbers in parentheses are absorption maxima at liquid helium temperatures.

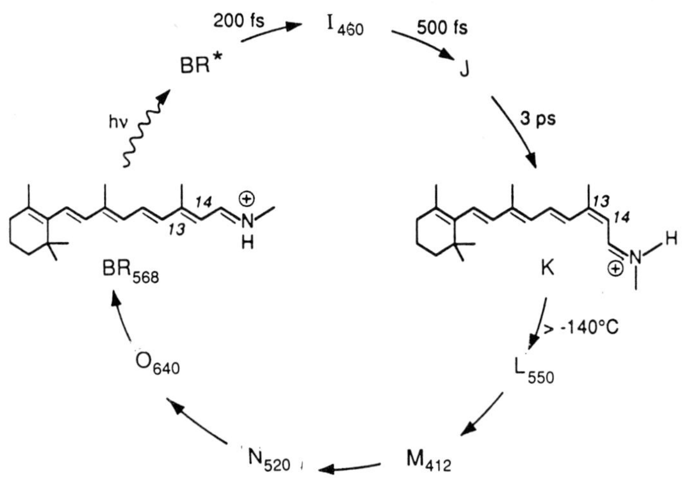

Fig. 2. Photocycle of bacteriorhodopsin. The subscripts denote the absorption maxima of each species.

EXPERIMENTAL

Ultrashort light pulses are obtained from an amplified colliding-pulse, mode-locked laser (50 fs centered at 620 nm) by chirping the pulse in an optical fiber followed by pulse compression using a combination of prisms and gratings.[1,2] By this method, pulses as short as 6 fs can be obtained with spectral widths of over 100 nm. More recent advances in laser design have allowed pulses with similar characteristics to be generated in the blue-green region of the spectrum.[14]

Although light pulses of this type have been used in a variety of experimental configurations to study molecular dynamics, we will focus on applications to pump-probe or transient absorption spectroscopy. Here, the pump beam both initiates the photochemistry and sets up a macroscopic polarization within the sample which is interrogated by the probe beam at a later time. The resulting changes in the probe can be monitored in one of two ways. The probe may be dispersed in a spectrograph and imaged onto a multichannel detector. In this configuration, advantage is taken of the full frequency spectrum of the probe light which can be measured with high spectral resolution. Alternatively, a narrow frequency range may be monitored by allowing the probe beam to pass through a filter placed before a photodiode detector. In both experiments, the effective time resolution is determined by the delay between the pump and probe pulses.

THEORY OF NONSTATIONARY STATE SPECTROSCOPY

Because of the short duration of the light pulses used in these experiments, it is possible not only to time-resolve exceedingly fast photochemical and photophysical processes but also to produce and probe non-stationary vibrational states of molecules. In the following paragraphs, we discuss how these non-stationary states arise and how their evolution can be monitored using pump-probe techniques.

If a molecule absorbs a pump pulse which is shorter than the timescale for vibrational motion, this interaction produces a vibrational wavepacket on the excited electronic state surface. The subsequent evolution of this wavepacket can be monitored using an ultrashort probe pulse. The interaction of the sample with the pump and probe pulses can be described using third-order perturbation theory in which the induced polarization, which is second order in the pump and first order in the probe, is calculated. However, it is possible to recast this third-order polarization response in terms of the first-order polarization response of the pump-perturbed system to the probe pulse.[15,16] This description relates pump-probe spectroscopy to ordinary linear absorption, with the difference that pump-probe spectroscopy represents linear spectroscopy of a non-stationary state $\Psi_{(1)}$ produced by the pump pulse. The interaction of this wavepacket with a probe pulse will typically produce time-dependent signals characteristic of either excited-state absorption or stimulated emission (Fig. 3). These spectral features are considered to arise in the following fashion. In the case of stimulated emission, the probe pulse causes a replica of the pump-induced wavepacket evolving on S_1 to be created on S_0 where it evolves under the influence of the ground-state vibrational Hamiltonian. The time-dependent dipole matrix element which couples these two wavepackets yields the induced polarization which radiates in the direction of the probe. The resulting signal is one of increased transmittance which is characteristic of stimulated emission. Signals due to excited-state absorption arise in a similar fashion when the probe pulse projects the initial S_1 wavepacket onto a higher-lying excited-state surface.

A non-stationary wavepacket may also be created on the ground-state surface due to the finite duration of the pump pulse.[17] In this process, the first interaction with the pump pulse raises the bra or ket wavepacket to the excited state where it evolves for some time before being coupled back down to the ground state by a second interaction with the pump field. Because the resulting ground-state wavefunction differs from the initial stationary ground-state wavefunction, the effect is to create a non-stationary wavepacket in the ground state. Time-dependent oscillatory features result which are superimposed on the differential transmittance spectra. In a molecule which contains a number of vibrational modes, several factors determine the contribution of each mode to the overall oscillatory response. To understand these factors, it is useful to consider a single vibrational mode having harmonic ground and excited state surfaces where the transition moment coupling these two surfaces is not coordinate dependent (Condon approximation). A large impulsive Raman signal results from creating a non-stationary wavepacket on the ground state which is strongly displaced with respect to the unperturbed ground-state wavefunction. This wavepacket acquires its displacement by propagating on the excited-state surface before being projected back to the ground state. In general, the momentum of the wavepacket in the excited state will be greatest when the minima of the ground and excited-state surfaces are the most displaced from one another so that the slope in the Franck-Condon region is large. Since this condition also results in large resonance Raman intensities, the modes which contribute to the oscillatory response are also those which are seen in the resonance Raman spectrum.[17,18] Because the total displacement attained by the wavepacket on the excited state will depend on the time of propagation, the strength of the impulsive Raman response also depends on the duration of the pump pulse. If the pump pulse is much shorter than the vibrational period, the wavepacket never moves far from the Franck-Condon region during the time that the pump is on. As a result, when this wavepacket is coupled back to the ground state via a second interaction with the pump field, it undergoes only small amplitude oscillations which do not strongly modulate the spectrum. In the limit of pulsewidths much larger than the period of the vibration, the final ground-state wavefunction that is built up will be stationary. In general, the pump pulse will be most effective in creating impulsive excitation in a particular mode when the pulsewidth is about 1/3 of the ground-state period.

In the following sections, experiments in which pump-probe spectroscopy is applied to photochemically active proteins are described. Extensive work has been done in

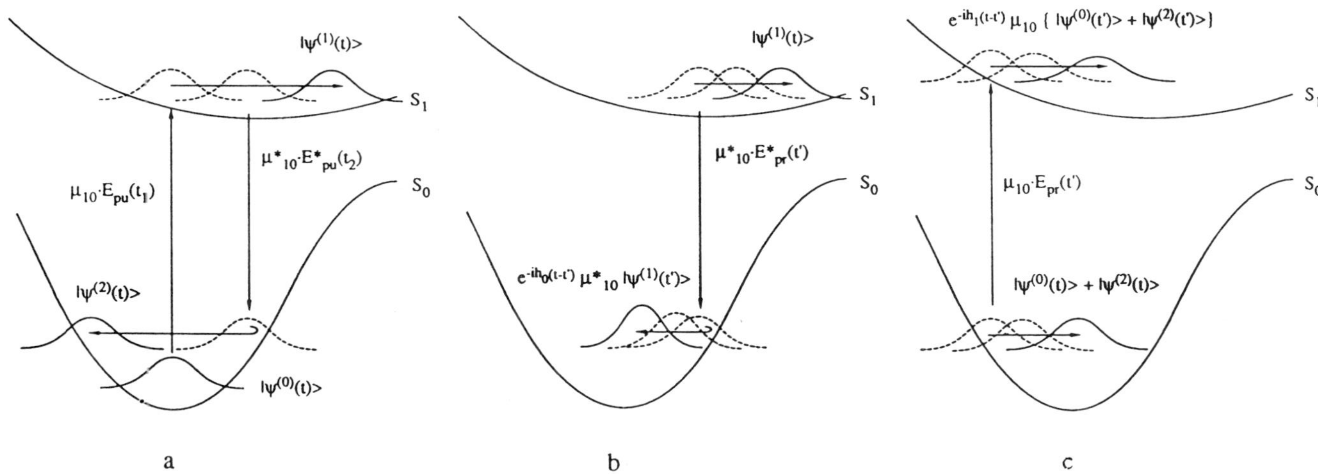

Fig. 3 (a) The components of the perturbed wavefunction following one and two interactions with the pump field. (b) Stimulated emission from a non-stationary excited-state wavepacket. (c) Absorption from a non-stationary ground-state wavepacket.

applying the wavepacket formalism to model both the early time evolution of the pump-probe signals in bacteriorhodopsin as well as the oscillatory behavior of these signals when the sample is excited with a very short pump pulse. The spectra which have been obtained for bacteriorhodopsin will then be compared to those of rhodopsin.

DYNAMIC ABSORPTION SPECTRA OF RHODOPSIN AND BACTERIORHODOPSIN

Because its absorption band lies in a region conveniently accessed by femtosecond lasers, bacteriorhodopsin was one of the first photochemically active molecules to be studied using short-pulse transient absorption techniques. The primary photoproduct of bacteriorhodopsin, denoted K, has been trapped at low temperature and its absorption spectrum was found to be red-shifted from that of ground state bacteriorhodopsin.[19] Numerous room-temperature transient absorption experiments have identified a precursor to K, denoted J, which is formed on the sub-picosecond time scale.[20-23] Spectroscopic evidence suggests[24,25] that J is a "hot" or vibrationally excited form of K which relaxes to form K on a ~ 3-6 picosecond time scale.[21,26-28]

Multichannel Spectra

Figure 4 presents differential transmittance spectra obtained using a 60-fs pump pulse centered at 620 nm and a 6-fs probe pulse which interrogates the sample at various time delays relative to the pump.[23] Four distinct features contribute to

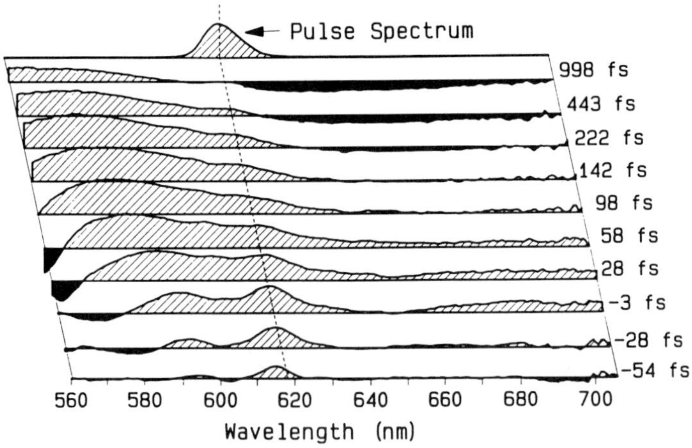

Fig. 4. Femtosecond pump-probe spectra of bacteriorhodopsin. The 60-fs pump and the 6-fs probe pulses are centered at 620 nm. The spectra are labeled with the time delay between the pump and probe pulses.

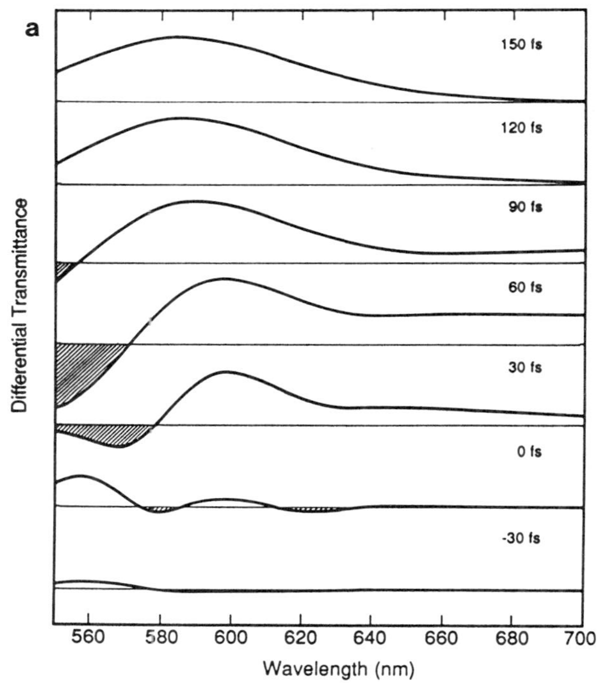

Fig. 5(a) Differential transmittance spectra of bacteriorhodopsin calculated using the surfaces of Fig. 5(b). Spectra are labeled with the time delay between the pump and probe pulses (from Reference 15).

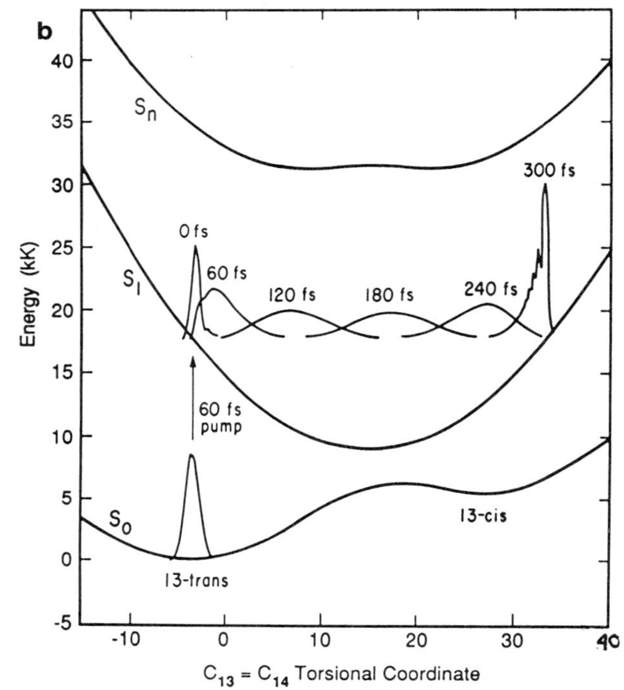

Fig. 5(b) Model potential surfaces and wavepacket evolution for the excited-state isomerization of bacteriorhodopsin. The S_0 and S_1 surfaces are 50 cm^{-1} harmonic potentials (from Reference 15).

the overall signal, the largest being a broad region of increased probe transmittance centered at 568 nm which represents the bleach of the ground-state bacteriorhodopsin absorption. Overlapping the ground state bleach at early times is a region of increased absorbance to the blue of 580 nm and a region of increased transmittance to the red of 650 nm. The blue-shifted feature is assigned to excited-state absorption from the initially populated S_1 excited state to higher lying excited states S_N. The increased transmittance to the red arises from stimulated emission between S_1 and the ground state S_0. At later times (> ~200 fs), an absorbance grows in whose maximum lies at 620 nm. This feature is characteristic of the ground-state J photoproduct. On a similar timescale, unreacted bacteriorhodopsin returns to the ground state thereby partially filling in the bleach.

We will focus first on the early-time (< 150 fs) spectral features which represent evolution of the wavepacket prepared by the pump pulse on the torsional reactive coordinate in the excited state. Later, we will discuss the kinetics of the return of the molecule to the ground state reactant and product wells.

The differential spectra given in Fig. 5(a) were calculated using numerical methods to propagate the wavepacket, produced by a pump pulse of finite duration, on an anharmonic 50 cm^{-1} surface which represents the excited-state reaction coordinate for the isomerization (Fig. 5(b)).[15] The S_0 - S_1 energy gap is determined by the absorption maximum of bacteriorhodopsin while the separation of the S_1 and S_N states as a function of the torsional coordinate was chosen to reproduce the observed excited-state absorption features. In this model, the excited-state absorption and stimulated emission features result from evolution of the same wavepacket on the S_1 surface. The differences between them, in terms of wavelength maxima and breadth, reflect both the initial slopes of the electronic surfaces as well as the behavior of the energy gaps between surfaces as a function of motion along the torsional coordinate. The rapid disappearance of both excited-state absorption and stimulated emission from the spectral window determined by the probe pulse reflects the rapid and irreversible motion of the wavepacket away from the Franck-Condon region and towards the 90 degree geometry in the excited state.

The femtosecond dynamic absorption spectra of rhodopsin are shown in Fig. 6.[29] Three main contributions to the overall signal can be identified. At early times (~ 0 fs) an absorption appears at 500 nm which vanishes within 40 fs to reveal a broad hole characteristic of a bleach of the ground-state rhodopsin absorption; an induced absorption near 570 nm appears within 50 fs, attains its maximal amplitude within 200 fs, and shifts slowly to the blue on a 1-3 ps timescale. Based on the results of low-temperature absorption spectra of the photoproduct of rhodopsin as well as picosecond transient absorption measurements, we assign this red-shifted absorption to the primary ground-state photoproduct of rhodopsin which is analogous to the product J which is formed from photolysis of bacteriorhodopsin.

The rapid disappearance of the excited-state absorption band in the pump-probe spectrum of rhodopsin is reminiscent of the behavior of the corresponding feature in the transient absorption spectrum of bacteriorhodopsin. This behavior is once again suggestive of rapid motion of the initially prepared wavepacket away from the Franck-Condon region along the excited-state torsional coordinate. However, in contrast to what was observed for bacteriorhodopsin, there is no readily identifiable stimulated emission signal in the differential transmittance spectra of rhodopsin. The absence of stimulated emission in these spectra could be the result of several processes including motion of the wavepacket away from the Franck-Condon region being even more rapid in rhodopsin than in bacteriorhodopsin, spreading of the

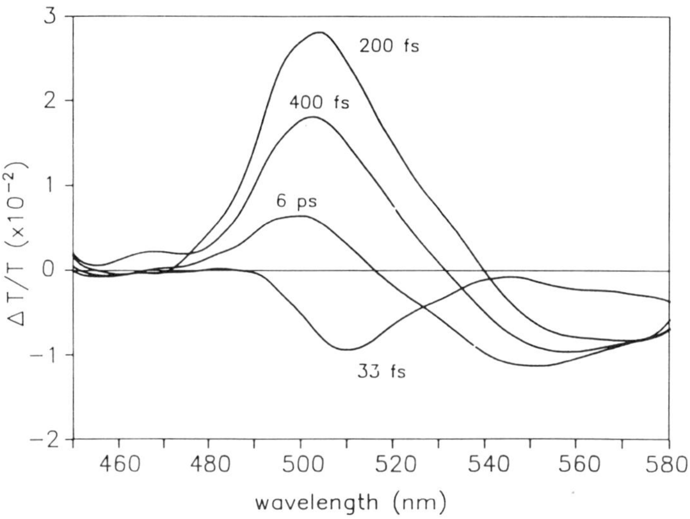

Fig. 6. Differential pump-probe spectra of rhodopsin. The 35-fs pump pulse and the 10-fs probe pulse were both centered at 500 nm. The spectra are labeled according to the time delay between the pump and probe pulses.

wavepacket due to motion on a dissipative excited-state surface, or loss of radiative coupling to the ground state due to a change in the electronic character of S_1 as the molecule changes geometry. Experiments are currently in progress which will begin to address this issue by expanding the wavelength range of the probe. Our expectation, based on the results of experiments on bacteriorhodopsin, is that the stimulated emission will be strongly red-shifted due to a narrowing of the S_0-S_1 energy gap as the molecule evolves towards the 90 degree geometry prior to crossing non-radiatively to the ground state.

Single Wavelength Traces

Although the multichannel spectra provide much of the information needed to understand the dynamics of the isomerization process, there are several advantages to obtaining single wavelength traces such as those shown for bacteriorhodopsin[18] (Fig. 7) and rhodopsin[29] (Fig. 8). For example, these traces clearly reveal an oscillatory component to the bands which is less obvious in the multichannel spectra. Single-wavelength traces for bacteriorhodopsin (Fig. 7) were obtained in three regions corresponding to predominantly ground-state absorption (580 nm), photoproduct absorption (650 nm), and a wavelength which is intermediate between the two (620 nm). The oscillations observed in these traces arise from a stimulated impulsive Raman process induced by the 12-fs pump pulse. Since the pump pulse duration is comparable to the period of the highest frequency Raman active modes of bacteriorhodopsin, (900 - 1550 cm^{-1})[30] the contribution of these frequencies to the overall oscillatory response is large.

Single wavelength traces of rhodopsin in the region of ground-state absorption (500 nm), photoproduct absorption (570 nm) as well as at intermediate wavelengths are shown in Fig. 8. In these traces, oscillations of very low-frequency (100-200 fs) can be seen which are damped after two or three recurrences. Two of the possible sources for these oscillations are wavepacket motion along a torsional coordinate in the excited or ground state, or an impulsive Raman process similar to what was observed for bacteriorhodopsin. The dominance of very low-frequency oscillations in the differential transmittance spectra of rhodopsin may result from the fact that longer (35 fs) pump pulses were used which would be less effective than shorter pulses at producing an impulsive excitation response in high-frequency skeletal modes. It is also true that the Raman spectrum of rhodopsin reveals numerous highly displaced, low-frequency torsional modes.[31] Absolute Raman intensity analysis of rhodopsin[31] has revealed that the displacements of the minima of the ground and excited state along several torsional coordinates are greater than the displacements in the higher frequency Raman active modes. This observation suggests another reason why low-frequency modes dominate the oscillatory response observed in rhodopsin but not that of bacteriorhodopsin, namely, that low-frequency torsions are much more highly displaced in rhodopsin than in bacteriorhodopsin.

KINETICS

In this section, we will discuss the time course of the differential transmittance signals that arise from product formation and ground state recovery. The focus will be on what the overall kinetics of the reaction can reveal about the mechanism of the isomerization process.

The kinetics of ground-state recovery in bacteriorhodopsin are measured by the time course of the signal at 568 nm (Fig. 7) which is the wavelength corresponding to the maximum of absorbance of ground state bacteriorhodopsin. The

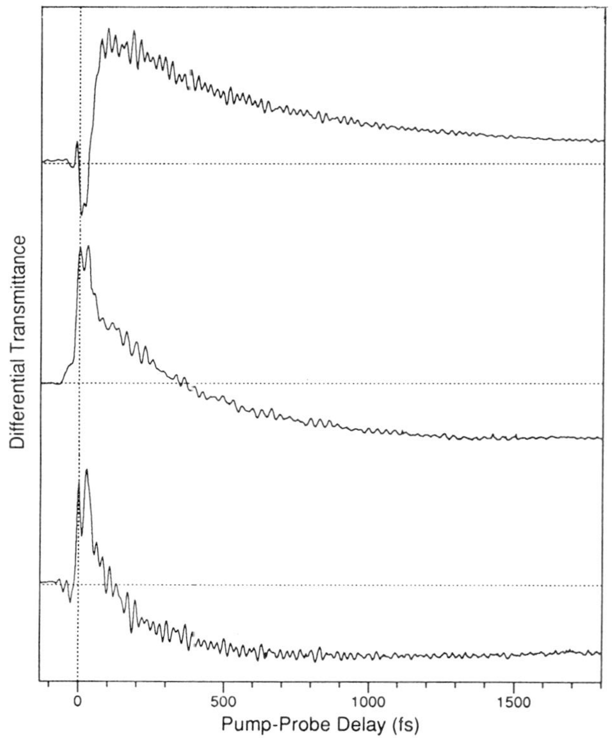

Fig. 7. Differential transmittance spectra of bacteriorhodopsin as a function of time delay between the 12-fs pump and 12-fs probe pulses recorded at 568 nm (top), 620 nm (middle), and 656 nm (bottom) (from Reference 18).

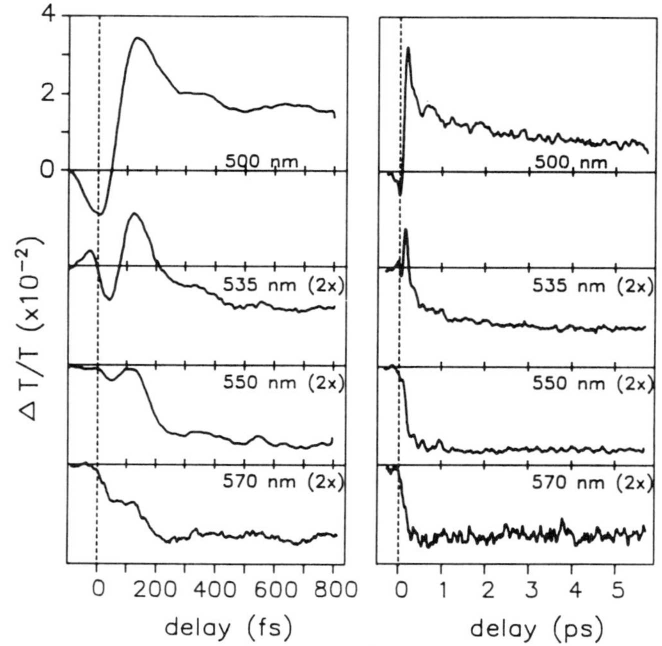

Fig. 8. Differential transmittance spectra of rhodopsin at various wavelengths following absorption of a 35-fs pump pulse at 500 nm using a 10-fs probe pulse centered at 500 nm. (from Reference 29).

recovery shows what appear to be single exponential kinetics with a time constant of ~500 fs. This rate agrees with numerous other measurements of the same process with lower time resolution.[20-23]

It is somewhat more difficult to determine the rate of photoproduct formation in bacteriorhodopsin because it is not possible to find a wavelength at which there is not a contribution to the measured signal from stimulated emission as well. The kinetics of the stimulated emission signal are also complex - experiments by Dobler et al. have shown that the signal at bluer probe wavelengths decays more rapidly than that at redder wavelengths (200 fs vs. 500 fs time constants).[22] The single wavelength trace at 656 nm shown in Fig. 7 shows contributions from three processes, namely, the fast component of the stimulated emission which is the positive signal at times earlier than 70 fs, a negative going signal which reflects the growth of photoproduct J absorption on the order of 500 fs, and a decay of J absorption as a second intermediate, K, is formed on the 3 ps timescale (see below).

To analyse the kinetics of photoproduct formation in the pigment rhodopsin, we examine the evolution of the absorption at 570 nm (Fig. 8). This trace reveals that the photoproduct absorption reaches a maximum at ~200 fs and that no further changes occur in this wavelength region out to 6 ps. The ground state bleach at 500 nm recovers on a similar timescale.

Although our understanding of the kinetics of these processes is not yet complete, it is interesting to compare the apparent rates of photoproduct formation in bacteriorhodopsin and rhodopsin. While bacteriorhodopsin's J photoproduct appears with a time constant of ~500 fs, the formation of the primary photoproduct in rhodopsin is essentially complete in only 200 fs. The more rapid formation of product in the case of rhodopsin is consistent with the more intense torsional slopes observed in the Franck-Condon region for rhodopsin[31] as well as its lower fluorescence quantum yield.[32]

The longer time (> 500 fs) evolution of the feature due to J in the transient absorption spectrum of bacteriorhodopsin shows a gradual blue shift of ~20 nm that occurs on the time scale of a few picoseconds. This can be observed in the multichannel spectrum (Fig. 6) and as a slow decay in the absorption signal at 620 nm (Fig. 7). This process has been identified with vibrational cooling of an initially hot population on the ground state surface which ultimately leads to the formation of the next identifiable intermediate in the photocycle, K. Picosecond transient Raman experiments have been performed on bacteriorhodopsin[24,25] which show that the initially formed photoproduct is vibrationally excited. The decay of the anti-Stokes signal, which arises from hot ground state molecules, follows the kinetics of the J -> K transition[25] that was originally identified by transient absorption methods.[21,26-28]

The absorption maxima of the photoproduct of rhodopsin also shifts towards shorter wavelengths between 200 fs and 3 ps. This process has also been observed in subsequent ~300 fs time resolution transient absorption spectra on rhodopsin by Yan et al..[33] They offer an alternative explanation for this finding, namely, that the photoproduct forms on a ~3 ps time scale from an excited-state that has a lifetime of ~200 fs. However, comparison of the kinetics measured for rhodopsin with the results for bacteriorhodopsin argues that the spectral evolution which occurs on a ~3 ps time scale is best interpreted as vibrational cooling in the ground-state manifold.

CONCLUSIONS

Numerous breakthroughs in the area of laser design and construction have opened up new avenues for research in photochemistry and photobiology. As a result, the timescales for the isomerization reactions of two important biological photoreceptors, bacteriorhodopsin and rhodopsin, have been determined. Our understanding of how these rapid and efficient reactions occur has also been advanced by the development of theoretical methods for the interpretation of femtosecond pump-probe spectra.

REFERENCES

1. R. L. Fork, C. H. Brito Cruz, P. C. Becker, C. V. Shank, Opt. Lett. 12, 483 (1987).
2. C. V. Shank, Science 233, 1276 (1986).
3. T. Yoshizawa and G. Wald, Nature 197, 1279 (1963).
4. G. Wald, Science 162, 230 (1968).
5. A. Cooper, Nature 282, 531 (1979).
6. G. A. Schick, T. M. Cooper, R. A. Holloway, L. P. Murray, R. R. Birge, Biochem. 26, 2556 (1987).
7. N. Bennett, M. Michel-Villaz, H. Kuhn, Eur. J. Biochem. 127, 97 (1982).
8. L. Stryer, Biopolymers 24, 29 (1985).
9. L. W. Hayness, A. R. Ray, K.-W Yau, Nature 321, 66 (1986).
10. A. L. Zimmerman, D. A. Baylor, Nature 321, 70 (1986).
11. E. Racker, W. Stoeckenius, J. Biol. Chem. 249, 662 (1974).
12. D. C. Todd, J. M. Jean, S. J. Rosenthal, A. Ruggiero, D. Yang, G. R. Fleming, J. Chem. Phys. 93, 8658 (1990).
13. R. J. Sension, S. T. Repinec, R. M. Hochstrasser, J. Chem. Phys. 93, 9185 (1990).
14. R. W. Schoenlein, J.-Y. Bigot, M. T. Portella, C. V Shank, Appl. Phys. Lett. 58, 801 (1991).
15. W. T. Pollard, S. Y. Lee, R. A. Mathies, J. Chem. Phys. 92, 4012 (1990); W. T. Pollard, H. L. Fragnito, J.-Y. Bigot, C. V. Shank, R. A. Mathies, Chem. Phys. Lett. 168, 239 (1990).
16. W. T. Pollard and R. A. Mathies, in Annual Review of Physical Chemistry, 1992 (in press).
17. W. T. Pollard, S. L Dexheimer, Q. Wang, L. A. Peteanu, C. V. Shank, R. A. Mathies, J. Phys. Chem., in press.

18. S. L. Dexheimer, Q. Wang, L. A. Peteanu, W. T. Pollard, R. A. Mathies, C. V. Shank, Chem. Phys. Lett. 188, 61 (1992).
19. R. H. Lozier, R. A. Bogomolni, W. Stoeckenius, Biophys. J. 15, 955 (1975).
20. E. P. Ippen, C. V. Shank, A. Lewis, M. A. Marcus, Science 200, 1279 (1978).
21. A. V. Sharkov, A. V. Pakulev, S. V. Chekalin, Y. A. Matveetz, Biochim. Biophys. Acta 808, 94 (1985).
22. J. Dobler, W. Zinth, W. Kaiser, D. Oesterhelt, Chem. Phys. Lett. 144, 215 (1988).
23. R. A. Mathies, C. H. Brito Cruz, W. T. Pollard, C. V. Shank, Science 240, 777 (1988).
24. T. L. Brack, G. H. Atkinson, J. Phys. Chem. 95, 2351 (1991).
25. S. J. Doig, P. J. Reid, R. A. Mathies, J. Phys. Chem. 95, 6372 (1991).
26. M. C. Nuss, W. Zinth, W. Kaiser, E. Kolling, D. Oesterhelt, Chem. Phys. Lett. 117, 1 (1985).
27. J. W. Petrich, J. Breton, J. L. Martin, A. Antonetti, Chem. Phys. Lett. 137, 369 (1987).
28. H.-J. Polland, M. A. Franz, W. Zinth, W. Kaiser, E. Kolling, D. Oesterhelt, Biophys. J. 49, 651 (1986).
29. R. W. Schoenlein, L. A. Peteanu, R. A. Mathies, C. V. Shank, Science 254, 412 (1991).
30. S. O. Smith, M. S. Braiman, A. B. Myers, J. A. Pardoen, J. M. L. Courtin, C. Winkel, J. Lugtenburg, R. A. Mathies, J. Am. Chem. Soc. 109, 3108 (1987) and references therein
31. G. R. Loppnow, R. A. Mathies, Biophys. J. 54, 35 (1988).
32. A. G. Doukas, M. M. Junnarkar, R. R. Alfano, R. H. Callender, T. Kakitani, B. Honig, Proc. Natl. Acad. Sci. USA 81, 4790 (1984).
33. M. Yan, D. Manor, G. Weng, H. Chao, L. Rothberg, T. M. Jedju, R. R. Alfano, R. H. Callender, Proc. Natl. Acad. Sci. USA 88, 9809 (1991).

APPROXIMATE METHODS FOR TIME EVOLUTION OF WAVE PACKETS

H.–D. Meyer, U. Manthe, J. Kucar and L.S. Cederbaum

Theoretische Chemie, Physikalisch Chemisches Institut
Universiät Heidelberg, Im Neuenheimer Feld 253
W–6900 Heidelberg, Germany

1. INTRODUCTION

The time–dependent picture allows for a vivid physical interpretation of numerical calculations. The relation to classical mechanics is more direct. For a certain class of problems the time–dependent approach has shown to be more efficient computational wise than the time–independent one. Despite the efficiency of the time dependent methods[1] it is not possible – except for special cases – to propagate a wave packet of more than three degrees of freedom. The development of reliable and systematically extendable approximate propagation schemes is therefore desired. There are two approximate methods that have been used to some extend in the past. The first one is the gaussian wave packet propagation of Heller[2] and the second one is the time–dependent Hartree (TDH) method[3]. Both approaches are quite simple but of rather limited quality. The development of the time–dependent rotated Hartree (TDRH) method[4] was our first attempt in the direction proposed above. TDRH is in some sense a generalization of both the above methods. TDRH becomes exact if the (rotated) Hamiltonian is either harmonic or separable. TDRH is able to recover an important fraction of the correlation that is missing in the TDH approach, TDRH however is more difficult to implement numerically.

Our second attempt to arrive at a reliable approximate propagation scheme was the development of a multi–configurational time–dependent Hartree (MCTDH) method[5]. Multiconfigurational approaches have been tried before[6 7], however, the method to be outlined below is more general than the previous attempts. It has been found that MCTDH is superior to TDRH. MCTDH is easier to implement and yields – as we will show below – excellent results for certain classes of problems. In the following we will therefore restrict ourselves to the discussion of MCTDH.

Time-Dependent Quantum Molecular Dynamics, Edited by J. Broeckhove
and L. Lathouwers, Plenum Press, New York, 1992

2. OUTLINE OF THEORY

The MCTDH method was introduced in a recent publication[5]. Here we will give a brief review using a slightly different notation. The MCTDH–ansatz for a time–dependent multi–configurational wave function reads

$$\Psi(x_1,..,x_f,t) = \sum_{j_1=1}^{n_1} \cdots \sum_{j_f=1}^{n_f} A_{j_1..j_f}(t) \prod_{\kappa=1}^{f} \varphi_{j_\kappa}^{(\kappa)}(x_\kappa,t). \tag{1}$$

Here x_1 to x_f denote the coordinates of the f degrees of freedom to be considered. The symbol A denotes the expansion coefficients and φ are the time dependent expansion functions which we will call "single–particle functions" or "orbitals". The index κ labels the f degrees of freedom and j_κ runs over the N_κ single–particle functions of the κ–th degree of freedom. The expansion (1) is not unique. Linear transformations among the orbitals of the κ–th degree of freedom leave the wave function (1) invariant if the inverse transformation is applied to the κ–th index of A. To arrive at unique formulas at least for the time evolution of Ψ one has to introduce additional constrains. We thus require

$$<\varphi_n^{(\kappa)} | \varphi_m^{(\kappa)}> = \delta_{nm} \tag{2}$$

and

$$<\varphi_n^{(\kappa)} | \frac{\partial}{\partial t} \varphi_m^{(\kappa)}> = <\varphi_n^{(\kappa)} | -i \cdot h_\kappa | \varphi_m^{(\kappa)}> \tag{3}$$

where the h_κ denote hermitian but otherwise arbitrary operators. ($\hbar=1$ is used throughout). In all our numerical calculations we have set $h_\kappa \equiv 0$, i.e.

$$<\varphi_n^{(\kappa)} | \dot{\varphi}_m^{(\kappa)}> = 0 . \tag{4}$$

This choice is not only the simplest, but it also leads to an evident interpretation. Using the constraint (4) the single particle functions will change in time as little as possible. It must be emphasized that the MCTDH wave function Ψ in independent of the choice of the h_κ. The numerical performance of the propagation, on the other hand, will depend on this choice.

Before we discuss the MCTDH equations of motion we have to introduce some definitions. The multi–index J represents the set of all j_κ

$$J = (j_1,j_2,..,j_f) \tag{5}$$

and a configuration ϕ_J is defined by

$$\phi_J(x_1,x_2,..,x_f,t) = \varphi^{(1)}_{j_1}(x_1,t)\cdot\varphi^{(2)}_{j_2}(x_2,t)\cdot\ldots\cdot\varphi^{(f)}_{j_f}(x_f,t) \tag{6}$$

Next we introduce "single–hole functions":

$$\Psi^{(\kappa)}_n = \sum_{j_1}\ldots\sum_{j_{\kappa-1}}\sum_{j_{\kappa+1}}\ldots\sum_{j_f} A_{j_1..j_{\kappa-1}n\,j_{\kappa+1}..j_f}\cdot \varphi^{(1)}_{j_1}\cdot\ldots\cdot\varphi^{(\kappa-1)}_{j_{\kappa-1}}\cdot\varphi^{(\kappa+1)}_{j_{\kappa+1}}\cdot\ldots\cdot\varphi^{(f)}_{j_f} \tag{7}$$

and mean fields

$$\langle H\rangle^{(\kappa)}_{nm} = \langle\Psi^{(\kappa)}_n|\,H\,|\Psi^{(\kappa)}_m\rangle\,. \tag{8}$$

The mean fields are operators on the κ–th degree of freedom. Finally we introduce the density matrix

$$\rho^{(\kappa)}_{nm} = \langle\Psi^{(\kappa)}_n|\Psi^{(\kappa)}_m\rangle = \sum_{j_1}\ldots\sum_{j_{\kappa-1}}\sum_{j_{\kappa+1}}\ldots\sum_{j_f} A^*_{j_1..j_{\kappa-1}n\,j_{\kappa+1}..j_f}\cdot A_{j_1..j_{\kappa-1}m\,j_{\kappa+1}..j_f} \tag{9}$$

and the projector on the space of κ–orbitals

$$P^{(\kappa)} = \sum_{j=1}^{n_\kappa} |\varphi^{(\kappa)}_j\rangle\langle\varphi^{(\kappa)}_j| \tag{10}$$

Splitting the Hamiltonian into an uncorrelated and a correlated part

$$H = \sum_{\kappa=1}^{f} h_\kappa + H_R \tag{11}$$

(the h_κ are the operators that appeared in eq.(3)) and employing the Dirac–Frenkel variational principle

$$<\delta\Psi | i\cdot\frac{\partial}{\partial t} - H | \Psi> = 0 \tag{12}$$

we finally arrive at the MCTDH equations of motion

$$i\cdot\dot{A}_J = \sum_L <\Phi_J | H_R | \Phi_L> A_L \, . \tag{13}$$

and

$$i\cdot\dot{\underline{\varphi}}^{(\kappa)} = h_\kappa \underline{\varphi}^{(\kappa)} + (1-P^{(\kappa)})\cdot \underline{\underline{\rho}}^{(\kappa)-1}\cdot <\underline{\underline{H}}_R>^{(\kappa)}\cdot\underline{\varphi}^{(\kappa)} \, . \tag{14}$$

In the latter equation we have indicated vectors and matrices by underlining in the usual way. Note that $\underline{\varphi}^{(\kappa)}$ denotes a vector of functions, $\underline{\underline{\rho}}^{(\kappa)}$ a matrix of numbers and $<\underline{\underline{H}}>^{(\kappa)}$ a matrix of single–particle operators. As already mentioned above, we have set $h_\kappa = 0$ and $H_R = H$ in the calculations reported below.

The eigenvalues of the hermitian positive–definite matrix $\rho^{(\kappa)}$ are called "natural populations" and the corresponding eigenvectors may be used to transform the κ–set of single–particle functions to the so called "natural orbitals". A natural population indicates the importance of the corresponding natural orbital in the expansion of the MCTDH wave function. Small natural populations hence indicate convergence. The natural populations allow us to judge the degree of convergence (with respect to the number of single–particle functions) without comparing to another calculation. This is a very important and helpful property of the MCTDH method.

A straightforward evaluation of the mean fields requires f–fold integrations making the usefulness of the MCTDH approach questionable. However, the evaluation of multi–dimensional integrals can be avoided if the Hamiltonian is written as a sum of single–particle operators.

$$H = \sum_{n=1}^{s} c_n \prod_{\kappa=1}^{f} h_n^{(\kappa)} . \tag{15}$$

Each $h_n^{(\kappa)}$ acts only on the κ–th degree of freedom. The restricted form (15) of the Hamiltonian can be fulfilled in most cases, at least after performing a new fit to the interaction potential.

In order to give an estimate of the numerical effort of a MCTDH calculation we have counted the numerical operations. For sake of simplicity of the argument let us ignore the κ–dependence and assume that there are n single–particle functions for each degree of freedom. Each of these orbitals is represented by N basis functions or grid points. The numerical effort of a MCTDH calculation then becomes

$$\text{effort} = c_1 \cdot f \cdot n \cdot N^2 + c_2 \cdot f^2 \cdot n^{f+1} . \tag{16}$$

(There are additional terms with different scaling behavior but these terms are small compared to the two terms above). The numerical values of the prefactors c_1 and c_2 are independent of f, n and N. They grow linearly with s, the number of expansion terms of the Hamiltonian (see eq.(15)). The first term in the sum varies linearly with the number of degrees of freedom, f, whereas the second term grows exponentially with f. For small values of n, e.g. for systems with little correlation, the first term dominates due to the large value of N. Then the numerical effort of the MCTDH–computation increases only linearly with the number of degrees of freedom, f.

The second term will dominate for large values of n or when treating many degrees of freedom. Despite the exponential scaling in this case a large gain factor with respect to an exact calculation is achieved. This gain factor is proportional to $(N/n)^{f+1}$. Since n is typically ten times smaller than N, the gain factor reaches enormous values for a large number of degrees of freedom.

Another aspect of the MCTDH–scheme should be mentioned. The number of basis functions (or grid points), N, used to represent the single particle functions enters only in the first term of eq.(16). Hence the use of large basis sets or grid sizes is not prohibitive for MCTDH. This is important for the treatment of systems including heavy atoms or high excitation energies. In most cases tricks like the introduction of absorbing boundaries or other truncation procedures are not necessary when using MCTDH.

3. PHOTO DISSOCIATION OF NOCl

In an initial exemplary MCTDH–calculation we had studied a model system of Henon–Heiles type[5]. Now we will test the MCTDH–scheme by applying it to a realistic problem, the photo dissociation of NOCl following an excitation to the S_1–state. This process has been investigated experimentally by several authors[8] and numerically exact wave packet dynamics studies have been performed by Untch, Weide and Schinke[9]. Due to the simple structure of the dissociation process this system should be easily described within a MCTDH–approach. Since numerically exact results are accessible, the accuracy and numerical efficiency of the MCTDH–approach can be judged by comparison with these data.

After a vertical transition the initial wave function is propagated on the S_1–surface. The time integration is stopped 50 fs after the S_0–S_1–excitation. Then the dissociation process is considered as completed and the product distributions are evaluated. The determination of the absorption spectrum $\sigma(\omega)$ is performed by a Fourier–transform of the autocorrelation function $<\Psi(0)|\Psi(t)>$:

$$\sigma(\omega) \sim \omega \cdot \mathrm{Re}\left[\int_0^\infty <\Psi(0)|\Psi(t)> e^{i \cdot \omega \cdot t}\, dt \right] . \tag{17}$$

Fig.1 displays the absorption spectrum calculated by different MCTDH–calculations. For reasons of comparison the spectrum resulting from a numerically exact calculation is also displayed in this Figure (broken line).

Within the Hartree–approximation, i.e. employing only one single–particle function per degree of freedom, one finds a structureless photo dissociation spectrum (Fig.1a). Excitations in the NO–vibration could not be resolved, since the Hartree–ansatz does not allow to assign a different energy distributions to every NO–vibrational state. Only the center and width of the absorption band are described correctly.

Including two single–particle functions per degree of freedom the description improves (Fig.1b). The low energy part of the spectrum, corresponding to the 0→0 transition of the NO–vibration, is reproduced correctly. Also structures, resulting from the 0→1 and 0→2 transitions, are found, but they are less pronounced compared to the exact spectrum. The description has improved to a semiquantitive level, since different energy distributions in the other degrees of freedom could now be assigned to different NO–vibrational excitations. The existence of only two single–particle functions per degree of freedom still describes the correlational effects incompletely. Therefore differences in details between the MCTDH and the exact result are still present.

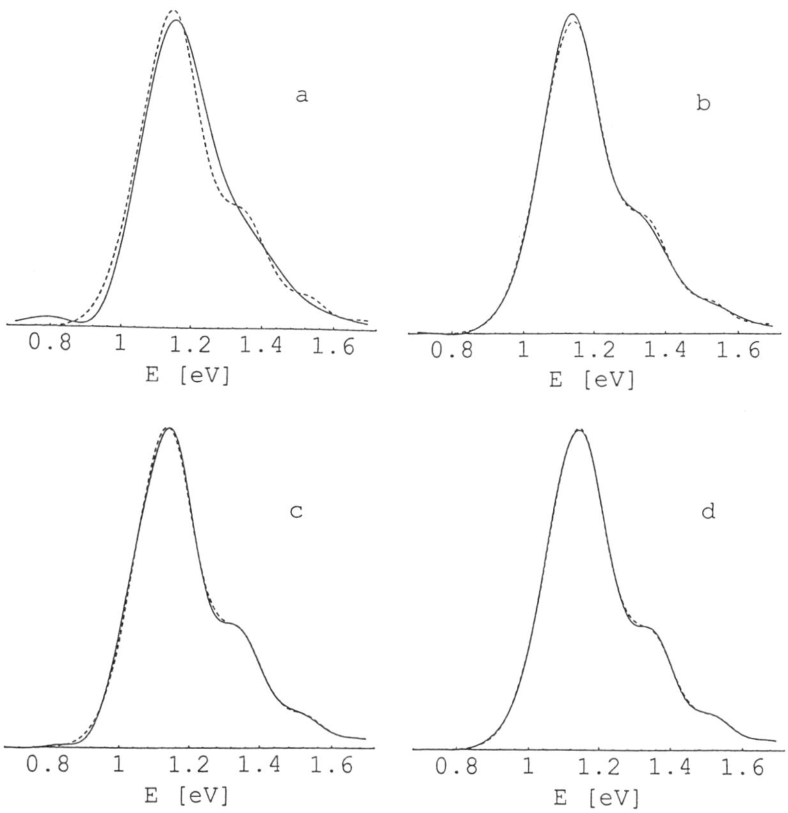

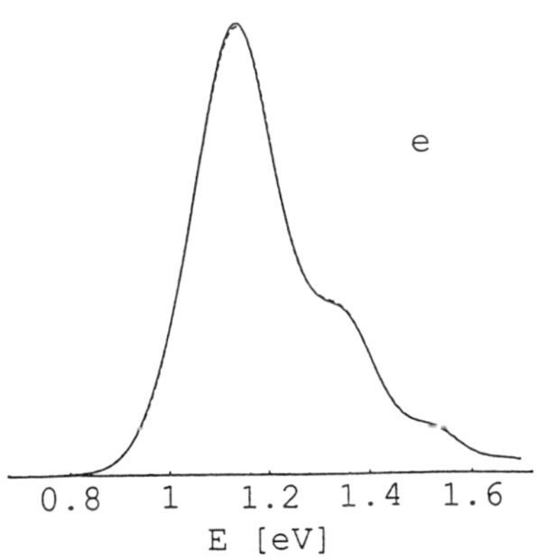

Fig. 1
Photoabsorption spectra of NOCl computed by using n = 1, 2, 3, 4, and 5 functions per degree of freedom. The spectrum of the exact calculation is indicated by a dashed line.

After inclusion of three or more single–particle functions per degree of freedom, a nearly quantitative agreement between the MCTDH and the exact result is reached (Fig.1c). Then each NO–vibrational excitation can be described by an independent set of single–particle functions. Since these excitations are the major source of correlational effects, all important correlational effects are now correctly included in the MCTDH–treatment. Calculations including four (Fig.1d) and five (Fig.1e) single–particle functions per coordinate demonstrate the convergence of the MCTDH–spectrum toward the exact one.

Comparing the numerical effort enormous differences are found between exact and MCTDH–calculations. The MCTDH–calculations with one, two ,three, four or five single–particle functions per degree of freedom require 5, 10, 20, 35 or 58 minutes of CPU–time on our IRIS–4D320 workstation (as discussed in section 3 the numerical effort increases linearly with the number of single–particle functions n for small n's. For larger n the n^4 scaling law becomes increasingly important.). On the other hand, the numerically exact calculation needs about 6 days of CPU–time on the same computer. The comparison demonstrates impressively the efficiency of the MCTDH–approach. Even the most expensive MCTDH–calculation that reproduces all exact results within graphical resolution requires over two orders of magnitude less CPU–time than the conventional exact calculation. Also the amount of memory required is drastically reduced by the MCTDH–approach.

Up to now we have judged the accuracy of the MCTDH–results by comparing them with the results of an exact calculation. But future studies may investigate systems by MCTDH–calculations for which exact calculations are impossible because of the numerical effort involved. Therefore it is necessary to develop criteria for estimating the accuracy that do not refer to other calculations. The inspection of the population of the natural single–particle functions (i.e. the eigenvalues of the single–particle density matrix) defined in section 2 provides such a criterion. Fig.2 displays the populations of the natural single–particle functions for the dissociation coordinate as a function of time. There the result of the calculation including five single–particle functions per degree of freedom is shown. Calculations employing fewer single–particle functions give similar results. The major difference is the absence of the weakest occupied single–particle functions, while approximately equal populations are found for the other ones.

One observes a rapidly decreasing population of the natural single–particle functions with increasing number. (The natural populations of the other degrees of freedom decrease even faster). The occupation probability of the first natural single–particle function is close to one (the numbering is in order of decreasing populations). This indicates that even a Hartree–approach may give a useful

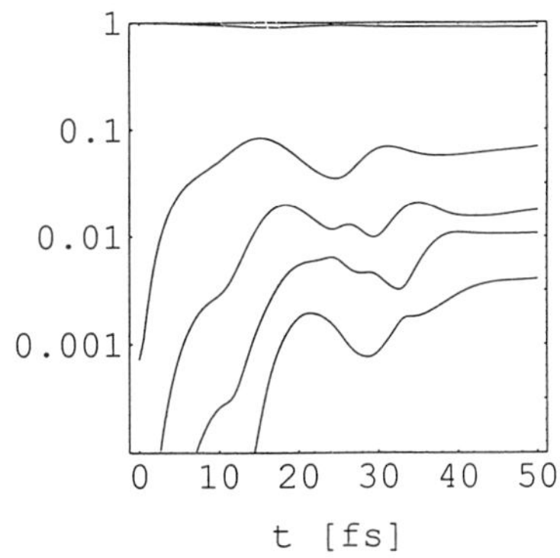

Fig. 2 Natural populations for the dissociative coordinate.

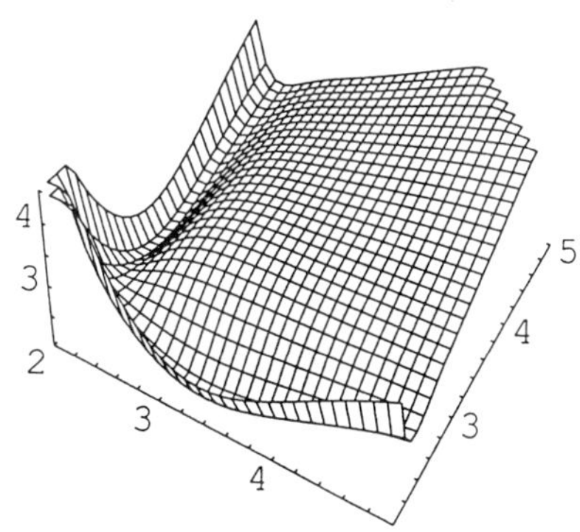

Fig. 3 Potential energy (in e.V.) of the B_2 state of NO_2 versus the two N–O distances (in a.u.). The bending angle is fixed at $\theta = 133^0$

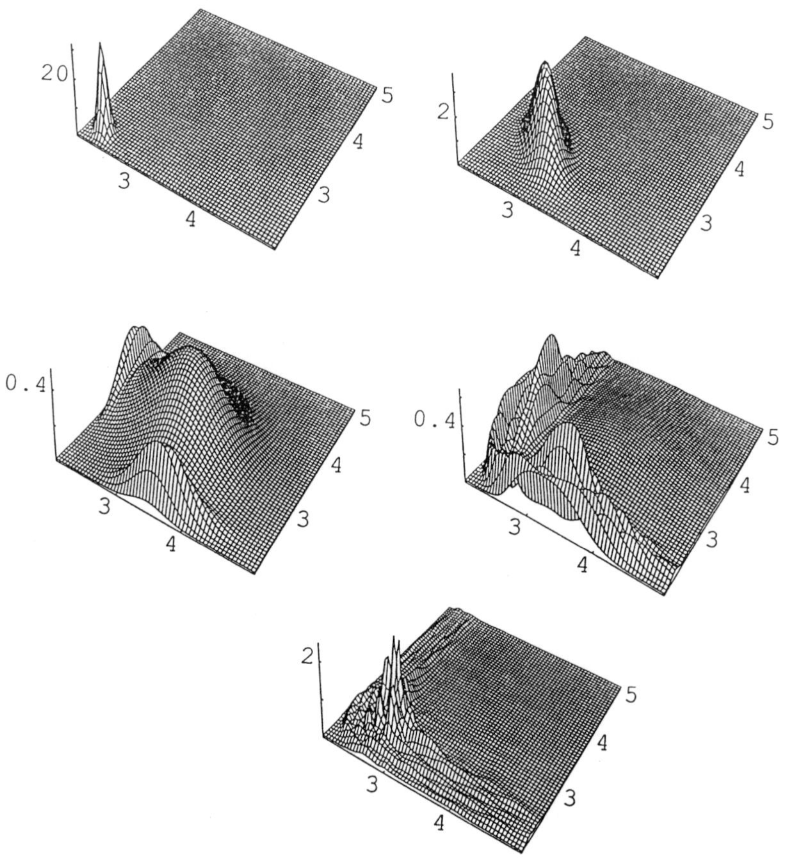

Fig. 4 Probability density versus the two N–O distances at times t = 0, 12, 24, 26 and 48 fs. (From left to right and from top to bottom)

description of the wave function. The population of the second natural single–particle function is below 10% everywhere. The contribution of the second single–particle function is of course neglected in the Hartree–approximation, implying that the error in the Hartree–wavefunction is of the order of 10%. Investigating the natural single–particle functions in other MCTDH–calculations, one also finds a rapid decrease of their population with increasing number. These functions give decreasing contributions to the description of the system demonstrating the convergence of the MCTDH–wavefunction.

4. THE ABSORPTION SPECTRUM OF NO_2

In the last section we have studied a system that was ideally suited for a MCTDH treatment. In this section we turn to discuss a problem that is among the hardest ones which can be treated nowadays by the time–dependent approach. We have studied the nuclear dynamics of the NO_2–molecule on its B_2 Born–Oppenheimer surface. This problem is difficult because i) one is investigating saddle–point dynamics, ii) the accessible volume in phase space is rather large and iii) the system is chaotic and the wave function will eventually fill the entire energetically allowed phase space. One hence cannot expect the wave function to be representable by a small number of configurations.

We have studied the photo absorption spectrum of the $A_1 \rightarrow B_2$ excitation of NO_2 using virtually the same techniques as above. However, when computing the spectrum by fourier transform of the autocorrelation function (see eq.(17)) we have introduced a simple but extremely helpful modification. We write the autocorrelation function as

$$\begin{aligned} \langle\Psi(0)|\Psi(t)\rangle &= \langle\Psi(0)|e^{-iHt}|\Psi(0)\rangle = \langle e^{iHt/2}\,\Psi(0)|e^{-iHt/2}\,\Psi(0)\rangle \\ &= \langle\Psi(t/2)^{*}|\Psi(t/2)\rangle \end{aligned} \tag{18}$$

where the last step follows because $\Psi(0)$ is real. This simple trick allowed us to stop the propagation after 50 fs whereas the conventional way would have required 100 fs because only then the autocorrelation function essentially vanishes.

The Born–Oppenheimer surfaces of both the A_1 and the B_2 state of NO_2 are taken from a recent publication of Hirsch and Buenker[10]. The non adiabatic coupling between these states is neglected. Fig.3 shows the potential surface of the B_2–state and Fig.4 depicts the wave packet as it evolves in time. One notices that the wave

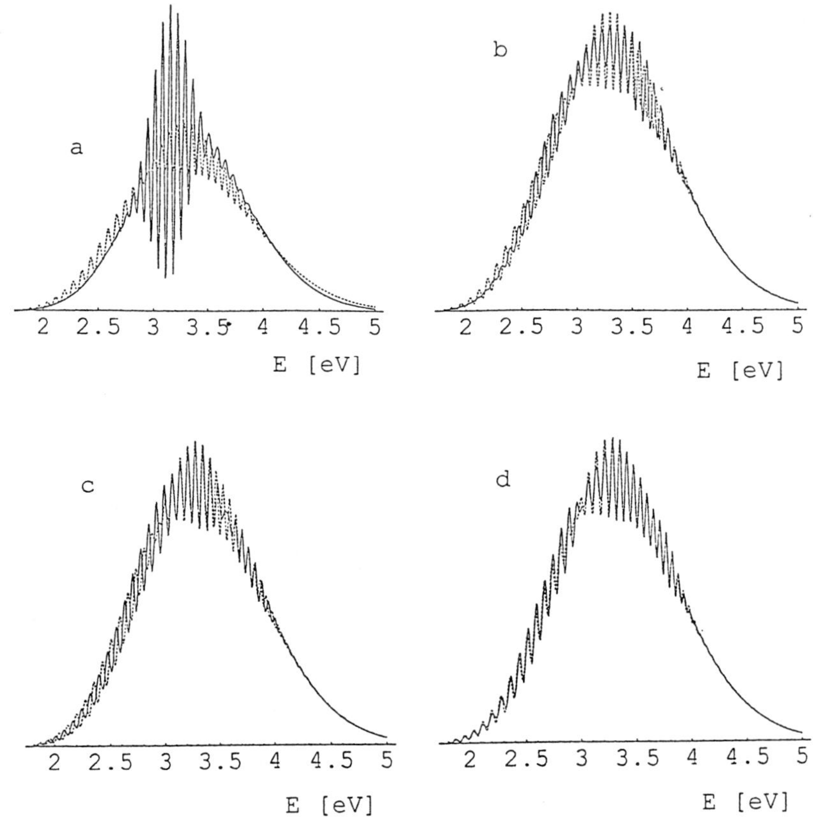

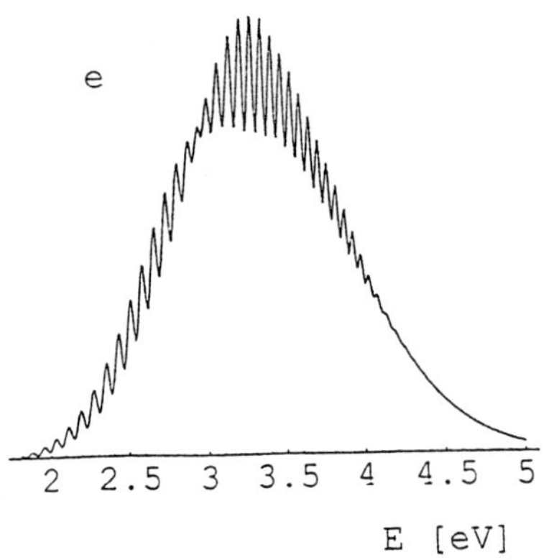

Fig. 5
Photoabsorption spectrum of NO_2 computed by using n = 1, 2, 4, 8, 12 functions per degree of freedom. The result of the n = 16 calculation is shown for comparison by a dashed line.

function rapidly acquires a very complicated structure. Fig.5 shows the absorption spectra computed by using n = 1, 2, 4, 8, and 12 single–particle functions for each degree of freedom. The spectrum obtained by employing n = 16 single–particle functions is indicated for comparison by a dashed line. Since the n = 12 and n = 16 calculations agree we conclude that we have reached convergence. Turning to the natural populations we see in Fig.6 that the smallest natural population is just above 10^{-2} when 16 functions are included. This supports our conclusion that convergence is reached.

We finally note that the computation times for the n = 1, 2, 4, 8, 12 and 16 functions calculations equaled 45min, 1.5h, 4h, 20h, 3 days and 9 days, respectively, on a IRIS 4D320 workstation. We estimate that a numerical exact propagation would take about 200 days on this workstation.

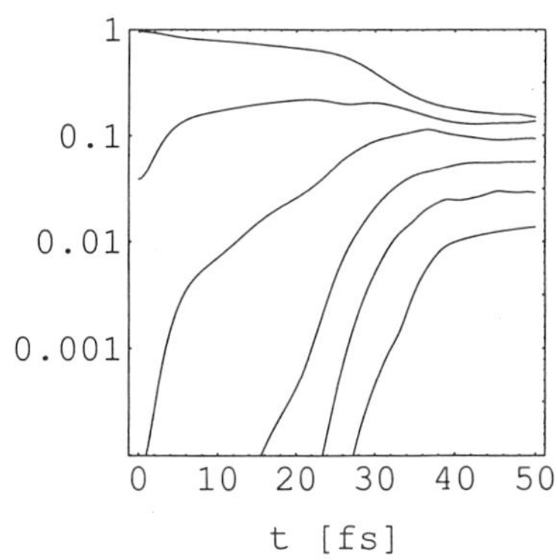

Fig. 6 Natural populations for the N–O–distance degree of freedom. Shown are the first, second, third, 4th, 8th, 12th and 16th eigenvalue of ρ versus time.

REFERENCES

1. C. Leforestier, R.H. Bisseling, C. Cerjan, M.D. Feit, R. Friesner, A. Guldberg, A. Hammerich, G. Jolicard, W. Karrlein, H.–D. Meyer, N. Lipkin, O. Ronccro and R. Kosloff, J. Comput. Phys. 94, 59, (1991)

2. E.J. Heller, J. Chem Phys. 62, 1544 (1975)

3. R.B. Gerber, V. Buch and M.A. Ratner, Chem. Phys. Lett. 91, 173 (1982); M. Messina and R.D. Coalson, J. Chem. Phys. 90, 4015 (1989)

4. J. Kucar, H.–D. Meyer and L.S. Cederbaum, Chem. Phys. Lett. 140, 525 (1987); J. Kucar, H.–D. Meyer and L.S. Cederbaum, J. Math. Phys. 29, 1417 (1988); J. Kucar and H.–D. Meyer, J. Chem. Phys. 90, 5566 (1989); J. Kucar and H.D. Meyer, Z. Phys. D 18, 325 (1991)

5. H.–D. Meyer, U. Manthe and L.S. Cederbaum, Chem. Phys. Lett. 165, 73 (1990)

6. N. Makri and W.H. Miller, J. Chem. Phys. 87, 5781 (1987)

7. R. Kosloff, J. Phys. Chem. 92, 2087 (1988)

8. A.E. Bruno, U.Brühlmann and J.R.Huber, Chem. Phys. 120, 155 (1988); A. Ticktin, A.E. Bruno, U. Brühlmann and J.R. Huber, Chem. Phys. 125, 403 (1988); A. Ogai, C.X.W. Qian, L. Iwata and H. Reisler, Chem. Phys. Lett. 146, 367 (1988); Y.Y. Bai, A. Ogai, C.X.W. Qian, L. Iwata, G.A. Segal and H. Reisler, J. Chem. Phys. 90, 3903 (1989); A. Ogai, C.X.W. Qian and H. Reisler, J. Chem. Phys. 93, 1107 (1990)

9. A. Untch, K. Weide and R. Schinke, J. Chem. Phys. 95, 6495 (1991)

10. G. Hirsch and R.J. Buenker, Mol. Phys. (in press)

ELECTRONIC ENERGY AND CHARGE TRANSFER IN SLOW ATOMIC COLLISIONS: A TIME-DEPENDENT MOLECULAR ORBITAL APPROACH

David A. Micha and Keith Runge

Departments of Chemistry and of Physics
University of Florida
Gainesville, Florida 32611, USA.

INTRODUCTION

We describe in what follows a *Time-Dependent Molecular Orbital (TDMO) theory* recently developed with two main objectives:

• to provide insight on electronically diabatic phenomena due to nuclear motions,

and

• to implement computational methods for many-electron, polyatomic systems.

By following the time-evolution of molecular orbitals, we hope to gain insight on the electronic charge fluctuations that occur during interactions. This insight is lost when the usual Born-Oppenheimer separation of electronic and nuclear motions is carried out, insofar as electronic variables are eliminated and nuclei are described as moving on potential energy surfaces. The present approach does not require the intermediate calculation of the PES's, but it nevertheless takes advantage of the large difference between electronic and nuclear masses, to construct MOs that describe the atomic conformation of the total system as it evolves due to nuclear motions.

Among the several approaches we could follow to describe both electrons and nuclei, we have chosen to work with an approach that can be extended to molecular systems with many degrees of freedom. This requires that we describe the nuclei in some classical limit, while we maintain a quantal description of the electrons; as a result, we shall be dealing with a practical approach in which classical and quantal degrees of freedom are coupled. It is desirable to do this starting from a general quantal treatment which allows us to take a classical limit for some degrees of freedom.

Time-Dependent Quantum Molecular Dynamics, Edited by J. Broeckhove
and L. Lathouwers, Plenum Press, New York, 1992

Phenomena of interest to us are:

- electron transfer , and
- electronic excitation and state polarization,

paying special attention to their time-evolution during interactions.

Our present studies are aimed at describing slow atomic collisions, where total kinetic energies are typically below about 100 eV, and the velocities of nuclei are as a result less that 0.1 au's. This is the range of interest in chemical applications, where one is dealing with complex molecular systems and averaged properties, such as thermal reaction rates. The same approach can however be applied with some modifications to a much higher range of kinetic energies, going up to 10,000 eV, where electronic and nuclear velocities are more comparable. This regime has been of interest to physicists for quite a long time; results of experimental and theoretical work are available from them for simple ion-atom systems and detailed properties such as cross sections.[1–3] For this reason the results to be presented here relate to both the lower and the higher energies.

In what follows, electronic states are obtained in the time-dependent Hartree-Fock (TDHF) approximation in the form originally proposed by Dirac, and later recast in terms of density matrices by McWeeney.[4] We do not therefore assume small density fluctuations as in the linearized versions introduced to describe responses to electromagnetic fields;[5,6] instead, we solve the TDHF equations in a way consistent with the strong couplings present in molecular interactions. They are furthermore coupled to nuclear classical motions in effective interaction potentials derived in an eikonal approximation.[7,8] Compared to other studies of electronically diabatic collisions, our approach differs in the that we explicitly solve for the electronic states during the interactions.

Molecular orbitals are described by time-dependent linear combinations of traveling atomic orbitals (LCTAO's), instead of the usual combinations of static atomic orbitals. This is necessary in a time-dependent treatment because as we shall see the standard expansions of Quantum Chemistry would lead to spurious couplings. The role of TAO's was clarified some time ago [9] and they have been successfully used by many authors for the higher collision energies. [10,11] Our use of MO-LCTAO's started some time ago in a study of extensions of the TDHF approach.[12,13]

The TDHF equations have been written in terms of the one-electron density operator, and have been solved by a local linearization in time with the use of a time-dependent reference density.[14] As we shall see, this provides a very convenient way to separate the calculation of the electron and nuclear dynamics, that occur over very different time scales. Other authors have also solved the TDHF equations for molecular systems, using numerical methods, [15–17] expansions in atomic and molecular basis sets, [18,19,2,3] and parametrizations of coherent states.[20,21]

We have performed many calculations to ascertain the validity of our approach. They relate to several physical systems, most of them consisting of ion-atom colliding pairs, [13,14] and some involving interactions of ions with solid metal surfaces [22]. They explored choices of effective potentials for the nuclear motions, and several electronic basis sets analyzed for the two-center one-electron problem. Here we shall only present a few results to illustrate the theoretical approach. We shall use examples where an electron with quantum numbers (*nlm*) may be transferred between two atoms *A* and *B*, in a process

$$A^+ + B(nlm) \rightarrow A^+ + B(n'l'm')$$
$$\rightarrow A(n''l''m'') + B^+ \ .$$

The derivations are however quite general, and the procedures followed in the applications, as well as the conclusions, are relevant to many-electron, polyatomic systems.

A recurring problem in approaches based on the TDHF approximation is the extent to which this can be trusted to provide correct transition probabilities. We shall at the end briefly summarize an extension of TDHF which uses its wavefunctions as trial functions within a time-dependent variational functional for transition amplitudes; earlier results[13] indicate that variational procedure appears very promising in the calculation of corrections to TDHF transition probabilities.

COUPLING OF ELECTRONIC AND NUCLEAR MOTIONS

The eikonal decomposition

The system Hamiltonian operator and the total wavefunction can be written as

$$H = -\frac{\hbar^2}{2M}\frac{\partial^2}{\partial \mathbf{R}^2} + H_{\mathbf{R}}(\frac{\hbar}{i}\frac{\partial}{\partial \mathbf{X}}, \mathbf{X})$$

$$\Psi(\mathbf{X}, \mathbf{R}) = \chi(\mathbf{X}, \mathbf{R}) exp[iS(\mathbf{R})/\hbar]$$

where $\mathbf{X}$ and $\mathbf{R}$ are the electronic and nuclear coordinates. The second term in the Hamiltonian describes the electronic system for fixed nuclei. The function S is real, while χ is complex-valued; in what follows we shall choose S for convenience to be an action integral approximating the quantal phase of the wavefunction; the phase of χ can compensate in principle for deviations from the correct phase.

For total energy E, we place the wavefunction in the time-independent Schroedinger equation to obtain

$$[\frac{1}{2M}(\frac{\hbar}{i}\frac{\partial}{\partial \mathbf{R}} + \frac{\partial S}{\partial \mathbf{R}})^2 + H_{\mathbf{R}} - E]|\chi(\mathbf{R})\rangle = 0$$

which is valid for any S . A specific equation for S follows by projecting the above equation on χ, and taking the real part of the result, which gives [7,8]

$$\frac{1}{2M}(\frac{\partial S}{\partial \mathbf{R}})^2 + V_{qu}(\frac{\partial S}{\partial \mathbf{R}}, \mathbf{R}) = E$$

$$V_{qu} = \langle\chi(\mathbf{R})|H_{\mathbf{R}}|\chi(\mathbf{R})\rangle / \langle\chi(\mathbf{R})|\chi(\mathbf{R})\rangle$$
$$+ V^{(1)} + V^{(2)}$$

where $V^{(1)}$ and $V^{(2)}$ contain $\partial\chi/\partial\mathbf{R}$ and $\partial^2\chi/\partial\mathbf{R}^2$ respectively, and the first term can be identified as the Ehrenfest potential for the nuclear motions.

Defining the momentum

$$\mathbf{P}_{\mathrm{R}} = \partial S/\partial\mathbf{R}$$

trajectories of nuclei follow, introducing the time t, from

$$H_{qu} = \mathbf{P}_{\mathrm{R}}.\mathbf{P}_{\mathrm{R}}/(2M) + V_{qu}(\mathbf{P}_{\mathrm{R}}, \mathbf{R})$$

$$d\mathbf{R}/dt = \partial H_{qu}/\partial\mathbf{P}_{\mathrm{R}}$$

$$d\mathbf{P}_{\mathrm{R}}/dt = -\partial H_{qu}/\partial\mathbf{R}$$

for initial $(\mathbf{P}_{\mathrm{Ra}}, \mathbf{R}_{\mathrm{a}})$ and give

$$S(\mathbf{R}) = S(\mathbf{R}_{\mathrm{a}}) + \int_{\mathbf{R}_{\mathrm{a}}}^{\mathbf{R}} d\mathbf{R}.\mathbf{P}_{\mathrm{R}} \quad .$$

The TDHF—eikonal approximation

We next introduce the following two simplifying assumptions:

- choose for χ

$$\chi(\mathbf{X}, \mathbf{R}) = \chi_I(\mathbf{R})\Phi_I(\mathbf{X}; \mathbf{R})$$

where Φ_I is a TDHF determinant, for an initial electronic state I , and $\chi_I(\mathbf{R})$ measures the divergence of trajectory bundles, and

- neglect $V^{(1)}$ and $V^{(2)}$ in a short-wavelength (eikonal) approximation to obtain the effective potential

$$\mathcal{V}_I(\mathbf{R}) = \langle\Phi_I(\mathbf{R})|H_{\mathrm{R}}|\Phi_I(\mathbf{R})\rangle/\langle\Phi_I(\mathbf{R})|\Phi_I(\mathbf{R})\rangle$$

and from it an average effective potential

$$\mathcal{V}(\mathbf{R}) = \sum_I f_I \mathcal{V}_I(\mathbf{R})$$

with chosen fractional weights f_I, adding over energetically accessible states I.

The TDHF states are usually acceptable to describe short-time events, and to calculate expectation values of single-electron properties. Improved treatments of the electronic states could be based on superpositions of electronic configurations, time-dependent multiconfigurational HF states, or time-dependent coupled cluster states, or on the variational improvement of TDHF we shall describe in a following section.

MOLECULAR ORBITALS AS LINEAR COMBINATIONS OF TRAVELING ATOMIC ORBITALS

The TDHF equation for the density operator is

$$\hat{F}^{\gamma}\hat{\rho}^{\gamma} - \hat{\rho}^{\gamma}\hat{F}^{\gamma} = i\hbar\partial\hat{\rho}^{\gamma}_{.}/\partial t$$

$$\hat{\rho}^{\gamma}(t) = \sum_{occ\ i} |\psi_i^{\gamma}(t)\rangle\langle\psi_i^{\gamma}(t)|$$

where $\hat{\rho}$ is the electron density operator,

$$\hat{F} = \hat{H}^{(1)} + \hat{G}^{\gamma}[\hat{\rho}^{\alpha}, \hat{\rho}^{\beta}]$$

is the Fock operator written as a sum of a one-electron term plus the self–consistent potential, and the ψ_i^{γ} are MO's for electron spin $\gamma = \alpha, \beta$.

If these MO's were expanded, as usual in quantum chemistry, in a basis of static atomic orbitals $\{\chi_{\mu}\}$, and then replaced in the TDHF equation, we would find spurious asymptotic couplings appearing in the form of $\langle\chi_{\mu}|\partial\chi_{\mu'}/\partial t\rangle$ terms where the two AO's are located at the same nucleus. To avoid this problem, MO's are instead expanded as linear combinations of traveling atomic orbitals (TAO's) ξ_{μ},

$$\psi_i^{\gamma}(\vec{r}, t) = \sum_{\mu} \xi_{\mu}(\vec{r}, t)c_{\mu i}^{\gamma}(t)$$

$$\xi_{\mu}(\vec{r}, t) = \chi_{\mu}(\vec{r})T_m(\vec{r}, t)$$

where χ_{μ} is an AO centered at nuclear position $\vec{R}_m(t)$, and

$$T_m(\vec{r}_j, t) = \exp\{im_e[\vec{v}_m(t)\cdot\vec{r}_j - \int_{t_a}^{t} dt' v_m^2(t')/2)]\}$$

is an electron translation factor.[9–11]

The density operator in this basis is

$$\rho^{\gamma}(t) = \sum_{\mu\nu} |\xi_{\mu}\rangle P_{\mu\nu}^{\gamma}(t)\ \langle\xi_{\nu}|$$

where $P_{\mu\nu}$ is the $\mu\nu$–element of the density matrix $\mathbf{P}^{\gamma}$ satisfying

$$\mathbf{S}^{-1}\mathbf{F}_{\mathrm{T}}^{\gamma}\mathbf{P}^{\gamma} - \mathbf{P}^{\gamma}(\mathbf{F}_{\mathrm{T}}^{\gamma})^{\dagger}\mathbf{S}^{-1} = \mathrm{i}\dot{\mathbf{P}}^{\gamma}$$

with $\mathbf{S}^{-1}$ the inverse of the overlap matrix of the TAO basis, and $\mathbf{F}_T^{\gamma}$ is the matrix of the modified Fock-like operator in

$$\langle\xi_{\mu}|\hat{F}_T^{\gamma}|\xi_{\nu}\rangle = \int d^3r\ \xi_{\mu}^{*}T_n\hat{F}^{\gamma}T_n^{-1}\xi_{\nu}$$

The rate of change of the density matrix is given by a set of first-order coupled differential equations in *t*. The solution to these equations is problematic as *the density matrix may oscillate significantly.*

THE LOCAL INTERACTION PICTURE

Given the starting values $\rho_0 = \rho(t_0)$ and $F_0 = F(t_0)$, and the time-evolution operator

$$U_0(t) = exp[-iF_0(t-t_0)/\hbar]$$

the density operator

$$\rho^{(0)}(t) = U_0(t)\rho_0 U_0(t)^\dagger$$

describes *electronic relaxation* at $t > t_0$ for *nuclei fixed* at $\mathbf{R}_0$.

For *moving nuclei*, with the Fock operator change $\Delta F(t) = F(t) - F_0$ giving their driving energy, we define the *local interaction picture* of an operator A by means of

$$A_L(t) = U_0(t)^\dagger A(t) U_0(t)$$

and in particular introduce the density change

$$\sigma_L(t) = \rho_L(t) - \rho_0$$

satisfying

$$i\dot{\sigma}_L(t) = D_L(t) + \Delta F_L(t)\sigma_L(t) - \sigma_L(t)\Delta F_L(t)$$

$$D_L(t) = \Delta F_L(t)\rho_0 - \rho_0 \Delta F_L(t)$$

Insofar both $\Delta F_L(t)$ and $\sigma_L(t)$ go to zero linearly with vanishing nuclear displacements, we find that retaining only the driving term $D_L(t)$ leads to a *local linearization in time* of the differential equation for $\sigma_L(t)$, which can be readily integrated to obtain

$$\rho(t) = \rho^{(0)}(t) - i\ U_0(t)[\int_{t_0}^{t} dt'\ D_L(t')]U_0(t)^\dagger\ .$$

LOCAL LINEARIZATION IN TIME OF THE MATRIX TDHF EQUATIONS

The operators $\rho^{(0)}(t)$ and $\sigma_L(t)$ have matrix representations $\mathbf{P}^{(0)}(t)$and $\mathbf{Q}_L(t)$ in the basis of TAO's. For fixed nuclei (and fixed spin γ) the density matrix $\mathbf{P}^{(0)}(t)$ satisfies

$$\mathrm{i}\dot{\mathbf{P}}^{(0)}(\mathrm{t}) = \mathbf{W}_0\mathbf{P}^{(0)}(\mathrm{t}) - \mathbf{P}^{(0)}(\mathrm{t})\ \mathbf{W}_0^\dagger$$

where $\mathbf{W} = \mathbf{S}^{-1}\mathbf{F}_\mathrm{T}$ and $\mathbf{W}_0 = \mathbf{W}(\mathrm{t}_0)$, or equivalently

$$\mathbf{P}^{(0)}(t) = \mathbf{U}_0(t)\mathbf{P}_0\mathbf{U}_0(t)^\dagger$$
$$\mathbf{U}_0(t) = exp[-i\mathbf{W}_0(t-t_0)]\ .$$

The equation for the density matrix change in the linearized approximation is then simply

$$\mathrm{i}\dot{\mathbf{Q}}_L(\mathrm{t}) = \mathbf{D}_L(t)$$

where the right hand side is the local interaction picture form of

$$\mathbf{D}(t) = \mathbf{\Delta W}(t)\mathbf{P}(t)^{(0)} - \mathbf{P}(t)^{(0)}\Delta\mathbf{W}(t)$$

This driving term **D** contains therefore contributions from both the *nuclear motions and electron density relaxation.*

The original change in the density matrix is given by

$$\mathbf{Q}(t) = \mathbf{U}_0(t)\mathbf{Q}_L(t)\mathbf{U}_o(t)^\dagger$$

and can be explicitly constructed at time t_1 from the integrals

$$\Gamma_{k\nu\rho l}(t_1, t_0) = \int_{t_0}^{t_1} exp[-i(w_k - w_l)(t_1 - t')]D_{\nu\rho}(t')dt'$$

where the w_k are eigenvalues of the $\mathbf{W}_0$ matrix. Therefore, one may integrate the equations step by step from beginning to end *without encountering the problem of rapid variation* in the density matrix, as $\mathbf{Q}(t)$ is constructed analytically within each time interval.

If the electron-electron interaction is absent one can write $\mathbf{W}_0 = \mathbf{S}_0^{-1}\mathbf{H}_0$ in terms of the one-electron Hamiltonian matrix **H** at times t_0 and t .

TIME-DEPENDENT ATOMIC PROPERTIES

Atomic populations

Atomic populations provide insight on the time evolution of the interacting atoms even though they can not be defined in a unique way. Two useful definitions are based on the Mulliken and Lowdin decompositions [23] of density functions. In the Mulliken decomposition we write the density operator as the sum

$$\rho(t) = \sum_a \rho_a^M(t)$$

$$\rho_a^M(t) = \sum_{\mu\epsilon a}\sum_{\nu}[|\xi_\mu\rangle P_{\mu\nu}(t)\langle\xi_\nu| + h.c.]/2$$

$$N_a^M = \sum_{\mu\epsilon a}\sum_{\nu}[P_{\mu\nu}(t)S_{\nu\mu} + c.c.]/2.$$

where N_a is the ocupation number of center a.
Alternatively we can introduce the orthonormal (travelling) Lowdin orbitals

$$|\eta_\lambda\rangle = \sum_\mu |\xi_\mu\rangle(\mathbf{S}^{-1/2})_{\mu\lambda}$$

and the decomposition

$$\rho(t) = \sum_a \rho_a^L(t)$$

$$\rho_a^L(t) = \sum_{\lambda \epsilon a} \sum_{\mu} (|\eta_\lambda\rangle P_{\lambda\mu}^L \langle \eta_\mu| + h.c.)/2$$

$$N_a^L = \sum_{\lambda \epsilon a} \sum_{\mu,\nu} ((\mathbf{S}^{1/2})_{\lambda\mu} P_{\mu\nu} (\mathbf{S}^{1/2})_{\nu\lambda} + c.c.)/2$$

These two decompositions give numerically different results, but one expects that they will provide qualitatively similar descriptions. It is nevertheless of interest to find out which one would be more appropriate for describing atomic interactions.

Atomic polarization parameters

Two useful measures of atomic orbital polarizations relate to *orientation* , in which magnetic substates *m* have differing populations, and *alignment* , where magnetic substates |*m*| have differing populations. They can be measured by parameters defined in terms of atomic angular momenta. [24]

Introducing the spherical tensor operators $L_q^{(n)}$, $-n \leq q \leq n$, derived from the electronic orbital angular momenta $\vec{L}$ *and* L_z with quantum numbers (*l,m*), the orientation vector $O_q^{(1)}$ and alignment tensor $A_q^{(2)}$ are, for center *a* ,

$$O_{q,a}^{(1)}(l_a) = \frac{1}{\sqrt{(l_a+1)l_a}} Re[tr(\rho_a L_{q,a}^{(1)})]/N$$

$$A_{q,a}^{(2)}(l_a) = \frac{\sqrt{6}}{(l_a+1)l_a} Re[tr(\rho_a L_{q,a}^{(2)})]/N$$

For excitation of a one-electron target, choosing the z-axis along the projectile incoming direction and the (xz)-plane as the trajectory plane, the alignment for the AO with $(nlm) = (2p_m)$ follows from

$$L_{0,a}^{(2)} = (3L_{z,a}^2 - \vec{L}_a^2)/\sqrt{6}$$

and is given at time *t*, for a collision with impact parameter *b*, by

$$A_{0,a}^{(2)}(t;b) = N_a(2p_\perp, t; b)/2 - N_a(2p_\parallel, t; b)$$

in terms of the populations of p-orbitals perpendicular and parallel to the initial velocity of the projectile atom.

Cross sections

Given an average effective potential, the inelastic differential cross section for impact parameter *b* and the transition $I \rightarrow J$ can be written as

$$(\frac{d\sigma}{d\Omega})_{I\rightarrow J} = (\frac{d\sigma}{d\Omega})_{AEP}\ P_{I\rightarrow J}[b(\Theta)]$$

$$P_{I\rightarrow J} = \langle \Phi_J | \rho_I | \Phi_J \rangle_{t_f}$$

in terms of the elastic differential cross section in the AEP, and of the transition probability at the final time.

For a one-electron target, and the transition $\mu \to \nu$, the cross sections follow from the populations at the final time, insofar as $\Phi_J = \xi_\nu$ and $\rho_I = \rho_\mu$, so that

$$(\frac{d\sigma}{d\Omega})_{\mu\to\nu} = (\frac{d\sigma}{d\Omega})_{AEP}\ N_\nu^{(\mu)}[t_f; b(\Theta)]$$

COMPUTATIONAL PROCEDURE

Effective interaction potentials

For initial state I , the average potential in the TDHF approximation is

$$\begin{aligned}\mathcal{V}_I(\mathbf{R}) &= V_{cc}(\mathbf{R}) + tr[\rho_I(H^{(1)} + F)/2]/N \\ &= V_{cc}(\mathbf{R}) + \sum_{\mu,\nu} P^I_{\mu\nu}(\mathbf{R})[H_{\nu\mu}(\mathbf{R}) + F_{\nu\mu}(\mathbf{R})]/(2N)\end{aligned}$$

where the first term is the sum of core-core interactions, and the second term is the electronic energy from TDHF. Forces are obtained taking the gradients of *all* its terms dependent on $\mathbf{R}$.

Average effective potentials follow from

$$\mathcal{V}(\mathbf{R}) = \sum_I f_I \mathcal{V}_I(\mathbf{R})$$

for given fractional occupations f_I . The advantage of using these potentials is that, because they do not depend on the initial state, the corresponding trajectories satisfy microscopic reversibility.

For a one-electron target, using the eigenstates of $\mathbf{W}(t)$, we have chosen a set of weights $\{f_\lambda, \lambda = 1\ to\ N_B\}$ so that

$$\mathcal{V}(\mathbf{R}) = \sum_\lambda f_\lambda \mathcal{V}_\lambda(\mathbf{R}) = V_{cc}(\mathbf{R}) + N_A^{-1}\sum_{j=1}^{N_A} w_j$$

where N_A is the number of energetically accessible states. The average of MO energies properly accounts for the screening effect of electrons during the interaction at long distances, and it is easily constructed as one integrates the TDHF equations.

Atomic basis sets

The type of atomic basis sets to be used in atomic collisions must be carefully considered, because requirements are different from those in quantum chemistry, which deals with electronic structure for fixed nuclei usually around equilibrium positions. In collisions, and particularly while describing electron transfer, it is important to properly describe the exponential tails of AOs and their kinks at the nuclei. It is not safe to assume that the standard linear combinations of Gaussian orbitals (GO's) in the literature [23] would be applicable in our problems. For these reasons we have chosen to first obtain results with hydrogenic orbitals, written as combinations of Slater–type orbitals (STO's), and then to compare these results with linear combinations of Gaussians, with parameters chosen to describe each whole hydrogenic orbital.

Numerical procedure

The numerical calculations require integration of the differential equations for the nuclear trajectories in the effective potentials, coupled to the integrated form of the TDHF density matrices. Time can be advanced in variable step sizes, to minimize the total number of steps between initial and final conditions. We choose the step $\Delta t = t_1 - t_0$ so that the relative change in the magnitude of the density matrix is within both upper and lower tolerances, by requiring that

$$\epsilon_{min} \leq ||\sigma||/||\rho^{(0)}|| \leq \epsilon_{max}$$

We further expand the integrand in the linearized expression for the density change around $t_{1/2}$ to obtain $||\sigma|| = \mathcal{O}(\Delta t)^2$ and its error to higher order in Δt .

To advance both trajectories and densities over time, we start from values at $t_{-1/2}$ and t_0 ,

$$[(\mathbf{R}, \mathbf{P}, \rho^{(0)}, \mathcal{V})_{-1/2} \ , (\mathbf{R}, \mathbf{P}, \rho, \mathcal{V})_0]$$

to calculate

$$(\mathbf{R}, \mathbf{P}, \rho^{(0)}, \mathcal{V})_{1/2}$$

using the Runge-Kutta 4th order algorithm [25] for the Hamiltonian equations, and then use local linearization to get

$$(\mathbf{R}, \mathbf{P}, \rho, \mathcal{V})_1$$

This provides the starting values for the next step in time.

The efficiency of the algorithm used to advance the trajectory can be easily improved, but does not have a noticeable effect on computation times because most of the time goes into the calculation of the Hamiltonian matrix elements. It is also worth pointing out that because the time intervals are restricted to small increments by the rapid changes of the electronic densities, there are always enough points along trajectories to calculate forces numerically. Thus, although there are now efficient ways to analytically generate forces with Quantum Chemical packages, there is no need to do this along the trajectories.

APPLICATIONS

Atomic systems

The following atom-atom pair collisions have been studied,

$$\begin{aligned} H^+ + H(1s\alpha) &\rightarrow H^+ + H(nlm\alpha) \\ &\rightarrow H(n'l'm'\alpha) + H^+ \end{aligned}$$

from 2 to 2,000 eV in the LAB frame, [14]

$$He^{+2} + He^{+}(1s\alpha) \rightarrow He^{+2} + He^{+}(nlm\alpha)$$
$$\rightarrow He^{+}(n'l'm'\alpha) + He^{+2}$$

from 200 to 2,500 eV in the COM frame, [26] and

$$He^{+2} + H(1s\alpha) \rightarrow He^{+2} + H(nlm\alpha)$$
$$\rightarrow He^{+}(n'l'm'\alpha) + H^{+}$$

at 500 and 8,000 eV in the LAB frame. [26]

The same approach, with a space partitioning procedure suitable for extended systems, was applied to the atom-metal surface collision events [22]

$$Na^{+} + W(011) \rightarrow Na + W(011)^{+}$$

from 1.0 to 100.0 au in the LAB frame.

Atomic basis sets

Our first calculations were done with a hydrogenic (1s)/nucleus, which suffices for $H^{+}+H$ below velocities of about 0.25 au's , where electronic excitation is improbable. This was followed by a study where the hydrogenic orbital was fit by a series of n Gaussians, for increasing n, to ascertain the accuracy of the GO expansions.[26] The advantage of using Gaussians is that there are thoroughly tested quantum chemistry packages, such as VMOL,[27] which can be used to generate all the one- and two-electron integrals needed to calculate the matrix elements of the Fock operator.

The more recent calculations were done with hydrogenic (1s, 2s, 2p)/nucleus, which allows for calculations of a number of ionic collisions provided the net core charges are similar. They were also compared with a limited set of our calculations using combinations of GO's.

Finally, in the case of scattering of an ion by a metal surface, we introduced localized orbitals in the metal, properly accounting for the presence of the surface, by constructing generalized Wannier functions at each metal center; to this basis we added STOs for pseudopotentials representing the core of the projectile ions (3s and 3pxyz STOs for the Na ion).[22]

Effective interaction potentials

The accuracy of interaction potentials becomes critical as collision energies are lowered, and electronic rearrangements leading to screening become more pronounced during collisions. We have used a zero-potential (ZP) model (or straight line potential) at the higher velocities, above about 0.20 au. For lower velocities, down to about 0.10 au's, we used the Coulomb potential (CP) due to nuclear repulsion.

This was tested against results with average effective potentials (AEP) which include screening (also called SCP's) and were obtained from the eigenenergies of the electronic Hamiltonian. These are certainly needed at velocities below about 0.05 au's, but it is of interest to numerically investigate how high one can go above these values before screening effects can be neglected.

Observables

In order of increasing demands on the quality of the models and the numerical accuracies, we have calculated: (a) integral cross sections for elastic and inelastic electron transfer, direct and exchange excitation; (b) differential cross sections for direct collisions and electron transfer; and (c) alignment parameters.

These studies were supplemented by calculations of time-dependent properties such as occupation numbers and orbital alignments versus time; although these properties can not be uniquely defined during collisions, they provide useful insight into their mechanisms.

Discussion of selected results

We describe here some results for electronic charge and energy transfer in ion-atom collisions over a wide range of energies. We concentrate on the simplest system, $H^+ + H$, to comment on features common to many of the systems we mentioned above. For $H^+ + H$, we had previously found that our results at a laboratory energy of 1 keV, with the ZP potential, were in good agreement with early results by Gallaher and Wilets [28]. Calculations have been done from 2 eV to 2 keV using the AEP and CP potentials and have been compared to other theoretical and experimental results.[29,30] Our results are shown in Fig. 1, where the full line gives our calculations, and the points are the experimental results; excellent agreement is found over three orders of magnitude in collision energies. Probabilities of electron transfer versus impact parameters have also been obtained over a wide range of energies and have been confirmed by recent results;[21] at the higher energies it is sufficient to work with the ZP potential, but as energies are lowered this, and also the CP potential become inaccurate. Results at 10 eV with the CP and SCP potentials, shown in Fig. 2, indicate why the latter (which includes screening) is needed at low energies; screening of the Coulombic nuclear repulsion allows for closer encounters at high impact parameters and, correspondingly, larger transfer probabilities.

Expansions with GO's were tested comparing the result for a basis of a hydrogen 1s orbital/center to results obtained using contracted Gaussians (1s-nG) for n=3,4,5,6 . At the larger impact parameter region all of the 1s-nG results are close to the hydrogen 1s result. At intermediate impact parameters the 3G result deviates somewhat from the 1s exact result, and at very small impact parameters all the nG results deviate from the hydrogen 1s, with 6G being most accurate.

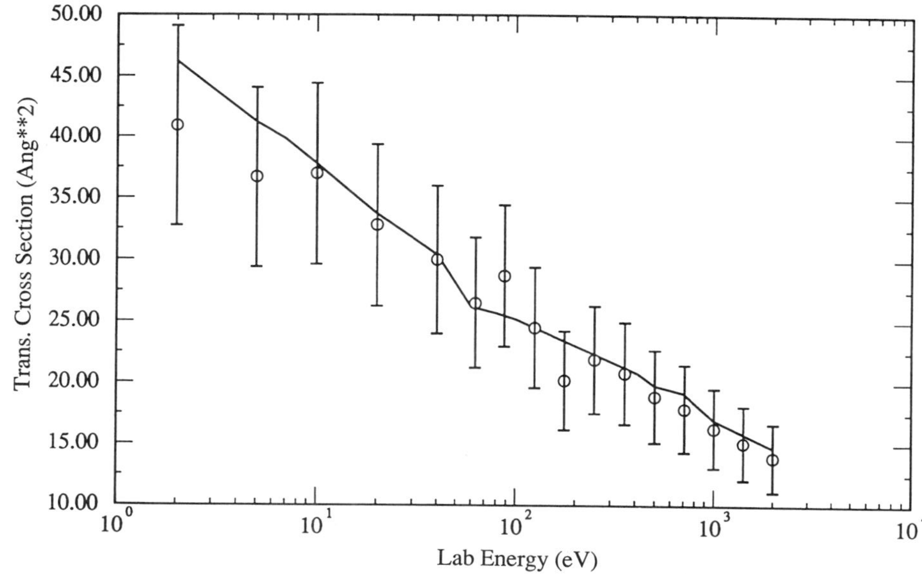

Fig. 1. Electron transfer integral cross section vs. laboratory collision energy for p + H . Full line: present results; circles : experiment.

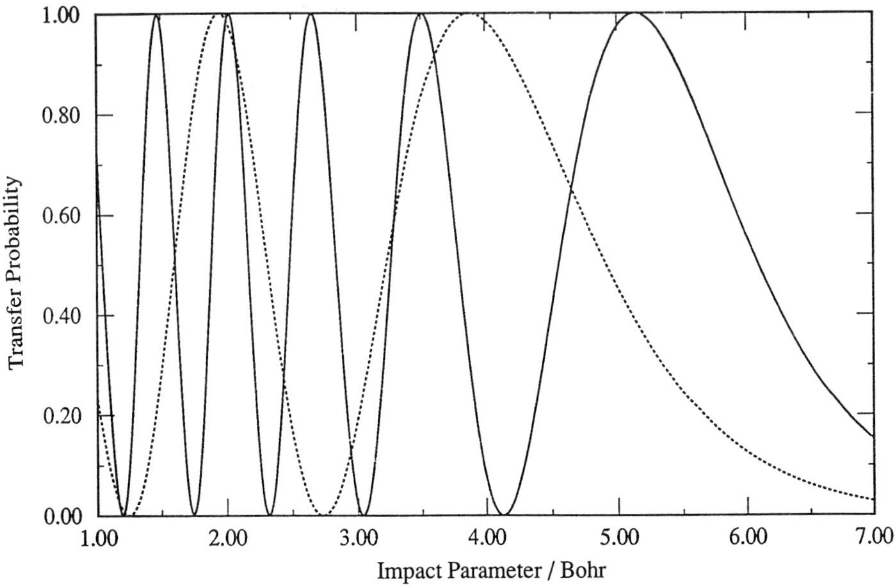

Fig. 2. Electron transfer probability vs. impact parameter for p + H collisions at a laboratory energy of 10.0 eV. Full line: SCP; dot line: CP from nuclear repulsion.

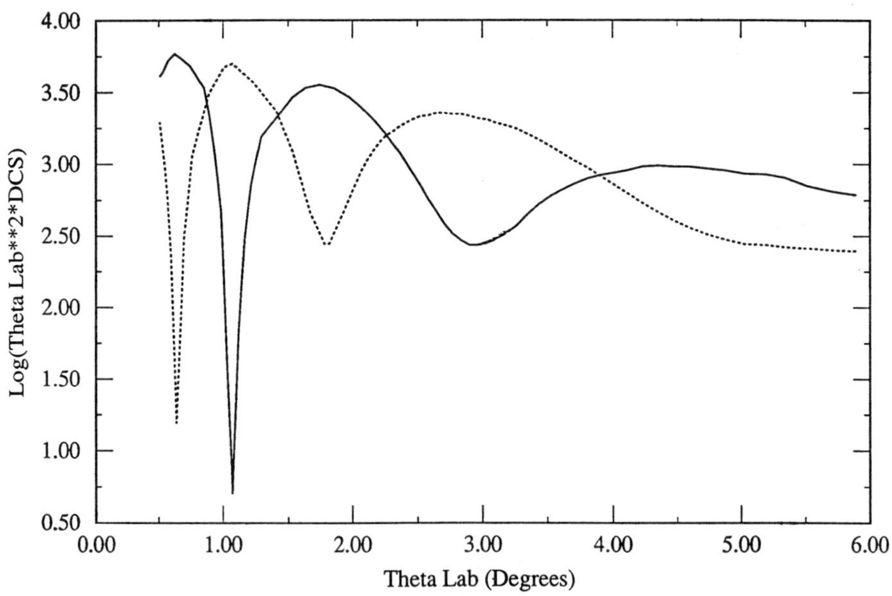

Fig. 3. Differential cross section for p + H in the laboratory frame at a collision energy of 500.0 eV. Full line: direct elastic scattering; dot line: exchange elastic scattering.

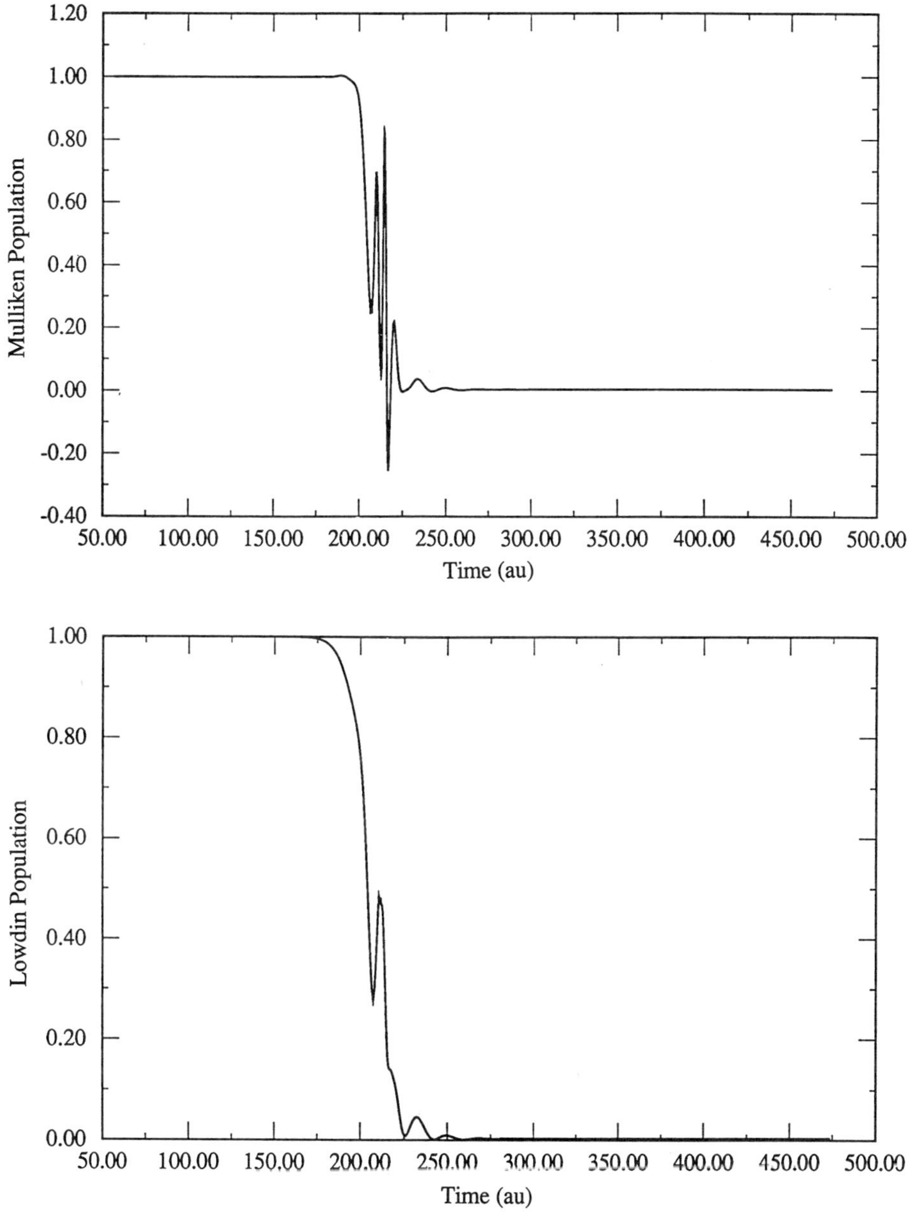

Fig. 4. Electron populations of the target atom 1s orbital vs. time for p + H at 2.0 keV in the LAB frame, for the impact parameter b = 0.2 au . Upper panel: Mulliken population; lower panel: Lowdin population.

An AO basis set consisting of 1s, 2s, 2px, 2py, and 2pz hydrogenic functions /center has been used for direct and exchange excitations, and alignment parameters, and for total charge transfer cross section in this system. Figure 3 shows differential cross sections in the LAB frame at a collision energy of 500 eV, for both direct elastic scattering and for total electron exchange, versus LAB angles. The dephasing of the oscillations and the spacing of peaks agree very well with experimental results [31] and other calculations.[32] Our calculations were done with both the CP and the AEP (or screened Coulomb) potential, and show that the second is needed to obtain good differential cross sections, even though the CP potential would have been acceptable for integral cross section calculations.

Additional insight could be extracted from the time-dependence of atomic populations. Figure 4 shows Mulliken and Lowdin populations for the target 1s orbital, at an energy of 2,000 eV and an impact parameter of 0.2 au's. It is seen that while the time evolutions of the two quantities are qualitatively similar, the Lowdin populations remain always positive, as populations should, but that the Mulliken populations become negative at intermediate times. This has also been found at other energies and impact parameters. On this basis it appears that the Lowdin population analysis is better suited for analyses of time evolving properties.

The system $He^{2+} + H$ has also been studied with the same basis of 5 hydrogenic orbitals/center. Electron transfer occurs in this case with higher probability into the electronically excited states of He^{+} with n = 2, that are resonant in energy with n = 1 of H.

Some time ago we also did extensive calculations on collisions in the two electron system $He^{2+} + He$, which can lead to one– and two– electron transfer. Results in the range of 10 — 100 keV were obtained directly from TDHF wavefunctions, and also from a variational procedure which uses TDHF trial functions.[13] Comparing exact and TDHF results, with a minimal basis set, it became clear that the latter could not be relied upon because the two final charge states were averaged by the TDHF description, giving unphysical results. The variational procedure was however very successful in correcting for the shortcomings of the TDHF description. We briefly describe in what follows how the variational procedure could be used to improve the numerical values of properties we have recently calculated.

VARIATIONAL IMPROVEMENT OF TDHF PROBABILITIES

Standard TDHF Probabilities

Transition probabilities can be calculated by evolving an electronic wavefunction from an initial time to a final time at which it is projected on a final state, or it can be extracted from a wavefunction starting at the final time, evolved backward in time, and projected on the initial state. In our case, we can calculate forward in time the TDHF function $|\Phi_I^{(+)}(t)\rangle$ *from the initial state* $\Phi_I^{(a)} = \Phi_I^{(+)}(t_a)$ at $t = t_a$, and project at the final time t_b on the final state $\Phi_J^{(b)}$ to find the transition amplitude

$$S_{JI}^{(+)} = \langle \Phi_J^{(b)} | \Phi_I^{(+)}(t_b) \rangle$$

or we can calculate backward in time *from the final state* $\Phi_J^{(b)}$ at t_b and find the amplitude

$$S_{JI}^{(-)} = \langle \Phi_J^{(-)}(t_a) | \Phi_I^{(a)} \rangle$$

The time evolution of these electronic wavefunctions are determined by the classical trajectories of the nuclei in the effective potentials chosen for the forward and backward motions. When one uses the Ehrenfest potentials, one finds that trajectories in the forward and backward directions are different, and one loses microscopic reversibility. Even if one chooses to work with an average effective potential, common to both forward and backward motions, the problem of loss of reversibility remains as long as one is working only with approximate electronic wavefunctions. The loss of microscopic reversibility may be considered a warning about loss of accuracy in the calculation of transition probabilities. A possible action is of course to improve the quality of the electronic states; another way, particularly convenient in our case, is to use already calculated TDHF functions as trial functions in a variational procedure leading to corrections.

A Phase-invariant Variational Procedure

We have proposed to use TDHF trial wavefunctions in a variational functional of the Demkov-type, modified to avoid unphysical phase effects, that gives corrected transition amplitudes between specific *initial and final* states.[12]

To apply the procedure to transition amplitudes, construct the trial functions

$$|\Psi_I^{(\pm)}(t)\rangle = |\Phi_I^{(\pm)}(t)\rangle exp[-i\gamma_I^{(\pm)}(t)]$$

$$\gamma_I^{(\pm)}(t) = \alpha^{(\pm)}(t) + \int_{t_{1,2}}^{t} dt'\, \Delta_I^{(\pm)}(t')$$

$$\Delta_I^{(\pm)}(t) = \langle \Phi_I^{(\pm)}(t) | (H - i\partial_t) | \Phi_I^{(\pm)}(t) \rangle$$

where $\alpha^{(\pm)}$ includes the action integral from the nuclear trajectories. With these trial functions, calculate the new transition amplitudes

$$\mathcal{S}_{JI} = \frac{1}{2}(\langle \Phi_J^{(b)} | \Psi_I^{(+)}(t_b) \rangle + \langle \Phi_J^{(-)}(t_a) | \Phi_I^{(a)} \rangle) \\ -i\mathcal{A}_{JI}$$

$$\mathcal{A}_{JI} = \int_{t_a}^{t_b} dt\, [\langle \Psi_J^{(-)} | \mathcal{D}_t \Psi_I^{(+)} \rangle_t + \langle \mathcal{D}_t \Psi_J^{(-)} | \Psi_I^{(+)} \rangle_t]/2$$

$$\mathcal{D}_t = H - i\partial_t$$

Here the first two terms give the average of the forward and backward forms of the transition amplitudes obtained from the original TDHF wavefunctions; the third term is a transition action, defined so that it is independent of the trial functions phases (as expected on physical grounds). The total sum satisfies microscopic reversibility by construction.

This functional leads to errors of second order in the deviations of the trial functions from the accurate wavefunctions; the transition action therefore provides a correction to the original transition amplitudes. This procedure was successfully applied to $He^{2+} + He$,[13] and appears well suited for other electronic energy and charge transfer calculations.

CONCLUSION

Summarizing the formalism and experience gained from calculations carried out so far, we can state the following conclusions.

• The density matrix version of TDHF gives a compact description, physical insight, and simple (linear) equations.

• A basis of TAO's eliminates spurious couplings; the ETF's can be omitted in integrals at low nuclear velocities, but are needed at high velocities.

• Local linearization in time of the TDHF equations allows for separation of time scales without approximations.

• The eikonal approximation for nuclear trajectories, with average effective potentials, allows for screening effects and leads to time-reversible trajectories.

• TDHF calculations provide results in good agreement with more extensive calculations and with experiments, in several collisional studies involving electron transfer and excitation.

• The Lowdin population analysis provides more physically acceptable values of properties during interactions.

• The TDHF results can be improved with a phase-invariant variational functional for transition amplitudes, that leads to state-to-state results satisfying microscopic reversibility and energy conservation.

ACKNOWLEDGEMENTS

The authors thank Dr. John Stanton at the University of Florida for his advice on the use of the VMOL computer package. This work has been supported by the National Science Foundation of the USA, through its Chemistry Division. One of us (DAM) did part of this work at the Institute for Theoretical Atomic and Molecular Physics, Harvard University and Smithsonian Center for Astrophysics, supported by an NSF Grant.

REFERENCES

1. R. McCarroll, in "Atomic and Molecular Collision Theory," F. A. Gianturco, Ed., Plenum Press, New York, 1980, Chap. 2.
2. M. Kimura and N. F. Lane, Adv. At. Molec. and Opt. Phys. 26:79 (1990).

3. W. Fritsch and C. D. Lin, Phys. Rep. 202:1(1991).
4. a)P. A. M. Dirac, Proc. Cambridge Philos. Soc. 26:376 (1930); b) R. W. McWeeney, J. Mol. Structure 93:1 (1983).
5. A. Dalgarno and G. A. Victor, Proc. Roy. Soc. London Ser. A291:292 (1966).
6. A. D. McLachlan and M. A. Ball, Rev. Mod. Phys. 36:844 (1963).
7. D. A. Micha, J. Chem. Phys. 78:7138 (1983).
8. J. M. Cohen and D. A. Micha, J. Chem. Phys., to appear (1992).
9. D. R. Bates and R. McCarroll, Proc. Roy. Soc. London Ser. A245:175 (1958).
10. J. B. Delos, Rev. Mod. Phys. 53:287 (1981).
11. A. Macias and A. Riera, Phys. Rep. 90:299 (1982).
12. D. A. Micha and B. Gazdy, Phys. Rev. A36:539 (1987).
13. a) B. Gazdy and D. A. Micha, Phys. Rev. A33:4446 (1986); b) B. Gazdy and D. A. Micha, Intern. J. Quantum Chem. S20:689 (1986); c) B. Gazdy and D. A. Micha, Phys. Rev. A36:546 (1987).
14. K. Runge, D. A. Micha and E. Q. Feng, Intern. J. Quantum Chem. S24: 781 (1990).
15. C. Bottcher, Phys. Rev. 48:85 (1982).
16. K. C. Kulander, K. R. Sandhya-Devi, and S. E. Koonin, Phys. Rev. A25:2968 (1982).
17. K. R. Sandhya-Devi, and J. D. Garcia, J. Phys. B16:2837 (1983).
18. D. Rapp and D. Dinwiddie, J. Chem. Phys. 57:4919 (1972).
19. W. Stich, H. J. Ludde, and R. M. Dreizler, Phys. Lett. 99A:5 (1983); J. Phys. B18:1195 (1985).
20. J. W. Negele and H. Orland "Quantum Many-Particle Systems", Addison-Wesley, Redwood City, California, (1988).
21. E. Deumens, A. Diz, H. Taylor and N. Y. Ohrn, J. Chem. Phys., to appear (1992).
22. E. Q. Feng, D. A. Micha and K. Runge, Intern. J. Quantum Chem. 50:545 (1991).
23. A. Szabo and N. S. Ostlund "Modern Quantum Chemistry"McMillan Publ. Co., New York (1982), Chap. 3.
24. R. N. Zare, "Angular Momentum" J. Wiley, New York (1988), Chap. 5.
25. W. H. Press, B. P. Flannery, S. A. Teukolsky, and W. T. Vetterling "Numerical Recipes", Cambridge Univ. Press, Cambridge, England (1986), Chap. 15.
26. K. Runge and D. A. Micha, work to be published.
27. VMOL is a local version of MOLECULE, a vectorized Gaussian integral computer program, authored by J. Almlof and P. R. Taylor.
28. a). L. Wilets and D. F. Gallaher, Phys. Rev. 147: 147 (1966); b) D. F. Gallaher and L. Wilets, Phys. Rev. 169: 139 (1968).
29. M. W. Gealy and B. Van Zyl, Phys. Rev. A36:3091 (1987).
30. J. H. Newman, J. D. Cogan, D. L. Ziegler, D. E. Nitz, R. D. Lunden, K. A. Smith, and R. F. Stebbings, Phys. Rev. A25:2976 (1982).
31. J. C. Houver, J. Fayeton, M. Abignola, M. Barat, Phys. Rev. Lett. 28:1433 (1972).
32. C. Gaussorgues, C. Le Sech, F Masnou-Seeuws, R McCarroll, and A. Riera, J. Phys. B8: 253 (1975).

CUMULATIVE AND STATE-TO-STATE REACTION PROBABILITIES VIA A DISCRETE VARIABLE REPRESENTATION — ABSORBING BOUNDARY CONDITION GREEN'S FUNCTION

William H. Miller and Tamar Seideman

Department of Chemistry, University of California, and
Chemical Sciences Division, Lawrence Berkeley Laboratory
Berkeley, California 94720-9989 USA

I. INTRODUCTION

This paper reports some significant recent progress in the quest for a *direct* means of computing rate constants for chemical reactions. For a bimolecular reaction, for example, the rate constant k can be expressed compactly in terms of the *cumulative reaction probability*[1] (CRP) N(E),

$$k(T) = [2\pi\hbar Q_r(T)]^{-1} \int_{-\infty}^{\infty} e^{-E/kT} N(E) \;, \tag{1.1a}$$

which itself is defined as the sum of state-to-state reaction probabilities (the square moduli of S-matrix elements) over all energetically allowed quantum states of reactants (with quantum numbers n_r) and products (with quantum numbers n_p),

$$N(E) = \sum_{n_r,n_p} |S_{n_p,n_r}(E)|^2 \;. \tag{1.1b}$$

Q_r in Eq. (1.1a) is the reactant partition function, and E is the total energy of the molecular system. (In some cases, typically for unimolecular reactions, the microcanonical rate is the quantity of interest, and it is also given in terms of the CRP,

$$k(E) = [2\pi\hbar\rho_r(E)]^{-1} N(E) \;, \tag{1.1c}$$

where ρ_r is the density of reactant states per unit energy.) By a "direct" way of calculating the rate one means a way that by-passes Eq. (1.1b), i.e., that avoids having to obtain the S-matrix (which is tantamount to solving the complete reactive scattering problem). Transition state theory is such a "direct" approach, but it is inherently approximate; we seek a procedure capable of arbitrary accuracy.

For almost ten years there has been a *formal* solution to the problem; i.e., the CRP can also be expressed as[2]

Time-Dependent Quantum Molecular Dynamics, Edited by J. Broeckhove
and L. Lathouwers, Plenum Press, New York, 1992

$$N(E) = \frac{1}{2}(2\pi\hbar)^2 \, tr[F\delta(E-H)F\delta(E-H)] \, , \tag{1.2a}$$

where H is the total Hamiltonian of the molecular system and F is a flux operator. The microcanonical density operator is usually obtained from the Green's function (actually an *operator*) with outgoing wave boundary conditions,

$$\delta(E-H) = -\frac{1}{\pi} \, ImG^+(E) \, , \tag{1.2b}$$

where

$$G^+(E) = \lim_{\varepsilon \to 0} (E+i\varepsilon-H)^{-1} \, . \tag{1.2c}$$

Eq. (1.2a) is rigorously equivalent to Eq. (1.1b) but clearly has no explicit reference to asymptotic quantum states of reactants or products. There have been some earlier attempts[3-5] based on Eq. (1.2), and though they have been successfully applied only to one dimensional examples, they have at least shown that it is in principle possible calculate N(E) "directly" via Eq. (1.2).

The progress[6] described in the present report is based on using a discrete variable representation (DVR)[7,8] of the Hamiltonian and flux operators in Eq. (1.2), and the use of absorbing boundary conditions (ABC)[9,10] to obtain a well-behaved representation of the Green's function in Eq. (1.2c) The desirable features of a DVR are well recognized: no integrals need to be computed in constructing the Hamiltonian matrix, and the latter is extremely sparse, thus allowing the efficient use of iterative methods for the linear algebra (matrix inversion or diagonalization) involved. DVR methods[11] are presently widely used in essentially all molecular problems which call for an L^2 basis set. Absorbing boundary conditions have been used by several groups in the past,[9,10] primarily in time-dependent wave-packet propagation schemes. Goldberg and Shore[9a] and Leforestier and Wyatt[9b] employed absorbing boundaries in models of laser induced dissociation, and Kosloff *et al.*[10a,b] and Neuhauser, Baer *et al.*[10c-h] have used ABC to study reactive scattering. For the present application, ABC provides a simple means of imposing the outgoing wave boundary conditions necessary for the proper and well-behaved representation of the Green's function, without having to include information regarding the asymptotic region (i.e., reactants and products), as is necessary, for example, in a variational representation of $G(E^+)$.[12] ABC provide, furthermore, a convenient means of limiting the DVR grid to the interaction region, where the reaction outcome is determined.[6]

It is shown below, moreover, that *state-to-state* reaction probabilities, i.e., the S-matrix itself, can also be obtained from this DVR-ABC Green's function, and perhaps of even more interest, that one can obtain *partial* state resolution, e.g., the total reaction probability from a given initial reactant state,

$$P_{n_r}(E) \equiv \sum_{n_p} |S_{n_p,n_r}(E)|^2 \, . \tag{1.3}$$

Section II first discusses the cumulative reaction probability, and then Section III shows how state-specific information can also be obtained from the DVR-ABC Green's function.

II. CUMULATIVE REACTION PROBABILITY[6]

Bold-face quantities below are matrices (or vectors) in a discrete variable (i.e., grid point) representation (DVR). The DVR of the outgoing wave Green's function is thus

$$\mathbf{G}(E) \equiv (E\mathbf{I}+i\varepsilon-\mathbf{H})^{-1} . \tag{2.1}$$

$\varepsilon \equiv \Gamma/2$ is an absorbing potential which is zero (or essentially so) in the relevant interaction region and gradually turned on at its edges so as to absorb all flux exiting the interaction region. Since ε is a potential energy-like operator, its DVR is a diagonal matrix. The reader is referred to earlier work[8,6] on the specifics of the DVR that has been used in the present work. Fig. 1 shows a contour plot of the potential energy surface for the $H+H_2 \rightarrow H_2+H$ reaction, along with contours of the absorbing potential $\varepsilon(\mathbf{q})$. The solid points are the DVR grid points.

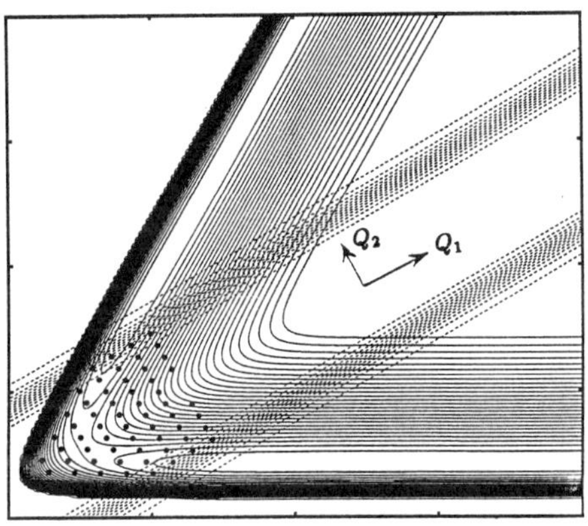

Figure 1. Contour plot of the potential energy surface (solid lines) for the collinear $H+H_2 \rightarrow H_2+H$ reaction; dashed lines are contours of the absorbing potential $\varepsilon\equiv\Gamma/2$. The solid points are the DVR grid points.

For future reference, we note that since

$$(E\mathbf{I}+i\varepsilon-\mathbf{H})\cdot\mathbf{G} = \mathbf{1} , \tag{2.2a}$$

the real and imaginary parts of $\mathbf{G} \equiv \mathbf{G}_r+i\mathbf{G}_i$ satisfy the following reactions

$$(\mathbf{H}-E\mathbf{1})\cdot\mathbf{G}_i - \varepsilon\cdot\mathbf{G}_r \tag{2.2b}$$

$$(\mathbf{H}-E\mathbf{1})\cdot\mathbf{G}_i\cdot(\mathbf{H}-E\mathbf{1}) = -\,\varepsilon - \varepsilon\cdot\mathbf{G}_i\cdot\varepsilon . \tag{2.2c}$$

The flux operator in Eq. (1.2a) is defined most generally as the Heisenberg time derivative of a step function which separates reactants and products,

$$F = \frac{i}{\hbar} [H,h(f(\mathbf{q}))] \ , \tag{2.3a}$$

where h is a step function

$$h(\zeta) = \begin{cases} 1, \zeta>0 \\ 0, \zeta<0 \end{cases} , \tag{2.3b}$$

and f(**q**) defines a *dividing surface* in the space of coordinates $(q_1, q_2, ...) = \mathbf{q}$ via the equation $f(\mathbf{q}) = 0$. Since the constant E commutes with h(f(**q**)), Eq. (2.3a) can also be written as

$$F = \frac{i}{\hbar} [\widetilde{H},h(f)] \ , \tag{2.4a}$$

where

$$\widetilde{H} = H\text{-}E \ . \tag{2.4b}$$

The DVR of the flux operator is thus

$$\mathbf{F} = \frac{i}{\hbar} [\widetilde{\mathbf{H}},\mathbf{h}] = \frac{i}{\hbar} (\widetilde{\mathbf{H}}\bullet\mathbf{h} - \mathbf{h}\bullet\widetilde{\mathbf{H}}) \ , \tag{2.4c}$$

where **h** is a diagonal matrix, equal to 0 or 1, respectively, for a grid point on the reactant or product side of the dividing surface.

Therefore the DVR of Eq. (1.2a) for the CRP is

$$N(E) = \frac{1}{2} (2\pi\hbar)^2 \left(\frac{i}{\pi\hbar}\right)^2 \mathrm{tr}[(\widetilde{\mathbf{H}}\bullet\mathbf{h} - \mathbf{h}\bullet\widetilde{\mathbf{H}})\bullet\mathbf{G}_i\bullet(\widetilde{\mathbf{H}}\bullet\mathbf{h} - \mathbf{h}\bullet\widetilde{\mathbf{H}})\bullet\mathbf{G}_i] \ , \tag{2.5a}$$

where $\mathbf{G}_i$ is the imaginary part of **G**. Multiplying out the operator product inside the trace and utilizing cyclic permutation of various factors gives

$$N(E) = -4\ \mathrm{tr}(\widetilde{\mathbf{H}}\bullet\mathbf{G}_i\bullet\mathbf{h}\bullet\widetilde{\mathbf{H}}\bullet\mathbf{G}_i\bullet\mathbf{h}) + 4\ \mathrm{tr}(\widetilde{\mathbf{H}}\bullet\mathbf{G}_i\bullet\widetilde{\mathbf{H}}\bullet\mathbf{h}\bullet\mathbf{G}_i\bullet\mathbf{h}) \ , \tag{2.5b}$$

which can also be written as

$$N(E) = 4\ \mathrm{tr}[\widetilde{\mathbf{H}}\bullet\mathbf{G}_i\bullet\mathbf{h}\bullet\widetilde{\mathbf{H}}\bullet\mathbf{G}_i\bullet(1\text{-}\mathbf{h})] - 4\ \mathrm{tr}[(\widetilde{\mathbf{H}}\bullet\mathbf{G}_i\bullet\widetilde{\mathbf{H}}\bullet\mathbf{h}\bullet\mathbf{G}_i\bullet(1\text{-}\mathbf{h})] \ . \tag{2.5c}$$

Employing Eq. (2.2b,c) and various cyclic permutations of the matrices in the trace, and noting that ε and **h** commute and that $\mathbf{h}\bullet(1\text{-}\mathbf{h}) = 0$, leads to the following remarkably simple result

$$N(E) = \mathrm{tr}\ (\Gamma_r\bullet\mathbf{G}_r\bullet\Gamma_p\bullet\mathbf{G}_r) + \mathrm{tr}\ (\Gamma_r\bullet\mathbf{G}_i\bullet\Gamma_p\bullet\mathbf{G}_i)$$

$$= \mathrm{tr}\ (\Gamma_r\bullet\mathbf{G}\bullet\Gamma_p\bullet\mathbf{G}^*) \ , \tag{2.6a}$$

where

$$\Gamma = 2\varepsilon,$$

and

$$\Gamma_r = (1-\mathbf{h})\bullet\Gamma \qquad (2.6b)$$

$$\Gamma_p = \mathbf{h}\bullet\Gamma\,. \qquad (2.6c)$$

Γ_r and Γ_p are the absorbing potentials in the reactant and product regions. Since Γ is a diagonal matrix, the matrix element version of Eq. (2.6a) reads,

$$N(E) = \sum_{j,j'} \Gamma_r(\mathbf{q}_j)\Gamma_p(\mathbf{q}_{j'})|G_{j,j'}|^2\,, \qquad (2.6d)$$

where $\{\mathbf{q}_j\}$ are the DVR grid points.

Equation (2.6) also has the desirable feature of being explicitly *independent of the location of the dividing surface*, provided only that it lies somewhere between the reactant and product absorbing regions. It is well known[13] that the reaction rate, if evaluated exactly, is independent of the choice of the dividing surface, but all expressions used to date for the canonical[14-19] and microcanonical[3-5] rate constants contain an explicit choice of f($\mathbf{q}$). The present form manifestly shows this independence. Eq. (2.6d) also lends itself to a simple qualitative interpretation. The summand of Eq. (2.6d) can be thought of as a transition probability from grid point $\mathbf{q}_j$ in the reactant absorbing region to grid point $\mathbf{q}_{j'}$ in the product absorbing region. The *cumulative* result, N(E), is then the sum of the transition probabilities over all reactant and product grid points, in analogy with Eq. (1.1b) which is a sum of reaction probabilities over asymptotic quantum states of reactants and products. The practical advantages of Eq. (2.6d) are that the reactant and product absorbing strips lie immediately outside the interaction region, rather than in the asymptotic regions, and that the sum is over simple grid points, rather than over quantum states.

Finally, although Eqs. (2.6) have been derived utilizing special features of a DVR, the trace structure in Eq. (2.6a) is clearly invariant to the choice of representation. It thus holds for any basis set representation of the matrices involved in case for some applications a choice other than a DVR might be preferred.

Figure 2 shows the CRP obtained[6a] for the collinear $H+H_2 \rightarrow H_2+H$ reaction, and Fig. 3 that for the three-dimensional (J=0) version of the reaction.[6b] The fact that N(E) does not increase monotorically with energy in Fig. 2 is an indication of transition state theory - violating dynamics in the collinear reaction, i.e., flux that re-crosses any dividing surface that one might construct.

This is seen not to happen for the three-dimensional reaction, another manifestation of the fact that TST is a better description of the dynamics the higher dimensionality.[20]

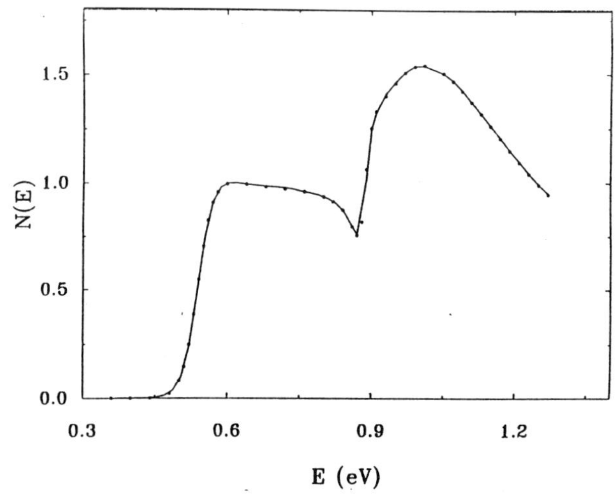

Figure 2. Cumulative reaction probability for the collinear $H+H_2$ reaction.

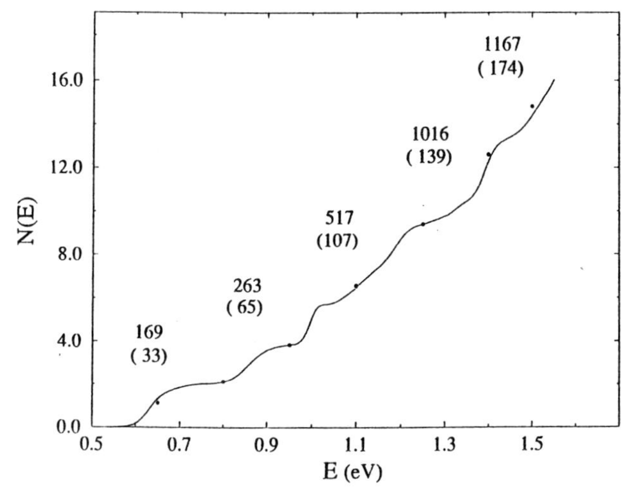

Figure 3. Cumulative reaction probability for the three-dimensional $H+H_2$ reaction (J=0). The numbers indicate the total number of grid points required in various energy regions (the numbers in parentheses being the number of this total that are located in the absorbing regions).

III. STATE-SPECIFIC REACTION PROBABILITIES

Let Φ_{n_r} and Φ_{n_p} be asymptotically incoming radial waves (free or distorted) for reactants and products, respectively, with channel quantum numbers n_r and n_p. With the specific normalization used in S-matrix Kohn variational reactive scattering calculations,[21] the explicit asymptotic form of Φ for an A+BC collinear reaction, for example, is

$$\Phi_{n_r}(r,R) \sim \phi_{n_r}(r) v_{n_r}^{-1/2} e^{-ik_{n_r}R} , \tag{3.1}$$

where $\phi_{n_r}(r)$ is the initial vibrational state of BC, R is the A-BC translational coordinate, and $R' = R_{AB\text{-}C}$ and $r' = r_{AB}$, respectively. A formally exact expression for the S-matrix element for the $n_r \rightarrow n_p$ transition is then

$$S_{n_p,n_r} = S^0_{n_p,n_r} + \frac{i}{\hbar}\left[\langle\Phi_{n_p}|H-E|\Phi_{n_r}\rangle + \langle\chi_{n_p}|G(E^+)|\chi_{n_r}\rangle\right] , \tag{3.2}$$

where

$$\chi_{n_r} = (H-E)\,\Phi_{n_r} \tag{3.3a}$$

and

$$\chi_{n_p} = (H-E)\,\Phi_{n_p} . \tag{3.3b}$$

(Note: throughout this paper wavefunctions inside a bra symbol $<|$ are not complex-conjugated unless explicited so indicated.) $S^0_{n_p,n_r}$ is the amplitude of the $n_r \rightarrow n_p$ transition contained in the asymptotic form of the wavefunction Φ_{n_r}, but since we are considering a *reactive* process, so long as Φ_{n_r} and Φ_{n_p} are non-reactive distorted waves, $S^0_{n_p,n_r}$ will be zero. Furthermore, since Φ_{n_r} and Φ_{n_p} are typically cutoff before they enter the reaction region, the Born term (the second term in Eq. (3.2)) is also (rigorously) zero. With the above two observations, Eq. (3.2) becomes

$$S_{n_p,n_r} = \frac{i}{\hbar}\langle\chi_{n_p}|G(E^+)|\chi_{n_r}\rangle . \tag{3.4}$$

Since Φ_{n_r} and Φ_{n_p} satisfy the correct asymptotic scattering boundary conditions (cf. Eq. (3.1)), χ_{n_r} and χ_{n_p} are short range, L^2 functions, so that the DVR-ABC representation of $G(E^+)$ is possible. The DVR of Eq. (3.4) thus reads

$$S_{n_p,n_r} = \frac{i}{\hbar}\,\mathbf{\Phi}^T_{n_p}\cdot(\mathbf{H}-E\mathbf{I})\cdot\mathbf{G}(E)\cdot(\mathbf{H}-E\mathbf{I})\cdot\mathbf{\Phi}_{n_r} , \tag{3.5}$$

where $\mathbf{G}(E)$ is the DVR-ABC Green's function of Eq. (2.1), $\mathbf{\Phi}_{n_r}$ and $\mathbf{\Phi}_{n_p}$ are vectors in grid point space

$$(\Phi_{n_r})_j = w_j^{1/2}\Phi_{n_r}(\mathbf{q}_j) \tag{3.6a}$$

$$(\Phi_{n_p})_j = w_j^{1/2}\Phi_{n_p}(\mathbf{q}_j) . \tag{3.6b}$$

T in Eq. (3.5) denotes vector transpose, and $\{w_j\}$ are the (multidimensional) DVR weights associated with $\{\mathbf{q}_j\}$. (It is possible to express the DVR of the vector χ_{n_r} as

$$\chi_{n_r} = (\mathbf{H}\text{-}\mathbf{E}\mathbf{I})\bullet\Phi_{n_r} \tag{3.6c}$$

because the ABC Green's function $\mathbf{G}(E)$ decays exponentially outside the interaction region, and similarly for χ_{n_p}.) Using Eq. (2.2) in Eq. (3.5) gives the final result

$$S_{n_p,n_r} = -\frac{i}{\hbar}\Phi^T_{n_p}\bullet\varepsilon_p\bullet\mathbf{G}(E)\bullet\varepsilon_r\bullet\Phi_{n_r}\,, \tag{3.7a}$$

where we used the fact that

$$-\frac{i}{\hbar}\Phi^T_{n_p}\bullet\varepsilon\bullet\Phi_{n_r} = 0$$

since Φ_{n_r} and Φ_{n_p} do not overlap, and replaced by ε by ε_r or ε_p since Φ_{n_r} and Φ_{n_p} are non-zero only in the reactant and product absorbing regions, respectively. Since the absorbing potential ε is diagonal, the explicit matrix expression corresponding to Eq. (3.7a) is

$$S_{n_p,n_r} = -\frac{i}{\hbar}\sum_{j,j'}\Phi_{n_p}(\mathbf{q}_j)\,w_j^{1/2}\,\varepsilon(\mathbf{q}_j)G_{j,j'}(E)\varepsilon(\mathbf{q}_{j'})w_{j'}^{1/2}\Phi_{n_r}(\mathbf{q}_{j'})\,. \tag{3.7b}$$

Equation (3.7) is a remarkably simple expression for state-to-state S-matrix elements which parallels Eq. (2.6) for the cumulative reaction probability. Thus having calculated matrix elements $\mathbf{G}(E)$ between grid points in the reactant and product absorbing regions, one may compute the cumulative reaction probability using Eq. (2.6d) or individual S-matrix elements using Eq. (3.7b). It should be noted that with the present formulation it is not necessary to "cutoff" the incoming wavefunctions Φ_{n_r} and Φ_{n_p} in order to guarantee regularity at small R (as it is variational calculations[21]) because the absorbing potentials eliminate the small R contribution from Eqs. (3.7).

Whether or not the DVR-ABC approach will prove more efficient than the variational approach[8,21] for the study of discrete-discrete transitions remains to be tested, and is likely to be system and energy dependent. The present method is, however, potentially of considerable advantage for the treatment of cases where *dissociative* channels are energetically allowed, including the case where the final state of interest is a dissociative state. In the latter case, Eqs. (3.2)-(3.7) all apply, and the DVR-ABC procedure still generates the correct Green's function. The final state Φ_{n_p} is of the form

$$\Phi_{n_p} \propto \exp[-ik(q_1 \sin\alpha + q_2 \cos\alpha)]\,, \tag{3.8}$$

where $k = \sqrt{2mE/\hbar^2}$, and α labels the partitioning of the translational energy among the fragment A, B and C. The treatment of such processes by the variational method would require discretizing the continuum (of dissociative states) which is considerably more involved.

Finally, from expressions given in ref. 13, the total reaction probability from a given initial state of the reactants,

$$P_{n_r}(E) \equiv \sum_{n_p} |S_{n_p,n_r}(E)|^2 , \qquad (3.9a)$$

can be expressed as

$$P_{n_r}(E) = \langle \Psi^*_{n_r} | F | \Psi_{n_r} \rangle , \qquad (3.9b)$$

where Ψ_{n_r} is the exact solution of the Schrödinger equation. Standard manipulations of formal scattering theory gives

$$\Psi_{n_r} = G(E^+) i\varepsilon\, \Phi_{n_r} , \qquad (3.10)$$

so that with the flux operator F taken in the form of Eq. (2.4c), the DVR of Eq. (3.9b) is

$$P_{n_r} = \frac{i}{\hbar} \Phi^{T*}_{n_r} \bullet (-i\varepsilon) \bullet \mathbf{G}^* \bullet (\widetilde{\mathbf{H}} \bullet \mathbf{h} - \mathbf{h} \bullet \widetilde{\mathbf{H}}) \bullet \mathbf{G} \bullet (i\varepsilon) \bullet \Phi_{n_r} . \qquad (3.11)$$

Since

$$\widetilde{\mathbf{H}} \bullet \mathbf{G} \equiv (\mathbf{H} - E\mathbf{I}) \bullet \mathbf{G} = -\mathbf{1} + i\varepsilon \bullet \mathbf{G} , \qquad (3.12)$$

Eq. (3.11) becomes

$$P_{n_r}(E) = \frac{2}{\hbar} \Phi^{*T}_{n_r} \bullet \varepsilon_r \bullet \mathbf{G}^* \bullet \varepsilon_p \bullet \mathbf{G} \bullet \varepsilon_r \bullet \Phi_{n_r} , \qquad (3.13)$$

where we have used the fact that $\mathbf{h} \bullet \varepsilon_r = 0$. This result, Eq. (3.13), parallels the completely state-selected and -detected expression, Eq. (3.7), and the cumulative expression, Eq. (2.6c).

All these results can be summarized by introducing the *projection matrix* for a specific quantum state,

$$\mathbf{Q}_{n_r} = \varepsilon_r \bullet \Phi_{n_r} \Phi^{*T}_{n_r} \bullet \varepsilon_r / 2\hbar \qquad (3.14a)$$

$$\mathbf{Q}_{n_p} = \varepsilon_p \bullet \Phi_{n_p} \Phi^{*T}_{n_p} \bullet \varepsilon_p / 2\hbar . \qquad (3.14b)$$

Then the completely state-specific, "half" state-specific, and cumulative reaction probabilities, respectively, can be expressed as

$$P_{n_p,n_r}(E) \equiv |S_{n_p,n_r}(E)|^2$$

$$= 4\, \mathrm{tr}(\mathbf{G} \bullet \mathbf{Q}_{n_p} \bullet \mathbf{G}^* \bullet \mathbf{Q}^*_{n_r}) \qquad (3.15a)$$

$$P_{n_r}(E) \equiv \sum_{n_p} P_{n_p,n_r}(E)$$

$$= 4\, \mathrm{tr}(\mathbf{G} \bullet \varepsilon_p \bullet \mathbf{G}^* \bullet \mathbf{Q}^*_{n_r}) \qquad (3.15b)$$

$$N(E) \equiv \sum_{n_r} P_{n_r}(E) \equiv \sum_{n_r,n_p} P_{n_p,n_r}(E)$$

$$= 4\ \mathrm{tr}(\mathbf{G}\bullet\varepsilon_p\bullet\mathbf{G}^*\bullet\varepsilon_r)\ . \tag{3.15c}$$

The projection matrices of Eq. (3.14) thus formally obey the summation relations

$$\sum_{n_r} \mathbf{Q}_{n_r} = \varepsilon_r \tag{3.16a}$$

$$\sum_{n_p} \mathbf{Q}_{n_p} = \varepsilon_p \tag{3.16b}$$

IV. CONCLUDING REMARKS

The DVR-ABC Green's function provides an extremely compact and efficient procedure for evaluating the cumulative reaction probability. It can also be used to obtain partially or completely state-specific reaction probabilities, though it remains to be seen just how practically useful this latter feature will be.

ACKNOWLEDGMENT

This research has been supported by the Director, Office of Energy Research, Office of Basic Energy Sciences, Chemical Sciences Division of the US Department of Energy under Contract No. DE-AC03-76SF00098 and by the National Science Foundation, Grant CHE-8920690.

REFERENCES

1. W. H. Miller, J. Chem. Phys. **62**:1899 (1975).
2. W. H. Miller, S. D. Schwartz, and J. W. Tromp, J. Chem. Phys. **79**:4889 (1983).
3. D. Thirumalai, B. C. Garrett, and B. J. Berne, J. Chem. Phys. **83**:2972 (1985).
4. C. W. McCurdy and B. C. Garrett, J. Chem. Phys. **84**:2630 (1986).
5. T. Seideman and W. H. Miller, J. Chem. Phys. **95**:1768 (1991).
6. a) T. Seideman and W. H. Miller, J. Chem. Phys. in press.
   b) T. Seideman and W. H. Miller, J. Chem. Phys. in press.
7. a) J. V. Lill, G. A. Parker, and J. C. Light, Chem. Phys. Lett. **89**:483 (1982).
   b) J. C. Light, I. P. Hamilton, and J. V. Lill, J. Chem. Phys. **82**:1400 (1985).
   c) J. V. Lill, G. A. Parker, and J. C. Light, J. Chem. Phys. **85**:900 (1986).
   d) Z. Bacic and J. C. Light, J. Chem. Phys. **85**:4594 (1986); **86**:3065 (1987).
   e) R. M. Whitnell and J. C. Light, J. Chem. Phys. **90**:1774 (1989).
   f) S. E. Choi and J. C. Light, J. Chem. Phys. **92**:2129 (1990).

8. D.-T. Colbert and W. H. Miller, J. Chem. Phys. **96**:1982 (1992).

9. a) A. Goldberg and B. W. Shore, J. Phys. B **11**:3339 (1978).

b) C. Leforestier and R. E. Wyatt, J. Chem. Phys. **78**:2334 (1983).

10. a) C. Cerjan, D. Kosloff, R. Kosloff, and T. Teshef, Geophysics **50**:705 (1985).

b) R. Kosloff and D. Kosloff, J. Comput. Phys. **63**:363 (1986).

c) D. Neuhauser and M. Baer, J. Chem. Phys. **90**:4351 (1989).

d) D. Neuhauser and M. Baer, J. Chem. Phys. **91**:4651 (1989).

e) D. Neuhauser, M. Baer, R. S. Judson, and D. J. Kouri, J. Chem. Phys. **93**:312 (1990).

f) M. Baer, C. Y. Ng, and D. Neuhauser, Chem. Phys. Lett. **169**:534 (1990).

g) D. Neuhauser, J. Chem. Phys. **95**:4927 (1991).

h) I. Last and M. Baer, Chem. Phys. Lett. **189**:84 (1992); I. Last, D. Neuhauser, and M. Baer, J. Chem. Phys. **96**:2017 (1992).

11. Other pointwise approaches are the pseudodospectral method suggested by Friesner [Chem. Phys. Lett. **116**:39 (1985); J. Chem. Phys. **85**:1462 (1987)], and the collocation method of Yang and Peet [Chem. Phys. Lett. **153**:93 (1988); J. Chem. Phys. **90**:1746 (1989)].

12. W. H. Miller and B.M.D.D. Jansen op de Haar, J. Chem. Phys. **86**:6213 (1987).

13. W. H. Miller, J. Chem. Phys. **61**:1823 (1974).

14. a) R. Jaquet and W. H. Miller, J. Phys. Chem. **89**:2139 (1985).

b) K. Yamashite and W. H. Miller, J. Chem. Phys. **82**:5475 (1985).

c) J. W. Tromp and W. H. Miller, J. Phys. Chem. **90**:3482 (1986).

d) J. W. Tromp and W. H. Miller, Faraday Disc. Chem. Soc. **84**:441 (1987).

15. R. E. Wyatt, Chem. Phys. Lett. **121**:301 (1985).

16. a) T. P. Park and J. C. Light, J. Chem. Phys. **85**:5870 (1986).

b) T. P. Park and J. C. Light, J. Chem. Phys. **88**:4897 (1988).

c) T. P. Park and J. C. Light, J. Chem. Phys. **91**:974 (1989).

d) M. Founargiotakis and J. C. Light, J. Chem. Phys. **93**:633 (1990).

e) T. P. Park and J. C. Light, J. Chem. Phys. **94**:2946 (1991).

17. a) G. Wahnstrom and H. Metiu, Chem. Phys. Lett. **134**:531 (1987); J. Phys. Chem. **92**:3240 (1988).

b) G. Wahnstrom, B. Carmeli, and H. Metiu, J. Chem. Phys. **88**:2478 (1988).

c) A. E. Depristo, K. Haug, and H. Metiu, Chem. Phys. Lett. **155**:376 (1989).

18. a) R. Lefebvre and N. Moiseyev, J. Chem. Phys. **93**:7173 (1990).

b) N. Ron, N. Moisseyev, and R. Lefebvre, in preparation.

19. P. N. Day and D. G. Truhlar, J. Chem. Phys. **94**:2045 (1991).

20. S. Chapman, S. M. Hornstein, and W. H. Miller, J. Am. Chem. Soc. **97**:892 (1975).

21. J. Z. H. Zhang, S. I. Chu, and W. H. Miller, J. Chem. Phys. **88**:6233 (1988).

TIME EVOLUTION OF ELECTRONS AND NUCLEI IN MOLECULAR SYSTEMS

Yngve Öhrn, Erik Deumens, Agustin Diz, Ricardo Longo, Juan Oreiro, and Hugh Taylor

Quantum Theory Project, Departments of Chemistry and Physics, University of Florida, Gainesville, FL 32611.

1. Introduction

Chemical change on the most fundamental level happens because of the dynamics of electrons and nuclei. Such dynamics is studied by solving the time-dependent Schrödinger equation. There is a growing literature of methods for accomplishing this difficult task by what can be referred to as "direct approaches". Such numerical methods have two key features. One is the representation of the wavefunction, e.g. by expansion coefficients in a basis set, or by discretization on a grid of points. This in turn determines how one evaluates the action of the operators, in particular the Hamiltonian $\hat{H} = \hat{K} + \hat{V}$ with its kinetic energy operator $\hat{K}$ and its potential energy operator $\hat{V}$, on the wavefunction. The other feature is the actual time propagation algorithm, i.e. how the solution is marched over a time interval, given its value at the initial time.

One method is the Pseudospectral Fourier Approximation [1] , which has been introduced rather recently into molecular dynamics, although for some time it has been used in several areas of physics, e.g. fluid dynamics, optics, and electron microscopy. This approach employs a grid representation of the wavefunction in coordinate space and applies the Discrete Fourier Transform (DFT) to obtain a momentum space representation. Computer codes performing Fast Fourier Transform (FFT) are generally available making possible efficient ways to switch from coordinate space to momentum space and back allowing for fast evaluation of the action of the necessary operators as the kinetic energy and the potential energy terms of the Hamiltonian. Error analysis and collocating functions other than plane waves are available within this technique. Appropriate functions for radial [2], and angular coordinates [3] have been implemented.

There are various time integration algorithms currently in use with this pseudospectral method in molecular dynamics. We distinguish four, viz. the second-order difference (SOD) method, the split operator (SO) method, the Short-time Iterative Lanczos (SIL) method and the Chebychev Expansion (CE) method.

In the SOD [4] applied to the Schrödinger equation the wavefunction is computed at successive short time steps through a combination of a forward and a backward time step, which eliminates the second order terms so the error is of third order in the time step. In the other three methods one focuses on the propagator. In the SO scheme [5] one makes a symmetric decomposition of the exponential operator that defines the exact propagator

$$\begin{aligned} \hat{U}(\epsilon) &= \exp\left(-i\hat{H}\epsilon\right) \\ \psi(\epsilon) &= \hat{U}(\epsilon)\psi(0) \\ \hat{U}(\epsilon) &\approx \exp(-i\frac{\epsilon}{2}\hat{K})\exp(-i\epsilon\hat{V})\exp(-i\frac{\epsilon}{2}\hat{K}) \end{aligned} \tag{1}$$

and introduces a third order error term. This is a short-time method that requires successive applications to propagate over a finite time interval.

The SIL propagation formula [6] is

$$\mathbf{U}(\epsilon) \approx \exp[-i\epsilon\mathbf{\Lambda}(\hat{H}, \psi(0))] \tag{2}$$

where $\mathbf{U}$ is now a matrix operator in the Krylov subspace generated by Hamiltonian and the initial wavefunction. The matrix Λ is the tridiagonal Lanczos matrix representing the Hamiltonian in the Krylov space [7]. The exponentiation is usually performed by diagonalizing the Lanczos matrix and working with the diagonal eigenvalue matrix. The length of time dictates the size of the Krylov space needed for a predefined accuracy [8]. In general one will opt for short time steps in order not to lose the advantage of the Lanczos reduction, i.e. iterating the algorithm is more efficient than using long time steps. The CE method [9] approximates the exact propagator by a Chebychev expansion

$$\hat{U}(t) = \sum_{n=1}^{N} a_n(t) T_n(-i\hat{H}_R) \tag{3}$$

where the Hamiltonian needs to be renormalized so that its spectrum coincides with the domain of the Chebychev polynomials T_n. this is a long-time method. The convergence requirement on the CE expansion is such that the number of terms does not decrease significantly for smaller t-values. Thus, for efficiency t should be large.

Should one want to consider time-dependent Hamiltonians, there are straightforward ways to extend the SO, SOD, and SIL methods to such cases, while the CE method would seem not to have this flexibility. For short time methods further approximations are usually implied as e.g. the use of short-time averaged Hamiltonians and disregard of the time ordering.

The specific merits of various methods to evaluate the action of the Hamiltonian on the wavefunction and to time propagate the solution have been reviewed [10]. The computational effort involved in these methods are presently such that applications involving at most a few degrees of freedom can be attempted. They are used to study quantum dynamics of nuclear motion on a potential energy surface or on a small number of coupled potential energy surfaces. The SO offers limited accuracy and has been used in studies of reactions [11], and absorption spectra [12]. The SOD has been used extensively because of its ease of implementation and satisfactory level of accuracy. Applications include the determination of eigenspectra [13], non-adiabatically coupled systems [14], photodetachment spectra [15], and systems with time-dependent Hamiltonians [16]. A significant number of applications of the CE method has been made among other problems to atom-diatom collisions [17] also using a multiconfigurational SCF approach [18], photodissociation [19], and to the computation of energy levels [20]

In contrast to these "direct methods" we have developed a general approach to solve approximately the time-dependent Schrödinger equation for a system of electrons and atomic nuclei by invoking the Time-Dependent Variational Principle (TDVP). This approach employs a wavefunction of a given form for the participating particles and generates a set of coupled first-order differential equations in time for the wavefunction parameters. Since a given wavefunction can be parameterized in a variety of ways, the resulting dynamical equations obtain different appearance, and might be ill behaved numerically, depending on the particular choice of parameters. With an appropriate choice of complex parameters [21] the dynamical equations that govern their time evolution obtain the form of Hamilton's classical equations [22] for the time derivatives of coordinates and conjugate momenta. The dynamics is then said to be taking place in a generalized phase space.

This means that this approach attempts in one step what traditionally has been divided into at least three sizable problems, i.e. (i) the calculation of a potential energy surface, (ii) the determination of an analytical representation of such a surface, and (iii) the dynamics on such a surface. This is possible because of existing efficient codes for the determination of energy gradients necessary to evaluate the instantaneous forces between participating particles.

The actual time propagation or integration of this system of a large number of coupled first-order differential equations can be performed in a variety of ways. We have used several standard methods available in the literature [43]. Depending on the particular problem one method provides better stability of the solution than another and one must have available a variety of techniques.

To solve a first order differential equation numerically, is basically very simple: one approximates the derivative with the difference formula and one finds a value for the variable at the next time step. This is known as the Euler method. Its major disadvantage is that in order to achieve high accuracy, very small time steps have to be taken. One can discern three major methods:

1. *Runge-Kutta* methods propagate a solution over an interval by combining the information from several Euler-style steps and then use the information obtained to match a Taylor series expansion to some higher order.
2. *Richardson extrapolation* uses the powerful idea of extrapolating a computed result to the value that *would* have been obtained if the step size had been much smaller than the one chosen. In particular, extrapolation to zero step size is the goal. When combined with a particular way of taking individual steps (the *modified midpoint method*) and a particular kind of extrapolation (rational function approximation) Richardson extrapolation produces the Bulirsch-Stoer method.
3. *Predictor corrector* methods store the solution along the way, and use these results to extrapolate the solution one step advanced. They then correct the extrapolation using derivative information at the new point. These techniques are best for very smooth functions.

In our computer code we have made available three methods, which represent these categories:

1. DE: modified divided difference form of the Adams-Pece formulas and local extrapolation. It adjusts the order and step size to control the local error. This method is fully explained in the text by Shampine and Gordon [42].
2. GEARB is a variant of the GEAR package. It uses Adams methods or, optionally, backward differentiation formulas, also known as the stiff methods of Gear [43]. The package can use functional iteration (no partial derivatives needed), or the chord method with an analytic Jacobian supplied by the user or with the Jacobian constructed approximately by numerical differentiation.

3. ODEINT uses Richardson extrapolation with the Bulirsch-Stoer method [43]. This method is extremely accurate, but we found that it requires many more points than any of the previous methods. We therefore use it only to verify a part of the evolution, when we have reason to doubt the accuracy of the faster, simpler methods.

All the above methods adapt the step size to the scale of the solution, starting from a small guess and gradually increasing the step size if the estimated error indicates that the predicted solution is accurate.

The dynamical equations for applications of this approach, which include applications to atom-atom, ion-atom, atom-molecule, and molecule-molecule collisions as well as a variety of bound state time evolution problems, can be integrated with one of these standard methods.

This general approach can be implemented at different levels depending on the choice of wavefunction or model state for the electrons and the atomic nuclei. The simplest *ab initio* model derives from using a single Slater determinant for the electrons and treating the nuclei as classical particles. The resulting dynamical equations have been worked out in detail [22] and we repeat in the next section the basic results. In the following section we give some of the general properties of the approach. We discuss in particular how the approximate dynamics relates to the exact dynamics in terms of conservation laws. In the fourth section we give some results for some small systems in order to illustrate this dynamical approach for electrons and nuclei together.

There exist several different attempts to treat electronic and nuclear dynamics via time-dependent formalisms. The method introduced by Car and Parinello [44] and reviewed by Remler and Madden [45] is a sometimes successful, phenomenologically based approach that uses density functional theory for the electrons and classical dynamics on an average surface for the nuclei. They add "fictitious kinetic energy terms" to obtain dynamical equations for the variables describing the electronic state. This method has been used for systems in condensed phases and when used for geometry optimization has the same characteristics as simulated annealing. The Time-Dependent Hartree-Fock (TDHF) approximation for electrons used with a classical nuclear dynamics on an average potential energy surface has been used by Runge, Micha, and Feng [46], Field, Hartke and Carter [47] and Meyer and Miller [48]. Introducing the electron hopping model to allow for several electronic states [49] has also been successful. The electron-nuclear coupling in the dynamics of these methods occurs only through the average surface; the electrons feel the nuclei through the instantaneous Fock operator. As a result the electron momenta are not treated correctly, which shows up mainly in higher non-chemical energy regimes. This deficiency is sometimes remedied by the introduction of electron translation factors (ETF) [50], particularly in the close coupling approaches. The close coupling method used primarily in the study of atomic collisions involving one or two electrons [36] is usually using prescribed trajectories for the nuclei.

One of the major differences of our theory with these methods is the explicit consideration of the generalized phase space, as shown in the next section, and the explicit construction of its metric made possible by the use of coherent state parameterization of the wavefunctions. In the methods briefly reviewed above one implicitly assumes the phase space to be canonical, i.e. flat. In some cases this may be correct, but in others this assumption warrants further investigation. For instance, in cases where ETFs are important the necessary correction terms are even sometimes referred to as "unphysical". In the END theory the geometry of the phase space is correctly accounted for at the outset.

2. Theory

The Slater determinant electronic wavefunction is ultimately defined in terms of a set of

K localized atomic spin orbitals. These orbitals are given as linear contractions of Gaussians and can in principle be centered anywhere in space, but preferably they are centered on the nuclei. These orbitals are non-orthogonal and we shall refer to them as *atomic* orbitals.

It is meaningful to divide the set of K spin orbitals N *occupied* and K-N *unoccupied* spin orbitals. We will use a superscript • for quantities referring to the occupied set and ∘ for the unoccupied set, while quantities referring to either have no superscript.

The set of atomic spin orbitals $\{\phi_i\}_{i=1...K}$ is partitioned into occupied and unoccupied as $\phi = (\phi^\bullet \quad \phi^\circ)$. Most manipulations with determinants are conveniently carried out in an orthonormal basis $\{\psi_j\}_{j=1...K}$ and the relation to the atomic basis can be expressed as $\psi = \phi W$ or in more detail

$$(\psi^\bullet \quad \psi^\circ) = (\phi^\bullet \quad \phi^\circ)\begin{pmatrix} W^\bullet & W' \\ W'' & W^\circ \end{pmatrix}, \tag{4}$$

where the columns of W represent the new basis vectors. The transformation matrix W satisfies the equation

$$W^\dagger \Delta W = I \tag{5}$$

i.e. $WW^\dagger = \Delta^{-1}$ and where the atomic metric matrix Δ has the elements

$$\Delta_{ij} = \langle \phi_i | \phi_j \rangle. \tag{6}$$

The model wavefunction can then be expressed as $|z\rangle = \det\{\chi_i(x_j)\}$ where the dynamical spin orbitals are

$$\chi_i(x) = \psi_i(x) + \sum_{j=N+1}^{K} \psi_j(x) z_{ji} \qquad (1 \le i \le N) \tag{7}$$

in terms of the orthonormal basis, with z_{ji} the complex time-dependent parameters and x the combined position space and spin variable. We note that the dynamical orbitals are nonorthogonal and that the particular form of the resulting Slater determinant is the one obtained by Thouless [23], which has the form of a coherent state [24]. The unnormalized determinant in this form depends only on the parameters z , not on z^*, a property of coherent states. This choice leads to dynamical equations that exhibit a simple phase space structure. We have developed the dynamical equations in terms of the nonorthogonal atomic basis [25], which results in a similar overall form albeit a somewhat more involved detailed structure of the dynamical equations. The virtual dynamical spin orbitals are

$$\chi_j = \psi_j - \sum_{i=1}^{N} z_{ji}^* \psi_j, \qquad (N+1 \le j \le K) \tag{8}$$

for the orthogonal basis, while for the nonorthogonal atomic basis we have the occupied dynamical spin orbitals

$$\chi_i = \phi_i + \sum_{j=N+1}^{K} \phi_j z_{ji}, \qquad (1 \le i \le N) \tag{9}$$

and the virtuals

$$\chi_j = \phi_j - \sum_{i=1}^{N} v_{ji}^* \phi_i, \qquad (N+1 \le j \le K). \tag{10}$$

The matrix v is defined as

$$v = (\Delta'' + \Delta^{\circ} z)(\Delta^{\bullet} + \Delta' z)^{-1} \tag{11}$$

in terms of the submatrices of the metric

$$\Delta = \begin{pmatrix} \Delta^{\bullet} & \Delta' \\ \Delta'' & \Delta^{\circ} \end{pmatrix}. \tag{12}$$

Note that the virtual or unoccupied space is kept orthogonal to the space of occupied dynamical spin orbitals in either basis. Provisions must be made for reordering of orbitals during the time evolution.

The TDVP now yields a set of coupled differential equations in time that can be expressed as [22]

$$\begin{aligned} i\sum_{\beta} C_{\alpha\beta}\dot{\zeta}_{\beta} &= \frac{\partial E}{\partial \zeta_{\alpha}^{*}} \\ -i\sum_{\beta} C_{\alpha\beta}^{*}\dot{\zeta}_{\beta}^{*} &= \frac{\partial E}{\partial \zeta_{\alpha}} \end{aligned} \tag{13}$$

where the ζ—parameters denote both the electronic z—parameters and the nuclear coordinates $Z_k = (R_k + iP_k)/\sqrt{2}$ with positions $\vec{R}_k = (R_{kx}, R_{ky}, R_{kz})$ and momenta $\vec{P}_k = (P_{kx}, P_{ky}, P_{kz})$. The energy or the generator of time translations is

$$E = E(\zeta^*, \zeta) = \sum_{k=1}^{n} \frac{\vec{P}_k^2}{2M_k} + \frac{1}{2}\sum_{k,l} \frac{Q_k Q_l}{|\vec{R}_k - \vec{R}_l|} + \frac{\langle z|H_e|z\rangle}{\langle z|z\rangle} \tag{14}$$

where H_e is the electronic Hamiltonian, M_k the nuclear mass, and Q_k the atomic number of nucleus k. The dynamic metric is defined as

$$C_{\alpha\beta} = \frac{\partial^2 \ln S}{\partial \zeta_{\alpha}^{*} \partial \zeta_{\beta}} \tag{15}$$

with $S = S_{el} S_{cl}$ where the electronic metric is

$$S_{el}(z^*, z) = \langle z|z\rangle = \det\{I^{\bullet} + z^{\dagger} z\} \tag{16}$$

for an orthonormal basis and

$$S_{el}(z^*, R, P, z, R', P') = \langle z|z\rangle = \det\{\Delta^{\bullet} + \Delta' z + z^{\dagger}\Delta'' + z^{\dagger}\Delta^{\circ} z\} \tag{17}$$

for the nonorthogonal atomic basis. The classical nuclear metric S_{cl} satisfies $\ln S_{cl}(Z^*, Z) = \sum_{l} Z_l^* Z_l$. The metric S_{el} involves the overlap of two determinantal states at different geometries when the atomic basis is used, and when electron translation factors are included in this basis the overlap also depends on the nuclear momenta.

Introducing the generalized Poisson bracket $\{f, g\}$ of two functions f and g, depending on the parameters ζ and ζ^*, as

$$\{f,g\} = (\partial f/\partial \zeta^{T} \quad \partial f/\partial \zeta^{\dagger}) \begin{pmatrix} -i\mathbf{C}^{-1} & \mathbf{0} \\ \mathbf{0} & i\mathbf{C}^{*-1} \end{pmatrix} \begin{pmatrix} \partial g/\partial \zeta^{*} \\ \partial g/\partial \zeta \end{pmatrix}, \tag{18}$$

enables us to write the dynamical equations in the standard form

$$\begin{aligned} \dot{\zeta} &= \{\zeta, E\} \\ \dot{\zeta}^{*} &= \{\zeta^{*}, E\}. \end{aligned} \tag{19}$$

The dynamical equations can then be expressed in the matrix form as

$$\begin{pmatrix} iC & 0 & 0 & 0 \\ 0 & iI & 0 & 0 \\ 0 & 0 & -iC^* & 0 \\ 0 & 0 & 0 & -iI \end{pmatrix} \begin{pmatrix} \dot{z} \\ \dot{Z} \\ \dot{z}^* \\ \dot{Z}^* \end{pmatrix} = \begin{pmatrix} \partial E/\partial z^* \\ \partial E/\partial Z^* \\ \partial E/\partial z \\ \partial E/\partial Z \end{pmatrix} \tag{20}$$

in terms of an electron basis that is fixed in space, which cannot be accurate other than in the hypothetical limit of a complete basis. The form of the dynamical equations for the case of the electronic basis being centered on the nuclei can, of course, be derived straightforwardly, but also obtained from Eq. (20) by a transformation among the parameters (dynamical coordinates) z and Z [22]. It becomes

$$\begin{pmatrix} iC & 0 & iC_R & iC_P \\ 0 & -iC^* & -iC_R^* & -iC_P^* \\ iC_R^\dagger & -iC_R^T & C_{RR} & -I + C_{RP} \\ iC_P^\dagger & -iC_P^T & I - C_{PR} & C_{PP} \end{pmatrix} \begin{pmatrix} \dot{z} \\ \dot{z}^* \\ \dot{R} \\ \dot{P} \end{pmatrix} = \begin{pmatrix} \partial E/\partial z^* \\ \partial E/\partial z \\ \partial E/\partial R \\ \partial E/\partial P \end{pmatrix} \tag{21}$$

where we also use the (real) nuclear position and momentum parameters. The notation for the various matrix blocks is such that

$$C = \frac{\partial^2 \ln S(z^*, R, P, z, R, P)}{\partial z^* \partial z} \tag{22}$$

for the purely electronic part,

$$C_X = \left. \frac{\partial^2 \ln S(z^*, R, P, z, R', P')}{\partial z^* \partial X'} \right|_{R'=R, P'=P} \tag{23}$$

for the electronic-nuclear coupling parts, and

$$C_{XY} = \left. \frac{\partial^2 S(z^*, R, P, z, R', P')}{\partial X \partial Y'} \right|_{R'=R, P'=P} \tag{24}$$

for the purely nuclear part, where X and Y stand for R or P. The overall phase factor of the evolving state $\exp(i\gamma)$ must satisfy the evolution equation [22]

$$\dot{\gamma} = -\mathrm{Im}\Big(\sum_\alpha \dot{\zeta}_\alpha \frac{\partial \ln S(\zeta^*, \zeta)}{\partial \zeta_\alpha}\Big) - E(\zeta^*, \zeta). \tag{25}$$

3. Conservation Laws

We employ laboratory fixed coordinate axes and describe the nuclear degrees of freedom in Cartesian form. This means that the dynamics includes the overall translational and rotational motions, which have to be separated out when necessary. The TDVP has rigorous conservation laws associated with it, when the Poisson bracket of the expectation values of a conserved operator and the energy (Hamiltonian) as defined in Eq. (18) equals the expectation value of the commutator of that operator with the quantum mechanical Hamiltonian H. The existence of these conservation laws makes it possible to avoid all transformations to internal coordinates.

As a simple example let us first consider the conservation of energy. We know from the Hamiltonian structure of the TDVP equations (19) that

$$\partial E/\partial t = \{E, E\} = 0 \tag{26}$$

By construction, in using the TDVP, the time derivative of the state vector is a variation in the manifold of states, and thus

$$\begin{aligned}\frac{d}{dt}\langle\zeta|H|\zeta\rangle &= \langle\frac{d}{dt}\zeta|H|\zeta\rangle + \langle\zeta|H|\frac{d}{dt}\zeta\rangle \\ &= \langle\delta\zeta|H|\zeta\rangle + \langle\zeta|H|\delta\zeta\rangle \\ &= i\hbar\langle\delta\zeta|\frac{d}{dt}\zeta\rangle - i\hbar\langle\frac{d}{dt}\zeta|\delta\zeta\rangle \\ &= i\hbar\langle\frac{d}{dt}\zeta|\frac{d}{dt}\zeta\rangle - i\hbar\langle\frac{d}{dt}\zeta|\frac{d}{dt}\zeta\rangle = 0\end{aligned} \tag{27}$$

Hence the energy is conserved both when viewed as the value of the Hamiltonian function in the generalized phase space and as the expectation value of the quantum mechanical Hamiltonian operator.

The conservation of total momentum and total angular momentum can be argued in similar ways and we limit our considerations to the total momentum. It becomes straightforward to do this by considering the classical description of the nuclei as the narrow width limit of a Gaussian wave packet description [22]. Invoking the coherent state parameterization of the Gaussians results in a form where all nuclear coordinates $\vec{R}$ enter through the combination $\vec{R}-\vec{Z}$, where $\vec{Z}$ is a vector with three complex components, being the coherent state parameters of the wave packet. Distinguishable nuclei are then represented by a product of such boosted wave packets [22], which constitutes one factor of the total system evolving state vector $|z, Z, Z^*\rangle$, the Thouless determinant for the electrons being the other.

The translation of the entire system is generated by the transformation

$$\begin{aligned}\vec{r} &\to \vec{r} + \vec{a} \\ \vec{R} &\to \vec{R} + \vec{a}\end{aligned} \tag{28}$$

of electron and nuclear coordinates. With the electron basis (Gaussian orbitals) centered on the moving nuclei we find that such a translation is accomplished by the parameter transformation

$$\begin{aligned}z &\to z \\ \vec{Z} &\to \vec{Z} + \vec{a}\end{aligned} \tag{29}$$

The derivation of the conservation law from the TDVP then proceeds by considering the action of the conserved operator on the evolving state as the variation. In the case of total momentum we get for κ kinds of nuclei (*N(k)* of kind *k*), and *N* electrons

$$\hat{\vec{P}}_{tot}|z, Z, Z^*\rangle = \frac{\hbar}{i}[\sum_{k=1}^{\kappa}\sum_{l=1}^{N(k)} \nabla_{\vec{R}_{lk}} + \sum_{i=1}^{N} \nabla_{\vec{r}_i}]|z, Z, Z^*\rangle \tag{30}$$

It can be seen that this is a state in the tangent plane of the manifold of states, i.e.

$$\begin{aligned}\hat{\vec{P}}_{tot}|z, Z, Z^*\rangle &= \lim_{\vec{a}\to\vec{0}} \frac{\hbar}{i\vec{a}} \sum_{k=1}^{\kappa}\sum_{l=1}^{N(k)} [|z, \vec{Z}_{kl} + \vec{a}, \vec{Z}^*_{kl} + \vec{a}\rangle - |z, \vec{Z}_{kl}, \vec{Z}^*_{kl}\rangle] \\ &= \frac{\hbar}{i}\sum_{k=1}^{\kappa}\sum_{l=1}^{N(k)} [\nabla_{\vec{Z}_{kl}} + \nabla_{\vec{Z}^*_{kl}}]|z, Z, Z^*\rangle = \frac{\hbar}{i}|z, \delta Z, \delta Z^*\rangle = \frac{\hbar}{i}|\delta\psi\rangle.\end{aligned} \tag{31}$$

Again using ζ to denote the collection of all parameters we write the conservation law as

$$\begin{aligned}\frac{d}{dt}\langle\zeta|\hat{\vec{P}}_{tot}|\zeta\rangle &= \langle\frac{d}{dt}\zeta|\hat{\vec{P}}_{tot}|\zeta\rangle + \langle\zeta|\hat{\vec{P}}_{tot}|\frac{d}{dt}\zeta\rangle \\ &= -\frac{\hbar}{i}\langle\frac{d}{dt}\zeta|\delta\zeta\rangle + \frac{\hbar}{i}\langle\delta\zeta|\frac{d}{dt}\zeta\rangle \\ &= \langle\zeta|H|\delta\zeta\rangle - \langle\delta\zeta|H|\zeta\rangle \\ &= -\frac{i}{\hbar}\langle\zeta|[\hat{\vec{P}}_{tot}, H]_-|\zeta\rangle = 0,\end{aligned} \tag{32}$$

where in going from line two to three we have used the TDVP equations derived from the stationarity of the quantum mechanical action [22]. This establishes the expectation value of the total momentum operator as a constant of motion associated with the translational symmetry. From the conservation laws for total momentum, it follows that the expectation value of the total momentum of an isolated system is constant in time. It further follows by applying the Ehrenfest relations that the expectation value of the CM moves on a straight line with constant velocity. As long as only simple expectation values and the associated correlation functions are considered, this suffices to ensure that only the dynamics of the internal degrees of freedom are observed. We then have a description in the laboratory frame, with its computational and conceptual simplicity, showing only the internal dynamics.

4. Applications

This basic model has been applied [22] to proton hydrogen atom collisions for energies between 0.02 to 4000 eV with excellent agreement with experiments for charge transfer cross sections, excitation cross sections, and alignment ratios. Bound state problems have also been studied. For instance, a short time evolution of the water molecule [22] from a modest distortion of the electronic ground state geometry yields a nicely resolved vibrational spectrum.

In the proton hydrogen collisions the hydrogen atom in its ground state is initially placed at rest at the origin of the coordinate system and the proton at a distance of 50 a.u. away and given different impact parameters and initial momentum corresponding to the collision energy. The system is allowed to evolve until the nuclei again are separated by 50 a.u..

For collision energies from 10 eV to 4 keV, an approximate n=1 and n=2 hydrogenic basis is used [22]. A comparison between the results with this larger basis and a smaller basis of a single 1s state per center is made at 10 eV and the difference in total cross-section is found to be less than 2%. Thus, for energies from 0.02 eV to 10 eV only the 1s functions are used. Impact parameters are chosen differently for various energy ranges: from 0.1 to 7.9 a.u in steps of 0.2 a.u. for energies from 100 eV to 4 keV; from 0.1 to 9.9 a.u. in steps of 0.2 a.u. for energies of 10 eV to 40 eV; and 0.0 to 13.9 a.u. in steps of 0.1 a.u. for energies of 0.02 eV and 0.18 eV. When only the 1s basis is used, the initial and final separations are taken to be 30 a.u.. Figure 1 shows our calculated total transfer cross sections compared to experiments over six orders of magnitude of collision energy.

Newman *et al.* [26] measure total transfer cross-sections for energies between 0.18 eV and 300 eV. For energies below 100 eV, they use protons colliding with deuterium atoms. Theoretical work by Hunter and Kuriyan [29] at energies below 10 eV indicates that below 1 eV appreciable differences appear between the total transfer cross-section of a proton on deuterium and that of the proton on hydrogen.[1] The former process being slightly endothermic (ΔE = 0.0037 eV), this is to be expected at some point. We study proton-deuterium and D^+-hydrogen collisions.

[1] Detailed trajectory as well as quantum scattering calculations on a given potential for proton + hydrogen atom and proton + deuterium atom have been performed by S. K. Knudson and W. R. Thorson, *Can. J. Phys.* **48**, 313 (1970) and by J. P. Davis and W. R. Thorson, *Can. J. Phys.* **56**, 996 (1978). The graphical representations of their results are less suitable to quantitative comparison with our calculations than are the numerical results of of Hunter and Kuriyan.

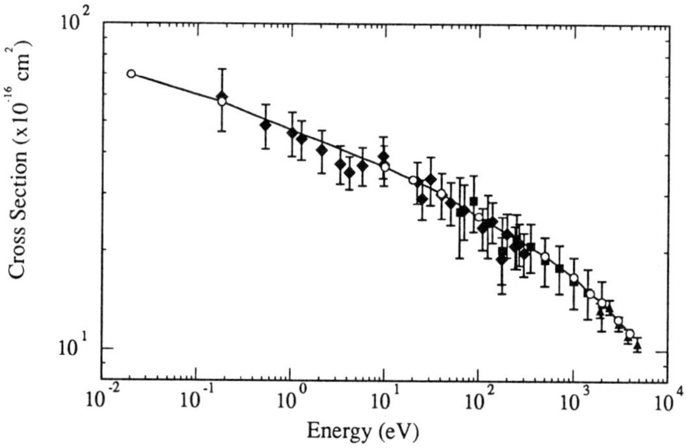

Figure 1. Total transfer cross sections. Theoretical calculations are the open circles joined by the solid line. Experiments are Ref. 26 (full diamonds); Ref. 27 (full squares); Ref. 28 (full triangles)

Table I. Total transfer cross-sections for collisions of a proton on hydrogen and deuterium, and that of a D^+ ion on hydrogen, for E < 1 eV. Results compared to those in Ref. 29 are also shown. (Units $x10^{-16}$ cm^2).

	E(eV)	$p+H \rightarrow H+p$	$p+D \rightarrow H+D^+$	$D^++H \rightarrow D+p$
Hunter&Kuriyan	0.02	72.63	60.78	72.17
Present work	0.02	69.55	69.23	69.25
Hunter&Kuriyan	0.2	55.88	50.64	51.31
Present work	0.18	57.32	56.89	–

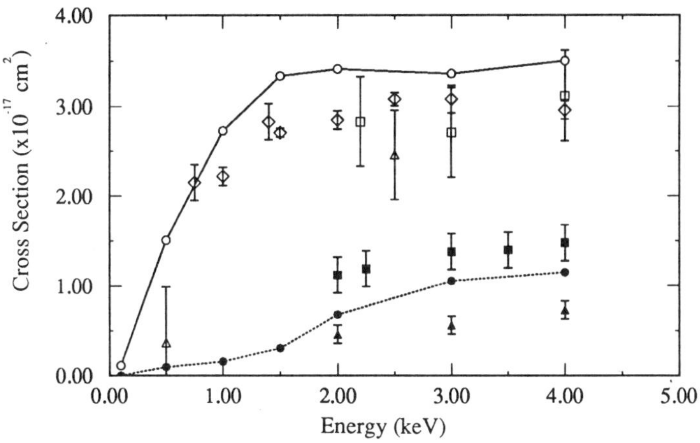

Figure 2. n=2 excitation cross sections. Total $2p$ excitation calculations are given by open circles joined by solid line; experimental results are: Ref. 30 (open diamonds); Ref. 31 (open squares); Ref. 32 (open triangles). $2s$ calculations are given by full circles joined by dotted line; experimental results are: Ref. 33 (full squares), and Ref. 34 (full triangles).

Our results, as well as those of Hunter and Kuriyan, for energies below 1 eV are shown in Table 1. The calculations with the deuterium atom and ion are done using the same basis as for the proton on hydrogen, except that the exponents in the basis are changed to reflect the change in the reduced mass of the deuteron-electron system.

Comparing our results with those of Hunter and Kuriyan [29] and noting that the experimental value of the cross-section at 0.18 ±0.06 eV for the $p+D \rightarrow H+D^+$ is $59.4^{+16.6}_{-9.7} \times 10^{-16}$ cm^2 [26], we find their result at 0.2 eV just within the error bars and considerably lower than ours. Our calculations show only a slight difference between the proton-hydrogen result and that of proton-deuterium. Even at 0.02 eV the difference is not significant. Since at 0.18 eV our result agrees closely with the experimental value, it would be interesting to see experiments repeated at these and perhaps even lower energies to verify if and when the isotopic effect becomes important. Due to obvious experimental difficulties with proton-hydrogen collisions, it would be easier and more interesting to measure proton-deuterium and D^+-hydrogen collisions. Hunter and Kuriyan report significant differences in the transfer cross-sections of these two processes for energies below 0.2 eV.

Figures 2 and 3 compare total $2p$ and $2s$ cross sections with experiments for excitation and transfer processes, respectively. Our total $2p$ transfer cross sections agree with experiments up to about 2 keV. Above that they are too large. For $2s$ transfer, our results are well above the existing experimental values. Other theoretical studies, in particular by Kimura and Thorson [35] and by Fritsch and Lin [36] do a better job of reproducing these experimental results. Our less accurate results are due to the omission of electron translation factors, which at these energies do make a difference since the incoming proton loses little momentum to the target for the impact parameters of importance in determining these cross sections.

The calculated excitation cross sections are very close to experimental values for the $2p$ state. This is so because of the small momentum transferred to the target for the important impact parameters, which makes ETF's not as important for excitation processes. On the other hand, our calculated $2s$ excitation cross-sections fall between the values of the two experiments performed at these energies. Other theoretical treatments [35, 36, 37] determine $2s$ excitation cross-sections below the experimental values.

The integral alignment provides a measure of the relative probabilities of excitation to the different $2p$ states of the target atom. The available experimental results are from Hippler

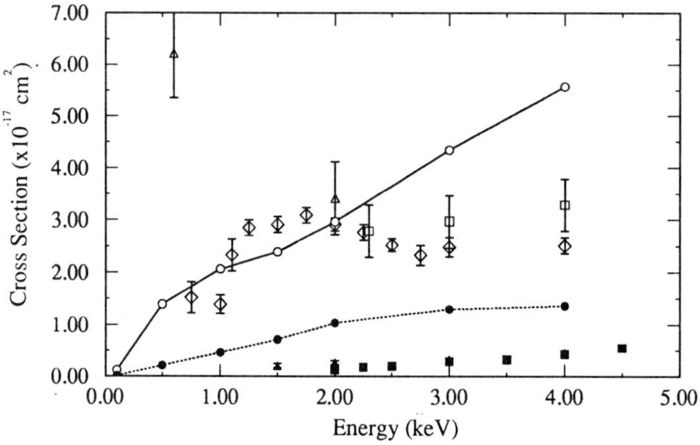

Figure 3. n=2 transfer cross sections. The symbols and lines have the same meaning as in Figure 2.

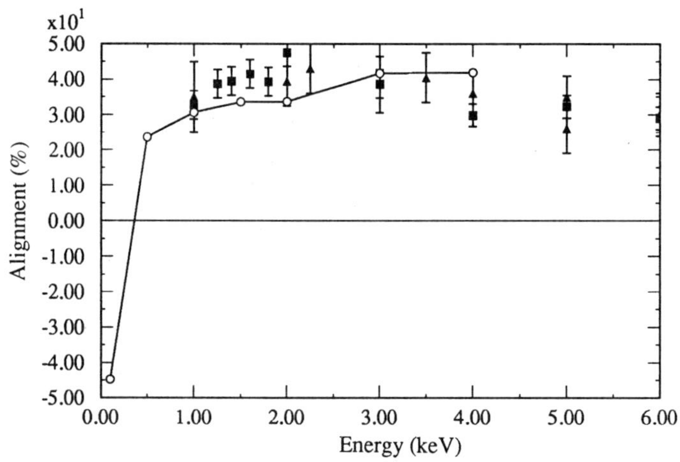

Figure 4. Computed integral alignment (open circles joined by solid line) and experimental results: Ref. 38 (full triangles) and Ref. 39 (full squares).

and collaborators [38, 39], starting from 1 keV and going up to energies of 25 keV. The integral alignment is computed as $A_{20} = (\sigma_1 - \sigma_0)/(\sigma_0 + 2\sigma_1)$ with $2\sigma_1$ being the total cross section to p-states aligned perpendicular to the initial direction of projectile motion and σ_0 the cross section to the p-state along this direction.

Figure 4 shows our results compared to experiment. In references [38, 39] the experiments are compared to theoretical work by Lüdde and Dreizler [37] and Fritsch and Lin [36],who employ straight line trajectories. As the energy drops to 1 keV, all these theoretical results converge to an alignment of 46%, just above the experimental error bars, and none predict the decrease in the alignment below 2 keV that experiments show. The straight line approximation appears to cause errors in the alignment at these energies, which are not present in our full electronic-nuclear dynamics treatment. Our results predict a sharp fall in the alignment as energy continues to drop, which warrants further experimental work.

These results also compares favorably with results using the Time-Dependent Hartree-Fock method [40], which considers electron-nuclear coupling and has been implemented with the so called eikonal [41] approximation for nuclear dynamics.

5. References

1. D. Gottlieb and S. A. Orszag, *Numerical Analysis of Spectral Methods: Theory and Applications,* SIAM, Philadelphia 1977.
2. R. Bisseling and R. Kosloff, *J. Comput. Phys.* **59**, 136 (1985).
3. F. L. Quéré and C. Leforestier, *J. Chem. Phys.* **92**, 247 (1990).
4. D. Kosloff and R. Kosloff, *J. Comput. Phys.*,**52**, 35 (1983).
5. M.D. Feit and J.A. Fleck, *Appl. Opt.*, **19**, 2240 (1980).
6. C. Leforestier, R. Bisseling, C. Cerjan, A. Guldberg, A. Hammerich, G. Jolicard, W. Karrlein, H. D. Meyer, N. Lipkin, O. Roncero, and R. Kosloff, *J. Comput.Phys.* **94**, 59 (1991).
7. J. K. Cullum and R. A. Willoughby, *Lanczos Algorithms for Large Symmetric Eigenvalue Computations* Vol.4 of *Progress in Scientific Computing*, Birkhäuser, Boston (1985).
8. J. Park and J. Light, *J. Chem. Phys.* **85**, 5870 (1986).
9. H. Tal-Ezer and R. Kosloff, *J. Chem. Phys.*, **81**, 3967 (1984).
10. R. Kosloff, *J. Phys. Chem.*, **92**, 2087 (1988).
11. R. Kosloff and D. Kosloff, *J. Chem. Phys.*, **79**, 1823 (1983).
12. S. L. Tang, D. G. Imre, and D. Tannor, *J. Chem. Phys.*, **92**, 5919 (1990).
13. M. D. Feit, J.A. Fleck, and A. Steiger, *J. Comput. Phys.* **47**, 412 (1982).
14. J. Avarellos and H. Metiu, *J. Chem. Phys.* **88**, 4957 (1988) and J. Broeckhove, B. Feyen, L. Lathouwers, F. Arickx, and P. Van Leuven, *Chem. Phys. Lett.* **174**, 504 (1990).
15. V. Engel, *J. Chem. Phys.***94**, 16 (1991).
16. S. Cheslowski, A. D. Bandrauk, and P. B. Corkum, *Phys. Rev. Lett.*, **65**, 2355 (1990).
17. Y. Sun, R. C. Mowrey, and D. J. Kouri, *J. Chem. Phys.* **87**, 339 (1987).
18. A. D. Hammerich, R. Kosloff, and M. A. Ratner, *Chem. Phys. Lett.* **171**, 97 (1990).
19. K. Kulander, C. Cerjan, and A. E. Orel, *J. Chem. Phys.* **94**, 2571 (1991).
20. R. Kosloff and H. Tal-Ezer, *Chem. Phys. Lett.*, **127**, 223 (1986) and D. Neuhauser, *J. Chem. Phys.***93**, 2611 (1990).
21. P. Kramer and M. Saraceno, *Geometry of the Time-Dependent Variational Principle in Quantum Mechanics*, Lecture Notes in Physics **140** Springer, New York (1981).
22. E. Deumens, A. Diz, H. Taylor, and Y. Öhrn, *J. Chem. Phys.* to appear (1992).
23. D. J. Thouless, *Nucl. Phys.***21**, 225 (1960); E. Deumens and Y. Öhrn, *J. Mol. Structure (Theochem)* **19**, 23 (1989).
24. A. M. Perelomov, *Commun. Math. Phys.* **26**, 222 (1972).

25. E. Deumens, A. Diz, and Y. Öhrn, *Coherent State Approach to Time Evolution with a Determinatal State* Technical Note T.F. 1146, Quantum Theory Project, University of Florida, Gainesville, FL 32611.
26. J. H. Newman et al., *Phys Rev.* **A 25**, 2976 (1982).
27. M. W. Gealy and B. Van Zyl, *Phys. Rev.* **A 36**, 3091 (1987).
28. G. W. McClure, *Phys Rev.* **148**, 47 (1966).
29. G. Hunter and M. Kuriyan, *Proc. Roy. Soc.* Ser. **A 353**, 575 (1977); *ibid.* **A358**, 321 (1977).
30. T. Kondow, R. J. Girnius, Y. P. Chong, and W. L. Fite, *Phys. Rev.* **A 10**, 1167 (1974).
31. T. J. Morgan, J. Geddes, and H. B. Gilbody, *J. Phys.* **B 6**, 2118 (1973).
32. R. F. Stebbings, R. A. Young, C. L. Oxley, and H. Ehrhardt, *Phys Rev.* **138**, A1312 (1965).
33. T. J. Morgan, J. Stone, and R. Mayo, *Phys. Rev.* **A 22**, 1460 (1980).
34. J. Hill, J. Geddes, and H. B. Gilbody, *J. Phys.* **B 12**, 2875 (1975).
35. M. Kimura and W. R. Thorson, *Phys. Rev.* **A 24**, 1780 (1981).
36. W. Fritsch and C. D. Lin, *Phys. Rev.* **A 26**, 762 (1982); *ibid.* **A 27**, 3361 (1983).
37. H. J. Lüdde and R. M. Dreizler, *J. Phys.* **B 15**, 2703 (1982).
38. R. Hippler, Coherence and orientation in few-electron ion-atom collisions, in *Electronic and Atomic Collisions* Eds. H. B. Gilbody, W. R. Newell, F. H. Read, and A. Smith, p241, Elsevier, New York 1988.
39. R. Hippler, H. Madeheim, W. Harbich, H. Kleinpoppen, and H. O. Lutz, *Phys Rev.* **A 38**, 1662 (1988).
40. D. A. Micha and B. Gazdy, *Phys. Rev.* **A36**, 539 (1987); *ibid* **A36**, 546 (1987)
41. D. A. Micha, *J. Chem. Phys.* **78**, 7138 (1983).
42. L. F. Shampine and M. K. Gordon, *Computer Solution of Ordinary Differential Equations: The Initial Value Problem* Freeman, San Fransisco, 1975.
43. see e.g. A. C. Hindmarsh, *Gear: Ordinary differential equation solver.* Technical Report, Lawrence Livermore Laboratory, Berkeley CA, 1975, UCID-30059; W. H. Press, B. P. Flannery, S. A. Teutolsky, and W. T. Vetterling, *Numerical Recipes: The Art of Scientific Computing,* Cambridge University Press, New York 1986
44. R. Car and M. Parinello, *Phys. Rev. Letters* **55**, 2471 (1985).
45. D. K. Remler and P. A. Madden, *Mol. Phys.* **70**, 921 (1990).
46. K. Runge, D. Micha, and E. Feng, *Int. J. Quantum Chem.* **S24**, 781 (1990).
47. M. J. Field, *J. Chem. Phys.* **96**, 3214 (1992); B. Hartke and E. A. Carter, *Chem. Phys. Letters* **189**, 358 (1992).
48. H.-D. Meyer and W. H. Miller, *J. Chem. Phys.***70**, 3214 (1979).
49. see e.g. J. C. Tully and R. K. Preston, *J. Chem. Phys.***55**, 562 (1971); R. D. Coalson, *J. Chem. Phys.***86**, 6823 (1987).
50. J. B. Delos, *Rev. Mod. Phys.* **58**, 287 (1981).

TIME DEPENDENT INTRAMOLECULAR QUANTUM DYNAMICS FROM HIGH RESOLUTION SPECTROSCOPY AND LASER CHEMISTRY

Martin Quack

Laboratorium für Physikalische Chemie
ETH Zürich (Zentrum)
CH–8092 Zürich, Switzerland

INTRODUCTORY SURVEY

The present paper reports a survey of the efforts of our research group concerning time dependent intramolecular quantum dynamics as derived from experiments in laser chemistry after infrared multiphoton excitation and high resolution infrared spectroscopy. The emphasis is on the conceptual background and recent results.

Time dependent intramolecular dynamics is at the origin of the primary processes in chemical reaction kinetics[1-4]. Explicitly time dependent descriptions of intramolecular dynamics and chemical reaction dynamics arise naturally in classical mechanical trajectory calculations for the motion of the nucleï in polyatomic molecules, macromolecules or condensed phases[5-7]. These use classical mechanics for the heavy particles and interaction potentials either from solutions of the quantum electronic Schrödinger equation in the adiabatic or Born-Oppenheimer approximations or from simple models. In the earlier years of *quantum theoretical reaction dynamics*, including heavy particle motion, descriptions were usually either based on *time independent* scattering theory for bimolecular, direct processes[8-10] or on quasiequilibrium theories for unimolecular reactions or bimolecular processes involving intermediate collision complexes[11-13]. Indeed, these time independent quantum theories have been sufficient to describe traditionally measurable quantities in kinetics, such as state-to-state cross sections, specific and thermal rate constants.

A need for a detailed, explicitly time dependent quantum description arose particularly with the advent of coherent multiphoton exciation with various shaped pulses and observation on time scales from microseconds to femtoseconds[14,15]. The developments of short time laser experiments in the last decade[16-20] have generated an increasing interest in time dependent descriptions for these experiments but also more generally including time dependent scattering theory[21-29].

The experimental access to time dependent intramolecular dynamics is based on a variety of different approaches. Whe the field is still in its infancy, some most significant conclusions can be drawn from the work of the last decade. Table 1 presents a broad view of time scales of intramolecular primary processes. In the upper part of the table we show some physical primary processes ranging from elementary particles and cosmology to fast electron transfer and molecular vibrations. Of more direct relevance for chemical reactions are the redistribution processes shown in the lower part of the table (illustrated mostly with examples from our own work).

Time-Dependent Quantum Molecular Dynamics, Edited by J. Broeckhove and L. Lathouwers, Plenum Press, New York, 1992

Table 1. A Broad View of Time Scales for Primary Processes of Intramolecular Energy Transfer

Type of primary process	Time scale (reference)
I. Ultrafast primary processes	
Dynamics in the primordial, cosmic supermolecule	Planck time $t_P = \sqrt{(G \cdot hc^{-5})}$ $\approx 1.35 \times 10^{-43}$ s
Nucleon motion in nuclei	$t_n \simeq 10^{-20}$ s
Fast electron transfer and electronic relaxation	$t_e = 4m_e^{-1}\,\epsilon_0^2\,h^3\,e^{-4}$ $= h/E_h \simeq 1.52 \times 10^{-16}$ s
Nuclear vibrational motion in molecules ($\lesssim$ 3333 cm^{-1})	$t_v \gtrsim 10^{-14}$ s
II. Redistribution processes	
Strongly, preferentially coupled anharmonic vibrational modes (example CH-stretch bend resonances)	$t_{IVR} \gtrsim$ 100 fs (24, 30, 31)
"Ordinarily" coupled modes (example: CF_3 frame modes)	1 ps $\lesssim t_{IVR} \lesssim$ 10 ps (3, 31, 53)
Adiabatically decoupled vibrational modes, $\Delta\ell$ selection rule in rotation vibration coupling	10 ps $\lesssim t_{IVRR} \lesssim$ 1 ns(?)
examples: —C≡C—H in acetylenes,	(32, 41)
HF stretching in $(HF)_2$	(33)
Slower electronic relaxation of spin- or adiabatically well decoupled states	$t_e \gtrsim$ 1 ns (34)
Rotation – Vibration – Nuclear Spin coupling; Violation of nuclear spin symmetry conservation	1 ns $\lesssim t_{nsrv} \lesssim$ 1 s (35, 36, 37)
Transfer of intramolecular parity (by violation of space inversion symmetry due to parity violating weak nuclear force)	(38, 39) 1 s $\lesssim t_{PV} \lesssim$ 1 day (theoretical estimates only)
Violation of time reversal symmetry in molecular processes	? (>> 1 day?, unknown) (40)

Nontrivial energy transfer processes between various vibrational motions have been characterized lately. These depend in specific ways upon certain structural features of molecules. About the fastest vibrational redistribution process identified so far is due to the Fermi resonance coupling between CH stretching and bending motions in the sp^3 alkyl CH-chromophore, combined with Darling-Dennison resonances between the bending modes in molecules that do not possess a threefold symmetry axis. Here we observe redistribution times of about 100 fs and even somewhat less, that is less than ten vibrational periods of the high frequency mode. More typical anharmonic redistribution times are from 1 to 10 ps, whereas adiabatic decoupling of high frequency from low frequency modes leads to much longer redistribution times up to a nanosecond and perhaps even longer. Of course, efficient adiabatic decoupling is also known in many cases for the electronic degrees of freedom, where slow intramolecular electronic relaxation (internal conversion) is common, particularly when combined with the change of electronic spin state (intersystem crossing). The even weaker coupling between nuclear spin and the other molecular degrees of freedom leads to a broad range of long relaxation times for change of nuclear spin symmetry by violation of the corresponding approximate conservation rule[35,36]. Finally very long times apply to the still unobserved, but theoretically predicted intramolecular relaxation of parity by means of the parity violating weak nuclear force and to the purely hypothetical violation of time reversal symmetry.

When discussing different *time scales for primary processes* due to different coupling mechanisms we should distinguish these from the great variation in *time scales in chemical reactions* arising from a combination of primary processes and merely statistical factors. An obvious example is the Arrhenius expression (1) for the rate constant of a thermal unimolecular reaction

$$k(T) = A(T) \exp(-E_a/RT) \tag{1}$$

The exponential factor involving the Arrhenius activation energy E_a generates almost any reaction time one could think of, from femtoseconds to the age of the universe. However, this arises from the statistical factors involving many activation steps, each of which being fast. The same is true for the statistical expression for the specific rate constant of a unimolecuar reaction by vibrational predissociation at energy E and angular momentum J[11,12]

$$k(E,J) = \frac{W(E,J)}{h\ \rho(E,J)} \tag{2}$$

Here h is Planck's constant, W(E,J) the number of open adiabatic reaction channels and $\rho(E,J)$ the reactant state density. If the latter is high and the former low, reaction rates will be low, because even for a fast primary vibrational process, reaction will be slow, because it is based on a random process with a rare chance for escape. A third example for this statistical effect is the power law for the threshold energy depedence of the chemical rate constant after multiphoton excitation

$$k(st,I) \simeq B\, I\, E_0^{-2.7} \tag{3}$$

In the linear regime,[2,3] each individual primary step in multiphoton excitation depends linearly upon radiation intensity I, but many steps are needed to overcome the threshold energy (energy barrier E_0), which results in an additional statistical factor, which is weaker here than in the other examples. One may compare the relative effect of primary processes and statistical factors in chemical reactions to the escape rate of runners from a complicated forest. The individual speed of the runners corresponds to the rate of the primary process, which can be slower or faster (there are hares, hedgehogs and snails ...). The escape rate is secondly determined by the complexity of the path of the runners in the forest. For a complex "random" motion of the runners in the forest, the escape rate will be low, because of statistical factors. These statistical factors are reasonably well understood, whereas much less is known about the molecular primary quantum processes, on which we shall concentrate here. We note in passing, that "intelligent reaction control" generally will operate on the statistical factors ("by choosing an optimum path for the runners to increase the escape rate"). Less can be done about the primary processes, as they correspond to fundamental properties of the molecular hamiltonian (however some influence is also possible here with external fields). We note that on various levels certain tunneling processes can lead to great varieties of time scales, which are not specifically indicated in the table.

SLOW, NANOSECOND REDISTRIBUTION IN IR-LASER CHEMISTRY

Unimolecular reactions induced by monochromatic infrared radiation (URIMIR[14]) generally consist in first exciting molecules to high levels by many (typically ten to forty) photons, from where they can then react. Normally, the infrared chromophore, which is excited by the infrared laser and the reactive site are quite different. For example, when exciting CF_3I with a CO_2 laser, one uses the frequency range of the strong absorption assigned usually to the symmetric CF_3-stretching vibration[42] near 1075 cm^{-1}. However, the reaction observed consists in breaking the carbon iodine bond

$$CF_3I \xrightarrow{nh\nu} CF_3I^* \quad n \gtrsim 17 \tag{4a}$$

$$CF_3I^* \longrightarrow CF_3 + I \tag{4b}$$

Of course, the two processes generally occur simultaneously. A large body of data has resulted in the dogma, that upon decomposition on the picosecond to microsecond time scale the highly excited molecules, such as CF_3I^*, essentially absorb light and react from a "statistical" internally relaxed state, generated by fast intramolecular relaxation on the picosecond or subpicosecond time scale[43, 44]. The evidence for this comes both from optical measurements (direct, but often ambiguous) and from chemical, dynamical observations (usually ambiguous, but sometimes unambiguous, generally indirect). For instance, in the reaction (4) an ambiguity concerning the statistical nature of the redistribution occurs: The hypothetical reactions at the chromophore site (for instance breaking the CF bond) and the actually observed reaction of breaking the carbon iodine bond are energetically not equivalent. It thus may well be, that multiphoton excitation would largely favour reactions at the chromophore site but not sufficiently to beat the dominant energetic effect, favouring C-I bond fission Eq. (4b), even if the distributions in CF_3I^* were highly nonstatistical. Evidence from product translational energy distributions in molecular beam multiphoton dissociation of CF_3I [45] and from rate constants of unimolecular decay after multiphoton exciation[46] favour statistical interpretations. This evidence is not conclusive, because comparison is made with parametrized statistical theories, where adjustments in the fit to experiment can compensate nonstatistical effects, as has been discussed before[47, 48]. These ambiguitites can be circumvented by selecting a reaction, where the chromophore site and the reaction site are identical (or nearly so).

We shall quote here such an example from our own work, which seems so far the only example of this "symmetrical" type[49]:

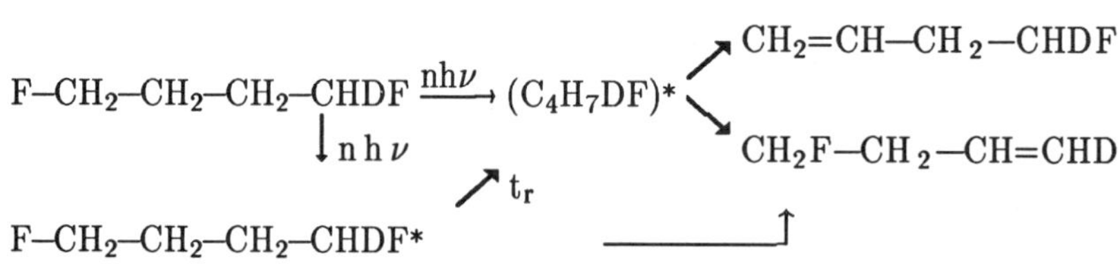

(5a)

(5b)

In this case one can excite the $-CHDF$ chromophore with a laser wavenumber at 930 cm^{-1} (the $-CH_2F$ chromophore absorbs at 1050 cm^{-1} only). Nevertheless, the reactive sites are almost identical at both ends of the butane, giving HF elimination with a very small secondary isotope effect. The products of reactions (5a) and (5b) can be easily distinguished by IR-spectroscopy. It is found that both products are formed in equal amounts on time scales of more than about 10 ps (timing by collisional quenching with inert gas). This can be compared with a "chromophore selectivity" of about 100:1, which is derived from the isotope selectivity in mixtures of ordinary fluorobutane and deuterofluorobutane. Thus one would conclude, that to a high degree of accuracy statistical redistribution occurs in this molecule on time scales of less than about 10 ps. Ambiguities still arise from the collisional timing in these experiments, which could be avoided in very short pulse experiments under collision free conditions (at low pressures or in molecular beams). Even then, observing just the *outcome* of the redistribution process is of course just *indirect* evidence. Furthermore, there is no reason to believe that a result of fast re-

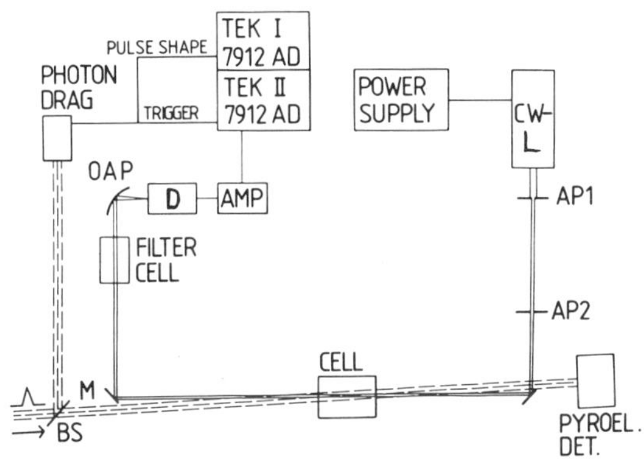

Fig. 1. Experimental scheme for relaxation-reaction measurements (see discussion in the text and refs. 19, 46, 50)

distribution that has been established for one or a few reactions under certain conditions is valid for all molecules under all conditions. Indeed, it seems unlikely that the dogma of universally fast, statistical redistribution is correct, although it is widely believed. We shall show here some recent evidence on slow intramolecular relaxation processes, that may give a blow to the dogma.

In order to study in detail relaxation and reaction processes in polyatomic molecules we have systematically developed a general scheme for kinetic spectroscopy, using shaped pulses from a CO_2 laser[19] and continuous detection with electronic devices allowing for nanosecond time resolution with various continuous light sources. The general scheme is shown in Figure 1. The continuous wave light source designated by CW-L can be a cw-CO_2 laser,[46] a mid infrared diode laser,[50] a near infrared diode laser or a monochromatic UV light source, allowing for great versatility. The probe beam passes through the reaction cell via a filter cell onto a detector D coupled to fast transient digitizers. The reaction cell is furthermore exposed to the exciting radiation from a pulsed CO_2 laser, which is not shown in the figure, the incoming beam being schematically drawn by —^— . The pulse shape of the beams ranging from about one to one hundred nanoseconds and various, controlled shapes is simultaneously measured by a fast photon drag detector (about nanosecond resolution). Using this device, we can measure the effective time dependent infrared absorption of reactant molecules and product atoms and molecules. We refer to the original papers for details[19, 46, 50]. The basic scheme of kinetic spectroscopy is, of course, very old, going back to the work of Norrish and Porter and Eigen in the fifties. Current work concentrates mostly on the picosecond to femtosecond domain, where *pulsed excitation and detection* is required. However, interesting observations are expected with continuous detection on the nanosecond time scale. In recent work we have found, for instance, a relaxation process of a few nanoseconds in the infrared absorption of CF_3I after coherent multiphoton excitation:

$$CF_3I \xrightarrow{nh\nu} CF_3I^* \xrightarrow{t_r} CF_3I^{**} \tag{6}$$

The relaxation time t_r of a few nanoseconds[46] is most striking and warrants some discussion. It is not due to reaction (which can also be observed at high excitation), it is becoming unmeasurably fast (on our time scale) if we probe only very highly excited CF_3I molecules for instance, after excitation with intense pulses of high fluence or by probing where only very "hot" molecules absorb, far on the low frequency side (around 1040 cm^{-1}) of the main absorption (around 1075 cm^{-1}). Under conditions of low pressure (< 1 mbar) and in the absence of pressure effects, we postulated a collision free relaxation[46]. Otherwise collision cross sections of more than 1000 $Å^2$ would have to be assumed, combined with an unusual absence of pressure dependence. This seems unlikely, although in early experiments on IR-multiphoton excitation of SF_6, even larger cross sections for collisional processes were postulated[51], but never confirmed. If the process is truly intramolecular, it could be dominated by just barely excited CF_3I, but we must consider that we have a coherent superposition from many steps in multiphoton excitation. It could secondly be a relaxation process involving nuclear spin, but the frequency- and fluence dependence would be unusual. It might thirdly be a coherent radiative effect from the excited molecules interacting with the probe laser, a rather less attractive interpretation. The most likely interpretation is clearly, that "slow" (i.e. nanosecond) internal vibrational (and rotational) redistribution has been observed here in real time. Small coupling matrix elements for ν_1 interacting with $\nu_3 + 3\nu_6$ (in $^{12}CF_3I$ [52]) and with $4\nu_6$ (in $^{13}CF_3I$ [53]) would seem to support this interpretation based on quite different evidence from high resolution spectroscopy. Ultimately, the proof must indeed come from simulations of the dynamics of IR-multiphoton excitation and IR-absorption of CF_3I based on a complete analysis of the high resolution spectra, similar to the wavepacket study for CX_3H outlined below. While at present this analysis cannot be carried out yet for CF_3I, the qualitative observation of nanosecond relaxation (even without knowledge of the time dependent wave function) is striking enough. Current analysis of spectroscopic couplings in CF_3I suggest both the existence of rather fast IVR for some modes and slow IVR for others, contradicting the dogma of *universally* fast IVR.

We may mention here two other pieces of evidence for slow, nanosecond IVR in polyatomic molecules. These come both from line shape analyses, not from real time observation. The first concerns vibrational predissociation linewidths for $(HF)_2$, which are small, even at very high excitation above threshold,[33] the second from CH stretching vibrational linewidth studies on substituted acetylenes R–C≡C–H, which, in agreement with earlier results suggesting adiabatic decoupling and long lifetimes,[32] have shown linewidths corresponding to nanosecond relaxation[41].

In summary of this chapter we may say, that there is experimental evidence becoming available on rich structures for intramolecular rovibrational redistribution processes including slow and fast kinetics. A drawback of kinetic relaxation measurements – be it on the nanosecond or femtosecond time scales – and also of lineshape evaluations is the lack of detailed information about exactly what molecular kinetic processes are observed, i.e. one knows *how fast* they are but not *what* they are. For this, additional information is necessary, although this point is often overlooked. In the next chapter we shall summarize an approach, which we consider one of the most powerful available, although tedious.

MOLECULAR SPECTRA AND MOLECULAR MOTION

What are the primary processes of molecular motion during and after coherent excitation, for instance IR-multiphoton excitation? Answering this question in terms of molecular quantum dynamics implies studying wave packet dynamics or, more generally, with a statistical initial state, the probability distributions for electrons and nuclei in their multidimensional coordinate and spin space[15]. It seems unlikely, that observation of some time dependent absorption cross sections, Raman scattering cross sections, or of reactant and product concentrations even with so called state-to-state resolution will provide enough information to obtain the complete dynamics. We have therefore undertaken over the last decade or so an indirect, systematic approach towards polyatomic molecular quantum dynamics. It consists of a conceptually simple two step procedure:

(i) Obtain from high resolution spectroscopy and additional information (such as ab initio potentials in the adiabatic approximation) the molecular "spectroscopic" states (or "eigenstates"), their wavefunctions, transition moments, and the underlying molecular hamiltonian H ("potential surface" in Born-Oppenheimer language, but really more general).

(ii) Solve the time dependent equation of motion for the matrix representation $\mathbf{U}$ of the time evolution operator U in the basis of molecular eigenstates, possibly including coupling to a time dependent radiation field (say, from a laser).

To the extent that this program can be carried out accurately, this solves the molecular motional problem in its most general form. In practice, of course, there arise some problems in detail. We shall discuss some of the experimental and theoretical aspects of the major steps involved and then present a summary of selected results. Figure 2 presents a more detailed scheme of the approach.

Starting from high resolution molecular spectra in the upper part of the scheme the first step of traditional spectroscopic line-by-line analysis consists in a least squares adjustment resulting in the "best" spectroscopic parameters of an effective hamiltonian. One way to look at these parameters is to consider the analogy with the paramters of simple analytical term formulae (the diagonal part of the effective hamiltonian), being generalized to non diagonal matrix representation, with analytical formulae for off diagonal matrix elements as well. To avoid a common misunderstanding, we should stress that these spectroscopic constants of the effective hamiltonian do not correspond in a simple way to properties of the real molecular hamiltonian, for instance potential constants. If common analytical formulae for anharmonic constants from perturbation theory are used,[54,55] the constants derived may be in error by a factor of two to five, as has been demonstrated in a case study on anharmonic couplings in the CH chromophore[30,31,56]. Thus, the effective hamiltonian does neither directly provide the molecular potential function nor the molecular wavefunctions associated with the observed spectroscopic lines. Nevertheless, effective hamiltonians provide a compact representation of the spectrum and allow for predictions of unobserved lines. They can also be made the starting point for automatic assignment procedures[58,59]. The simple analytical

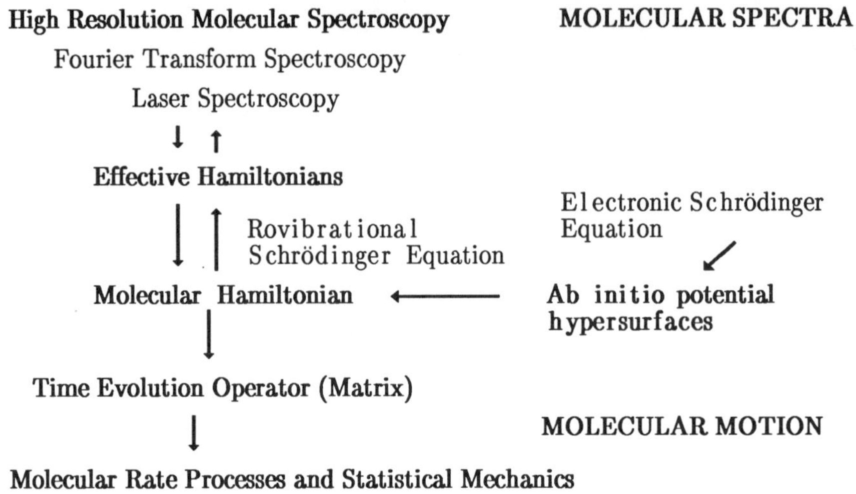

Fig. 2. Scheme of the combined experimental-theoretical approach "Molecular Spectra and Molecular Motion"

structure of the effective hamiltonian may also contain fundamental insight, although in general the relation to the true molecular hamiltonian is complex.

The next step in the scheme is to derive the "true" molecular hamiltonian. This step is more complicated than commonly believed and generally cannot be carried out by means of spectroscopic, empirical information only. In practice, additional information is used, for instance ab initio calculations for some aspects of the potential hypersurface in conjunction with the Born-Oppenheimer (or adiabatic) approximations. Even then, the need arises to calculate molecular energy levels and wavefunctions by solving a multidimensional revibrational Schrödinger equation for the polyatomic system

$$\hat{H}_{mol}\, \phi_k(r) = E_k\, \phi_k(r) \tag{7}$$

Here "r" designates the ensemble of all spin and space coordinates of all particles, in general, but for the rovibrational problem this is reduced to the space coordinates of the nuclei only. This still presents a formidable problem. Usually it is simplified by assuming either complete or adiabatic separability of some degrees of freedom from all the others. There is currently considerable activity in developing new methods in this field, quite a few of which being discussed at this conference (see, for instance, the paper by Light[57]). Table 2 A. summarizes methods used and developed in our laboratory in conjunction with the spectroscopic experiments (mostly FTIR-spectroscopy[31]). Usually, numerical solutions of the Schrödinger equation for an important subspace of the problem are combined with other vibrational-rotational motions by means of perturbation theory, separability assumptions etc. The number of dimensions treated numerically is shown for the examples (2-d, 3-d etc.). In the case of the Fermi resonance in CHX_3 symmetric top molecules the important part of the problem is 2-dimensional, for asymmetric tops CHX_2Y it is 3-dimensional. The straight-forward variational basis set expansion and grid techniques are generally limited to few dimensions. In contrast, Anderson[72] Quantum Monte Carlo techniques can be used for high dimensional systems and we have obtained accurate solutions for $(HF)_2$ with six vibrational and 3 rotational degrees of freedom. However, only ground states can be calculated in a direct way. For excited states additional aproximations are necessary, such as the newly developed quasiadiabatic channel quantum Monte Carlo method[68,69]. There is little doubt that future developments will bring further progress.

Table 2. Methods developed or used in the Zürich group (and some others) for solving the time independent and time dependent Schrödinger equations

A. Time independent problem (computation of H)		
Method	Examples	References
Vibrational, variational, basis set expansion in normal coordinates	2-d CHD_3, 2-d CHF_3, $CHCl_3$, $CH(CF_3)_3$ 2-d $CHBr_3$,	56 60, 61, 62 63, 64
Vibrational, variational, basis set expansion in internal coordinates (approximate)	2-d CHD_3, CHF_3, 2-d $CHCl_3$ 2-d $CHBr_3$	62, 65, 66 62 63
Vibrational-rotational quantum Monte Carlo and quasiadiabatic channel quantum Monte Carlo	$(HF)_2$ (9-d, 6-d, 5-d + 1-d)	67, 68, 69
Vibrational, numerical grid methods	CHD_2F (3-d)	70, 71

B. Time dependent problem of coherent radiative excitation (Computation of U)		
Method	Typical Application	References
Direct numerical integration of differential equations (in eigenstate basis or on grids)	Ultrashort pulses, strong, nonperiodic fields	21, 22, 23, 24, 25
Floquet approximation (combined with numerical integration for one period, H_{eff} = periodic)	Typical, coherent laser pulses ps to μs periodic field	14, 15, 23, 24
Quasiresonant approximation (QRA) $H_{eff} \neq f(t)$	Typical laser pulses, not too strong coupling with near resonant states,	14, 15, 73
Field free evolution similar	isolated molecules	30,83,84,85
Closed analytical solutions (special cases only)	Degenerate levels harmonic oscillators	14 74, 75
Quantum statistical Master equations	Very large numbers of irregularly coupled levels,	14, 15
cases A, B	intermediate coupling strength	76
case C (generalized)	low coupling strength, strong selection rules	14, 15, 77, 79
case D	high coupling strength	14, 15
Rate constants, first order kinetics	Typical laser pulses, only reaction rates needed	2, 3, 14, 15, 76, 77, 78

We shall turn now to the second step, the solution of the time dependent problem, assuming that the rovibronic, time independent problem has been solved by one of the techniques mentioned. This amounts to solving the differential equation for the time evolution matrix **U** in the basis ϕ_k of molecular eigenstates (rovibronic states)

$$i \frac{h}{2\pi} \frac{dU}{dt} = \mathbf{H}(t)\, \mathbf{U} \tag{8}$$

In the basis of the molecular eigenstates and a coupling to a periodic classical field, which is an excellent approximation to the laser fields in multiphoton excitation,[14] one has explicitly

$$i \frac{dU}{dt} = \{\mathbf{W} + \mathbf{V} \cos(\omega t + \eta)\}\, \mathbf{U} \tag{9}$$

with H_{mol} in the basis of eigenstates being diagonal

$$\mathbf{W} = 2\pi\, \mathbf{H}_{mol}/h \tag{10}$$

or

$$\hat{H}_{mol}\, \phi_k = \frac{h}{2\pi} W_{kk}\, \phi_k \tag{11}$$

For electric dipole coupling (common in IR-multiphoton excitation) one has with z-polarized radiation (electric field E_0)

$$V_{kj} = -\langle \phi_k \mid \boldsymbol{\mu}\, \mathbf{e}_z \mid \phi_j \rangle \left[\frac{2\pi}{h}\right] \mid E_0 \mid \tag{12}$$

$\boldsymbol{\mu}$ is the electric dipole moment, $\mathbf{e}_z$ the unit vector in the z-direction. For other couplings this is to be replaced by analogous expressions. The function $E_0(t)$ designates the laser field amplitude, which is assumed to be slowly varying compared to the period of the field. With lasers in the mid- and near infrared the period of the field is in the femtosecond range (~ 30 fs for CO_2 laser), whereas variations in E_0 fall in the picosecond to nanosecond range, typically. $E_0 = 0$ corresponds to the special case of field free evolution of the molecules. One may thus assume that the problem can be decomposed into pieces with constant E_0 each. These pieces are joined smoothly to provide a laser field envelope function f, which varies slowly in time. If this is not the case, one cannot make use of the periodicity of the field, but then this is also not necessary. Beause of the short times involved, one can then use one of the many schemes for direct numerical integration of the time-dependent coupled differential equations. In this case, the eigenstate expansion in our two step procedure is not necessary and may also not be the most useful starting point. It is a situation where radiative couplings are of similar strength as intramolecular couplings. While most recently a number of experiments fall in this class, it is not very common. It is often overlooked that direct integration methods will fail totally, numerically, when extended to "long" times into the picosecond and nanosecond domain of typical coherent laser excitation experiments.

For typical conditions of laser excitation in the longer time ranges, one may use the periodicity of the field by means of the Floquet-Liapounoff approximations[14]. If the coupling by the laser field is not too strong and if the importantly coupled states are fairly close to resonance at each step of coherent multiphoton excitation, one may use the quasiresonant approximation (QRA),[14] which is identical to the well known rotating wave approximation (RWA) in the special case of the simple two level problem. It is at the level of these approximations that the two step procedure with expansion of the time dependent problem in the basis of field free molecular eigenstates becomes essential. In this basis, it is possible to reduce the size of the basis, which for an exact solution would be infinite, to manageable numbers of coupled states. The quasiresonant approximation corresponds to a further transformation of this eigenstate basis to the quasiresonant basis[15]. We have critically discussed the merits of these techniques elsewhere[2,3]. Published program packages exist[73]. In very special cases, analytical solutions can be found[14,74,75].

The approaches discussed so far all attempt to solve the *complete* dynamical problem or statistical mechanical problem. In terms of the expansion in eigenstates we may write, using the **U** matrix from the solution of Eqs (8) and (9):

Solution of Schrödinger equation for the wavefunction $\Psi(t)$ (explicitly, with expansion in the eigenstate basis ϕ_k)

$$\Psi(t) = U(t,t_0)\,\Psi(t_0) \tag{13a}$$

$$\Psi(t) = \sum_k b_k(t)\,\phi_k(r) \tag{13b}$$

$$\mathbf{b}(t) = \mathbf{U}(t,t_0)\,\mathbf{b}(0) \tag{13c}$$

Solution of the Heisenberg equations of motion for the observables Q (particle coordinates, spins and momenta, matrix representations **Q**)

$$\mathbf{Q}(t) = \mathbf{U}^{\dagger}(t,t_0)\,\mathbf{Q}(t_0)\,\mathbf{U}(t,t_0) \tag{14}$$

Solution of the Liouville-von Neumann equation for the complete density matrix ***P***

$$\mathbf{P}(t) = \mathbf{U}(t,t_0)\,\mathbf{P}(t_0)\,\mathbf{U}^{\dagger}(t,t_0) \tag{15}$$

Even with the use of the QRA the ultimate problem in solving these equations is the size of the matrices, which would be of infinite order for an exact solution of real molecular problems. Good approximations require too large a basis to be tractable even for polyatomic molecules of modest size. One then may resort to statistical mechanical approximations. These can take the form of Bloch-type differential equations for *reduced density matrices* ρ for one mode or a subset of modes:

$$\frac{d\rho_{kk}}{dt} = f(\hat{H}, \rho_{jj}, T_1 \ldots) \tag{16a}$$

$$\frac{d\rho_{jk}}{dt} = f(\hat{H}, \rho_{\ell,m}, T_2 \ldots) \tag{16b}$$

Alternatively, one can use Pauli-type master equations for average or coarse grained populations p_k of molecular levels (involving very large numbers of quantum states):

$$\dot{\mathbf{p}} = \mathbf{K}\,\mathbf{p} \tag{17}$$

One may connect Eqs (16) and (17). We think that future accurate treatments of time dependent quantum molecular dynamics will always be based upon such approaches for larger polyatomic and macromolecular systems. For a discussion of both fundamental and practical aspects we refer to [1-4,14,15]. At the final level of simplification of the dynamics one just calculates the generally time dependent reaction rate constant, as mentioned at the end of table 2. For thermal reactions, this will be the thermal rate coefficient, depending on T, P etc. For laser induced reactions it is the time- and intensity dependent reaction rate coefficient.

Table 3 provides a summary of selected results on time dependent molecular quantum dynamics in specific systems obtained with these methods. For details of the great variety of quantum dynamical and kinetic results the reader should

Table 3 Selected Results on Time Dependent Molecular Quantum Dynamics (from work of Zürich group)

Method	Specific system, Calculated property	References
QRA	Irregular, statistical spectra, populations	14, 15
Direct Integration Floquet, QRA	harmonic and anharmonic oscillator (1-d), populations wavefunctions	23, 74, 75, 80
Analytical, closed expressions	harmonic oscillator, populations wavefunctions, densities	74, 75
QRA	Ozone (O_3), populations, reaction $O_3 \longrightarrow O_2 + O$	22, 80, 81, 82
Floquet and QRA	CHD_3, CHF_3 (populations, wavepackets)	24
Free evolution $H_{eff} \neq f(t)$	CHF_3, populations, wavepackets	30,83,85
	$CH(CF_3)_3$, populations	84
	CHD_2F, populations	70
	$CHClF_2$. $CHCl_2F$, $CH(CF_3)_2F$,	86
	populations, entropy	90, 31
	statistical spectra	89
Master equations cases B and C	CF_3I, populations, reactions	33
Master equations cases B and C	CF_3I, reaction	88
Master equations	CH_2O, reaction, populations CF_2Br_2, C_2F_5I, reaction, product energy distributions	48
Master equations rate constants	Various molecules with experiments, systematic studies	2, 3, 77, 78
Master equation rate constants	$C_4H_9F \longrightarrow C_4H_8 + HF$ reaction	49

consult the original literature cited in the table. We shall summarize here only one important aspect of a spectroscopic discovery, which is complementary to the kinetic nanosecond relaxation observed in the IR-multiphoton excitation of CF_3I, as discussed in the previous section. We have found that there is group specific preferential vibrational redistribution on the time scale of less than 100 fs for the CH stretching and bending modes in alkyl CH compounds. For C_{3v} symmetric top CHX_3 molecules two vibrational degrees of freedom are involved in this redistribution, with the vibrational angular momentum quantum number ℓ for the degenerate bending modes being a good, conserved quantum number on short time scales. For C_s or C_1 alkyl CH, this good quantum number is not present and thus there are three preferentially coupled modes giving redistribution on short time scales. The other vibrational degrees of freedom ($s_f \geq 6$) from the heavy frame do not seem to participate appreciably for $t \lesssim 1$ ps.

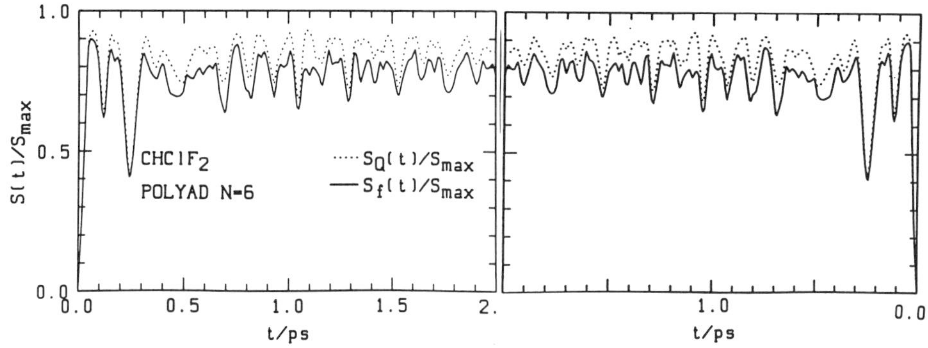

Fig. 3. Quantum evolution of partial, reduced entropy for the subset of strongly coupled CH modes in $CHClF_2$ with an initial state of six quanta of stretching from t = 0 to t = 2 ps and the mirror image evolution, formally, from t = 2 ps backwards to t = 0 ps as forward process (see discussion in text and references 90, 91.

For the subset of strongly coupled modes we have found a striking quantum statistical redistribution, which is sensitive to the strength of the anharmonic coupling and which, for the realistic molecular hamiltonians, leads to a nonclassical quasimicrocanonical spreading of the probability distribution for wavepackets under IR-multiphoton excitation or in isolated molecules[24, 85]. We believe that this phenomenon is important for the understanding of the origin of quantum irreversibility in large polyatomic molecules. In order to illustrate and quantify this we show in Fig. 3 the evolution of suitably defined quantum entropies for the subset of three strongly coupled CH modes in $CHClF_2$, initially excited with six quanta of pure CH-stretching[90, 91]. Within less than 100 fs 90% of the quasiequilibrium maximum entropy S_{max} for the subset of modes is attained, with subsequent fluctuations of an order of magnitude expected for such a small quantum system. To be sure, the irreversible increase of entropy is due only to a *de facto* symmetry breaking of time reversal symmetry in the real, dynamical situation of Fig. 3. If "at equilibrium at 2 ps" we inverted all momenta, we would observe the time reversed mirror image of the evolution shown in the right hand part of the figure, with a return to zero entropy at t = 0 (or t = 4 ps, if we continue to run time forward with reversed initial state at 2 ps). One of the more interesting open questions in our current investigations of molecular quantum dynamics concerns the *de lege breaking* of this time mirror image symmetry on long time scales, similar to the *de lege* breaking of space mirror image symmetries in chiral molecules due to the weak nuclear interactions, as might be possibly indicated by the CP violation in K^0-delay[4, 38-40]. Our investigations of the origin of the *de facto* symmetry breaking (Fig. 3) provide a framework for possible further investigations.

SUMMARY AND CONCLUSIONS

The primary processes of intramolecular quantum dynamics cover a wide range of time scales. They can be investigated by kinetic spectroscopy combined with coherent laser excitation. Using this technique, recent evidence points to the surprising existence of comparatively slow (nanosecond) vibrational redistribution processes. An alternative approach to intramolecular kinetics is based on the analysis of time independent spectral features. This idea has been used for a long time for instance in line shape evaluations for predissociation processes[4, 92] or in Weller's pioneering studies of fast proton transfer using the kinetics of fluorescence change[93]. Our experimental approach, using a combination of high resolution spectroscopy and quantum dynamical calculations (at various levels reviewed here) provides a new, deep insight into femtosecond to picosecond molecular dynamics[31].

It has given first answers to some long standing questions, such as mode specific anharmonic redistribution in dynamical functional groups. The old dogma of universally fast intramolecular vibrational redistribution towards microcanonical equilibrium in a few ps has been replaced by the experimentally established fact that some vibrational modes show, indeed, this behaviour, while others don't. The dynamical reasons for this can be understood semiquantitatively, even on an ab initio basis. Our spectroscopic approach may furthermore help to attack some of the most fundamental questions of molecular irreversibility de facto and de lege.

Acknowledgement: The work reviewed here results from enthusiastic contributions of many coworkers over the years, as indicated by the long list of references. Our work is supported financially by the Schweizerischer Nationalfonds and the Schweizerischer Schulrat. This paper is dedicated to Albert Weller on the occasion of his 70th birthday.

REFERENCES

1. M. Quack, Statistical mechanics and dynamics of molecular fragmentation, *Nuovo Cim.* B63: 358 (1981).
2. D.W. Lupo and M. Quack, IR-laser photochemistry, *Chem. Rev.* 87: 181 (1987).
3. M. Quack, Infrared laser chemistry and the dynamics of molecular multiphoton excitation, *Infrared Physics* 29: 441 (1989).
4. M. Quack, The role of quantum intramolecular dynamics in unimolecular reactions, *Phil. Trans. Roy. Soc. Lond.* A 332: 203 (1990).
5. D.L. Bunker, Classical trajectory methods, *Methods in Computational Phys.* 10: 287 (1971).
6. W.F. van Gunsteren and H.J.C. Berendsen, Computer simulation of molecular dynamics, methodology, applications and perspectives in chemistry, *Angew. Chem. Internat. Ed.* (English) 29: 992 (1990).
7. M.P. Allen and D.J. Tildesley, Computer Simulation of Liquids, Clarendon Press, Oxford (1987).
8. R.D. Levine, Quantum Mechanics of Molecular Rate Processes, Clarendon Press, Oxford (1969).
9. J.R. Taylor, Scattering Theory, John Wiley, New York (1972).
10. M.S. Child, Molecular Collison Theory, Academic Press, New York (1974).
11. M. Quack and J. Troe, Statistical methods in scattering, *in:* "Theoretical Chemistry: Advances and Perspectives" Vol. 6B: 199, D. Henderson ed., Academic Press, New York (1981).
12. M. Quack and J. Troe, Unimolecular reactions and energy transfer of highly excited molecules, *in:* "Gas Kinectics and Energy Transfer" Vol. 2: 175, P.G. Ashmore and R.J. Donovan, eds, The Chemical Society London (1977)
13. M. Quack and J. Troe, Current aspects of unimolecular reactions, *Internat. Reviews in Phys. Chem.* 1: 97 (1981).
14. M. Quack, Theory of unimolecular reactions induced by monochromatic infrared radiation, *J. Chem. Phys.* 69: 1282 (1978).
15. M. Quack, Reaction dynamics and statistical mechanics of the preparation of highly excited states by intense infrared radiation, *Advances in Chem. Phys.* 50: 395 (1982).
16. W. Kaiser (ed.), Ultrashort Laser Pulses and Applications, Springer Verlag, Berlin (1988).
17. G.R. Fleming, Chemical Applications of Ultrafast Spectroscopy, Oxford University Press, Oxford (1986).
18. M. Gruebele, G. Roberts, and A.H. Zewail, Femtochemistry of the reaction of IHgI, theory versus experiment, *Phil. Trans. Roy. Soc. Lond.* A332: 223 (1990).

19. M. Quack, C. Rüede and G. Seyfang, Generation of shaped pulses for IR-laser chemistry, *Spectrochim.Acta* 46A: 523 (1990).
20. R.A. Mathies, Analysis of femtosecond pump-probe spectra, at this workshop (in this volume, 1992).
21. R. Kosloff and D. Kosloff, Absorbing boundaries for wave propagation problems, *J. Comput. Phys.* 63: 363 (1986); H. Tal-Ezer and R. Kosloff, *J. Chem. Phys.* 81: 3967 (1984).
22. M. Quack and E. Sutcliffe, Quantum interference in the IR-multiphoton excitation of small asymmetric-top molecules: Ozone, *Chem. Phys. Lett.* 99: 167 (1983).
23. M. Quack and E. Sutcliffe, Primary photophysical processes in infrared multiphoton excitation: Wavepacket motion and state selectivity, *Infrared Phys.* 25: 163 (1985).
24. R. Marquardt and M. Quack, The wave packet motion and intramolecular vibrational redistribution in CHX_3 molecules under IR-multiphoton excitation, *J. Chem. Phys.* 95: 4854 (1991).
25. K.C. Kulander, Time dependent calculations for atoms and molecules in intense, pulsed laser fields, at this workshop (in this volume, 1992).
26. C. Leforestier, Time dependent study of the photodissociation of ozone, at this workshop (in this volume, 1992).
27. R. Schinke, Application of 3D wave packet dynamics to spectroscopy and photodissociation, at this workshop (in this volume, 1992).
28 Y. Öhrn, Time evolution of electrons and nuclei in molecular systems, at this workshop (in this volume, 1992).
29. W.A. Lester, Wave packet studies of atom molecule scattering, at this workhop (in this volume, 1992).
30. H.R. Dübal and M. Quack, Tridiagonal Fermi resonance structure in the IR spectrum of the excited CH-chromophore in CF_3H, *J. Chem. Phys.* 81: 3779 (1984).
31. M. Quack, Spectra and dynamics of coupled vibrations in polyatomic molecules, *Annu. Rev. Phys. Chem.* 41: 839 (1990).
32. K. von Puttkamer, H.R. Dübal and M. Quack, Time-dependent processes in polyatomic molecules during and after intense infrared irradiation. *Faraday Disc. Chem. Soc.* 75: 197 (1983).
33. K. von Puttkamer and M. Quack, Vibrational spectra of $(HF)_2$, $(HF)_n$ and their D-Isotopomers: Mode selective rearrangements and nonstatistical unimolecular decay. *Chem. Phys.* 139: 31 (1989).
34. J. Jortner, S.A. Rice, and R. Hochstrasser, Radiationless transitions in photochemistry, *Adv. Photochemistry* 7: 149 (1969).
35. M. Quack, Detailed symmetry selection rules for reactive collisions, *Mol. Phys.* 34: 477 (1977).
36. M. Quack, Detailed symmetry selection rules for chemical reactions, in: "Symmetries and properties of non-rigid molecules: A comprehensive survey", *Studies in Physical and Theoretical Chemistry* 23: 355, Elsevier Publ. Co., Amsterdam (1983).
37. M. Quack, On the densities and numbers of rovibronic states of a given symmetry species: Rigid and nonrigid molecules, transition states and scattering channels, *J. Chem. Phys.* 82: 3277 (1985).
38. M. Quack, On the measurement of the parity violating energy difference between enantiomers, *Chem. Phys. Lett.* 132: 147 (1986).
39. M. Quack, Structure and dynamics of chiral molecules, *Angew. Chemie Internat. Ed.* (English) 28: 571 (1989).
40. M. Quack, Die Symmetrie von Zeit und Raum und ihre Verletzung in molekularen Prozessen, Akademie der Wissenschaften zu Berlin, Jahrbuch 1990, W. de Gruyter Verlag.
41. K.K. Lehmann, B.H. Pate, and G. Scoles, Intramolecular vibrational redistribution on the nanosecond time scale, ν_1 and $2\nu_1$ spectra of substituted acetylene compounds, *Jerus. Symp.* 29: 17 (1991).
42. M. Quack and G. Seyfang, Absolute rate parameters for infrared photochemistry: $CF_3I \rightarrow CF_3 + I$. *J. Chem. Phys.* 76: 955 (1982).

43. H.S. Kwok and E. Yablonovitch, Collisionless intramolecular vibrational relaxation in SF_6, *Phys. Rev. Lett.* 41: 745 (1978).
44. P.A. Schulz, A.S. Sudbo, D.J. Krajnovitch, H.S. Kwok, Y.R. Shen, and Y.T. Lee, Multiphoton dissociation of polyatomic molecules, *Ann. Rev. Phys. Chem.* 30: 379 (1979).
45. A.S. Sudbo, P.A. Schulz, E.R. Grant, Y.R. Shen, and Y.T. Lee, Simple bond rupture reactions in multiphoton dissociation of molecules, *J. Chem. Phys.* 70: 912 (1979).
46. M. Quack, R. Schwarz, and G. Seyfang, Time resolved infrared spectroscopic observation of relaxation and reaction processes during and after infrared multiphoton excitation of $^{12}CF_3I$ and $^{13}CF_3I$ with shaped nanosecond pulses. *J. Chem. Phys.* 00: 000 (1992), in press.
47. M. Quack, Statistical models for product energy distributions in bimolecular reactions with metastable intermediates. *Chem. Phys.* 51: 353 (1980).
48. M. Quack, The role of intramolecular coupling and relaxation in IR-photochemistry, *in:* "Intramolecular Dynamics" (15th Jerus. Symp.): 371, J. Jortner and B. Pullmann, eds., Reidel, Dordrecht (1982).
49. M. Quack and H.J. Thöne, Absolute and relative rate coefficients in the IR-laser chemistry of bichromophoric fluorobutanes: Tests for inter- and intramolecular selectivity, *Chem. Phys. Lett.* 135: 487 (1987).
50. P. Dietrich, M. Quack, and G. Seyfang, Time resolved diode laser infrared absorption spectroscopy of the nascent HCl in the infrared laser chemistry of 1,2-dichloro-1,1-difluoroethane, *Chem. Phys. Lett.* 167: 535 (1990).
51. R.C.Sharp, E. Yablonovitch, and N. Bloembergen, Picosecond infrared double resonance studies on SF_6, *J. Chem. Phys.* 74: 5357 (1981).
52. H. Bürger, A. Rahner, A. Amrein, H. Hollenstein, and M. Quack, Free-jet high resolution FTIR spectroscopy of the complex structure of the ν_1-band of CF_3I near 9μm, *Chem. Phys. Lett.* 156: 557 (1989).
53. H. Bürger, K. Burczyk, H. Hollenstein, and M. Quack, High resolution FTIR spectra of $^{12}CF_3I$, $^{13}CF_3I$ and $^{12}CF_3{}^{79}Br$ near 1050 cm^{-1} and 550 cm^{-1}, *Mol. Phys.* 55: 255 (1985).
54. G. Amat, H.H. Nielsen, and C. Tarrago, Rotation and Vibration of Polyatomic Molecules, Marcel Dekker, New York (1982).
55. D. Papousek and M.R. Aliev, Molecular Vibrational-Rotational Spectra, Elsevier, Amsterdam (1982).
56. M. Lewerenz and M. Quack, Vibrational spectrum and potential energy surface for the CH-chromophore in CHD_3. *J. Chem. Phys.* 88: 5408 (1988).
57. J.C. Light, Quantum Dynamics using the DVR, at this conference (in this volume, 1992); Z. Bacic and J.C. Light, Theoretical methods for rovibrational states of floppy molecules, *Ann. Rev. Phys. Chem.* 40: 469 (1989).
58. D. Luckhaus and M. Quack, The far infrared pure rotational spectrum and the Coriolis coupling between ν_3 and ν_8 in $CH^{35}ClF_2$, *Mol.Phys.* 68: 745 (1989).
59. D. Luckhaus and M. Quack, The rovibrational analysis of the high-resolution IR spectrum of CD_2HF below 1200 cm^{-1}: An interacting tetrad of vibrational levels. *Chem. Phys. Lett.* 180: 524 (1991).
60. H.R. Dübal, T.K. Ha, M. Lewerenz, and M. Quack, Vibrational spectrum, dipole moment function and potential energy surface of the CH chromophore in CHX_3 molecules, *J. Chem. Phys.* 91: 6698 (1989).
61. H. Hollenstein, M. Lewerenz, and M. Quack, Isotope effects in the Fermi resonance of the CH-chromophore in CHX_3 molecules, *Chem. Phys. Lett.* 165: 175 (1990).
62. T.K. Ha, M. Lewerenz, R. Marquardt, and M. Quack, Overtone intensities and dipole moment surfaces for the isolated CH-chromophore in CHD_3 and CHF_3: Experiment and ab initio theory, *J. Chem. Phys.* 93: 7097 (1990).
63. J. Davidsson, J.H. Gutow, R.N. Zare, H. Hollenstein, R. Marquardt, and M. Quack, Fermi resonance in the overtone spectra of the CH-chromophore in $CHBr_3$. II. Visible spectra, *J. Phys. Chem.* 95: 1201 (1991).
64. A.J. Ross, H. Hollenstein, R. Marquardt, and M. Quack, Fermi resonance in the overtone spectra of the CH chromophore in bromoform. *Chem. Phys. Lett.* 156: 455 (1989).

65. T. Carrington jr., L. Halonen, and M. Quack, Fermi resonance in CHX_3: A Hamiltonian in symmetrized curvilinear internal coordinates, *Chem. Phys. Lett.* 140: 512 (1987).
66. L. Halonen, T. Carrington jr., and M. Quack, Investigation of Fermi resonances in CHX_3 molecules with an internal-coordinate Hamiltonian, J. Chem. Soc. Faraday Trans. 2, 84: 1371 (1988).
67. M. Quack and M.A. Suhm, Potential energy surface and energy levels of $(HF)_2$ and its D-Isotopomers, *Mol. Phys.* 69: 791 (1990).
68. M. Quack and M.A. Suhm, Potential energy surfaces, quasiadiabatic channels, rovibrational spectra and intramolecular dynamics of $(HF)_2$ and its isotopomers from quantum Monte Carlo calculations, *J. Chem. Phys.* 95: 28 (1991).
69. M. Quack and M.A. Suhm, Quasiadiabatic channels and effective transition state barriers for the disrotatory in-plane hydrogen-bond exchange motion in $(HF)_2$, Chem. Phys. Lett. 183: 187 (1991).
70. D. Luckhaus and M. Quack, Spectrum and dynamics of the CH chromophore in CHD_2F: I. Vibrational Hamiltonian and analysis of rovibrational spectra, *Chem. Phys. Lett.* 190: 581 (1992).
71. T.K. Ha, D. Luckhaus, and M. Quack, Spectrum and dynamics of the CH chromophore in CHD_2F: II. Ab initio calculations of the potential and dipole moment functions, *Chem. Phys. Lett.* 190: 590 (1992).
72. J.B. Anderson, A random walk simulation of the Schrödinger equation, H_3^+, *J. Chem. Phys.* 63: 1499 (1975).
73. M. Quack and E. Sutcliffe, URIMIR: Programs for the calculation of the quantum dynamics of IR multiphoton excitaion and dissociation (unimolecular reactions induced by monochromatic infrared radiation), Quantum Chemistry Program Exchange, Program 515, QCPE Bulletin 6: 98 (1986).
74. R. Marquardt and M. Quack, Molecular motion under the influence of a coherent IR-laser field, *Infrared Physics* 29: 485 (1989).
75. R. Marquardt and M. Quack, IR-multiphoton excitation and wavepacket motion of the harmonic and anharmonic oscillators: exact solutions and quasiresonant approximation, *J. Chem. Phys.* 90: 6320 (1989).
76. M. Quack, Master equations for photochemistry with intense infrared light, *Ber.Bunsenges.Phys.Chem.* 83: 757 (1979); Master equations for photochemistry with intense infrared light (III). The influence of molecular parameters in URIMIR, *Ber.Bunsenges.Phys.Chem.* 83: 1287 (1979).
77. M. Quack, Master equations for photochemistry with intense infrared light (IV), A unified treatment of case B and case C including nonlinear effects, *Ber.Bunsenges.Phys.Chem.* 85: 318 (1981).
78. D.W. Lupo and M. Quack, Absolute rate coefficients in IR-photochemistry: Simple estimates of the rate constant from molecular parameters. *Ber. - Bunsenges.Phys.Chem.* 91: 565 (1987).
79. M. Quack and E. Sutcliffe, Generalized quantum statistical model for molecular infrared multiphoton excitation, *Chem. Phys. Lett.* 121: 315 (1985).
80. M. Quack and E. Sutcliffe, On the validity of the quasiresonant approximation for molecular infrared multiphoton excitation, *J. Chem. Phys.* 83: 3805 (1985).
81. M. Quack and E. Sutcliffe, On the possibility of mode-selective IR-multiphoton excitation of ozone, *Chem. Phys. Lett.* 105: 147 (1984).
82. M. Quack and C. Sutcliffe, IR multiphoton excitation of small asymmetric top molecules: rotational and vibrational inversions for laser action in ozone, *Isr. J. Chem.* 24: 204 (1984).
83. H.R. Dübal and M. Quack, High resolution IR-spectrum of fluoroform: A close resonance, *Chem. Phys. Lett.* 80: 439 (1981).
84. J. Baggott, M.C. Chuang, R.N. Zare, H.R. Dübal, and M. Quack, Structure and dynamics of the excited CH chromophore in $(CF_3)_3CH$, *J. Chem. Phys.* 82: 1186 (1985).
85. R. Marquardt, M. Quack, J. Stohner, and E. Sutcliffe, Quantum-mechanical wavepacket dynamics of the CH group in symmetric top X_3CH compounds

using effective Hamiltonians from high-resolution spectoscopy, *J.Chem.Soc. Faraday Trans.* 2, 82: 1173 (1986).
86. H.R. Dübal and M. Quack, Vibrational overtone spectra and vibrational dynamics of $CFHCl_2$ and $(CF_3)_2$ CFH, *Mol. Phys.* 53: 257 (1984).
87. M. Quack, Photochemistry with infrared radiation. *Chimia* 35: 463 (1981).
88. M. Quack, E. Sutcliffe, P.A. Hackett, and D.M. Rayner, Molecular photofragmentation with many infrared photons: Absolute rate parameters from quantum dynamics, statistical mechanics, and direct measurement. *Faraday Disc. Chem. Soc.* 82: 229 (1986) and 82: 289 (1986) (replies).
89. M. Quack, On local and global vibrational states, *Faraday Disc. Chem. Soc.* 71: 359 (1981).
90. A. Amrein, H.R. Dübal, and M. Quack, Multiple anharmonic resonances in the vibrational overtone spectra of $CHClF_2$, *Mol. Phys.* 56: 727 (1985) and to be published.
91. M. Quack, Mode selective vibrational redistribution and unimolecular reactions during and after IR-laser excitation, *Jerusalem Symp.* 24: 47 (1991).
92. G. Herzberg, Molecular Spectra and Molecular Structure. III. Electronic spectra, Van Nostrand, New York (1966).
93. A. Weller, Quantitative Untersuchungen der Fluoreszenzumwandlung bei Naphtholen, *Ber.Bunsenges.Phys.Chem.* 56: 662 (1952); Zur Kinetik der Fluoreszenzumwandlung, *Z. Physik. Chem.* 15: 438 (1958).

MOLECULAR MODEL OF FAST ATOMIC COLLISIONS

A. Riera

Department of Chemistry, C-XIV
Universidad Autónoma de Madrid
Canto Blanco, 28049-Madrid

INTRODUCTION

Aside from its fundamental significance in dynamics, the study of atomic collision provides important data for other fields such as astrophysics, plasma physics, fusion reactor technology, laser physics, beam experiments using electron cyclotron resonance (ECR) sources and channeling and energy loss in solids. From a fundamental viewpoint, it is a privileged field where theory is able to yield similar accuracy as experiment, without the help of parametrized models. Schematically, some physical processes that can arise in the collision between two atomic species, either neutral or ionized, are :

$A + B \rightarrow A + B$	elastic
$A + B \rightarrow A^{*} + B$	inelastic
$A + B \rightarrow A^{+} + B^{-}$	charge exchange
$A + B \rightarrow A^{+} + B + e^{-}$	ionization

and it is now common, with coincidence experiments using energy gain (or loss) and photon emission spectroscopy, to measure state-to-state cross sections as well as alignment and orientation parameters for either one, or all, of those processes.

At high enough nuclear velocities, the interaction between the colliding species is sufficiently small as compared to the total energy, so that one can adequately describe the situation by using perturbative methods stemming from high energy physics.[1] On the other hand, these methods yield very inaccurate results at lower velocities where the interatomic forces cannot, even approximately, be considered as a perturbation. Consequently,

Time-Dependent Quantum Molecular Dynamics, Edited by J. Broeckhove
and L. Lathouwers, Plenum Press, New York, 1992

for the low velocity range, a different approach was devised, starting from the opposite assumption that those forces are overwhelmingly important. In a sense, which will be taken up in the present outline, the applications of this latter approach may be looked upon as a strive to extend to higher velocities the simple and fruitful idea that a slow collision between two atoms may be interpreted in terms of the properties of the molecular system formed by the colliding partners - hence the name of *molecular model*.

Originally, the model was introduced on the physical basis of the adiabatic theorem[2], and is often thougth of in such a framework : one may consider the system formed by two slowly colliding atoms in terms of an adiabatically slow change of the corresponding clamped-nuclei situation (Born-Oppenheimer approximation). However, the domain of validity of such an approach may be easily seen to be extremely limited. In particular, the method can only be employed to study elastic or resonant processes[3]. The theory was then naturally extended to cover inelastic and electron exchange processes too, by explicitly including non-adiabatic transitions between different electronic states - especially once it was realized that the coupling terms can be accurately calculated[4] without much difficulty. The ensuing close-coupling formalism, named[5] perturbed stationary states method was not new : it had already been derived in Born and Huang's[6] early treatment of vibrational problems. It relies for its justification on the completeness of the set of eigenfunctions of the electronic Hamiltonian. This completeness ensures the convergence to the exact results as the basis set of molecular wavefunctions is progressively enlarged..

A second important development of the method was the generalization of the kind of molecular states involved in the close-coupling expansion, through the introduction of the so-called diabatic states[3,7]. In this way, some awckward situations of states interacting, in slow collisions, with infinite Rydberg series, and/or with the ionizing continuum, could be offset in an elegant and computationally simple way. A further development was the introduction of translation factors, which provide the simplest way to enforce the boundary conditions of a scattering problem with a finite close-coupling expansion, and eliminate several practical difficulties of the method.

With the introduction of diabatic states and translation factors, the domain of validity of the model was recognized to cover the range of nuclear velocities that are sensibly less than those of the electrons that are "active" (that is, suffer transitions) during the collision. We thus had a molecular model of slow atomic collisions. For a long time, this boundary seemed the "natural" one for a method that was developed starting from the approximation of fixed nuclear positions. Furthermore, when nuclear and electronic velocities are of the same order, ionization occurs, and to account for this effect it seemed necessary to explicitly introduce the electronic continuum, which is a complicated, time-consuming step. One consequence of the failure to account for ionization was that molecular calculations did not reproduce the conspicuous feature of the rapid fall of the charge transfer cross sections at high velocities. Most further developments thus focused on this problem, hence on charge exchange processes, whose interest was enhanced by their importance in the treatment of fusion and astrophysical plasmas, as well as of ECR experiments. Some of these studies have led to an understanding of the mechanism of transitions to the ionizing continuum, and this has resulted in the "natural" boundary being crossed in very recent treatments[8,9]. We now have the satisfactory situation that the method can be used to treat fast processes, such that high energy methods[1] are also applicable. We have a molecular model of fast, as well as slow, atomic collisions.

The aim of the present work is limited to a theoretically-oriented presentation of the main aspects of the previous basic developments. Alternative high and intermediate-energy treatments of atomic collisons, such as non-perturbative atomic expansions will not be described. For the sake of conciseness, since good reviews and state-of-the-art reports are available[12-19], the formalism will be kept as simple as possible, comparison between theory and experiment will be left for the references, and these will be restricted to a (subjective) selection. Atomic units will be used throughout.

PERTURBED STATIONARY STATES METHOD

Quantal formalism

To simplify the presentation, a heteronuclear one-electron system will be explicitly considered in the formulation (but not necessarily in the discussion). We start from the Schrödinger equation, after extracting the motion of the total center of mass :

$$H\,\Psi(\underline{r},\underline{R}) = E\,\Psi(\underline{r},\underline{R}) \qquad (1)$$

where $\underline{R}$ denotes the relative nuclear vector; $\underline{r}$ the electronic one defined with respect to the nuclear center of mass; and H is the sum of the relative nuclear kinetic energy and the electronic Hamiltonian H_{el}, defined for clamped nuclear positions.

For a vibrational problem, eq. (1) must be solved subject to the condition that Ψ should vanish for $R \to \infty$. For a collisional problem with the electron initially belonging to atom A, the boundary condition for elastic and inelastic processes is :

$$\Psi(\underline{r}, \underline{R}) \underset{\rho\to\infty}{\sim} \exp(i\underline{k}_i\cdot\underline{\rho})\ \varphi_i^A + \sum_j \rho^{-1} \exp(ik_j\rho)\ f_{ij}(\Theta)\ \varphi_j^A \qquad (2)$$

with ρ the distance of the center of mass of atom A from nucleus B; φ_i^A the atomic wavefunction representing the entrance channel and φ_j^A an exit or reaction channel; k_j the linear momentum corresponding to the jth channel; and Θ the scattering angle. The condition for charge transfer is :

$$\Psi(\underline{r}, \underline{R}) \underset{\rho'\to\infty}{\sim} \sum_j \rho'^{-1} \exp(ik_j\rho')\ f_{ij}(\Theta)\ \varphi_j^B \qquad (3)$$

where ρ' stands for the distance between the center of mass of atom B and nucleus A. Eqs. (2) and (3) must be modified in the standard way[20] when both atomic systems are charged. The experimental quantities, which are the cross sections, can be simply written[20] as functions of the corresponding amplitudes $f_{ij}(\Theta)$ in eqs. (2) and (3), and the reduced masses for the initial and exit channels :

$$d\sigma_{ij}(\Theta)/d\Omega = (\mu_j/\mu_i)\ |f_{ij}(\Theta)|^2 \qquad (4)$$

for the $i \to j$ partial differential cross section, and :

$$\sigma_{ij} = \int d\sigma_{ij}(\Theta) \qquad (5)$$

for the corresponding total one.

Semiclassical formalism

Except at very small nuclear velocities, the relative linear momentum of the nuclei is a very large quantity when expressed in atomic units, and the wavefunctions $\Psi(\underline{r}, \underline{R})$ present strong oscillations, as may be seen from their asymptotic form in eqs. (2) and (3). One can extract a large part of these oscillations by defining a vector $\underline{k}$ such that $\hat{k} = \hat{k}_i$ and $k^2 = 2\ \mu\ E$, with $\mu = M_A M_B/(M_A+M_B)$ the reduced nuclear mass; one then formally expands the solutions of eq. (1) in powers of μ^{-1} :

$$\Psi(\underline{r}, \underline{R}) = \exp(i\underline{k}\cdot\underline{R})\ [\ \psi + O(\mu^{-1})] \qquad (6)$$

where most of the oscillations appear in the exponential factor. By substituting eq. (6) in (1), one obtains to lowest order the simplest semiclassical approximation, where the so-called eikonal, or impact parameter, equation[21-23] determines the wavefunction ψ in eq. (6) :

$$i\ \partial\psi/\partial t = H_{el}\ \psi \qquad (7)$$

Eq. (7) must be solved for each nuclear trajectory $\underline{R} = \underline{b} + \underline{v}\ t$, where a nuclear velocity $\underline{v} = \mu^{-1}\ \underline{k}$ and "time" $t = v^{-1}\ Z$ have been defined, with Z the nuclear coordinate along $\hat{k}$, and $\underline{b} \perp \hat{k}$, the impact parameter vector. In the one-electron case, eq. (7) may be considered as the dynamical equation for a time-dependent molecular orbital; otherwise, ψ involves all electronic coordinates - as well as the relative nuclear one $Z = \hat{k}\cdot\underline{R}$ through the parameter $t = Z/v$.

Expansion in powers of μ^{-1} of the boundary condition (2) yields the corresponding initial condition for eq. (7) :

$$\psi(\underline{r}, t) \underset{t\to-\infty}{\sim} \exp[-ip\underline{v}\cdot\underline{r} - (p^2/2)v^2 t] \exp(-iE_i^A t)\, \varphi_i^A \qquad (8)$$

where $p = M_B/(M_A+M_B)$ and E_i^A is the entrance channel energy.

Substitution of (6) in the Lippmann-Schwinger integral equations[22] together with eqs. (2) - (5) yields simple expressions[21-23] for the differential :

$$d\sigma_{ij}(\Theta)/d\Omega = \mu_i\mu_j\, v^2\, |\int db\, b\, J_m(b\mu_j v \sin\Theta)\, [t_{ij}(b)-\delta_{ij}]|^2 \qquad (9)$$

and total cross sections :

$$\sigma_{ij} = 2\pi \int db\, b\, |t_{ij}(b)-\delta_{ij}|^2 \qquad (10)$$

where J_m is a Bessel function of argument $m = |m_i - m_j|$, with m_i, m_j the projection of the initial and final angular momentum, respectively, on the $\hat{k}$ axis; and $t_{ij}(b)$ the $i\to j$ transition amplitude for the classical trajectory with impact parameter b :

$$t = \lim_{t\to\infty} \langle \exp[-ip\underline{v}\cdot\underline{r} - (p^2/2)v^2 t] \exp(-i\, E_j^A\, t)\, \varphi_j^A \mid \psi(\underline{r}, t)\rangle \qquad (11)$$

for excitation and :

$$t = \lim_{t\to\infty} \langle \exp[\, iq\underline{v}\cdot\underline{r} - (q^2/2)v^2 t] \exp(-i\, E_j^B\, t)\, \varphi_j^B \mid \psi(\underline{r}, t)\rangle \qquad (12)$$

with $q = 1-p$, for charge exchange.

Method

In the molecular approach we first solve the Born-Oppenheimer equation:

$$H_{el}\, \phi_n(\underline{r}, R) = E_n(R)\, \phi_n(\underline{r}, R) \qquad (13)$$

for $\underline{R}$ fixed. Except at extremely small collisional energies, small relativistic corrections to the electronic energies (such as spin-orbit interactions in light atoms) can be neglected. The solution of eq. (1) is then formally expanded in terms of the complete set of the stationary, or adiabatic, electronic wavefunctions :

$$\Psi(\underline{r}, \underline{R}) = \sum_n \chi_n(\underline{R})\, \phi_n(\underline{r}, R) \qquad (14)$$

where the sum in eq. (14) includes an integral over the ionizing continuum. Substitution in eq. (1) yields a system of coupled differential equations for the nuclear functions :

$$[-(2\mu)^{-1}\nabla_R^2 + E\,(R) - E\,]\,\chi_m(\underline{R}) = \sum_n\,[\,\mu^{-1}\langle\phi_m|\nabla_R|\phi_n\rangle\cdot\nabla_R + (2\mu)^{-1}\langle\phi_m|\nabla_R^2\,|\phi_n\rangle\,]\,\chi_n(\underline{R}) \qquad (15)$$

where we have assumed the electronic wavefunctions to be normalized.

In the semiclassical formalism, the molecular expansion reads :

$$\psi(\underline{r},t) = \sum_n a_n(t)\,\exp\{-i\int E_n[R(t)]\,dt\}\,\phi_n(\underline{r},[R(t)]) \qquad (16)$$

and substitution in eq. (7) yields a system of ordinary differential equations :

$$i\,da_m/dt = \sum_n a_n(t)\,\langle\phi_m|-i\partial/\partial t|\phi_n\rangle\,\exp\{-i\int(E_n-E_m)dt\} \qquad (17)$$

As will be expected, in practice the expansions (14), (16) must be truncated, and in the simplest possible approach one retains a single term; this yields the Born-Oppenheimer approximation. The restriction to one term is equivalent to neglecting all right-hand-side matrix elements, called non-adiabatic, or dynamical, couplings, in eqs. (15) and (17); these couplings are related through $\mu^{-1}\,\underline{k}\cdot\nabla_R = \underline{v}\cdot\nabla_R = \partial/\partial t$. Obviously, to represent a scattering event, the electronic wavefunction is chosen so that at infinite internuclear separations $\phi_i(\underline{r}, R=\infty) = \varphi_i^A$, it represents the entrance channel defined in eq. (2) (neglecting the tiny effect of the electronic mass in the definition of this atomic wavefunction). The validity of the Born-Oppenheimer approximation is ensured by the adiabatic theorem[2] at sufficiently low nuclear velocities, such that the collision is elastic.

A straightforward extension of the Born-Oppenheimer approximation, which involves a linear combination of adiabatic wavefunctions with constant coefficients in eqs. (14) and (16), is required for homonuclear systems. For these systems, a combination $2^{-1/2}\,(\phi_g \pm \phi_u)$ must be employed to represent the initial electronic function φ_i^A. The gerade and ungerade states of the linear combination are degenerate in the limit $R\to\infty$, but have different electronic energies at finite internuclear distances. According to eq. (15) the corresponding nuclear wavefunctions χ_g, χ_u will therefore evolve differently; similarly, in eq. (17) the corresponding coefficients a_g, a_u will also differ. This can give rise to resonant charge exchange (final state represented by φ_i^B) as well as to elastic (final state φ_i^A) scattering.

Another simple extension is required when the spin structure of the initial state differs from that of any adiabatic wavefunction. A typical example is provided by collisions involving two active electrons, such that the initial state is prepared as $A(\uparrow) + B(\downarrow)$; this is written as a sum of singlet $2^{-1/2}$ $[\varphi_i^A(\uparrow)\ \varphi_i^B(\downarrow) - \varphi_i^A(\downarrow)\ \varphi_i^B(\uparrow)]$ and triplet $2^{-1/2}$ $[\varphi_i^A(\uparrow)\ \varphi_i^B(\downarrow) + \varphi_i^A(\downarrow)\varphi_i^B(\uparrow)]$ wavefunctions. Because exchange forces are different for singlet and triplet states, the corresponding nuclear wavefunctions χ_s, χ_t in eq. (14), and the coefficients a_s, a_t in eq. (16), evolve differently, and the spin structure of the final state can differ from that of the initial one, giving rise to a spin change $[A(\downarrow) + B(\uparrow)]$ process. When only one of the colliding species is spin-polarized, the situation is slightly more complicated[24,25].

Non-trivial extensions of the Born-Oppenheimer approximation involve using more than the minimum number of terms in expansions (14) or (16). This gives rise to the so-called perturbed stationary states method, which usually consists in retaining in the expansion only the terms that are most liable to be involved in transitions (close-coupling approach). The ensuing system (15) or (17) are then solved following a series of standard steps that in the quantal case may be considered as an extension of the well-known partial wave expansion[2,5,20] for scattering by a central potential, and which are described in detail, for example, in refs. 12 and 14. In order not to considerably burthen this paper with algebra, these steps will only be very briefly summarized :

▶ The nuclear wavefunctions χ_n are factorized into angular and radial (R) components; correspondingly, to evaluate the coupling matrix elements, a transformation is made from the laboratory-fixed frame to a molecule-fixed one, and use is made of the symmetry of the electronic wavefunctions in this latter frame; the nuclear gradient couplings are also written in polar coordinates, as sums of angular and radial components.

▶ Integration over the nuclear angular wavefunctions is analytical, and eq. (15) is finally transformed into a system of second degree differential equations for the nuclear radial wavefunctions[12,14], which is numerically solved starting from a set of independent initial conditions.

▶ The scattering matrix[5,12,20] is defined from the asymptotic behavior of the solutions; the boundary conditions are enforced, and the amplitudes $f_{ij}(\Theta)$ are obtained and introduced in eqs. (4) and (5).

The treatment in the semiclassical case is simpler. Passage is also made from the laboratory-fixed to a molecule-fixed frame of reference for the $\partial/\partial t = \underline{v} \cdot \nabla_R$ coupling operator, and the symmetry of the molecular wavefunctions in the latter frame is employed. The system (17) must be integrated subject to the initial condition (8), and the transition amplitudes are obtained from the limits (11) and (12). Introduction of these amplitudes in (9) and (10) yields the cross sections.

Would a complete set of electronic molecular wavefunctions ϕ_n be employed in eqs. (15) and (17), the solutions obtained using the molecular method would be exactly equivalent to those of the quantal [eq. (1)] and semiclassical [eq. (7)] equations, respectively. As in the treatment of molecular vibrations, truncation of the expansion (14) results in an approximation for the exact solution. However, and unlike the vibrational case, here the truncation precludes, in general, an exact fulfillment of the scattering boundary conditions, which must, therefore, be approximated:

$$\Psi(\underline{r},\underline{R}) \underset{R\to\infty}{\sim} \exp(i\underline{k}_i \cdot \underline{R})\ \varphi_i^A + \textstyle\sum_j R^{-1} \exp(ik_j R)\ f_{ij}(\Theta)\ \varphi_j^A \tag{18}$$

$$\Psi(\underline{r},\underline{R}) \underset{R\to\infty}{\sim} \textstyle\sum_j R^{-1} \exp(ik_j R)\ f_{ij}(\Theta)\ \varphi_j^B \tag{19}$$

instead of eq. (2) and (3). The same holds in the semiclassical case, where one uses :

$$\psi(\underline{r},t) \underset{t\to-\infty}{\sim} \exp(-i\ E_i^A\ t)\ \varphi_i^A \tag{20}$$

instead of eq. (8). Similarly, the expression for the transition amplitudes of eq. (11) is approximated by :

$$t_{ij} = \lim_{t\to\infty} \langle \exp(-i\ E_j^A\ t)\ \varphi_j^A \mid \psi(\underline{r},t)\rangle \tag{21}$$

and of eq. (12) by :

$$t_{ij} = \lim_{t\to\infty} \langle \exp(-i\ E_j^B\ t)\ \varphi_j^B \mid \psi(\underline{r},t)\rangle \tag{22}$$

Even those simpler boundary conditions (18) - (20) can only be approximately imposed - and the amplitudes calculated - because in general the molecular wavefunctions remain coupled[12-14, 26, 27] in the limit $R \to \infty$. In particular, one finds that radial couplings (the radial components $\langle\phi_n|\partial/\partial R|\phi_m\rangle$ of the nuclear gradient couplings) often tend to a constant value in the limit of infinite internuclear separation. Since the effectiveness of the couplings increases with the nuclear velocity, this limitation leads to serious difficulties in the intermediate and high range of collisional energies. As it stands, the method is therefore limited to the energy region where the existence of those "residual" couplings is not a practical problem.

When applying the method, it is indispensible to consider first which molecular channels are liable to be accessible through non-adiabatic transitions from the entrance channel. This non-automatic step is performed through a detailed study of the molecular energies E_n and the dynamical couplings as functions of internuclear distance R. Consideration of eqs. (15) and (17) shows[14] that, as in any close-coupling treatment, transitions between two channels are the more effective the closer their energies and the larger the coupling betwen them.

Large couplings $\langle \phi_n | \nabla_R | \phi_m \rangle$ between two molecular wavefunctions occur whenever the second one ϕ_m acquires a sizeable component on the first ϕ_n as either the radial or angular component of $\underline{R}$ is changed. To exemplify the effect of such a variation, let us consider the case where, as the internuclear distance is increased from $R = R_0$ to $R = R_0 + \delta R$, ϕ_m changes from $\phi_m(R_0)$ to $\phi_m(R) \approx \phi_n(R_0)$. A typical example is provided by a sharp pseudo crossing between two energy curves E_m, E_n, where the corresponding wave functions ϕ_m, ϕ_n interchange their structure. Since both functions must remain orthogonal for all R, the transformation between them is unitary, so that we must have $\phi_n(R_0+\delta R) \approx -\phi_m(R_0)$. In an infinitesimally slow collision where the electronic state of the system is represented by ϕ_m its structure will undergo a drastic variation in the small interval δR where ϕ_m and ϕ_n exchange their characteristics. On the other hand, at high nuclear velocities the electronic cloud is unable to follow this change of character, so that if the electronic structure is represented by ϕ_m at $R = R_0$, at the distance $R = R_0 + \delta R$ it will keep the same character and will therefore be represented by $-\phi_n(R)$. This is described, in the molecular model, by a $\phi_m \to -\phi_n$ transition, occurring through the $\langle \phi_n | \partial/\partial R | \phi_m \rangle$ coupling, the radial component of $\langle \phi_n | \nabla_R | \phi_m \rangle$. A similar situation, concerning changes when the internuclear axis is rotated, applies to the rotational component. It can be seen that[14] radial couplings occur between wavefunctions of like (Λ_n) symmetry, and angular couplings between those with $\Lambda_n = \Lambda_m \pm 1$.

Typical cases of rotational couplings occur :

▶ In the neighbourhood of the united atom limit $R = 0$, when through rotation of the internuclear axis the wavefunction ϕ_m acquires a component on ϕ_n, as in $2p\sigma$ - $2p\pi$ interactions (using Mulliken's united-atom labels).

▶ In the asymptotic $R \to \infty$ limit between states that become degenerate in this limit, as in $\sigma 2p$ - $\pi 2p$ interactions (using Mulliken's separated-atoms labels).

Typical radial couplings are due to :

▶ Pseudocrossings between two molecular energies, where the corresponding wavefunctions exchange their properties, as in ionic-covalent avoided crossings.

▶ At large internuclear distances, partial hybridization or polarization of orbitals making up the atomic wavefunctions, which leads when R decreases to a delocalization of the wavefunctions.

▸ Contraction of the molecular wavefunctions at very small internuclear distances, which is often followed by a steep variation of the energies.

DIABATIC STATES

The names diabatic, near-diabatic, quasi-diabatic, P-diabatic, F-diabatic, M-diabatic, etc., have been used in the literature to label molecular functions $\phi_m^d(\underline{r}, R)$ that have a smoother R-dependence than that of the adiabatic ones $\phi_n(\underline{r}, R)$, which are, by definition, the eigensolutions of eq. (13)). When diabatic states are introduced in expansions (14) or (16), they present (electrostatic) couplings through H_{el}, which must then be incorporated in the dynamical equations (15) and (17). On the other hand, because of their smooth R-dependence, radial couplings between diabatic functions are smaller than those of the corresponding adiabatic ones; as a consequence, they are sometimes better suited to the treatment of collision processes, and this is the reason why they are introduced.

<u>Diabatic states as a technical device</u>

Let us first consider some simplifications that can be achieved by transforming from an adiabatic basis to a new one that presents no radial couplings at all. From eq. (15) we see that this implies the absence of first derivatives $\partial/\partial R$ of the nuclear radial wavefunctions in the corresponding system of differential equations for these functions[12-14]. This permits the use of numerical techniques such as Numerov-type or log-derivative methods to integrate this system.

Next, whatever method of integration is used, numerical problems often arise when radial couplings between adiabatic states present sharp, needle-like peaks, that are very difficult to integrate. These peaks are usually associated to pseudocrossings between energy curves, where the wavefunctions exchange their properties (see previous section). Then, a considerable simplification is reached by substituing diabatic for adiabatic wavefunctions, since the former have energies that cross rather than pseudocrossing, they present no radial couplings, and their electrostatic interaction is small and easy to integrate. In the same sense, when a sharp pseudocrossing is traversed with unit transition probability in a collisional treatment, so that the character of the electronic state is maintained (rather than follow the change described by the adiabatic wavefunction), one speaks of a diabatic behaviour, and one says that the system follows a diabatic energy correlation[28].

The simplest formalism that yields diabatic states with zero radial couplings starts from the adiabatic wavefunctions, and exactly integrates the couplings. It was proposed by Smith[29] that a unitary transformation be performed from the adiabatic to a diabatic basis :

$$\phi_m^d(\underline{r}, R) = \sum_n U_{mn}(R)\ \phi_n(\underline{r}, R) \tag{23}$$

where the summation runs over a selected set of molecular wavefunctions. The condition that the new molecular wavefunctions $\phi_m^d(\underline{r}, R)$ are not radially coupled :

$$\langle \phi_m^d | \partial/\partial R | \phi_n^d \rangle = 0 \tag{24}$$

leads to a set of differential equations for the matrix elements U_{mn} :

$$dU_{mn}/dR = \sum_j \langle \phi_j | \partial/\partial R | \phi_n \rangle\ U_{mj} \tag{25}$$

The system (25) is to be solved with an appropiate initial condition, such as $U_{mn}(\infty) = \delta_{mn}$. A description of the characteristics of Smith's definition, and its implementation by other authors, may be found in ref. 14. Some straighforward generalizations may be found in the literature. For example, Delos[13] proposed that adiabatic wavefunctions should be modified with translation factors (discussed below) before performing the transformation (23).

In practice, the most common application of the previous idea is the numerical diabatization of an isolated sharp radial coupling, of height $1/2\delta$, at the pseudocrossing between two molecular energies E_i and E_j, at an internuclear distance R_0. The simple transformation is chosen :

$$U_{ii} = U_{jj} = \cos\theta(R);\ \ U_{ij} = -\,U_{ji} = \sin\theta(R);\ \ U_{nm} = \delta_{nm}\ (n \neq i) \tag{26}$$

with :

$$\theta = (\pi/4) - (1/2)\tan^{-1}\left[(R - R_0)/\delta\right] \tag{27}$$

and applied in eq. (23) for the domain of internuclear distances $[R_0 - \Delta, R_0 + \Delta]$ such that $\Delta \gg \delta$, so that $\theta(R_0-\Delta) \approx 0$, $\theta(R_0+\Delta) \approx \pi/2$ and outside this domain the original adiabatic expansion can be employed. In the new diabatic basis we have:

$$\langle \phi_i^d | \partial/\partial R | \phi_j^d \rangle = \langle \phi_i | \partial/\partial R | \phi_j \rangle - d\theta/dR \tag{28}$$

$$\langle \phi_i^d | \partial/\partial R | \phi_n \rangle = \cos\theta\ \langle \phi_i | \partial/\partial R | \phi_n \rangle + \sin\theta\ \langle \phi_j | \partial/\partial R | \phi_n \rangle \tag{29}$$

$$\langle \phi_j^d | \partial/\partial R | \phi_n \rangle = -\sin\theta\ \langle \phi_i | \partial/\partial R | \phi_n \rangle + \cos\theta\ \langle \phi_j | \partial/\partial R | \phi_n \rangle \tag{30}$$

and :

$$d\theta/dR = 2^{-1}\ \delta/\left[(R - R_0)^2 + \delta^2\right] \tag{31}$$

It may be easly seen from eq. (26) that the diabatic energies cross instead of pseudocrossing, and coincide with the adiabatic ones outside the crossing. Furthermore, the pseudocrossing between E_i and E_j is often sharp enough so that the electrostatic coupling :

$$\langle \phi_i^d | H_{el} | \phi_j^d \rangle = 2^{-1} \sin 2\theta \, (E_j - E_i) \qquad (32)$$

can be neglected. Notice that in a sufficiently small neighborhood of the middle of the pseudocrossing this interaction can always be taken as being constant, and the behaviour of the diabatic energies at the crossing as linear, (Landau-Zener-Stückelberg linear model[15]). This exactly yields eq. (31) for the radial coupling; hence, eq. (28) eliminates the peak of this coupling for $R \simeq R_0$.

In the sixties and early seventies, the routine calculation of radial couplings was not standard, and collisional treatments required the use of states that could be assumed to have negligible radial interactions. Since eq. (25) could not be used because the right hand side was unknown, many approximate ways to construct diabatic wavefunctions were devised. All these methods were based on the the criterion that if one requires the molecular wavefunctions to smoothly change with R the ensuing radial couplings will be small. Since nowadays the calculation of these couplings presents no serious dificulty, we shall not dwell upon these approximate methods, with the exception of one that possesses a useful physical meaning.

As mentioned before, one often finds that sharp pseudocrossings between adiabatic energies occur between states that possess, outside these crossings, a markedly different *character*, and which exchange this character at the avoided crossing. The character involved can be a quasi-gerade or ungerade approximate symmetry[30], a ionic or covalent structure[31], or a configurational structure with approximate quantum numbers[32].

Now, since, in order to construct the adiabatic wavefunctions $\phi_n(\underline{r}, R)$, one usually employs a linear variational treatment for H_{el}, we can introduce a constraint in the treatment so that the ensuing wavefunctions $\phi_m^d(\underline{r}, R)$ possess a definite character; then, the energies of states with different characters will not pseudocross. More specifically, let us suppose that we are able to define a projector[33-35] Q_α onto a manifold of functions that possess the required character. Then, a variational treatment for $Q_\alpha H_{el} Q_\alpha$ will yield wavefunctions :

$$Q_\alpha H_{el} Q_\alpha \, \phi_{\alpha n}^d(\underline{r}, R) = E_{\alpha n}^d(R) \, \phi_{\alpha n}^d(\underline{r}, R) \qquad (33)$$

with the desired properties. This is to be carried out for as many charac ters Q_α, Q_β,... as desired, and is equivalent to solving the secular equation (block-diagonalization[14]) for each of the diagonal blocks of the Hamiltonian matrix :

$$\begin{matrix} Q_\alpha H_{el} Q_\alpha & Q_\alpha H_{el} Q_\beta & \cdots \\ Q_\beta H_{el} Q_\alpha & Q_\beta H_{el} Q_\beta & \cdots \\ \cdots & \cdots & \cdots \end{matrix} \tag{34}$$

instead of solving the secular equation for the whole matrix (which yields approximations to the adiabatic states). Then, the energies $E^d_{\alpha n}(R)$ and $E^d_{\beta n}(R)$ with $\alpha \neq \beta$ will not pseudocross. On the other hand, the energies $E^d_{\alpha n}(R)$ and $E^d_{\alpha m}(R)$ can, in principle, pseudocross, so that the method leads to states of a given character being diabatic with respect to those of a different character.

It should be mentioned that, in some cases, the Q_α projectors are only meaningful within a finite basis set; an example is provided by ionic-covalent crossings: since complete sets of wavefunctions can be formed using (an infinite set of) functions of either ionic or covalent type alone, the corresponding projectors Q_{ionic}, $Q_{covalent}$ cannot be defined onto the whole of Hilbert space because they would then coincide with the identity operator and thence with each other; see a discussion of this point in refs. 14 and 31. In this specific example, we have $Q_\alpha Q_\beta \neq 0$; in most other cases $Q_\alpha Q_\beta = 0$.

For each value of α, the solution of eq. (33) can, in practice, be done in three equivalent ways[36,37] :

▸ One can perform[30-32] separate calculations by restricting each basis set to exactly have the corresponding character.

▸ Alternatively[38], the matrix elements of the projected Hamiltonians $Q_\alpha H_{el} Q_\alpha$ can be constructed, using a general basis.

▸ When $Q_\alpha Q_\beta = 0$, one can employ the analogy[37,39] between matrix partitioning and Phillips-Kleinman pseudopotential theory[40].

The former way is the simplest one, but is limited to cases where the basis configurations have a definite character. The second approach is more general but can be time consuming, while the last technique is a fast procedure that consists in performing variational calculations for an effective Hamiltonian :

$$H_{eff} = H_{el} + \gamma (1 - Q_\alpha) \tag{35}$$

using a general basis, and where γ is a large real number (say, $\gamma \simeq 1000$). Eq. (36) is then repaced by :

$$
\begin{matrix}
Q_\alpha H_{el} Q_\alpha & Q_\alpha H_{el} Q_\beta & \cdots \\
Q_\beta H_{el} Q_\alpha & Q_\beta (H_{el} + \gamma) Q_\beta & \cdots \\
\cdots & \cdots & \cdots
\end{matrix}
\qquad (36)
$$

Then, solving the secular equation for the Hamiltonian of eq. (36) will yield, as lowest eigenvalues those of the projected Hamiltonian $Q_\alpha H_{el} Q_\alpha$. Similarly, by expanding the eigenfunctions of H_{eff} in powers of γ^{-1} it is easy to see[37] that they converge as $\gamma \rightarrow \infty$ to those of the projected Hamiltonian $Q_\alpha H_{el} Q_\alpha$. For another character β the same procedure can be set up by substituting β for α in eq. (35), and so on.

In the previous developments, the use of diabatic, rather than adiabatic, states in the molecular model of atomic collisions has been introduced mainly as a technical matter, which permits to simplify the numerical calculations, on the same footing as the particular algorithm that is chosen to solve the dynamical equations. On the other hand, it may also be argued that since diabatic states have smaller couplings, they describe the physical processes better. This is the case when transitions are strongly localized at pseudocrossings, or when the physical situation is well described by a simplified analytical model[15], which is usually written in terms of diabatic states. However, one often finds in quantitative treatments that couplings outside pseudocrossings are very important, and that in many mechanisms sizeable transitions occur in the absence of pseudocrossings. As a consequence, the discussion on the "best" definition of diabatic states has long ago been recognized as academical and subsided in the literature.

It was also soon realized that Smith's definition of "exact" diabatic wavefunctions in eq. (23) has a limited physical sense, since it depends on the manifold spanned by the adiabatic wavefunctions $\phi_m(\underline{r}, R)$ in the (truncated) expansion (14). Moreover, when it is applied to a complete set of functions, it yields[41,42] states that are so unphysical as not to change with R. Finally, while the adiabatic theorem[2] guarantees that, in the absence of resonant processes, a single adiabatic wavefunction $\phi_n(\underline{r}, R)$ describes a collisional event at sufficiently low nuclear velocities, it is false in general that there exists a high enough velocity that a single diabatic function $\phi_m^d(\underline{r}, R)$ has the same property. In other words, diabatic states, in the sense used here, are not the high-v counterparts of the adiabatic ones, and, unlike adiabatic states, they are therefore devoid of an "absolute" sense.

In spite of these aspects that would seem to confine the use of diabatic wavefunctions to a facultative numerical step, we shall now see that the "approximate" diabatic wavefunctions introduced above, have an important physical sense, and are in fact indispensible to treat a large number of collisions involving more than one active electron.

Diabatic states as a physical representation

To see the need of such a representation, one may follow Lichten's original[7] reasoning. Let us consider the case of resonant double charge exchange in He^{2+} + He collisions, and use a semiclassical formalism, that permits a simpler visualization of the process. The initial state He($1s^2$) + He^{2+} is represented at large R by an electronic wavefunction $1s_A^2$, which can be written as the sum of two adiabatic functions : $\phi_2(\underline{r}, R)$ and $\phi_3(\underline{r}, R)$, of structures $1s_A^2 - 1s_B^2$ and $1s_A^2 + 1s_B^2$, respectively [the ground state wavefunction $\phi_1(\underline{r}, R)$, of structure $1s_A 1s_B + 1s_B 1s_A$, is irrelevant to the present context]. At short distances, the wavefunctions $\phi_2(\underline{r}, R)$ and $\phi_3(\underline{r}, R)$ take up a molecular orbital $1s\sigma_g\ 2p\sigma_g$ and $2p\sigma_u^2$ structure, respectively, and the energy of the latter is found to avoid crossing with that of $\phi_4(\underline{r}, R)$, of structure $1s\sigma_g\ 2s\sigma_g$, which in turn pseudocrosses the energy of $\phi_5(\underline{r}, R)$, of structure $1s\sigma_g\ 3s\sigma_g$, and so on; schematically :

```
2p²
      .
        .
          .
            .  ionization threshold                        (37)
             /  v
                ^
                    v
                    ^
 ↑                      v
 E                      ^
                            v
                            ^   v (5 - 6)  pseudocrossing
                                ^  v(4 - 5)  pseudocrossing
                                   ^   (3 - 4) pseudocrossing

 R = 0                   R  →
```

This yields an infinite series of pseudocrossings between the Rydberg states of character $1s\sigma_g\ nl\sigma_g$, converging to the He_2^{3+} ($1s\sigma_g$) ionization threshold. The pseudocrossings are so sharp that transitions occur with near-unit probability at each one of them. As a result, to represent the collision, one needs to include an infinite set of intermediate, adiabatic states.

Lichten's idea was to introduce a diabatic state, represented by $\phi_3^d(\underline{r}, R)$, of character $2p\sigma_u^2$, whose energy crosses the infinite series of $1s\sigma_g\ nl\sigma_g$ levels, penetrates in the ionization continuum, and correlates, in the united atom R = 0 limit to a $2p^2$ state of Be^{2+}. A quantitative implementation[32] of this idea using the block-diagonalization method (i.e., solving eq. (33)) showed that an excellent description of the experimental results on resonant double charge exchange : $He(1s^2) + He^{2+} \rightarrow He^{2+} + He(1s^2)$ could be achieved with only three diabatic states. Obviously, the use of diabatic functions is, for this problem, indispensible rather than facultative, and the best way to construct these functions is the block-diagonalization, or partitionning, method described above. The physical sense of the diabatic states is that the $2p\sigma_u^2$ character of the $\phi_3^d(\underline{r}, R)$ wavefunction is preserved during the collisions.

An interesting question is the structure of the diabatic state for the region of internuclear distances where its energy lies inside the ionization continuum. Using the partitionning in eq. (34), to obtain that state one must solve the projected equation (33) to obtain functions of character $nl\sigma_u \, n'l'\sigma_u$ (hence, in particular, that of character $2p\sigma_u^2$), and the corresponding one :

$$Q_\beta H_{el} Q_\beta \, \phi_{\beta m}^d(\underline{r}, R) = E_{\beta m}^d(R) \, \phi_{\beta m}^d(\underline{r}, R) \tag{38}$$

for those states of character $1s\sigma_g \, nl\sigma_g$ or $1s\sigma_g \, \varepsilon\sigma_g$. Now, eqs. (33) and (38) may be recognized as being identical to the closed and open channel equations, respectively, in Feshbach's[43] treatment of autoionizing resonances, in the frozen core or static exchange approximation (i.e., disregarding the mutual distortion of core and outgoing electron wavefunctions in the ionization process) and neglecting second order terms. The operator Q_α is identical to Feshbach's closed channel projector Q, and Q_β to the opne channel (continuum) one P = 1 - Q, which is defined such that $P\phi_\varepsilon$. behaves asymptotically like ϕ_ε when either of the electronic coordinates tends to infinity. Hence, for energies higher than the ionization threshold, the block diagonalization method of eq. (34) merges[33] with the Feshbach partitionning

$$\begin{matrix} Q\,H_{el}\,Q & Q\,H_{el}\,P \\ P\,H_{el}\,Q & P\,H_{el}\,P \end{matrix} \tag{39}$$

in the treatment of autoionizing resonances. In fact, most applications of the pseudopotential technique of eqs. (35), (36) have been performed on the calculation of resonance parameters of doubly excited series lying above one[44] and several[45] ionization thresholds.

A state whose energy is embedded in a continuum, and is diabatic with respect to it, is autoionizing. In practice, many experimentally important collisions (such as of the type X^{n+} + He) have been seen[19,46,47] to proceed through molecular states that are autoionizing for some range of internuclear distances.

In the Feshbach theory, the autoionizing state may be ascribed a ionization lifetime τ that is given, in atomic units, as the inverse of its width. This width can be evaluated, to an excellent approximation, by the golden rule :

$$\Gamma_\alpha = 2\pi\, |\langle\phi_\alpha^d|Q_\alpha H_{el} Q_\beta|\phi_\beta^d\rangle|^2 \tag{40}$$

where $E_\beta^d(R) = E_\alpha^d(R)$ and the continuum wavefunction ϕ_β^d is δ-normalized with respect to the energy. Provided the collisional time is larger than the lifetime $\tau(R) = \Gamma^{-1}$, autoionization during the collision may be ignored[46,48]; oherwise, it must be taken into account. In many situations, the region where molecular states are autoionizing extends to $R \to \infty$ with $\tau(R{=}\infty) < \infty$, so that the corresponding atomic exit channels are autoionizing; then, electron emission after the collision can compete with, or even dominate, radiative decay from these channels; examples are provided by processes[49] such as $N^{5+}(1s^2) + He(1s^2) \to Ne^{3+}(1s^2 3l3l') + He^{2+}$

As final remarks on the block-diagonalization technique, it should be noted that radial couplings between diabatic states thus obtained may or may not be identically zero : they vanish, for example, in the specific example of Lichten's I have quoted, but they do not exactly vanish in ionic-covalent crossings. What the method guarantees is that radial couplings between the ensuing diabatic states are small. Next, I have mentioned the possibility that energies of states which are solutions of the same projected equation (33) pseudocross, and therefore are not, even approximately, diabatic. Only states which are solutions of different equations (33), (38) can be spoken of as diabatic with respect to each other. Obviously, the same holds for the states which are obtained by solving Feshbach's closed channel equation - a point that was taken up in ref. 14 to show that a simple extension of Feshbach's formalism does not provide a general definition of diabatic states. If it is desired to further diabatize the states so that the remaining pseudocrossings are eliminated, block-diagonalization can be set up in a nested way, as in the partitionning :

$$\begin{matrix} Q_\alpha H_{el} Q_\alpha & Q_\alpha H_{el} Q_\beta & Q_\alpha H_{el} P \\ Q_\beta H_{el} Q_\alpha & Q_\beta H_{el} Q_\beta & Q_\beta H_{el} P \\ P H_{el} Q_\alpha & P H_{el} Q_\beta & P H_{el} P \end{matrix} \tag{41}$$

An implementation of this procedure is presented in ref. 46 for Be^{4+} + He collisions. Alternatively, Smith's method can be applied, in the form of eqs. (23), (26)-(32), to the remaining pseudocrossings. Finally, it should be remarked that the use of the term diabatic in the literature is certainly not limited to either Smith's definition or to block-diagonalized states - cf. e.g. ref. 12, where any function that is not adiabatic is called diabatic.

TRANSITIONS TO THE CONTINUUM

In principle, expansions (14) and (16) include integrals over ionizing, continuum molecular wavefunctions $\phi_\varepsilon(\underline{r}, R)$ which are solutions of eq. (13). The calculation of dynamical coupling matrix elements corresponding to these exact continuum wavefunctions is a difficult numerical problem[50]. Fortunately, this calculation is often unnecessary : when a linear variational treatment is, as usual, employed to solve eq. (13) for the bound wavefunctions $\phi_n(\underline{r}, R)$, it also yields *discretized* approximations $\bar{\phi}_m(\underline{r}, R)$ to the continuum ones; these discretized wavefunctions are L^2-integrable functions that represent, within a viven volume[51], and up to a normalization factor, those exact non-L^2-integrable ones $\phi_\varepsilon(\underline{r}, R)$ for $\varepsilon = E_m$. Hence, they can be employed instead of the latter ones in many calculations. For instance, eq. (14) can be replaced by :

$$\begin{aligned} \Psi(\underline{r}, \underline{R}) &= \sum_n \chi_n(\underline{R})\, \phi_n(\underline{r}, R) + \int d\varepsilon\, \chi_\varepsilon(\underline{R})\, \phi_\varepsilon(\underline{r}, R) \\ &= \sum_n \chi_n(\underline{R})\, \phi_n(\underline{r}, R) + \sum_m \bar{\chi}_m(\underline{R})\, \bar{\phi}_m(\underline{r}, R) \qquad (42) \end{aligned}$$

where the n-summations are restricted to molecular energies E_n less than the first ionization threshold; and $\langle\phi_\varepsilon|\phi_{\varepsilon'}\rangle = \delta(\varepsilon-\varepsilon')$, $\langle\bar{\phi}_m|\bar{\phi}_n\rangle = \delta_{mn}$. Similar expressions may be written for the semiclassical wavefunction in eq. (16). The use of discretized wavefunctions permits to apply the usual bound-state techniques to calculate the discrete-continuum dynamical couplings.

In the limit of a complete variational basis set, both expansions in eq. (42) are exact, and hence describe the scattering wavefunction Ψ, within any finite volume. When an incomplete, finite basis set is employed, the second equation in (42) yields a quadrature procedure for the integral in the first equation. Alternatively, the so-called packed states[50-52], which are L^2-integrable linear combinations of the continuum wavefunctions, can be employed for the same purpose in eq. (42).

The use of a large number of discretized states to represent the ionizing continuum as well as the infinite series of Rydberg levels usually converging to it, is a very cumbersome procedure. This procedure is unavoidable when one aims at calculating the spectra of electron emission during a collision. On the other hand, it may sometimes be dispensed with when only total cross sections for elastic, excitation, charge exchange - even ionization - are required. To undestand how this can be carried out, one must know the mechanisms whereby the continuum states are coupled to the discrete ones, in the molecular model of atomic collisions.

In practice, two kind of ionization mechanisms operate, and these correspond to resonant and non-resonant processes. The resonant mechanism has been discussed in the previous section, and consists in an autoionizing adiabatic state immersed in an ionizing continuum. The problem is solved through the construction of a new state that is diabatic with respect to the continuum, according to Feshbach's theory. This state has a lifetime $\tau = \Gamma^{-1}$, given by eq. (40), before it decays into the continuum. Use of the block-diagonalization technique permits to define the diabatic state for energies below the ionization threshold, and thus to avoid introducing in eqs. (14) and (16) the infinite Rydberg series converging to that threshold. In addition, the state may be ascribed a lifetime for decay into this Rydberg quasi-continuum[14].

A totally different, non-resonant ionization mechanism occurs at high velocities, when the discrete state is not embedded in a continuum but is nevertheless closely coupled to it. To treat this problem, we may take, as is adequate at these velocities, the semiclassical approach of eqs. (6)-(12), and introduce the time-dependent counterpart[53] of Feshbach's[43] approach. As in the previous section, the state space is separated into two subspaces (notice the reversal of the roles of P and Q with respect to refs. 53 and 54, which is made here for the sake of consistency with the previous section) : the Q-subspace is the manifold spanned by the wavefunctions included in the truncated molecular expansion, and the P-subspace is its complement, and contains all ionizing contributions. Eq. (7) is then equivalent to :

$$i\, Q\, \partial/\partial t\, Q\, \psi = Q\, H_{el}\, Q\, \psi + Q\, \left(H_{el} - i\, \partial/\partial t\right) P\, \psi \qquad (43)$$

$$i\, P\, \partial/\partial t\, P\, \psi = P\, H_{el}\, P\, \psi + P\, \left(H_{el} - i\, \partial/\partial t\right) Q\, \psi \qquad (44)$$

with Q and P the corresponding projectors. The perturbed stationary states method amounts to solving instead of (43), (44) the projected equation :

$$i\, Q\, \partial/\partial t\, Q\, \psi_{approx} = Q\, H_{el}\, Q\, \psi_{approx} \qquad (45)$$

instead of eq. (7). The neglected last term in eq. (43) yields a probability leakage from Q to P space, which in turn influences the population of the Q functions trough the last term in eq. (44). Use of infinite order perturbation theory for eq. (44) yields :

$$i\, Q\, \partial/\partial t\, Q\, \psi = Q\, H_{el}\, Q\, \psi + Q\, V_{opt}\, Q\, \psi \qquad (46)$$

where the optical potential V_{opt}, whose explicit form is given in ref. 53, is nonlocal in time, expressing memory effects, and accounts for the probability flux from Q to P states, and in particular to continuum states leading to ionization.

Eq. (46) is exactly equivalent to eq. (7), and is as difficult to solve as this equation without introducing some approximation. One may then look for approximate ways to to account for the probability flux towards P-states due to V_{opt}. One possibility that has been recently proposed[55] is to add to the molecular basis a few (non-adiabatic) functions, each of which acts as a "probability flux absorber" for a selected Q-function ϕ_m. This addition may be considered as a "best augmentation" procedure that selects, from P-space, those functions that are most closely coupled to the Q-space ones. In terms of the adiabatic basis, absorber wavefunctions are linear combinations of continuum and higher excited states not included in the molecular basis. Their main advantage is that they can be determined without constructing these linear combinations.

To infer the form of the absorber function corresponding to a given adiabatic wavefunction ϕ_n, we can assume that at $t = t_0$ the system is exactly described by ϕ_n. At $t = t_0 + \delta t$ the exact electronic wavefunction has evolved according to eq. (7), while the stationary state approximation fulfills eq. (45). It can then be seen that the error funcion is :

$$\varepsilon(\underline{r}; t) \equiv \psi - \psi_{approx} = -\, i\, (\delta t)\, P\, \left(H_{el} - i\, \partial/\partial t\right) Q\, \phi_n + O(\delta t^2) \qquad (47)$$

Hence, when transitions from ϕ_n to P space just begin to be important, the first term in eq. (47) wil be responsible for them, and it is shown in ref. 50 that this term is the seeked probability absorber. Furthermore, with the definition of absorbers of absorbers, etc., a hierarchical basis can be formally set up so that the method can be rendered as exact as needed - at the expense, of course, or a considerable computational efort if one goes any further than first-generation absorbers.

The use of absorber states is a simple procedure that replaces the integral or summation of the second terms in eq. (42) by the addition of a few terms involving L^2-integrable functions. This presents considerable computational advantages. On the other hand, the method has the inconvenience that information is lost as to the particular continuum states where the probability flux eventually ends up; only a global information as to transitions towards P-space is obtained. This prevents, for example, the calculation of ionization cross sections as functions of the energy of the emitted electron.

Besides simplifying the calculations, the use of absorber states has permitted to explicitly study the mechanisms of non-resonant transition processes towards the molecular continuum. The main conclusions[55] of this study are in full agreement with the picture that emerges from classical trajectory simulations[56,57] and atomic three-center expansion calculations[58]. Roughly speaking, those transitions proceed via two mechanisms :

▸ Direct transitions from one or more bound molecular wavefunctions to continuum states. The relevant discrete-continuum couplings are due to the strong variation of the discrete molecular wavefunctions in the small-R region, as their structure passes from being one-centered at R = 0 to two-centered at slightly larger distances; this is accompanied by a considerable relative expansion of the electronic clouds as R increases. The specific range of internuclear distances involved depends on the particular state, being the larger the more excited (hence, more diffuse) this state is. As a consequence of this structural variation, the molecular wavefunction describing this state is radially coupled to those of many higher excited states and especially with those of the ionizing continuum. At close atomic encounters with slow nuclear velocities, the adiabatic behaviour of the electronic wavefunction describes, in the small-R region, a contraction followed by an expansion of the electron cloud as the nuclei approach and recede from each other. On the other hand, at high velocities the cloud cannot follow these sudden changes, with the result that non-adiabatic couplings cause the electron to be impulsively promoted to positive energy states.

▸ Indirect transitions, giving rise to a ladder mechanism, in which probability flows, at large distances and through radial couplings, from a bound molecular state towards a higher excited one; then from this one, at larger distances, to an even higher excited one (and back towards the original state); and so on until the ionization continuum is reached. Physically, this mechanism represents a process whereby a small part of the electronic cloud is left stranded in the region between the nuclei as they recede from each other (saddle-point electrons[57]) - a process that is reminiscent, though not identical[59] to Wannier two-electron threshold ionization. This process is represented in the molecular model in terms of transitions to increasingly diffuse states. Because of the large distances involved, indirect transitions give an important contribution to the total ionization cross section. In particular, from the previous explanation one may expect the ionization threshold to be entirely dominated by the indirect transition mechanism.

The distinction between both processes is not a sharp one, since it is the (direct) transitions to the continuum from each rung of the ladder that lead to ionization in the indirect process, and the two mechanisms overlap. It was noticed in ref. 55 that, in order to reproduce direct ionization, absorber states have energies that are immersed in the continuum at short internuclear distances. On the other hand, to reproduce indirect ionization, absorber states were found to approximate, at large internuclear distances, the adiabatic ones whose energies lie just above the highest excited ones included in the molecular basis, and are more closely coupled to them. In this way, they tend to reproduce the first-missing rung of the ladder leading to the ionizing continuum. This is an important finding, because it suggests that to account for indirect transitions to the continuum, such that only the large-R structure of the absorber states is involved, these absorbers may be approximated by selected Rydberg states. This amounts to just augmenting the molecular basis with a few more adiabatic states, which act as an approximation to the absorber ones, and allow for probability leakage towards the ionization continuum. The success of their selection can be ascertained by repeating the procedure by further augmenting the basis set with closely coupled Rydberg states : probability will again flow towards these new states introduced. One even finds that, at sufficiently high velocities such that ionization dominates charge exchange, the exit probabilities through these approximate absorber states may be employed[9] to evaluate the cross section for indirect ionization.

As a conclusion, in order to describe all processes that occur during an atomic collision at high energies, one must reproduce the probability flux towards the ionization continuum. This means that to accurately evaluate partial cross sections for a given set of exit channels, one must include[60] in the molecular basis, besides the wavefunctions that dissociate yielding these channels, selected representatives of the manifold lying next higher in energy. The final probability corresponding to this manifold then represents genuine exit though the corresponding atomic exit channels and also probability flux towards higher lying, and ionizing, states.

In order to implement these ideas to apply the molecular model at intermediate and high nuclear velocities, one has to cope with a fundamental limitation of the perturbed stationary states approach, which is studied in the following section.

TRANSLATION FACTORS

Reaction coordinates and translation factors

A knowledge of the mechanisms of transitions to the continuum is very useful when applying the molecular model at energies such that ionization cannot be neglected. On the other hand, it is not directly useful when continuum wavefunctions are required, not to describe a ionizing process but simply to cope with the impossibility to enforce the scattering boundary conditions with a finite molecular expansion - in particular, to obtain the cross sections in spite of the existence of residual couplings at infinite internuclear separation. To do away with this difficulty, a departure from the perturbed stationary states approach is needed.

It may be worth mentioning here that the existence of residual couplings at infinite internuclear separation is not the only practical difficulty that arises when truncating the perturbed stationary states expansion (14). As is easily seen, dynamical ∇_R couplings in eq. (15) depend on the origin of electronic coordinates :

$$\nabla'_R = \nabla_R + \delta \nabla \qquad (48)$$

with δR the distance from the new to the old origin, so that $\underline{r} = \underline{r}' + \delta \underline{R}$. Since this origin is at the center of nuclear mass, one finds large isotopic dependences in the couplings; a typical example is g-u couplings in HD^+. The effect of this origin dependence on the exit transition probabilities cancels in the limit of a complete set of functions in eq. (14), but it does not cancel for a truncated expansion, thus leading to unphysical isotopic effects in calculations. A similar difficulty arises in the close-coupling treatment of vibrational problems, and in the semiclassical treatment of atomic collisions. Furthermore, in this latter formalism the exact solution of eq. (7) is easily seen[27] to be invariant with respect to respect to a change of origin of electronic coordinates. While this invariance is preserved by a complete expression (16), it is destroyed when the expansion is truncated. In simplified calculations, these origin dependences of the calculated cross sections often remain concealed by the procedure of employing model[15], rather than exact, interactions. These simplified models are usually defined in terms of some diabatic basis such that non-adiabatic couplings (including their origin dependence) are assumed to be negligible. This practical "solution" is not adequate in either accurate treatments or at sufficiently high nuclear velocities such that the origin-dependence of the couplings is large.

In present-day calculations, the standard solution to these difficulties is to modify the molecular (diabatic or adiabatic) wavefunctions, by multiplying them by phases, depending both on the electronic and nuclear coordinates, called electron translation factors (ETF's). To avoid misunderstandings that have plagued the literature, it should be stated from the start that ETF's are *not* indispensible to treat collisional problems. In fact, their use in the quantal formalism is only an approximation to the general solution that was first proposed by Mitleman[61] : to substitute R in expansion (14) by a mass-weighted reaction coordinate[12-14,61,62] $\underline{\xi}(\underline{r},\underline{R})$, such that : $\underline{\xi} \to \underline{\rho}, \underline{\rho}'$ in the limits $\rho, \rho' \to \infty$, respectively. Then the boundary conditions (2), (3) can be trivially imposed. Furthermore, the reaction coordinate is chosen to depend on the origin of electronic coordinates so that the new couplings are independent of this location.

All (flawless) formal theories that have claimed to do away with the abovementioned dificulties of the perturbed stationary states approach without recourse to ETF's involve one form or another of reaction coordinates. If one further wishes to guarantee the convergence of the molecular treatment, the reaction coordinate must be the same for all states, so that the new expansion reads :

$$\Psi(\underline{r},\underline{R}) = \sum_n \chi_n(\underline{\xi}) \quad \phi_n(\underline{r}, R = \xi) \tag{49}$$

where the $R = \xi$ equation must be interpreted as both quantities being numerically identical; in this way, the completeness of the molecular wavefunctions for a fixed R value guarantees convergence of expansion (49). For an application of this method see, e.g. ref. 63. In practice, the use of reaction coordinates is hindered by the selection of an appropiate analytical form for ξ, and by the appearance of cross terms $\nabla\xi \cdot \nabla_i$ in the expression for the Hamiltonian H in eq. (1), leading to new coupling terms in eq. (15). In addition, the mass ratios involved make reaction coordinates less appropiate for atomic collisions than for chemical reactions[12]. A simpler approach is obtained by expanding ρ, ρ' in powers of μ^{-1} :

$$\underline{\rho} = \underline{R} - M_A^{-1} \, \underline{r}_A + O(\mu^{-2}) \tag{50}$$

$$\underline{\rho}' = \underline{R} + M_B^{-1} \, \underline{r}_B + O(\mu^{-2}) \tag{51}$$

where M_A, M_B are the nuclear masses. Retaining first-order terms and substitution in eqs. (2), (3) yields :

$$\Psi(\underline{r},\underline{R}) \underset{\rho\to\infty}{\sim} \exp(i\underline{k}_i\cdot\underline{R}) \exp(-i\, M_A^{-1}\, \underline{k}_i\cdot\underline{r}_A)\, \varphi_i^A + \sum_j R^{-1} \exp(ik_jR) \exp(-i\, M_A^{-1}\, \underline{k}_j\cdot\underline{r}_A)\, f_{ij}(\theta)\, \varphi_j^A \tag{52}$$

$$\Psi(\underline{r},\underline{R})_{\rho'\to\infty}^{\sim} \sum_j R^{-1} \exp(ik_jR) \exp(i\, M_B^{-1}\, \underline{k}_j\cdot\underline{r}_B)\, f_{ij}(\theta)\, \varphi_j^B \tag{53}$$

These approximate boundary conditions may be compared to eqs. (18) and

(19), respectively, where only zeroth-order terms in μ^{-1} were retained. To fulfill them, it suffices to substitute expansion (14) by :

$$\Psi(\underline{r},\underline{R}) = \sum_n \chi_n(\underline{R}) \exp(iU_n) \, \phi_n(\underline{r},R) \tag{54}$$

where each wavefunction ϕ_n has been multiplied by a phase $\exp(iU_n)$ that equals $\exp(-i\, M_A^{-1}\, \underline{k}_n \cdot \underline{r}_A)$ near nucleus A and $\exp(i\, M_B^{-1}\, \underline{k}_n \cdot \underline{r}_B)$ near B. More specifically :

$$U_n(\underline{r},\underline{R}) \underset{\rho\to\infty}{\sim} - \bar{M}_A^{1} \;\; \underline{k}_n \cdot \underline{r}_A \tag{55}$$

$$U_n(\underline{r},\underline{R}) \underset{\rho'\to\infty}{\sim} M_B^{-1} \;\; \underline{k}_n \cdot \underline{r}_B \tag{56}$$

Hence, in the asymptotic regions ρ, $\rho' \to \infty$, the factor $\exp(iU_n)$ describes the following of the nuclear translational movement by the electron, and is appropiately called an electron translation factor. The analytical form of U_n is further chosen to depend on the origin of electronic coordinates such that all new couplings are origin-invariant; moreover, from eqs. (54), (55) and (56), it can be shown that the new dynamical couplings vanish in the limit $R \to \infty$, hence in each asymptotic region ρ, $\rho' \to \infty$.

Since the semiclassical formalism was derived from an μ^{-1} expansion [eq. (6)], it is not surprising that the solution to the boundary condition problem provided by ETF's becomes an exact one in this formalism. Expansion (16) is modified :

$$\psi(\underline{r},t) = \sum_n a_n(t) \exp\{-i\int E_n[R(t)]\, dt\} \exp(iU_n) \, \phi_n(\underline{r},[R(t)]) \tag{57}$$

and $\exp(iU_n)$ equals $\exp(-i\, p\, \underline{v} \cdot \underline{r}_A)$ near nucleus A and $\exp(\, i\, q\, \underline{v} \cdot \underline{r}_B)$ near B, with p R the distance of the origin of electronic coordinates from nucleus A, and q = 1 - p. More specifically :

$$U_n(\underline{r},\underline{R}) \underset{\substack{R\to\infty \\ r_A \text{ fixed}}}{\sim} - p\, \underline{v} \cdot \underline{r}_A \tag{58}$$

$$U_n(\underline{r},\underline{R}) \underset{\substack{R\to\infty \\ r_B \text{ fixed}}}{\sim} q\, \underline{v} \cdot \underline{r}_B \tag{59}$$

Then, the initial condition (8) can be exactly fulfilled, and one is able to extract the transition amplitudes (11) and (12). As in the quantal case, the functions U_n are chosen to depend on the origin of electronic coordinates in such a way that all new couplings are origin-invariant.

In either formalism, when more than one electron is involved, the antisymmetry condition on the wavefunction Ψ or ψ can be fulfilled by taking U_n to be a sum over all electrons, of the form :

$$U_n = \sum_j u_n(\underline{r}_j, \underline{R}) \tag{60}$$

Let us now turn to the physical interpretation of ETF's. Since the molecular wavefunctions $\phi_n(\underline{r}, R)$ are usually taken to be real-valued, they represent a state of the system with a zero net electronic flux. It is not usually recognized that an electronic charge density that changes with R without such a flux yields a self-contradictory picture; this entails a limitation of the description of molecular structure provided by the Born-Oppenheimer approximation, which is the lack of correlation between electronic and nuclear motions; this limitation affects the description of vibrational as well as collisional states[64-66]. Multiplication of the molecular wavefunctions by an ETF $\exp(iU_n)$ yields a travelling wavefunction $\phi'_n(\underline{r}, R) = \phi_n(\underline{r}, R) \exp(iU_n)$, which represents the same charge density as $\phi_n(\underline{r}, R)$, but with a net flux :

$$\underline{J}_n = \mathrm{Im}(\phi'^*_n \nabla \phi'_n) = \nabla U |\phi_n|^2 = \nabla U |\phi'_n|^2 \qquad (61)$$

and therefore provides a more appropiate picture of the electronic motion in a scattering event. This flux or probability current density[2] describes the following of the nuclei by the electronic cloud, and may be intrepreted as the product of an effective velocity ∇U and a probability density $|\phi_n|^2$. For example, in the asymptotic limit $R \to \infty$, from eqs. (55), (56), (58) and (59) we see that ∇U yields the pertinent relative nuclear velocity, and $\underline{J}_n$ the corresponding (translationally uniform) flux. Since curl $\underline{J}_n = \underline{0}$, not all effective electronic velocities can be represented in this way. While one can represent, for instance, a translationally uniform flux, or a rotation with constant angular momentum, its irrotational property precludes the representation of any flux involving a net circulation (or turbulence, using the standard[2] fluid mechanics analogy) : a typical case which cannot be described in this way is that of a uniform rotation, such as a $2p_z - 2p_x$ transition in the united atom $R = 0$ limit; on the other hand, this uniform rotational flux is easily represented through breakdown of the Born-Oppenheimer approximation by rotational transitions between the two ($2p\sigma$, $2p\pi$) relevant molecular states.

We now consider the determination of ETF's, since the plane-wave forms of eqs. (55), (56), (58) and (59) only refer to the asymptotic $R \to \infty$ limit. Two additional obvious conditions are that $U_n \to 0$ as $k_n \to 0$ [or $v \to 0$]; and that U_n should be continuous in all variables, since an exponential of a discontinuous function cannot be differentiated twice, even in the framework of the theory of generalized functions or distributions; this precludes, for example to employ, in the semiclassical formalism, the simplest "intuitive" form of a step function $U_n = [\mathrm{sign}(\underline{R}\cdot\underline{r})+(q-p)/2]\ \underline{v}\cdot\underline{r}$.

To prove the usefulness of further general criteria for ETF's has been a difficult task, because the effect of the translation factors is closely linked with the mechanisms of the collisional processes, and these can strongly differ from one system to another, and at low and high nuclear velocities. Furthermore, most choices require considerable computational efforts, with the result that the analytical forms employed for ETF's have often been determined by practical reasons rather than by ab initio conditions. Nevertheless, some criterion is needed, since results obtained with different choices can significantly vary from each other[67]. A sensible one seems to consider ETF's as *convergence factors* for the modified molecular expansions (54) or (57), so as to diminish, as much as possible, the size of these expansions. A suitable analytical form may be chosen for U_n, containing (at least) an adjustable parameter, which can be determined from the condition that convergence of the modified expansion be optimal. Furthermore, since the exact enforcement of this condition is still very difficult, it has been proposed that it be substituted by a minimization of the couplings between the wavefunctions included in eqs. (54), (57) (which may be called Q-functions as in the previous section) and the infinite set of discarded functions (called P-functions). More specifically, in the framework of the semiclassical theory, it has been proposed to minimize either selected Q-P couplings[68], or a (Euclidean or weighted) norm of all couplings[69] :

$$N = \| P \left(H_{el} - i\, \partial/\partial t\right) Q \| \qquad (62)$$

[cf. eqs. (44), (45)]. This quantity (62) can be analytically evaluated[64]; also, in many applications, a simplified version of this norm[65] is sufficient. By minimizing the norm (62), the effect of the discarded second term in eq. (43) is also minimized. The application of these criteria to the quantum mechanical formalism is straightforward. The norm minimization technique has been employed in a large number of applications, and for several analytical forms for U_n. I shall now consider the two most popular choices for the ETF's.

Common translation factors

One additional condition that may seem desirable for ETF'S is that they should preserve the very reason to introduce the molecular expansion : the completeness of the set $\{\phi_n(\underline{r}, R)\}$ for each value of R. Thus, although one never envisages to employ a complete expansion, it is reassuring to know that by increasing the basis one will eventually converge, and converge to the exact result. According to this criterion, the modified wavefunctions $\{\phi_n \exp(iU_n)\}$ should form a complete set in the limit whereby all solutions of eq. (13), including the ionizing ones, are included. The only completeness proof presented[70] so far refers to the particular case of a common (state-independent) translation factor[71] (CTF) $U_n \equiv U$ in the semi-classical formalism. Starting from the completeness of the set $\{\phi_n\}$, which enables any wavefunction to be expanded in terms of these functions in any finite domain of the electronic coordinates, and choosing $\psi \exp(-iU)$ as one such particular wavefunction, one obtains an expansion of ψ in terms of the set $\{\phi_n \exp(iU)\}$ - hence this set is complete.

Roughly speaking, for any given molecular state we can consider two ranges of internuclear distances, which we may call molecular ($R < \Xi_n$, where the wavefunction ϕ_n is delocalized) and atomic ($R > \Xi_n$, where ϕ_n is localized). For the latter, we have for a heternonuclear one-electron system, either :

$$\phi_n \underset{R\to\infty}{\sim} \varphi_n^A (\underline{r}_A) \tag{63}$$

or

$$\phi_n \underset{R\to\infty}{\sim} \varphi_n^B (\underline{r}_B) \tag{64}$$

to an excellent approximation (neglecting the effect of the electronic mass in the solution of the atomic Schrödinger equations). Then, we have, from eqs. (58), (59) :

$$\exp(iU)\ \phi_n \underset{R\to\infty}{\sim} \exp(-i\ p\ \underline{v}\cdot\underline{r}_A)\ \varphi_n^A (\underline{r}_A) \tag{65}$$

if (63) holds, or

$$\exp(iU)\ \phi_n \underset{R\to\infty}{\sim} \exp(i\ q\ \underline{v}\cdot\underline{r}_B)\ \varphi_n^B (\underline{r}_B) \tag{66}$$

if (64) holds. Thus, from eqs. (65), (66), when $R \gg \Xi_n$ the CTF becomes identical to a plane-wave translation factor (PWTF). Hence, at large R the CTF molecular approach becomes identical to an expansion over atomic wave-functions modified by PWTF's. We notice that the radius Ξ_n defining the division between atomic and molecular domains is strongly state-dependent, steeply increasing for progressively excited states.

In the CTF formalism the (truncated) molecular expansions reads :

$$\psi(\underline{r},t) = \exp(iU) \sum_n a_n(t) \exp\{-i\textstyle\int E_n[R(t)]\,dt\}\ \phi_n(\underline{r},[R(t)])$$
$$\equiv \exp(iU)\ \tilde{\psi}(\underline{r},t) \tag{67}$$

In most calculations, U is taken to be of the form :

$$U = f(\underline{r},\underline{R})\ \underline{v}\cdot\underline{r} \tag{68}$$

with f an interpolating, or switching, function between the values :

$$f(\underline{r}, \underline{R}) \underset{\substack{R\to\infty \\ r_A \text{ fixed}}}{\sim} -p \tag{69}$$

$$f(\underline{r}, \underline{R}) \underset{\substack{R\to\infty \\ r_B \text{ fixed}}}{\sim} q \tag{70}$$

and an origin dependence :

$$f(\underline{r}', \underline{R}) = f(\underline{r}, \underline{R}) + (p - p') \tag{71}$$

so that all couplings can be shown to be origin independent.

If we now evaluate the net electronic flux corresponding to eq. (67) we obtain :

$$\underline{J} = \mathrm{Im}(\psi^* \nabla \psi) = \nabla U \, |\psi|^2 + \mathrm{Im}(\tilde{\psi}^* \nabla \tilde{\psi}) \tag{72}$$

Hence, the CTF introduces an overall electronic probability flux [first term in eq. (72)], in addition to the one that can be described by the transitions betwen the molecular wavefunctions [second term in eq. (72)]. If we interpret, as usual, $|\psi|^2$ as the electronic probability density, ∇U represents an effective velocity field. This field permits, for example, to easily describe the electronic fluxes that appear in eqs. (8), (11) and (12) . On the other hand, as mentioned above, not all effective fields can be described through the gradient ∇U, and the second term in eq. (72) is indispensible for this purpose. Notice that in this approach the effective field is chosen to be common to all molecular channels, but this does not imply that the CTF introduces a common flux; in fact, because of interference effects in eq. (72), one cannot define individual fluxes for those channels.

Apart from its convergence properties, the use of CTFs has important computational advantages too, and it is the most popular molecular approach used. In particular, the functions $\{\phi_n \exp(iU_n)\}$ are orthogonal, which simplifies the integration of the dynamical equations, and the new couplings, involving ∇U, $\nabla^2 U$ and $\partial U/\partial t$, can be treated (e.g., interpolated) as the usual, dynamical or electrostatic, ones. On the other hand, since the method relies on the completeness of the set $\{\phi_n \exp(iU)\}$ to converge to the exact solution that describes the scattering event, in order to treat a collisional process (charge exchange, excitation), the method must also reproduce all the other ones simultaneously taking place. This is obviously an asset of the method, but may require large basis sets, of the same order of those of an atomic expansion of the scattering wavefunction. In particular, when ionization occurs, allowance for loss of probability flux towards continuum wavefunctions must be made, as explained in the previous section.

Plane-wave translation factors

For state-dependent factors, the arbitrariness in their determination of ETF's is enhanced, and the question of completeness is an open problem. In practice, calculations have been practically limited to the simplest choice, which are PWTF's. Their use was suggested in the pioneer work of Bates and McCarroll[26]. In this approach, one takes for all internuclear distances the large-R limits of eqs. (55), (56), (58) and (59) :

$$U_n = - M_A^{-1} \underline{k}_n \cdot \underline{r}_A \text{ (quantal) ; } U_n = - p\, \underline{v} \cdot \underline{r}_A \text{ (semiclassical)} \tag{73}$$

when eq. (63) holds and :

$$U_n = M_B^{-1} \underline{k}_n \cdot \underline{r}_B \text{ (quantal) ; } U_n = q\, \underline{v} \cdot \underline{r}_B \text{ (semiclassical)} \tag{74}$$

when eq. (64) holds. This ascribes a constant effective electronic velocity to each molecular state. The molecular PWTF model has the advantage that, even with small expansions, it has been found[72,73] to yield accurate values for the charge transfer cross sections at intermediate velocities, and in particular the conspicuous fall of these cross sections at higher velocities[8].

As mentioned above [eqs. (65) and (66)], in the atomic region of internuclear distances, the PWTF and CTF approaches coincide. This is not so, however, for the molecular region of internuclear distances, with the consequence that, at intermediate and high velocities, the transition mechanisms are rather different.[8,9]

Introduction of the modified expansions (54), (57) - with U_n given by eqs. (73), (74) - in the dynamical equations (1), (7), yields coupling matrix elements that are of two different kinds :

▸ Those of a direct type arise between two wavefunctions which have the same asymptotic limit (63) or (64), hence the same PWTF.

▸ Those of the charge, or momentum, transfer type involve two functions with different asymptotic limits, hence different PWTF's. As a consequence, from eqs. (73) and (74) their integrands contain phases such as $\exp(i\underline{v} \cdot \underline{r})$ that describe a change in the translational momentum of the electron. For nuclear velocities $v > 1$ a.u. ($> 10^6$ m sec^{-1}), these phases cause the charge transfer couplings to strongly oscillate with R and hence decrease on the average.

The decrease in the effect of the charge transfer couplings at high v cause a drop in the corresponding cross section for $v \gg 1$ a.u.. This mechanism of the charge exchange process is very similar to that of an atomic model, whereas it strongly differs from that of a CTF approach, where the fall of the cross section is accounted for by less efficient transitions due to a modification of the electronic energies, and by probability flow towards ionizing channels.

At high velocities, the PWTF approach suffers from two important drawbacks :

▶ From the computational viewpoint, the use of a non-orthogonal basis $\{\phi_n \exp(iU_n)\}$ is numerically awkward and requires a time-consuming interpolation[69] of the couplings.

▶ From a theoretical point of view, there is no ascription of PWTF's to the continuum solutions of eq. (13) that represent ionizing wavefunctions.

The first difficulty is of a technical nature, and has resulted in the applications of the method to be very limited. In particular, there have been no applications to systems with more than one electron. The second difficulty is a fundamental one because not only it hints to the possibility that the definitions (73), (74) be rather arbitary for very diffuse excited states, but it also precludes the construction of a complete set $\{\phi_n \exp(iU_n)\}$, which would justify introducing the molecular expansions (54) and (57) to represent scattering events. This poses the question of the reason for the success of the approach. To simplify the discussion of this point I shall employ a semiclassical formalism.

At low collisional energies, the PWTF's are close to unity, and therefore do not affect much the completeness of the set of molecular wavefunctions $\{\phi_n\}$. This explains the success of the method at low nuclear velocities. To understand the workings of the method at higher energies, it may be reasoned that the total cross sections are often dominated by transitions at distances where the wavefunctions ϕ_n are only partly delocalized, and can be written in terms of polarized atomic functions :

$$\phi_{n,A} \propto \varphi_n^A(\underline{r}_A) + \eta_n(R)\, \varphi_\nu^B(\underline{r}_B) \tag{75}$$

or

$$\phi_{\mu,B} \propto \varphi_\mu^B(\underline{r}_B) + \eta_m(R)\, \varphi_\omega^A(\underline{r}_A) \tag{76}$$

with η_n exponentially decreasing with R. Then, when η_n, η_m are small, one may expect that the convergence of the PWTF molecular expansion (57):

$$\psi(\underline{r},t) = \sum_n \alpha_n(t) \exp(-i\, p\, \underline{v}\cdot\underline{r}_A - i \int E_n\, dt)\, \phi_{n,A}(\underline{r}) + \sum_m \beta_m(t) \exp(i\, q\, \underline{v}\cdot\underline{r}_B - i \int E_m\, dt)\, \phi_{m,B}(\underline{r}) \tag{77}$$

will share some properties of that of an atomic expansion :

$$\psi(\underline{r}, t) = \sum_j \alpha'_j(t) \exp(-i\, p\, \underline{v}\cdot\underline{r}_A - i\, E_j^A\, t)\, \varphi_j^A(\underline{r})$$

$$+ \sum_k \beta'_k(t) \exp(i\, q\, \underline{v}\cdot\underline{r}_B - i\, E_k^B\, t)\, \varphi_k^B(\underline{r}) \qquad (78)$$

In expansion (78) the set $\{\varphi_j^A(\underline{r}_A),\ \varphi_k^B(\underline{r}_B)\}$ is overcomplete in the limit when one includes the continuum wavefunctions of both atomic Hamiltonians, and so is the set $\{\varphi_j^A(\underline{r}) \exp(-i\, p\, \underline{v}\cdot\underline{r}_A),\ \varphi_k^B(\underline{r}) \exp(i\, q\, \underline{v}\cdot\underline{r}_B)\}$. Overcompleteness usually results in a faster convergence than simple completeness - consider for example, the relative performances of two-center (overcomplete) and one-center (complete) expansions in molecular calculations - and only leads to difficulties when overlap effects become so strong that they introduce quasi-linear dependences. The first terms of the molecular basis $\{\phi_{n,A}(\underline{r}) \exp(-i\, p\, \underline{v}\cdot\underline{r}_A),\ \phi_{m,B}(\underline{r}) \exp(i\, q\, \underline{v}\cdot\underline{r}_B)\}$ are similar to those of the previous atomic one, but the set is not complete, since we do not know how to include continuum wavefunctions to achieve this completeness. Hence, one may expect similar convergence properties for the first terms of both expansions (77) and (78), whereas the similarity will deteriorate for the highly excited states, with only eq. (77) converging to the exact result.

From the previous reasonings, we may conclude that when charge exchange and inelastic cross sections are determined by transitions occuring at distances such that eq. (75) and (76) hold, the molecular PWTF expansion may be considered as a roughly interpolative method between two approaches (molecular without ETF's and atomic with PWTF) that are convergent. In particular, like the atomic expansion, the method reproduces[8] the rapid fall of total charge exchange cross sections as a consequence of momentum transfer between wavefunctions of type $\exp(-i\, p\, \underline{v}\cdot\underline{r}_A)\ [\varphi_n^A(\underline{r}) + \eta_n(R)\, \varphi_\nu^B(\underline{r}_B)]$ and $\exp(i\, q\, \underline{v}\cdot\underline{r}_B)\ [\varphi_m^B(\underline{r}_B) + \eta_m(R)\, \varphi_\omega^A(\underline{r})]$. This partaking of both convergent properties results in the method being able to describe the collisional events with a small basis, although (especially as v increases) there must appear a limit to the accuracy of the method when augmenting the number of terms in the expansion. At high collisional energies, the cross sections are determined by transitions occurring at such small distances that the molecular states are strongly delocalized. Then, one may expect the molecular method modified with PWTF to be increasingly inaccurate to describe the collision of two atomic species. In particular, while it reproduces the fall of charge exchange cross sections, their relative error may be expected to increase with the nuclear velocity.

CONCLUDING REMARKS

The present work contains an outline of some characteristics of the present-day application of the molecular approach to atomic collisions. It is envisaged as a complement to the many reviews that are available, by presenting an outlook that, while not original, will not be usually found in these reviews. A considerable simplification and shortening of the text has been achieved by explicitly considering one-electron systems in most equations, and by not spelling out all contributions to the non-adiabatic couplings. On the other hand, in order to facilitate the connection with other areas in atomic and molecular physics, the discussion has not been restricted to the semiclassical formalism.

The subjects that have been covered here may be considered as resulting from successive extensions from very low to low, then to intermediate and high collisional energies of the original idea that a collision between two atoms at low energies can be interpreted in terms of the properties of the corresponding molecular system. These extensions have involved including first non-adiabatic transitions between the molecular states, then diabatic states and electron translation factors. For a satisfactory solution to these points, it has proved indispensible to understand the role of transitions to the ionizing continuum. With these tools, the molecular approach is able to reach the domain of collisional energies where high energy perturbational methods are accurate, and ionization dominates.

However, there is also a different way to interpret these developments. Consider, for example, a collision between a fully ionized atom X^{Z+} with $Z > 2$ and a He atom (a process of interest in the treatment of nuclear fusion plasmas as well as in astrophysical research). Since the energy of the He atom lies higher than that of the $X^{(Z-1)+}(1s)$ ion, the molecular state that dissociates, at infinite internuclear separation, into X^{Z+} + He is embedded in the continuum of $XHe^{(Z+1)+} + e^{-}$, and therefore must possess a ionization lifetime $\tau(R)$ that is given, to a good approximation, by eq. (40); of course, as $R \to \infty$ we shall have $\tau \to \infty$ exponentially, and the system is no longer autoionizing. However, since $\tau(R) < \infty$ for finite R, this means that even the simplest approximation to describe the collision of an X^{Z+} ion an a He atom requires constructing a state that is diabatic with respect to the ionizing continuum. For collisions with interaction times much shorter than τ, the autoionizing character of the entrance molecular channel can be ignored, but for sufficiently slow collisions this is not so. From this example, we may conclude that, especially at very small nuclear velocities the concept of diabatic states is required for a general description of collisional events.

A similar reasoning applies to the introduction of ETF's. When applying a given method, it is obviously a desideratum that one knows that it is firmly grounded, even if one must introduce approximations in the course of its implementation. However, in the application of the perturbed stationary states, or Born-Huang expansion method to represent scattering events, one knows that, no matter how small the nuclear velocity, if one imposes the boundary conditions (18) -(20) at an arbitrarily large internuclear distance $R = \Xi$, and then varies the value of Ξ, provided calculations are performed with unlimited precision, the cross sections will, in general, diverge[13] as $\Xi \to \infty$, because of the existence of residual couplings - hence, of transitions - for arbitarily large internuclear separations. It is because we employ a limited precision that the problem is often (rather trivially) unimportant at low energies. This practical "solution" is obviously awckward from the theoretical point of view. The energy region where the problem passes from being "academical" to a real stumbling-block depends on the process, on the accuracy required, and on the precision employed in the calculations. To cope with this difficulty, the formalism must be modified with the introduction of reaction coordinates - or, to what is usually an excellent approximation, translation factors.

An related, awckward description is also obtained for vibrational states with non-negligible amplitudes[66], although the solution for this vibrational problem involves an electron "translation" operator[64-66] rather than an ETF. Also, at extremely low energies the influence of the electronic mass in the difference between molecular and electronic energies at infinite internuclear separaction may become of importance[74], and reaction coordinates may again be needed.

The previous reasonings show that at least part of the developments explained here may also be interpreted as being required for a fully cogent presentation of the molecular expansion; this refers, in particular, to the Feshbach formalism and to electron translation factors or reaction coordinates. On the other hand, in applications those developments may often be dispensed with : to treat many collisional problems, the Feshbach formalism is unnecessary, and ETF's can often be approximated $\exp(iU_n) \approx 1$ at low velocities.

Acknowledgements

This work was partially supported by DGICYT project No. PB90-0213. Many conclusions of the present work were drawn from discussions with L. F. Errea, A. Macías, F. Martín and L. Méndez.

References

1. Dž. Belkić, R. Gayet and A. Salin, Phys. Rep. 56:279 (1979)
2. A. Messiah, "Quantum Mechanics II", North Holland, Amsterdam (1965)
3. W. Lichten, Phys. Rev. 131:229 (1963)
4. A. Macías and A. Riera, J. Phys. B 10:861 (1977), 11:1077 (1978)
5. N. F. Mott and H. S. W. Massey, "The Theory of Atomic Collisions", Oxford Univesity Press, Oxford (1965)
6. M. Born and K. Huang, "Dynamical Theory of Crystal Lattices", Oxford University Press, London (1954)
7. W. Lichten, Phys. Rev. 139:A27 (1965), 164:131 (1967)
8. L. F. Errea, J. M. Maidagan, L. Méndez and A. Riera, J., Phys. B 24:L387 (1991)
9. L. F. Errea, C. Harel, H. Jouin, J. M. Maidagan, L. Méndez, B. Pons and A. Riera, Phys. Rev. A (in press)
10. W. Fritsch, Nucl. Instr. and Meth. B 23:9 (1987)
11. B. H. Bransden, C. J. Noble and J. Chandler, J. Phys. B 16:4191 (1983)
12. B. C. Garrett and D. G. Truhlar, Theoret. Chem. Advan. and Perspectives 6A:215 (1981)
13. W. Thorson and J. B. Delos, Phys. Rev. A 18:117 (1978); J. B. Delos, Rev. Mod. Phys. 53:287 (1981)
14. A. Macías and A. Riera, Phys. Rep. 90:299 (1982)
15. E. E. Nikitin and S. Ya. Umanskii, "Theory of Slow Atomic Collisions", Springer-Verlag, Berlin (1984)
16. R. K. Janev and L. P. Presnyakov, Phys. Rep. 70:1 (1981)
17. P. T. Greenland, Phys. Rep. 81:131 (1982)
18. R. K. Janev and H. Winter, Phys. Rep. 117:265 (1985)
19. C. Harel and A. Salin, "Electronic and Atomic Collisions", North Holland, Amsterdam (1988)
20. C. J. Joachain, "Quantum Collision Theory", North-Holland, Amsterdam (1975)
21. T. Adachi and T. Kotani, Prog. Theor. Phys. 39:430 (1968)
22. L. Wilets and S. J. Wallace, Phys. Rev. 169:84 (1968)
23. R. McCarroll and A. Salin, J. Phys. B 1:163 (1968)
24. H. Kleinpoppen, Phys. Rev. A 3:2015 (1971)
25. S. Osimitsch, W. Jitschin, H. Reihl, H. Kleinpoppen, H. O. Lutz, O. Mó and A. Riera, Phys. Rev. A 40:2958 (1989)
26. D. R. Bates and R. McCarroll, Proc. Roy. Soc. A245:175 (1958)
27. A. Riera and A. Salin, J. Phys. B 9:2877 (1976)
28. M. Barat and W. Lichten, Phys. Rev. A 6:211 (1972)
29. F. T. Smith, Phys. Rev. 179:111 (1969)
30. A. Macías, A. Riera and M. Yáñez, Phys. Rev. A 23:2941 (1981)
31. F. Borondo, A. Macías and A. Riera, J. Chem. Phys. 74:6126 (1981)
32. V. López, A. Macías, R. D. Piacentini, A. Riera and M. Yáñez, J. Phys. B 11:2889 (1978)
33. T. F. O'Malley, Adv. Atom. Mol. Phys. 7:223 (1971)
34. A. Macías, F. Martín, A. Riera and M. Yáñez, Phys. Rev. 36:4203 (1987)
35. A. Russek and R. J. Furlan, Phys. Rev. A 39:5034 (1989)
36. L. Lipsky and A. Russek, Phys. Rev. 142:59 (1966)
37. F. Martín, O. Mó, A. Riera and Yáñez, J. Chem. Phys. 87:6635 (1987)
38. M. C. Bacchus-Montabonel and P. Vermeulin, Comput. Phys. Comm. 30:163 (1983), F. Spiegelmann and J. P. Malrieu, J. Phys. B 17:1259 (1984); see also list of references in R. Cimiraglia, these proceedings.
39. F. Martín, O. Mó, A. Riera and Yáñez, Europhys. Lett. 4:799 (1987)
40. J. C. Phillips and L. Kleinman, Phys. Rev. 116:287 (1959)
41. B. Andersen and S. E. Nielsen, Mol. Phys. 21:523 (1971)
42. H. Gabriel and K. Taulbjerg, Phys. Rev. A 10:741 (1974)
43. H. Feshbach, Ann. Phys. (N.Y.) 5:357 (1958), 19:287 (1962)

44. H. Bachau, P. Galan, F. Martín, A. Riera and Yáñez, At. Data and Nucl. Data Tables 44 : 305 (1990); H. Bachau, F. Martín, A. Riera and Yáñez, At. Data and Nucl. Data Tables 48: 167 (1991)
45. H. Bachau, P. Galan, F. Martín, A. Riera and Yáñez, J. Phys. B 23: L83 (1990); F. Martín, H. Bachau, P. Galan, A. Riera and Yáñez, J. Chem. Phys. 94: 5011 (1991)
46. F. Martín, A. Riera and M. Yáñez, J. Chem. Phys. 86: 6927 (1987)
47. C. Harel and H. Jouin, Europhys. Lett. 11: 121 (1990)
48. H. Bachau, Europhys. Lett. 3: 559 (1987)
49. M. Boudjema, P. Moretto-Capelle, A. Bordenave-Montesquieu, P. Benoit-Cattin, A. Gleizes, H. Bachau, P. Galan, F. Martín, A. Riera and M. Yáñez, J. Phys. B 22: L121 (1989)
50. G. Bandarage and W. R. Thorson, Phys. Rev. A 37: 692, 716 (1988)
51. D. F. Gallagher and L. Wilets, Phys. Rev. 169: 139 (1968) and many other papers; for an elementary example illustrating the method see A. Riera, Am. J. Phys. 59: 62 (1991)
52. D. A. Micha and R. D. Piacentini, Phys. Rev. A 25: 204 (1982)
53. H. J. Lüdde and R. M. Dreizler, J. Phys. B 22: 3243 (1989)
54. A. Riera, Phys. Rev. A 30: 2304 (1984)
55. L. F. Errea, L. Méndez and A. Riera, Chem. Phys. Lett. 164: 261 (1989), Europhys. Lett. 13: 43 (1990), Phys. Rev. A 43: 3578 (1991)
56. R. E. Olson, Phys. Rev. A 23: 4397 (1986)
57. G. Bandarage and A. Parson, Phys. Rev. A 41: 5878 (1990)
58. T. G. Winter and C. D. Lin, Phys. Rev. A 29: 3071 (1984)
59. R. E. Olson, T. J. Gay, H. G. Berry, E. B. Dale and V. D. Irby, Phys. Rev. Lett. 59: 36 (1987)
60. L. F. Errea, J. M. Gómez-LLorente, L. Méndez and A. Riera, J. Phys. B 20: 6089 (1987)
61. M. H. Mittleman, Phys. Rev. 188: 231 (1969)
62. M. H. Mittleman and H. Tai, Phys. Rev. A 8: 1880 (1973)
63. M. Gargaud, R. McCarroll and P. Valiron, J. Phys. B 20: 1555 (1987)
64. N. V. Cohan and H. F. Hameka, J. Chem. Phys. 45: 4392 (1966)
65. T. H. Walnut and L. A. Nafie, J. Chem. Phys. 67: 1491 (1977)
66. A. Riera, J. Mol. Struct. (Theochem) (in press)
67. L. F. Errea, L. Méndez and A. Riera, Phys. Rev. A 28: 3357 (1983)
68. J. Rankin and W. R. Thorson, Phys. Rev. A 18: 1990 (1978)
69. L. F. Errea, L. Méndez and A. Riera, Phys. Rev. A 39: 2404 (1989)
70. L. F. Errea, L. Méndez and A. Riera, J. Phys. B 15: 101 (1982)
71. S. B. Schneiderman and A. Russek, Phys. Rev. 181: 311 (1969)
72. G. J. Hatton, N. F. Lane and T. G. Winter, J. Phys. B 12: L571 (1979)
73. T. G. Winter and G. J. Hatton, Phys. Rev. A 21: 793 (1980)
74. J. B. Delos and W. R. Thorson, J. Chem. Phys. 70: 1774 (1979)

CONTROL OF PHOTOCHEMICAL BRANCHING: NOVEL PROCEDURES FOR FINDING OPTIMAL PULSES AND GLOBAL UPPER BOUNDS

David J. Tannor* and Vladimir Kazakov

Department of Chemistry and Biochemistry
University of Notre Dame
Notre Dame, IN 46556

Vladimir Orlov

Department of Chemistry
University of Chicago
5735 South Ellis Avenue
Chicago, IL 60637

Abstract: Given the great progress in femtosecond laser technology – with amplitude, frequency and phase control of the electric field – the *systematic* design of optical pulse sequences to control microscopic atomic and molecular motion has become an inexorable topic of study. The problem may be formulated variationally as the search for the optimal pulse sequence (i.e. complex electric field amplitude, $\varepsilon(t)$) which prepares a wavepacket such as to maximize the expectation value of an appropriate projection operator at some final time. For control of photofragmentation, for instance, the projection operator is equal to 1 for a particular chemical arrangement channel and 0 elsewhere. The optimization must be performed subject to the constraint that the wavepacket satisfy the time dependent Schrodinger equation in the presence of the field, leading to a mathematical structure of the type found in Optimal Control Theory (OCT). The first part of the paper provides the physical motivation for the definition of the objective functional. Then, the original objective, subject to the constraint of the TDSE, is transformed into an unconstrained, or modified objective and partitioned in a novel way following the work of Krotov. The partitioned form for the objective functional leads to a great deal of insight into the problem, as well as to a variety of remarkable new results. It is first used to introduce and contrast the necessary and sufficient conditions for optimal pulses. The partitioned objective is then used to derive a novel and remarkably efficient iterative algorithm to find optimal pulses. The new algorithm, which can be said to derive from the sufficient conditions of optimality, explicitly includes both linear and quadratic dependence of the objective on the field and the wavefunction, and as such is able to take macrosteps in the field at every iteration. In contrast, the usual gradient type methods include only the linear

*Alfred P. Sloan Foundation Fellow 1991-93

dependence of the objective on the field and the wavefunction; as a result they require a line search at each iteration and converge very slowly. Numerically, the new method several times more efficient than gradient type methods for the systems we have studied. In the last part of the paper the sufficient conditions for global optimality are used to obtain the first known upper bound on control of chemical reactions, subject to a penalty on the field energy. Although the upper bound is less than one (one represents 100% yield in the desired chemical arrangement channel) the discrepancy vanishes for zero penalty, consistent with 100% control of the channel from which the wavepacket exits.

1 Introduction

Is it possible to take advantage of the unique properties of lasers – monochromaticity, high power, and phase coherence – to bring about selective and energetically efficient photochemical reactions? In the last six years a new approach has emerged in which laser pulse shapes and pulse sequences, designed to achieve a desired chemical objective, are calculated systematically. In 1985, Tannor and Rice [1](see also [2] and [3]) formulated the problem of the search for optimal optical pulses, subject to suitable constraints, as a problem in the calculus of variations. Their formulation was based on a perturbation theory expression for the time dependent wavepacket amplitude, and led to a complicated integral equation expression for the optimal pulse(s). A formal and computational advance was achieved in 1988 by Rabitz and coworkers, [7],[8],[9],[10] who recognized the utility of optimal control techniques, used widely in engineering, for the calculation of optimal pulses. Optimal control theory (OCT) is essentially the extension of the classical calculus of variations to problems with differential equations constraints. In the context of molecular control, the time dependent Schrödinger equation (TDSE) is introduced as an additional constraint into the variational equations. The OCT methodology, in addition to providing a simplified algorithm for obtaining optimal fields numerically, has no difficulty in handling fields of arbitrary strength. This results in a significant generalization of the perturbative formalism, not only because greater chemical yields can be obtained with strong fields, but also because many new mechanisms are available to strong fields that are not accessible in the weak field regime. It is worth noting that most of the components of the TDSE/OCT formalism developed by Rabitz et al. were discovered independently by Kosloff et al. [4], who developed an intuitively based iteration procedure to optimize pulses. In fact, the first iterative method developed in that work prefigures some of the features of the new method presented in Section 4 below.

Despite the preliminary application of OCT technology to the search for optimal pulses, many issues are unresolved. The first set of issues are numerical. If OCT is to be a useful tool the optimization algorithms must converge rapidly. For instance, in the calculations reported in [5] the product yield had not converged, even after 300 iterations! Clearly, this slow convergence severely limits the utility of OCT techniques for chemically interesting models, i.e. 3 degree of freedom systems are not conveniently studied, even with supercomputers. Moreover, if convergence is never attained it is hard to answer a host of interesting numerical questions having to do with the structure of the solution space. For instance, do different initial guesses converge to the same final pulse, i.e. are there many local maxima of the objective or just one global maximum? Are there any theoretical upper limits on quantum controllability? Do fully converged optimal pulses have any interesting mathematical properties? For instance, are they eigenfunctions of some generalized operator? Do optimal pulses corresponding to different exit channels (or even corresponding to the same exit channel) obey a type of orthogonality relation? The second set of issues are physical. For OCT to be useful the pulse sequences which come out must be of interest to experimentalists, at least as regards their qualitative features, if not their quantitative details. To date, a good fraction of the pulse sequences calculated by OCT are so complex that they are extremely unlikely to be synthesized in the laboratory, and even qualitative analysis of their mechanism has not been attempted.

To a great extent the physical and numerical issues are intertwined: intuitively, there would seem to be some correlation between an OCT problem which is robust numerically (i.e. converges quickly, many different initial guesses for the field converge to the same final field) and a solution which will be robust in the laboratory. Robust pulses, in turn, seem likely to have an interesting underlying physical mechanism.

This article provides a preliminary report of the application of a set of unconventional methods for solving OCT problems, developed by Krotov and coworkers [16],[15],[19], to the problem of control of photochemical branching. Whereas standard methods of OCT are based on necessary conditions for optimality, Krotov's methods are based on sufficient conditions. In Section 4 a numerical method for iterative optimization based on Krotov's methods is presented, and contrasted with the usual gradient and conjugate gradient methods. Preliminary experience indicates that the method is several times more efficient than gradient type methods. Section 5 presents the first known upper bound on quantum control of chemical reaction, derived using Krotov's sufficient conditions for optimality. The upper bound obtained has the structure of a Carnot upper bound on efficiency. It is strictly less than one for any nonzero penalty on the electric field (and goes to one in the limit of zero penalty). This upper bound is low enough to be physically meaningful and is consistent with the empirical numerical evidence to date, that there is no limit (other than the normalization of the wavefunction) to the amount of amplitude obtainable in a desired exit channel. The ability to find global upper bounds to the objective (and, in principle, to iteratively reduce these upper bounds) provides a potential diagnostic of how close the objective has come to the global maximum. To the extent that the upper bound is meaningful, comparing this bound with a converged calculated objective provides a measure both of the efficiency of the pulse as well as some measure of how well conditioned is the problem. The new methods presented in this article, therefore, 1) provide a much more efficient route than previously available to calculate pulses corresponding to local maxima of the objective, and 2) a plausible measure of how close the resulting objective is to the global maximum. These two methodological developments, taken together, are viewed as important advances toward a preliminary goal of a global mathematical characterization of the problem of control of photochemical process. It is believed that such a mathematical characterization is a valuable step toward the ultimate goal: of calculating pulse sequences whose qualitative and quantitative features are meaningful for laboratory experiments.

This article may be viewed in some sense as a sequel to Ref [5]. That reference focussed on the conventional OCT formulation, and the physical interpretation of the iteratively obtained pulses. The word "optimal pulses" was carefully avoided in that reference because of the convergence problems referred to above, and the focus was on good, but not optimal pulses and their physical interpretation. This article therefore picks up precisely where that article left off, i.e. with a report of the sought after method for efficient calculation of converged optimal pulses and some characterization of the global optimum. Our implementation of these new methods was accomplished very recently. A complete study of the physical and mathematical properties of the optimal pulses is planned for publication elsewhere. It is believed that the new methodologies presented here, by facilitating wavepacket optimal control calculations to a great degree, cannot help but become a tool of widespread interest and utility to wavepacket practitioners.

2 Formulation of the Optimal Control Problem

The time dependent Schrödinger equation for a two electronic state system with transition dipole moment μ can be written as:

$$i\hbar\frac{\partial}{\partial t}\begin{pmatrix}\psi_a(t)\\ \psi_b(t)\end{pmatrix}=\begin{pmatrix}H_a & \mu\varepsilon(t)\\ \mu\varepsilon^*(t) & H_b\end{pmatrix}\begin{pmatrix}\psi_a(t)\\ \psi_b(t)\end{pmatrix} \tag{1}$$

The subscripts a and b refer to the ground and excited electronic state indices, respectively. $H_{a/b}$ refers to the Born-Oppenheimer Hamiltonian for the ground/ excited electronic state, respectively. The two electronic states are coupled by the transition dipole moment, μ, which interacts with the electric field $\varepsilon(t)$ associated with the laser pulse. Complex values of the field are considered admissible, in keeping with the spirit of the rotating wave approximation in which the complex conjugate of the active part of the field is neglected on the grounds that it is off resonant. $\psi_{a/b}$ is the wavefunction (wavepacket) associated with the ground/excited Born-Oppenheimer potential energy surface.

One seeks to maximize

$$J \equiv \lim_{T\to\infty} < \psi(T)|P_\alpha|\psi(T) > \tag{2}$$

where P_α is a projection operator for the excited electronic state in the population inversion study, or for chemical channel α in the bond breaking study (here, α takes on two values, referring to arrangement channels A+AB and AA+B; in general, in a triatomic molecule ABC, α takes on three values, 1,2,3, referring to arrangement channels A+BC, AB+C and AC+B). The quantity J is a *functional* of $\varepsilon(t)$, and the problem of maximizing J with respect to $\varepsilon(t)$ falls into the class of problems belonging to the calculus of variations. It has been the practice, starting with the original work in the perturbative regime [1], to introduce a constraint on the energy of the field:

$$||E||^2 \equiv \int_0^T dt|\varepsilon(t)|^2 \tag{3}$$

where E is a constant. This leads to a maximization problem for the modified objective functional,

$$\bar{J} \equiv < \psi(T)|P_\alpha|\psi(T) > -\lambda\int_0^T dt|\varepsilon(t)|^2 \tag{4}$$

where λ is a Lagrange multiplier. Alternatively, it is possible to let the norm of the field vary, but put a penalty on the size of the norm. In this case, one treats λ as fixed, the higher the value of λ the higher the penalty. Finally, it is possible to remove the constraint on the norm of the field completely. This leads to what is known as a singular optimal control problem. Note that here and henceforth we omit the limit $T \to \infty$ in anticipation of the use of OCT techniques operating within a finite time constraint. It is possible in principle to remove this constraint on the final time, although that has not been attempted here.

In the OCT formulation, the time dependent Schrödinger equation written as a 2x2 matrix in an Born-Oppenheimer basis set, is introduced into the objective functional with a Lagrange multiplier, $\chi(x,t)$. The modified objective functional may now be written as:

$$\bar{J} \equiv < \psi(T)|P_\alpha|\psi(T) > -2Re\int_0^T dt < \chi(t)|\frac{\partial}{\partial t} - \frac{H}{i\hbar}|\psi(t) > - \lambda\int_0^T dt|\varepsilon(t)|^2 \tag{5}$$

where

$$H \equiv \begin{pmatrix} H_a & \mu\varepsilon(t) \\ \mu\varepsilon^*(t) & H_b \end{pmatrix} \tag{6}$$

It is clear that as long as ψ satisfies the time dependent Schrödinger equation $\bar{J} = J$ for any $\chi(x,t)$.

Integrating by parts it is possible to rewrite $\bar{J}$ as follows:

$$\bar{J} = <\psi(T)|P_\alpha|\psi(T)> -2Re<\chi|\psi>|_0^T + \int_0^T dt 2Re[<\chi(t)|\frac{H}{i\hbar}|\psi(t)> + <\dot{\chi}|\psi>] - \lambda|\varepsilon(t)|^2 \quad (7)$$

Both of the above expressions for $\bar{J}$ have the following motivation. In the original problem $\psi(T)$ and $\varepsilon(t)$ are not independent. They are connected through the equation $\psi(T) = \int_0^T dt \frac{H[\varepsilon]}{i\hbar}\psi(t) + \psi_0$. By adding this equation to the original (with a time dependent Lagrange multiplier, $\chi(t)$) we have made the variation in $\psi(T)$ independent of the variation in $\varepsilon(t)$ and $\psi(t)$. In the end, as in any use of Lagrange multipliers, there will be an equation for $\chi(t)$ such that the original constraint is satified (i.e. $\bar{J} = J$). The advantage of the second form of the modified objective is that $\dot{\psi} = \frac{\partial\psi}{\partial t}$ has disappeared from $\bar{J}$.

Following Krotov [16], we now partition $\bar{J}$ in the following suggestive form:

$$\bar{J} = G + \int_0^T dt R \quad (8)$$

where

$$G \equiv <\psi(T)|P_\alpha|\psi(T)> -2Re<\chi|\psi>|_0^T \quad (9)$$

$$R \equiv 2Re[<\chi(t)|\frac{H}{i\hbar}|\psi(t)> + <\dot{\chi}|\psi>] - \lambda|\varepsilon(t)|^2 \quad (10)$$

This way of dividing the modified objective functional is a special case of a more general procedure developed by Krotov [16]. The new expression for $\bar{J}$ can be conceptualized as follows. The objective functional has been partitioned into two independent parts, a terminal part (G) and a time integral (integrand R). To maximize $\bar{J}$ one proceeds to maximize G and R independently. Note that if R is maximized at each time t the integral of R will be maximized.

Eqn 8-10 will be the jumping off point for the rest of the paper. In the next Section we will use this form to distinguish between necessary and sufficient conditions for optimal control. In the following Section it will be the basis for the derivation of a novel numerical algorithm for calculating optimal pulses (based on the necessary conditions for optimality). In the final Section it provide the basis for a derivation of the first global upper bounds for control of photochemical products.

3 Necessary and Sufficient Conditions for Optimality

In this Section the partitioned form of the objective functional will be used to introduce and contrast the necessary and sufficient conditions for optimality. The key observations about eqs. 8-10 at this point are: 1) Taking the variation of G with respect to $\psi(T)$, and R with respect to $\psi(t)$ and $\varepsilon(t)$, and setting them equal to zero yields a set of equations, which together with the original constraint equation provides the *necessary* conditions for a maximum value of the objective ($\delta\bar{J} = 0$). However, what these equations really specify are the conditions for a stationary point of the objective, in complete analogy with the condition of vanishing first derivative in the ordinary calculus. These conditions are therefore consistent not only with a maximum, but also with a minimum or a saddle point of the objective. Moreover, they are strictly local conditions and cannot be used to distinguish local from global maxima or minima.

2) Imagine now that G is truly maximized (i.e. globally) with respect to $\psi(T)$ and R is truly maximized with respect to $\psi(t)$ and $\varepsilon(t)$. If the resulting $\psi(T)$, $\psi(t)$ and $\varepsilon(t)$ satisfy the TDSE with initial conditions $\psi(0) = \psi_0$ the *sufficient* conditions for a global

maximum of the objective are satisfied. This is because the global maximum found by suspending the constraint of the TDSE (i.e. by maximizing G and R separately) can only be greater than or equal to the global maximum of the constrained problem. It should be noted that the assumption that G and R have been globally maximized stands in direct contrast with the strictly local assumptions in the necessary conditions, that G and R are at a stationary point.

3) It seems as if some miracle is required for the TDSE to be fortuitously satisfied. In other words, how do we know that the sufficient conditions for optimality are ever necessary? Let us consider the sufficient conditions of optimality from another perspective. For any $\chi(t)$, $\bar{J} \geq \tilde{J}$ where $\tilde{J} \equiv \max_\varepsilon J[\varepsilon]$, i.e. one has an upper bound to the global maximum. Now, one can proceed to minimize $\bar{J}$ with respect to χ to find the lowest upper bound. If the $\min_\chi \bar{J} = \tilde{J}$ then the global upper bound to the original problem has been found and ψ will satsify the TDSE (more precisely, the integrals of the two will be equal). In practice, $\tilde{J}$ is not known. However, one can compare $\bar{J}$ with the highest value of the objective obtained in the original problem. If they agree, to within some tolerance, one can safely assume that the global maximum has been found.

Given the flexibility of choosing χ (in Krotov's original work [16] he allows for a much more general class that of so-called 'dual' functions than we consider here) it does not seem as unlikely to attain the condition $\min_\chi \bar{J} = \tilde{J}$. Choosing a good dual function is an art as well as a science.

4) The sufficient conditions for optimality can be used to obtain *analytical global upper bounds* on the objective. If it is not convenient to find the maximum of G and R, but if they can be bounded in a meaningful way, one may still obtain a meaningful upper bound to J. This approach will be used in Section 5 to obtain global upper bounds on control of photochemical products.

3.1 Necessary Conditions

As discussed above, we can take the variation of $\bar{J}$ with respect to $\psi(T)$, $\psi(t)$ and $\varepsilon(t)$ independently. This leads to the following necessary conditions for optimality:

The requirement that $\frac{\delta \bar{J}}{\delta \psi(t)} = 0$ leads to the equation:

$$i\hbar \frac{\partial \chi}{\partial t} = H\chi \tag{11}$$

while the requirement that $\frac{\delta \bar{J}}{\delta \psi(T)} = 0$ leads to the equation:

$$\chi(x, T) = P_\alpha \psi(x, T) \tag{12}$$

The condition that $\frac{\delta \bar{J}}{\delta \varepsilon} = 0$ leads to the equation:

$$\tilde{\varepsilon}(t) = \frac{i}{\lambda}[< \psi_b|\mu|\chi_a > - < \chi_b|\mu|\psi_a >] \tag{13}$$

Finally, it is required that the original constraint equation be satisfied, i.e.

$$i\hbar \frac{\partial \psi}{\partial t} = H\psi \tag{14}$$

$$\psi(x, 0) = \psi_0(x) \tag{15}$$

The five equations, 11-15 are the necessary conditions for optimality for the two surface control of photochemical reactions. Note that the dual function, χ, obeys the time dependent Schrödinger equation, subject to the boundary condition at the *final* time T that χ be equal to the projection operator operating on the Schrödinger wavefunction. The equation for the optimal field consists of two terms, both of which involve an overlap between the the dual function and the original wavefunction, on different surfaces, multiplied by i. For a deeper physical understanding of the set of equations 11-15 the reader is strongly encouraged to see Ref. [4].

Equations 11, 12, 14 and 15 lead to a double ended boundary value problem. ψ is known at $t = 0$, while χ is known at $t = T$. Given an initial *guess* for $\varepsilon(t)$ one can propagate ψ forward in time to obtain $\psi(t)$, and propagate χ backwards in time to obtain $\chi(t)$. However, the equation for the *optimal* $\tilde{\varepsilon}(t)$ requires prior knowledge of $\psi(t)$ and $\chi(t)$, which in turn requires knowledge of $\tilde{\varepsilon}(t)$. In general these equations must be solved iteratively. In Section 4 below a detailed comparison of two different iterative methods will be presented. It will be shown, both analytically and numerically, that an iterative scheme which incorporates immediate feedback of the field satisfying eqn 13 into the TDSE at the next time step – a 'local' scheme – is much more effective than the usual gradient type methods which store the expression of the gradient and use it 'globally' at the next iteration. A local iterative scheme was actually the first approach tried in [4] but it differed in two significant ways from the scheme introduced below. A local iterative scheme virtually identical to that described below was proposed earlier by Tannor [6] but never before tested.

3.2 Digression on the Hamiltonian Structure of the Necessary Conditions

There is an alternative formulation of the set of equations 11-15 that bears the same relationship to the original formulation as Hamiltonian mechanics does to Lagrangian mechanics [17], [20], [21]. A Control Hamiltonian, $\mathbf{H}$, is constructed and the optimal pulse is that for which the Hamiltonian is an extremal, i.e. $\frac{\partial \mathbf{H}}{\partial \varepsilon} = 0$. Algorithmically, the search for the optimal pulse can then be performed as a maximization of $\mathbf{H}$. One should not confuse the optimization Hamiltonian $\mathbf{H}$, with the physical Hamiltonian, H.

The optimization Hamiltonian is defined as $\mathbf{H} \equiv 2Re < \chi|H/i\hbar|\psi > -\lambda|\varepsilon(t)|^2$. This is the Legendre transform of the original objective, neglecting the boundary term $< \psi|P|\psi >$. Hamilton's equations take the form

$$\dot{\chi} = -\frac{\partial \mathbf{H}}{\partial \psi} = \frac{H}{i\hbar}\chi \tag{16}$$

$$\dot{\psi} = \frac{\partial \mathbf{H}}{\partial \chi} = \frac{H}{i\hbar}\psi \tag{17}$$

in agreement with eqns 11 and 14 above. The condition for the optimal field is

$$\tilde{\varepsilon} = \arg\max_{\varepsilon} \mathbf{H}(t, \varepsilon, \psi(\varepsilon)) = \frac{1}{\lambda} < \chi|\mu/i\hbar|\psi > \tag{18}$$

in agreement with eqn. 13. (Note that eqn. 18 is written as a single surface expression, i.e. only one potential surface is considered. For simplicity, this 'single surface' form for the optimal field is the one adopted for the remainder of this paper, unless otherwise specified).

It is straightforward to show that $\frac{d\mathbf{H}}{dt} = 0$ for $\varepsilon = \tilde{\varepsilon}(t)$, i.e. the Control Hamiltonian is time independent on the optimal trajectory. This relationship can be used as a diagnostic of how close a particular control is to one which satisfies the necessary conditions for optimality. Note the analogy to the relationship in classical mechanics, that if $\frac{\partial H}{\partial t} = 0$ then $\frac{dH}{dt} = 0$ (i.e. energy is conserved) on a trajectory for which the integral of the Lagrangian is an extremum (i.e. any physical trajectory) [18].

3.3 Sufficient Conditions

We now proceed to a discussion of the sufficent conditions for global optimality, based on Krotov's functionals R and G defined in the previous Section. The sufficent conditions for optimality rest on the independent maximization of G over $\psi(T)$, and R over $\psi(t)$ and ε. The function χ is initially taken as fixed and given.

Maximizing first R with respect to ψ we may write

$$\tilde{R}(t,\tilde{\psi}(\varepsilon),\varepsilon) = \max_{<\psi|\psi>=1,\forall\varepsilon\in R_1} R(t,\psi,\varepsilon) \tag{19}$$

where we have defined the optimal ψ as $\tilde{\psi}$ and the corresponding R as $\tilde{R}$.

Defining

$$\tilde{\varepsilon} = \arg\max_{\varepsilon} \tilde{R}(\varepsilon)(t,\tilde{\psi},\varepsilon) \tag{20}$$

we may write

$$L(\chi,\dot{\chi}) = \tilde{R}(t,\tilde{\psi}(\tilde{\varepsilon}),\tilde{\varepsilon}) \tag{21}$$

where the change in notation from R to L, following Krotov, is intended to suggest the correspondence of L to a Lagrangian, and to show explicitly the dependence on χ and $\dot{\chi}$. Maximizing $\int dtL(\chi,\dot{\chi},t)$ with respect to χ in general is a problem in the classical calculus of variations. Similarly, we define the maximum of G over $\psi(T)$:

$$\tilde{G}(\tilde{\psi}(T)) = \max_{\psi(T)} G(\psi(T)) = \max_{\psi(T)}[<\psi(T)|P|\psi(T)> -2Re<\chi(T)|\psi(T)>] + 2Re<\chi(0)|\psi(0)> \tag{22}$$

Now,

$$\tilde{J} \leq \bar{J} = \tilde{G}(\tilde{\psi}(T))) + \int_0^T dtL(\chi,\dot{\chi}) \tag{23}$$

If a function χ is found such that $\tilde{\psi}$ and $\tilde{\varepsilon}$ satisfy the TDSE then the sufficient conditions for global optimality are met.

To summarize, the similarities and differences between the necessary and sufficient conditions are as follows. Both approaches suspend the constraint that ψ satisfy the TDSE, and both approaches treat $\psi(T)$,$\psi(t)$ and $\varepsilon(t)$ as independent variables. In the necessary approach the first variation with respect to these variables is found, and is set to zero. In the sufficient approach both terms G and $\int dtR$ must be rigorously maximized. This necessarily includes quadratic (and potentially higher) powers of the dependence of the functional on the independent variables. In Section 5 below, to obtain a well conditioned maximization problem for these terms it is necessary to build into both the G and R terms the constraint that $||\psi|| = 1$. This corresponds to a partial, but not complete relaxation of the constraints on the time evolution of ψ. However, once this constraint is built in these terms can be maximized (or at least bounded) and meaningful upper bounds obtained to $\tilde{J}$.

4 Iterative Solution Scheme Based on Necessary and Sufficient Conditions for Optimality

In this Section we focus on algorithms for obtaining pulse sequences that satisfy the necessary conditions for optimality. The partitioning of $\bar{J}$ into $G + \int_0^T dtR$ will provide a unified framework for comparing two classes of iterative algorithms. The first is a gradient type method (the gradient of $\bar{J}$ with respect to ε is defined in an N dimensional space, where $N \approx 10^3$ is the number of time steps allotted for the pulse sequence.) Variations on this method include the conjugate gradient method, in which the step is taken in a direction orthogonal to all previous steps. Regardless of the choice of gradient or conjugate gradient method, this whole family of methods is based on the local expansion of the functional about the previous value of the control, in the same spirit as the necessary conditions for optimality.

The second algorithm for iterating on the control explicitly considers quadratic terms in the expansion of the functional on the control and the wavefunction. As a result, there is a macrostep in the control at every iteration. Conceptually, the algorithm is closely tied to the sufficient conditions for optimality. Empirically, we have found the the new method to be several times more efficent than gradient type methods, if the initial guess for the field is far from an extremum.

To compare the usual gradient based methods and the present iterative algorithm it is instructive to study the analytical expression for the difference in the objective $\bar{J}$ at the kth and $k+1^{st}$ iteration.

Let $\psi^{(k+1)} \equiv \psi^{(k)} + \Delta\psi$, $\varepsilon^{(k+1)} \equiv \varepsilon(k) + \Delta\varepsilon$ where $\Delta\psi$ and $\Delta\varepsilon$ are *not* assumed to be small. For simplicity, we consider a single surface Hamiltonian in the derivation.

Adapting a derivation from [15] ,[19], it is straightforward to show that

$$\bar{J}^{(k+1)} - \bar{J}^{(k)} = \Delta_1 + \Delta_2 + \Delta_3 \tag{24}$$

where

$$\begin{aligned} \Delta_1 &\equiv G(\psi^{(k+1)}(T)) - G(\psi^{(k)}) &(25)\\ &= 2Re[< \psi^{(k)}(T)|P_\alpha|\Delta\psi(T) > - < \chi^{(k)}(T)|\Delta\psi(T) >] + < \Delta\psi(T)|P_\alpha|\Delta\psi(T) &(26) \end{aligned}$$

$$\begin{aligned} \Delta_2 &\equiv \int_0^T dt[R(t,\psi^{(k+1)},\varepsilon^{(k+1)}) - R(t,\psi^{(k+1)},\varepsilon^{(k)})] &(27)\\ &= -2Re\lambda\int_0^T dt\varepsilon^{(k)*}(t)\Delta\varepsilon(t) - \lambda\int_0^T dt|\Delta\varepsilon(t)|^2 \\ &+ 2Re[\int_0^T dt < \chi^{(k)}(t)|\frac{\mu}{i\hbar}\Delta\varepsilon(t)|\psi^{(k)} > + \int_0^T dt < \chi^{(k)}(t)|\frac{\mu}{i\hbar}\Delta\varepsilon(t)|\Delta\psi^{(k)} >] &(28) \end{aligned}$$

$$\begin{aligned} \Delta_3 &\equiv \int_0^T dt[R(t,\psi,\varepsilon^{(k)}) - R(t,\psi^{(k)},\varepsilon^{(k)})] &(29)\\ &= 2Re[\int_0^T dt < \chi^{(k)}(t)|\frac{H^{(k)}}{i\hbar}|\Delta\psi(t) > - \int_0^T < \dot{\chi}^{(k)}(t)|\Delta\psi(t) >] &(30) \end{aligned}$$

Note that in deriving these expressions we have chosen $\chi \equiv \chi^{(k)}$ in the expression for $\bar{J}^{(k+1)}$ as well as in that for $\bar{J}^{(k)}$. Recall that we are free to choose χ in each case in any way we like: $\bar{J}$ will equal J regardless of this choice provided ψ satsifies the TDSE, and this choice is the most convenient for comparison of $\bar{J}^{(k+1)}$ with $\bar{J}^{(k)}$. (For simplicity, here and in the remainder of this section, the single surface TDSE is treated, so that only a single overlap term will be obtained in the gradient; the derivation needs to be modified only slightly for the original two surface problem). Making the choice:

$$\chi^{(k)}(T) = P_\alpha\psi^{(k)}(T) \tag{31}$$

$$\dot{\chi}^{(k)}(t) = \frac{H}{i\hbar}\chi^{(k)}(t) \tag{32}$$

we obtain

$$\Delta_1 =< \Delta\psi(T)|P_\alpha|\Delta\psi(T) > \tag{33}$$

$$\Delta_3 = 0 \tag{34}$$

Eq. 28 for Δ_2 remains unchanged.

Eqns 33,28, and 34 provide a convenient way of contrasting iterative methods for improving $\bar{J}$. The usual gradient type methods neglect the quadratic terms in $\bar{J}$: i.e. the terms that go as $\Delta\psi^2$,$\Delta\varepsilon^2$ and $\Delta\psi\Delta\varepsilon$. $\Delta\varepsilon$ is therefore chosen to maximize the first and third terms of Δ_2. The direction of $\Delta\varepsilon$ is therefore given by

$$\Delta\varepsilon(t) = \frac{1}{\lambda} < \chi^{(k)}(t)|\frac{\mu}{i\hbar}|\psi^{(k)}(t) > \tag{35}$$

For small steps in this direction it is clear that the objective functional must increase or stay the same at each new iteration. However, because the second order dependence of the objective on the field has been neglected, there is no a priori way of knowing the magnitude of the step which will maximize the objective in the direction of the gradient. In practice, the magnitude of $\Delta\varepsilon$ is usually determined empirically from a line search, i.e. $\varepsilon^{(k+1)}(t) = \varepsilon^{(k)} + \alpha\frac{\Delta\varepsilon}{||\Delta\varepsilon||}$ (α a positive real parameter). This was the approach adopted in Ref [5]. However, the neglect of the quadratic terms not only forces the need for a line search; it makes the method strictly local and leads to extremely poor rate of convergence in most cases. Some workers have reported rapid convergence to the optimal pulse using a conjugate gradient (CG) method [8] [13] [14]. In our experience we have not found this to be the case, unless the initial guess is close to the extremal.

In the 'local' iterative method [15],[19] the quadratic terms are not neglected completely; some, but not all of the terms in the second order expansion for the objective functional are included, with essentially no additional computational cost over gradient type methods. Since P is a positive semidefinite operator (its eigenvalues are either 1 or 0), Δ_1 is manifestly greater than or equal to zero. Δ_2 is rewritten in the following suggestive form:

$$\Delta_2 = -2Re\lambda\int_0^T dt\varepsilon^{(k)*}(t)\Delta\varepsilon(t) + \int_0^T dt|\Delta\varepsilon(t)|^2 + 2Re\int_0^T dt < \chi^{(k)}(t)|\frac{\mu}{i\hbar}|\psi^{(k+1)}(t) > \Delta\varepsilon(t) \tag{36}$$

This expression suggests the following algorithm such that Δ_2 will be greater than or equal to zero. Given $\chi^{(k)}(0)$ from the previous iteration calculate the overlap, $< \chi^{(k)}(0)|\frac{\mu}{i\hbar}|\psi^{(k+1)}(0) >$ with $\psi^{(k+1)}(0) = \psi_0$. This leads to a quadratic equation for $\varepsilon(0)$ which is then used to propagate ψ_0 until time Δt. At time Δt the overlap $< \chi^{(k)}(\Delta t)|\frac{\mu}{i\hbar}|\psi^{(k+1)}(\Delta t) >$ is calculated and a new quadratic equation solved to find $\varepsilon(\Delta t)$. This "local" iterative scheme is repeated until $t = T$. The solution of the quadratic equation of $\Delta\varepsilon(t)$ is very similar to that obtained in the usual gradient type methods:

$$\Delta\varepsilon(t) = \frac{1}{\lambda} < \chi^{(k)}(t)|\frac{\mu}{i\hbar}|\psi^{(k+1)}(t) > \tag{37}$$

The significant difference is the appearance of $\psi^{(k+1)}(t)$ in the expression for the overlap, i.e. the feedback from the overlap of ψ and χ is used immediately as input into the field at the next time step for the propagation of ψ. For emphasis, in gradient type methods the overlap is not used immediately but is stored and used for the next iteration. We refer to methods that use immediate feedback of the overlap into the gradient as "local" methods and those that store the gradient and use it at the next iteration as "global" methods.

We summarize here the main features of this local iterative scheme. 1) The method takes macrosteps in the control $\varepsilon(t)$ at every iteration. 2) Since the form of Δ_2 is strictly quadratic in $\varepsilon(t)$ the step size is found analytically as the solution of a quadratic equation in $\varepsilon(t)$ and no line search in the direction of the gradient (or conjugate gradient) is required. 3) The method is guaranteed to provide an improvement of $\bar{J}$ at every iteration, unless the necessary conditions for optimality are satisfied. It is interesting to note that the present local iterative method, with the immediate feedback of the overlap into the form of the pulse at the next time step, is similar in spirit to the original optimization scheme developed by Kosloff et al.[4] Shortly after the local iterative scheme was developed by Kosloff et al. a variant of that scheme was developed by Tannor [6], which is virtually identical to the present algorithm. However, the algorithm was not tested at that time and there was no basis for the belief that it would be numerically superior to the usual gradient methods. The existence of a rigorous mathematical proof that it is possible to take macrosteps at each iteration was indispensible to the success of the present numerical results, not only because it provided a rigorous motivation to expect improved performance; but because the numerical implementation, because it takes macrosteps in the field at every iteration, is extremely sensitive to numerical error. It is therefore extremely helpful to know that Δ_1 and Δ_2 must separately be greater than or equal to zero, and to know that Δ_3 must be maintained equal to zero to very high precision.

5 Global Upper Bounds for Control of Reactions Based on Sufficent Conditions for Optimality

As in the previous section we choose a function χ and define new functionals R and G:

$$R(t,\psi,\varepsilon) = 2Re < \chi|\frac{H}{i\hbar}|\psi > -\lambda|\varepsilon|^2 + 2Re < \dot{\chi}|\psi > \tag{38}$$

$$G(\psi) =< \psi|P_\alpha|\psi > -2Re < \chi|\psi > |_0^T \tag{39}$$

The new optimization is on $\bar{J} \equiv G + \int_0^T dtR$. It is clear that if ψ obeys the TDSE then $\bar{J}$ is equal to J, the original objective. The key now is to relax the constaint of the TDSE, and to perform an unresticted optimization on R at every instant of time with respect to both and ψ and ε and on G with respect to ψ. Not only does ψ not have to satisfy the TDSE; it does not even have to be equal in the two terms R and G! It is obvious that be removing this constraint the optimal value obtained must always be greater than or equal to that obtained with the constraint, i.e., $\tilde{J} \leq \bar{J}$.

What use is a result obtained if ψ doesn't satisfy the TDSE? Two answers. 1) The value of the functional obtained $\bar{J}$ is a global upper bound to the functional. Such upper bounds, if they are not too far from the true maximum, can be useful estimations (in the same sense that transition state theory, while providing an upper bound to a rate constant, also frequently provides a useful estimation.) 2) The upper bound to the objective so obtained is a functional of the adjoint function χ selected. One can now proceed to minimize this functional with respect to χ to obtain the best global upper bound. This minimization can be viewed as the solution of a "dual" OCT problem. If the global minimum obtained by solution of the dual problem is equal to the maximum of the original problem it is a global maximum; at this point, the TDSE is satisfied (or at least its time integral). Moreover, because of the relationship of the dual problem to the original, it is often the case that the same field which minimizes $G + \int dtR$ maximizes J. The general criteria under which the minimum of J with respect to χ is equal to $\tilde{J}$ is not perfectly understood. Developing a good choice for χ is as much an art as a science given the present state of the theory.

3) The minimization of the dual problem with respect to χ generally has the following attractive feature. The functional space is purely convex, i.e. there is only a single minimum, and it is the global minimum. This makes it, in general, a much more desirable system numerically than the original OCT problem.

Maximizing first R with respect to ψ we may write

$$\tilde{R}(t,\tilde{\psi},\varepsilon) = \max_{<\psi|\psi>=1,\forall\varepsilon\in R_1} R(t,\psi(\varepsilon),\varepsilon) \tag{40}$$

where we have defined the optimal ψ as $\tilde{\psi}$ and the corresponding R as $\tilde{R}$.

We find

$$\tilde{\psi} = (\dot{\chi} - H\chi)^*/||\dot{\chi} - H\chi|| \tag{41}$$

so that

$$\tilde{R}(t,\tilde{\psi},\varepsilon) = 2||\dot{\chi} - H\chi|| - \lambda|\varepsilon(t)|^2 \tag{42}$$

Proceeding to the maximization of $\tilde{R}$ over ε and defining

$$L(\chi,\dot{\chi}) = \max_{\varepsilon} \tilde{R}(t,\tilde{\psi},\varepsilon) \tag{43}$$

it is possible to find a closed (but relatively complicated) expression for L:

$$L(\chi,\dot{\chi}) = \max_{\varepsilon}[-\lambda\varepsilon^2 + 2\sqrt{c_1(\chi,\dot{\chi}) - 2c_2(\chi,\dot{\chi})\varepsilon + c_3(\chi,\dot{\chi})\varepsilon^2}] \tag{44}$$

where

$$c_1(\chi,\dot{\chi}) = ||\dot{\chi} - H_0\chi||^2 \tag{45}$$

$$c_2(\chi,\dot{\chi}) = Re(< \dot{\chi} - H_0\chi, H_1\chi >) \tag{46}$$

$$c_3(\chi,\dot{\chi}) = ||H_1\chi||^2 \tag{47}$$

(Throughout this section H_0 and H_1 are defined generically such that $\frac{H}{i\hbar} \equiv H_0 + \varepsilon H_1$ and ε is assumed to be real). Maximizing $\int dt L(\chi,\dot{\chi})$ with respect to χ is a problem in the classical calculus of variations. If, however, we restrict our attention to χ's which satisfy the TDSE,

$$\frac{\partial\chi}{\partial t} = H_0\chi + v(t)H_1\chi \tag{48}$$

($v(t)$ is some arbitrary function of time) we find the following simple expression for L:

$$L(\chi,\dot{\chi}) = ||H_1\chi||(2|v| + ||H_1\chi||/\lambda) \tag{49}$$

We turn our attention now to the maximization of G:

$$\tilde{G}(\tilde{\psi}(T)) = \max_{\psi(T),<\psi(T)|\psi(T)>=1} G(\psi(T)) \tag{50}$$

For convenience define

$$l(a) = \max_{\psi}[< \psi|P|\psi > -2Re < a|\psi >] \tag{51}$$

It is clear that

$$l(a) \leq \min_{\alpha} \max_{\psi}[< \psi|P|\psi > -2Re < a|\psi > -\alpha < \psi|\psi > +\alpha] \tag{52}$$

from the same general considerations about relaxation of constraints that were used above. Assuming $\alpha > 1$ the maximum is given when

$$\tilde{\psi} = \frac{Pa}{1-\alpha} + \frac{(P-1)a}{\alpha} \tag{53}$$

which yields

$$l(a) \leq \min_{\alpha>1}[\alpha + \frac{<\chi|P|\chi>}{\alpha - 1} + \frac{<\chi|1-P|\chi>}{\alpha}] \tag{54}$$

If one finds $\alpha > 1$ from the equation

$$\frac{<\chi|P|\chi>}{(1-\alpha)^2} + \frac{<\chi|1-P|\chi>}{\alpha^2} = 1 \tag{55}$$

then eqn 54 will be the exact value for $l(a)$. However, it is possible to proceed further analytically by assuming $Pa = 0$. Then

$$l(a) \leq \min_{\alpha>1}[\alpha + \frac{<a|1-P|a>}{\alpha}] = 1 + <a|a> \tag{56}$$

Now,

$$\tilde{J} \leq \bar{J} = l(\chi(T)) + 2Re <\chi(0)|\psi_0> + \int_0^T dt||H_1\chi||(2|v| + \frac{||H_1\chi||}{\lambda}) \tag{57}$$

subject to the constraints

$$\dot{\chi} = H_0\chi + v(t)H_1\chi \tag{58}$$

Choosing $v \equiv 0$, $P\chi(T) = 0$ we obtain

$$\tilde{J} \leq 1 + ||\chi(T)||^2 + 2Re(<\chi(0)|\psi_0>) + \int_0^T dt\frac{||H_1\chi||^2}{\lambda} \tag{59}$$

for any $\chi(t)$ obeying $\dot{\chi} = H_0\chi$ and $P\chi(T) = 0$. Now fix some $\tilde{\chi}(t)$ and take $\chi(t)) = b\tilde{\chi}(t)$. This yields

$$\tilde{J} \leq 1 + b^2||\tilde{\chi}(T)||2 + 2bRe(<\tilde{\chi}(0)|\psi_0>) + \frac{b^2}{\lambda^2}\int_0^T dt||H_1\chi|| \tag{60}$$

Minimizing with respect to the scalar b we get

$$\tilde{J} \leq 1 - \frac{[Re <\tilde{\chi}(0)|\psi_0>]^2}{||\tilde{\chi}(T)||^2 + \frac{1}{\lambda}\int_0^T dt||H_1\tilde{\chi}||^2} \tag{61}$$

Thus, for all initial conditions $\tilde{\chi}(0)$ not orthogonal to ψ_0 we have an estimate which is strictly lower than 1. For example, choosing

$$||\tilde{\chi}(T)|| = 1, \tilde{\chi}(0) = \psi_0 \tag{62}$$

we get

$$\tilde{J} \leq 1 - \frac{1}{1 + \frac{1}{\lambda}\int_0^T dt||H_1\chi||^2} \tag{63}$$

Equation 63 may be viewed as a quantum mechanical analog of the Carnot formula for engine efficiency. It is a global upper bound to $\tilde{J}$; the lower this upper bound the more useful it will be. One can now proceed to solve the dual optimal control problem to minimize $\bar{J}$ with respect to χ using numerical techniques similar to those discussed in Section 4 . As mentioned at the beginning of this section, the functional $\bar{J}$ often has the property that it is strictly convex, making it attractive to deal with numerically. There is some reason to hope that the minimum of the dual problem formulated in this way will be equal to the global maximum of the original problem, at which point the following properties may be expected to hold: 1) ψ will satisfy the TDSE 2) $\psi(0)$ and $\psi(T)$ in G will be continuous with their values in R. 3) The function $v(t)$ which minimizes the dual problem would then be equal to $\tilde{\varepsilon}$, the control field which leads to the global maximum of $J[\varepsilon]$.

Knowledge about the global maximum and methods for locating it have been an important goal in optimal control of chemical reactions, and indeed in all optimal control theory. The present formulation looks promising for the partial or complete realization of these goals. Numerical testing of this dual formulation of the problem of control of chemical reactions is planned for the near future.

Acknowledgement

DJT wishes to acknowledge the many significant contributions of his collaborators, Professors Ronnie Kosloff, Pierre Gaspard, Stuart Rice and Jozsef Somlöi, to this research. This work was supported by a grant from the US Office of Naval Research.

References

[1] D.J.Tannor and S.A.Rice, J.Chem.Phys. **83**, 5013 (1985).

[2] D.J.Tannor, R.Kosloff and S.A.Rice, J.Chem.Phys. **85**, 5805 (1986).

[3] D.J.Tannor and S.A.Rice, Adv.Chem.Phys. **70**, 441 (1988).

[4] R.Kosloff, S.A.Rice, P.Gaspard, S.Tersigni and D.J.Tannor, Chem.Phys. **139**, 201 (1989).

[5] D.J.Tannor and Y.Jin, in *Mode Selective Chemistry*, B. Pullman, J.Jortner, and R.D.Levine, eds.(Kluwer, 1991).

[6] D.J.Tannor, unpublished.

[7] A.P.Pierce, M.A.Dahleh and H.Rabitz, Phys.Rev.A. **37**, 4950 (1988).

[8] S.Shi, A.Woody and H.Rabitz, J.Chem.Phys. **88**, 6870 (1988).

[9] S.Shi and H.Rabitz, J.Chem.Phys. **92**, 364 (1990).

[10] S.Shi and H.Rabitz, Comp.Phys.Comm. **63**, 71 (1991).

[11] P.Brumer and M.Shapiro, Chem.Phys.Lett. **126**, 54 (1986).

[12] M.Shapiro and P.Brumer, J.Chem.Phys. **84**, 4103 (1986).

[13] W.Jakubetz, J.Manz and H.-J.Schreier, Chem.Phys.Lett. **165** 100 (1990).

[14] B.Just, J.Manz and J.E.Combariza, J.Phys.Chem. 95, 10351 (1991).

[15] V.F.Krotov and I.N.Fel'dman, Engrg. Cybernetics **21** (1983), 123 (1984).

[16] V.F.Krotov, Control and Cybernetics **17** 115 (1988).

[17] *Applied Optimal Control*, A.E.Bryson, Jr. and Y.-C. Ho (Hemisphere, 1975).

[18] *Classical Mechanics*, H. Goldstein (Addison-Wesley, 1950).

[19] V.A.Kazakov and V.F.Krotov, Automation and Remote Control, no. 5, 430 (1987).

[20] L.I.Rozonoér, Automation and Remote Control **20** 1288 (1959).

[21] L.I.Rozonoér, Automation and Remote Control **20** 1405 (1959).

TIME DEPENDENT QUANTUM MOLECULAR DYNAMICS: EXPERIMENTS AND THEORY

Howard S. Taylor

Department of Chemistry
University of Southern California
Los Angeles, CA 90089-0482

I. Overview

Over the past several years a method of analysis has been developed that uses classical mechanics and the associated ideas of non-linear dynamics to extract the basic motions of the nuclei in highly excited molecules and of electrons in highly excited atoms.[1, 2, 3] The method works hand in hand with, and uses as input, experimental and/or quantally computed theoretical spectra and cross sections. It is to be noted that often such calculated quantities do not easily exhibit the underlying nature of the dynamics. These quantal calculations can yield correct cross sections, spectra, and energy levels but the implicit averages and eigenfunction complexity make dynamic motions less than apparent. This classical method of analysis (CMMA) aims at extracting the detailed motions that when quantized, give rise to the dominant structure in the measured or computed "quantum" results. The CMMA excels at extracting dynamics:

(a) from chaotic spectra [e.g., highly excited rovibrational spectra in molecules and in excited atoms in fields].[4, 5]

(b) from transition state spectra.[6]

(c) from absorption spectra of photodissociating systems.[7,8]

(d) from high quality scattering and half scattering calculations and experiments. In all cases the information CMMA extracts is the motions the system undergoes when it forms a resonance complex.

(e) from stabilization and L^2 calculation of pseudo-discreet states and resonances in the continuum or quasi-continuum.

In this report systems exemplifying points b and e are reported upon.

It is probably useful to note again that CMMA which like many other theories, depends on having a potential surface available, does not compute cross

Time-Dependent Quantum Molecular Dynamics, Edited by J. Broeckhove and L. Lathouwers, Plenum Press, New York, 1992

sections, energy levels or spectra. It goes to the bottom line of the dynamics and extracts from measured or computed quantities the motions that cause many of their features.

II. The Model and the Methodology

The ability to simplify calculations while still extracting motions, implies that there must be an underlying physical model which justifies the limited computations. As such we will first review our generic model which will supply the justification for what we calculate. The calculational methods will then briefly be described. This is followed in the next section by studies of the FH_2^- and $ClHCl^-$ systems. Even though we are here describing our recent work on open systems our discussion will cover spectra of bound states in the highly excited, vibrational quasi-continuum as well as photodetachment spectra. The reason for this, as will be seen below, is the similarity between the low resolution of the bound spectra and resonances in a half-scattering process as exemplified by photodissociation and transition state spectroscopy.

II-A: The Model[1, 2, 3]

Central to the model is the idea that low resolution spectral features in chaotically excited spectra and resonance features in open systems are very often extremely difficult to relate directly to features of the static potential hypersurface. Moreover unlike the lower energy regular spectral region, the requirements that underlie all traditional spectral analysis do not exist. These requirements are first that in some global set of coordinates the Hamiltonian can be divided into a separable zero order part (H_0) that models the dominant motion and a perturbation. Secondly, it is assumed that perturbation theory converges when started from the zero order states of H_0. The separable problem then supplies the N, equal to the number of degrees of freedom, quantum numbers or constants of the motion that label the levels. Experimentalists seeking to assign spectra and extract dynamics then replace the relevant matrix elements in the formal perturbation series for the eigenvalues by parameters which are fit to the spectra. Theoreticians compare their computed eigenvalues to experiment and extract dynamics by comparing the wave functions to those of various H_0 choices. For molecular vibrations H_0 is usually one made of bounded oscillators. Clearly for open systems at energies where no obvious wells or barriers exist in the static potentials, such ideas can not be used to assign the motions of observed or computed resonances. For bounded highly excited small systems or at lower energies for big systems, the large number of states either defy computation or when computed do not project well on to any H_0 model. Here experimentalists found they were unable to assign levels in the above sense.[9]

The resolution of the problem for open systems was pointed out in the late 1970's and early 1980's when dynamic adiabatically separable potentials with oscillator type minima were found.[10, 11 12, 13, 14] This involved either slow motion along the reaction coordinate or some globally radial hyperspherical coordinate. Child and Pollak[15] further pointed out that periodic orbits mark the extrema of these dynamic potentials and Pollak and Wyatt[16, 17, 18] located a particular type of periodic orbit called the Resonant Periodic Orbit (RPO) and quantized it. Most of these attempts worked well in 2D colinear systems. In the bound region for atoms in fields the periodic orbit relation to low resolution and resonant structures in absorption spectra had long been known. It was in our work that these ideas were extended to molecules and the following more general picture that relates periodic orbits to scattering-like resonances and low resolution structure arose.[19] Generically, as the density of states increases as in highly excited smaller or even less excited bigger molecules, quasicontinua develop and bound systems act as open systems, in which local (i.e., in small regions of the static potential) adiabatic potentials exist.[20] These potentials trap trajectories and wave packets, and in, turn cause wave functions, as in resonant scattering problems, to have high amplitude (Scars)[21] in the local regions.[22] In the adiabatic wells regular motion then gives rise in the usual way to resonances. If these potentials could be found, the usual methods of analysis would give the resonant energies. In the bound region the low resolution spectral structure is due to local adiabatic resonance states adjacent to a quasicontinua as opposed to open systems, where true continua exist. From the local potentials point of view, all static potentials in which they are embedded are "big" and in this sense "open". The difference between open and closed systems only comes in the high resolution where one sees the mixed adiabatic plus quasicontinuum states, i.e., the exact eigenstates. This is why the same methods work for scattering and transition state and photodissociation resonances, and also for low resolution features in chaotic systems. The problem is that the potentials cannot generally be found because first, the fast and slow directions often do not lie along the analytically available coordinate systems, and because the adiabatically separable region is so local that one would need to know where it is ahead of time to successfully observe the separation. CMMA uses classical methods to find these locally adiabatic regions.

CMMA depends on a key additional feature of these adiabatic potentials, namely that the trapped parts of the trajectories mimic in their trapped motion regular (tori) structures and that these tori, as do all tori, have periodic orbits associated with them. Moreover the trapping and mimicked tori need not always be in the full dimension of the system and can occur in reduced dimension. This gives rise to reduced dimension tori (RDT) of which periodic orbits (P.O.) are 1D examples. The situation is not unlike Van der Waals systems like He..I_2, where the motional energy gets transferred from the 3D open I_2 - He motion to the I_2 reduced dimension bond direction causing Feshbach resonances. The P.O.'s can be stable or even unstable if either the P.O. is not unstable in all directions or if it is unstable it can be stable in some orthogonal direction in a local adiabatic sense. Importantly the P.O.'s lie at the bottom of, and therefore are markers for, the dynamic potential. Noteworthy for our strategy is that the Fourier transforms of the correlation functions of ensembles of trapped trajectories show the frequencies of these local potentials and signal a P.O. lying under the trapped trajectory. Outside the locally adiabatic region the trajectories are chaotic in bound systems and look like typical scattering trajectories in open systems. Quantally, the wave function has complicated nodal and usually low amplitude structure outside, and regular high amplitude structure inside, the local potentials.

II-B: The Methodology of Extracting Dynamics

The methodology we have to extract dynamics is essentially to get the frequencies of the motion first and the dynamic potentials last.[1, 3] As such, starting with an experiment or theoretically computed spectrum or cross section, which we call I (E), we form the time correlation function $C(t) \equiv |FTI(E)|^2$. Computing the Fourier transform of C(t), called S(w) the spectral density, yields frequencies, w. From the fact that C(t) is also equal to $|<\psi(o)\psi(t)>|^2$, $\psi(o)$ being the experimentally initially created wave packet, these frequencies are seen to be the packets return frequencies that are most often hidden in I (E). They are also energy spacings between resonances and low resolution spectral peaks.

In the classical limit[23] these frequencies are the return frequencies of the trapped trajectory as well as the frequencies of the reduced dimension tori or equivalently the frequencies of the P.O. and its associated stability eigenvalues. They are also the frequencies of the harmonic potential approximations to the adiabatic trapping potential.[24, 25]

To extract motions the exact classical (limit) analogue of S(w) is computed. This, the ensemble averaged spectral density, is given as the Fourier Transform of a phase space density correlation function,[1, 23]

$$S(w) = FT \int dpdq\, \rho[p(o), q(o)]\, \rho[p(t), q(t)] \quad (1)$$

Here the density ρ [p(o), q(o)] carries the experimental information about the transition region and initial state, as it is the function in phase space (Wigner or coherent state transform) that corresponds to the initial state wave function or the initially prepared packet. $\rho[p(t), q(t)]$ carries the dynamics in that it is obtained by replacing p(o) and q(o) in ρ(o) by their values at time t as determined by running a trajectory starting at p(o) and q(o) on the potential surface under study. $\rho[p(t), q(t)]$ should be the phase space analogue of a propagating packet. A deeper understanding of S(w) comes from expressing the phase space integral on a grid as,

$$S(w) = \sum_{i\,=\,\text{grid points}} \rho[p_i, q_i]\, FT\rho[p_i(t), q_i(t)] = \sum_{i\,=\,\text{trajectories}} S_i(w) \quad (2)$$

Noting that each grid point p_i, q_i is an initial condition of a trajectory given as p_i (t), q_i(t), we see that the integral in eq. (2) is the ρ [p_i, q_i] weighted sum of contributions from trajectories in the ensemble ρ [p(o), q(o)]. S_i(w) displays the contribution to each w of the ith trajectory.

At this juncture it is hoped, as has proven to be the case in the problems that we have worked on, that the classical S(w) at least resembles the one obtained from eq. 1 and we can proceed. If this is not the case the CMMA would fail.

Basically, we are hoping, at least for short times, that the ebb and flow of the classical ensemble resembles that of the quantum packet. As dictated by the correspondence principle, we have used functions of energy spacing and not the spectrum itself to compare the classical mechanics to the quantum mechanics. The spectra I(E) has no classical analog as no discrete energies exist classically.

Now, with the $S_i(w)$ in hand, one can search for each peak in S(w), for the trajectories that contribute most to the ith peak. In the open-system, nonchaotic problem, projections of these trajectories immediately reveal the trapped regions that produce the w peaks in S(w). For chaotic systems this is not the case, as all trajectories visit and contribute to the trapping and further, all projections of the trajectories look random and space filling. The projection in any two dimensional plane in phase space superimposes chaotic regions over trapped ones and hides the trapping region. As such, a detailed inspection of the chaotic trajectory must be made. This search is often aided by our knowledge of the energy and relative position of the transition region for the features that we are trying to explain in S(w). For the problems done so far such simple searching has found the trapped region. When this will not be the case a "time slicing" technique suggested by Pollak can be used.[26] Here the trajectory is repeatedly and randomly segmented. Each segment is tested for stability and only the more stable segments are retained. For model systems inspection of these segments has revealed the trapping region. The trapped regions of the trajectory are the regular motion that we seek to extract. This motion can be further refined by using standard techniques for searching for P.O.'s and RDT in the trapped region. The frequencies of the RDT and the P.O. and associated stability frequencies and eigenvectors immediately and simultaneously give the discussed local harmonic approximation to the trapping potentials and its localized normal modes.[24, 25] If insufficient experimental data is available to determine unambiguously the recurrences in S(w), quantization of the local motions can be carried out using any of several semiclassical periodic orbit quantization schemes. Stabilization computations can also be done by putting basis functions around the P.O. or RDT.[2] The results would be compared to I(E). Ultimately when accurate eigenstates are available, they can be inspected to see if they "scar" along the segments. Tennyson's quantum calculations H_3^+ and Na_3 yielded wave functions that were localized (scarred) on the reduced dimension tori found by our phase space search. Moreover, the bound state wave functions in the quasicontinuum appeared in a way that is reminiscent of scattering functions in a resonant region. That is, several similarly localized eigenstates were in the same energy region; an arrangement that makes long lived wave packet formation and motion on the reduced dimension torus possible. Conversely, the knowledge of the periodic orbits should also greatly aid basis and grid point placement in eigenstate calculations. The difficulty of these quantum calculations is clearly due to having to describe the quantum analogue of the chaotic, non-trapped space regions, where by the nature of chaos no basis is physical and computations can expect to converge slowly.

We complete the section on methodology by pointing out that if I(E) is not available, as is presently always the case for large systems and often the case in smaller systems, then quantum stabilization calculations[27] which are much simpler than full eigenstate computations, can produce an approximate I(E). This I(E), made of pseudoscattering, as well as stable resonance states, though incorrect in not giving detailed high resolution eigenvalues, does give the w peaks in S(w) that we wish to obtain. In this method,[28] first the possibly relevant parts of configuration space are divided arbitrarily into overlapping local regions. In each region a basis is laid down and a diagonalization carried out. This basis should be good enough to describe the stable eigenvalues of the local dynamic potentials if they are present in the region. As such among the many meaningless eigenvalues, some not immediately obvious ones are physical. If unit length stick spectra from all regions are superimposed (a histogram), since regions overlap, stable eigenvalues will show up as a cluster of "sticks". Moreover the stick spectra having the "local" dynamics serves as a sufficiently good I(w) to describe the still unknown adiabatic regions. Calculation of S(w) from a smoothed I(w) then reveals the frequencies of the P.O.'s and RDT. Inspection of the stable eigenstates, which are much less defaced by orthogonality requirements and than full eigenstates, locates the P.O.s. Hence an "acceptable" S(w) can be produced theoretically in a simpler "ab initio" way than would be required to compute the fully resolved spectrum. This gives some hope to treat bigger systems than triatoms.

When high quality eigenstates are available in the energy range and transition region of the complex, chaotic (check by running classical trajectories) dense spectra, the analogy to resonance scattering[19] lead others[29] to develop a method of locating the RDT and P.O.'s and as such, extracting the dynamics, from what often appears as very complex looking "chaotic" eigenfunctions and spectra. The method, starting with a transition "stick" spectra, smoothes the spectra progressively, seeking places where the spectral topology shows structure which is quasi-stable to changes in the smoothing parameters. Several such places may exist; the lower resolution ones revealing trapped motions that exist in more restricted regions of phase space nearer to the transition region and corresponding to motion taking place at shorter times. For each region of resolution a wave packet is constructed by adding together the eigenfunctions under the peak. In forming the packet, each eigenfunction is weighted by its transition intensity (stick height). In the cases studied each low resolution case had successive peaks separated by a frequency w, and the peaks yielded a successive series of wave packets which were localized in the same small region of space looked regular, and had the extra nodes expected of excited regular wave functions. To our mind these regular wave functions are those of the local adiabatic potential discusses in the previous section. The localized packets are viewed also as scars over RDT or P.O.s A search of the region covered by the packet and an S(w) study carried out, as discussed above, reveals the classical RDT or P.O.s that are the desired "dynamics" or "assignment". The method clearly is also a powerful one for interpreting computed eigenfunctions.

III: Applications of our Methodology

Over the past several years we have applied our methods to the systems ClHCl, ClDCl, FH_2, FDH, FD_2, IHBr, H_3^+, Na_3, HCN-HNC, C_6H_6 and O_3. We shall review here only the study of FH_2[30] and its isomers and the study of ClHCL[3]. These systems were studied in order to assign transitionstate photodetachment spectra. This elegant technique is at the forefront of chemical physics research and is the culmination of many years effort by a multitude of investigators. In particular, the observed resonances of both ClHCl[31] and FH_2[32] have a long history which we will not review for the sake of brevity. As such, we will here emphasize what we learned in our past studies of the transition state resonances.

III-A.: ClHCl

For ClHCl the key prior theoretical works, that could be compared to Neumark's experiments,[33] were a coupled channel hyperspherical reactive scattering computation by Schatz[34] and a transition region stick spectrum from an L^2 diagonalization and stabilization computation by Bowman.[35] From the examination of the stable wave functions and/or the scattering data, it became clear that the two higher energy peaks observed in Neumark's TS spectra were resonances of (002) and (004) antisymmetric stretch type and that the first of the observed peaks was not a resonance but was a threshold effect due to the wavefunction enhancement at the energy of the barrier at which the hydrogen crosses between the HCl channels. Pollak's work on Resonance Periodic Orbits (RPO: periodic orbits for which the central atom moves between the reactant and product channels) although not done in 3D for ClHCl properly indicated that such antisymmetric stretch resonances might be quantized using the RPO.[36] Schatz's scattering calculations saw additional small structures on the high energy side of the (002) asymmetric stretch resonance which was tentatively assigned to either bending (022) or symmetric stretch (202) combinations; the definitive situation being unclear.

In our work the L^2 stabilization type diagonalization, carried out by Bowman, was first used to give I(E). The quantum S(w) was then formed and, with increasing frequency, four peaks, a, b, and the pair c and c′, were found. The classically generated S(w) and $S_i(w)$ showed peaks a and b to be due to trajectories trapped near the RPO which when found and analyzed for stability was a stable P.O. Peak b was the antisymmetric stretch along the RPO and twice this frequency fit the (002) - (004) observed spacing. Stability analysis of the RPO showed that the two stability eigenvectors, which are orthogonal to the antisymmetric stretch were a symmetric stretch and a bend motion. Surprisingly they had degenerate ie., equal frequencies. This frequency was exactly that of peak "a" in S(w) and was the separation in Schatz's work, between his (002) peak and the structure tentatively assigned as either (022) or (202). Of course the degeneracy was the cause of this uncertainty. Note that the three modal frequencies define an adiabatic bound trapping potential which accounts for the spectra from the (002) line up. Moreover the assigned modes are at the P.O. and not at the saddle.. Such a situation of degenerate frequencies would invalidate usual adiabatic approximations as they usually assume $w_b << w_{ss}$. Upon deuteration these peaks in S(w) moved as expected but the analysis revealed that the RPO was now singly unstable, i.e., the bend motion was unstable and would therefore not be seen. As such the deuterated resonance was a 2D, RDT with only an asymmetric and symmetric stretch. Moreover as the energy decreased into the region of Neumark's first peak, the RPO became suddenly extremely unstable in all directions indicating in our way, that no resonance could exist in this region. Most interesting was that in the classical S(w) peaks c and c′ were merged and could be traced to two symmetrically placed "in channel periodic orbits" representing in both channels the HCl vibrations in the presence of the other Cl. The splitting into c and c′ in the quantum but not in the classical S(w) was clearly evidence of tunneling under the ridge, from channel to channel. Interestingly upon deuteration peak c disappears from the classical S(w) as the Franck-Condon (FC) transition regions shrinks and no longer overlaps the in-channel p.o. It is to be noted that except for line widths almost all the features of the observed resonances seen in both experiment and theory, could be and have been, explained by our CMA. Moreover many dynamic features not evident in either the previous theory or experiment have been extracted.

III-B.: FH_2

The reaction $F + H_2 \rightarrow HF + H$ is the basis of the most important and first chemical laser.[37] As such, the unveiling of the dynamics that occurs on the way from reactants to products has occupied both experimentalists and theorists in chemical physics for over a decade. It is clearly a central problem of the subject. In spite of this, as will be seen below, a complete qualitative understanding of many of the motions underlying resonances that make up the transition state has eluded theoretical and experimental investigations. On the experimental front, an incisive negative ion FH_2^- photodetachment experiment,[6] which accesses in its final state, the transition state, has seen many structures (Fig. 1) that potentially are resonances. The importance of resonances was first suggested by the backward to forward symmetry of the state-to-state differential cross sections in molecular-beam studies of the $F + H_2$ reaction and its deuterates.[32] Theoretically, quantum scattering calculations[38, 39] have been published contemporaneously that are in good accord with the photodetachment experiment, but which did not analyze the dynamic basis of any of the observed structure.

Our purpose will be to assign the transition-state resonance seen in both experiment and quantitative theory. By assignment we again mean associating each peak or series of peaks with a classical model which, when quantized, gives rise to the observed peaks.

It will be seen that all the previous theories were successful in analyzing resonance A in Fig. 1. This resonance is particularly important as it vibrationally adiabatically connects to the H_2 (v = 0) - to -HF (v = 2) population inversion. The previous theories fall into two categories. The first is the quantitative quantum-scattering-type-calculations and the second is the more qualitative and interpretive periodic obit analysis and the global adiabatic separation methods. It will be seen that these latter qualitative methods made some conceptual assumptions that were sufficiently limiting to cause them to unnecessarily miss predicting resonance C in Figure 1 and to underemphasize the role of resonances B and D.

We will model the transition-state spectra (TSS) of FH_2^-, FDH^- , and FD_2^- . The potential hypersurface used is 5SEC,[40] which unlike past surfaces gives the correct balance of final HF vibration-state populations. (For the nature of the surface and its differences from previous surfaces, see Ref. 39.) The initial state of the negative ion, ϕ_o, is taken as a Gaussian or the ground state of a harmonic oscillator in all three degrees of freedom. We have used the ab initio geometry and harmonic frequencies reported in Ref. 39. From ϕ_o (q), q being configuration-

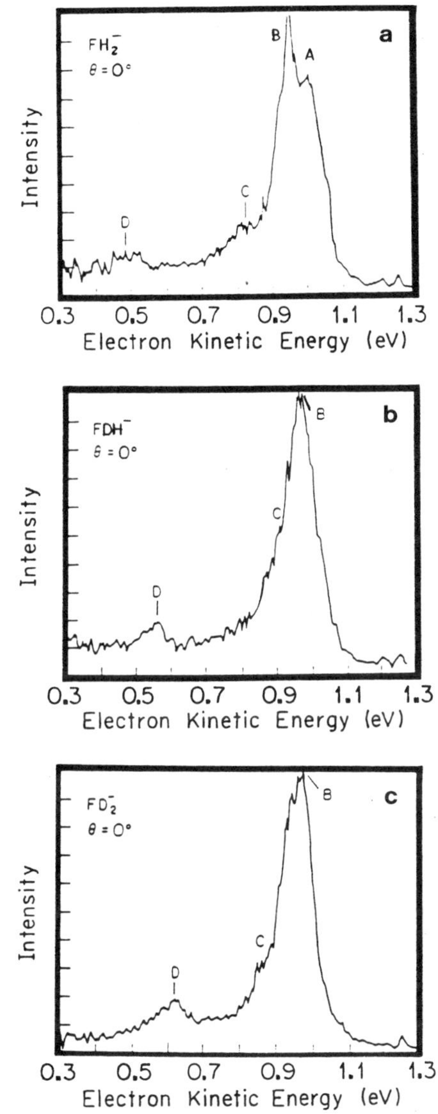

Fig. 1. (a) Photodetachment spectrum of FH_2 with the laser polarization angle $\phi = 0^0$ taken from Ref. 2. (b) Same for FDH
(c) Same for FD

space coordinates (scattering coordinates for the present work), a phase-space function $\rho(p(0), q(0))$,[23] i.e., the Wigner transform of ϕ_o (q), is constructed,

$$\rho(p(0), q(0)) = \prod_i \exp\left\{ -\frac{m_i w_i}{h} [q_i - q_i(0)]^2 - \frac{1}{m_i h w_i}[p_i - p_i(0)]^2 \right\}. \qquad (3)$$

Here the subscript i stands for each coordinate. The initial state is input into the problem via $\rho(p(0), q(0))$. Note here that as m_i increases, the area subtending the Gaussian ϕ_o and its Wigner transform $\rho(\mathbf{p}(0),\mathbf{q}(0))$ shrink in configuration space. A batch of 500 initial conditions $(\mathbf{p},(0),\mathbf{q}(0))$ is randomly selected according to the initial distribution ρ and propagated forward and backward in time for 512 fs, using the J = 0, three-dimensional (3D) Hamiltonian of the F + H_2 system. This is where the F + H_2 dynamics comes into the method. The potential is shown in Figure 2 in a collinear cut. We note it has no obvious wells or barriers to trap the trajectories. The ith trajectory yields $\mathbf{p}_j(t)$, $\mathbf{q}(t)$ which is then substituted into the now known $\rho(\mathbf{p},\mathbf{q})$ at each t to give $\rho(\mathbf{p}_i(t),\mathbf{q}_i(t))$. With this in hand, the function $S_i(w)$ of eq. 2 is created for each trajectory. We record for further use in a table of w values, whether or not trajectory i contributes strongly to a particular w value. Finally, the Monte Carlo evaluation of the ensemble $[\rho(\mathbf{p}(0), \mathbf{q}(0))]$ averaged classical spectral density S (w) is created using eq. 1. The results for FH_2^-, FDH^-, and FD_2^- are shown in Figure 3, where the arrows point to two emphasized frequencies. Note from Figure 1 that the lower frequency (dashed arrow) is the separation between peaks A and C (therefore, denoted as w_{AC}) and the higher one (solid arrow) is the separation between peaks B and D (therefore, denoted as w_{BD}). Although the positions of the dashed arrows are not clear in Figure 3 obtained from the batch of 500 trajectories, the w_{AC} peak in Fig. 3(a) does show up with a small intensity when a swarm of 1000 trajectories is used; this is shown as the inset to Fig. 3(a). Our confidence is further strengthened by noting the change in w_{BD} upon deuteration, going from Fig. 3(a) to 3(c), and by comparing to that from Fig. 1. Similar behavior for w_{AC} is not quite discernable and the reason for this will be discussed below.

At this point, we look up in the "w table" which trajectories contribute most to w_{BD}. We repeat this for w_{AC}. A generic trajectory contributing to w_{BD} is shown in Fig. 4(a) and that to w_{AC} in Fig. 4(b). Clearly, trapping is seen in both cases. w_{BD} is the in-channel frequency and w_{AC} the hydrogen exchange frequency. In each trapped region a PO search is carried out and the result is given in Fig. 4(c) and Table I. Figure 4(c) shows the BD periodic orbit at the experimental energies of peak B and peak D, and the AC periodic orbit at the energies of peak A and peak C. The transition region is also shown and it is to be noted that peak A is well to its left. For the deuterated species, since the w_{BD} appears clearly in S(w) [Figs 3(b) and 3(c)], the same procedure as above finds the PO with frequency w_{BD}. Now Figs 3(b) and 3(c) run for 500 trajectories do not show the lower frequency peak. Even a 1000 trajectory run was unconvincing. Since we saw no reason that POs A and C should disappear upon a small mass change and since we expected them to be near POs A and C for FH_2, we instituted a PO search which found the deuterated counterparts of POs A and C. Note since peak A is not seen in the deuterated species in Fig. 1, we had to estimate its probable energy for use in Fig. 4(d) and 4(e), by reducing the energy of peak C by the frequency of the PO C. As will be discussed shortly, the spatial positions of

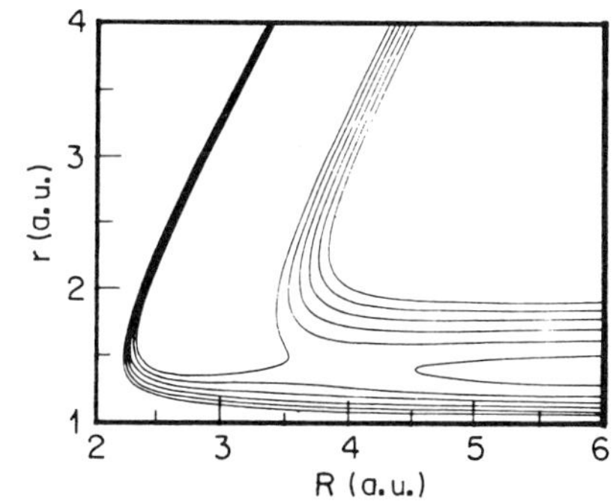

Fig. 2. The colinear cut of the 5 SEC potential, R is the distance of the F atom to the center of mass of the H_2 diatom and r is the distance between the two hydrogens. They are in a. u. The values of contours are 0.04,19.034, 049,0.64, and 0.79 eV.

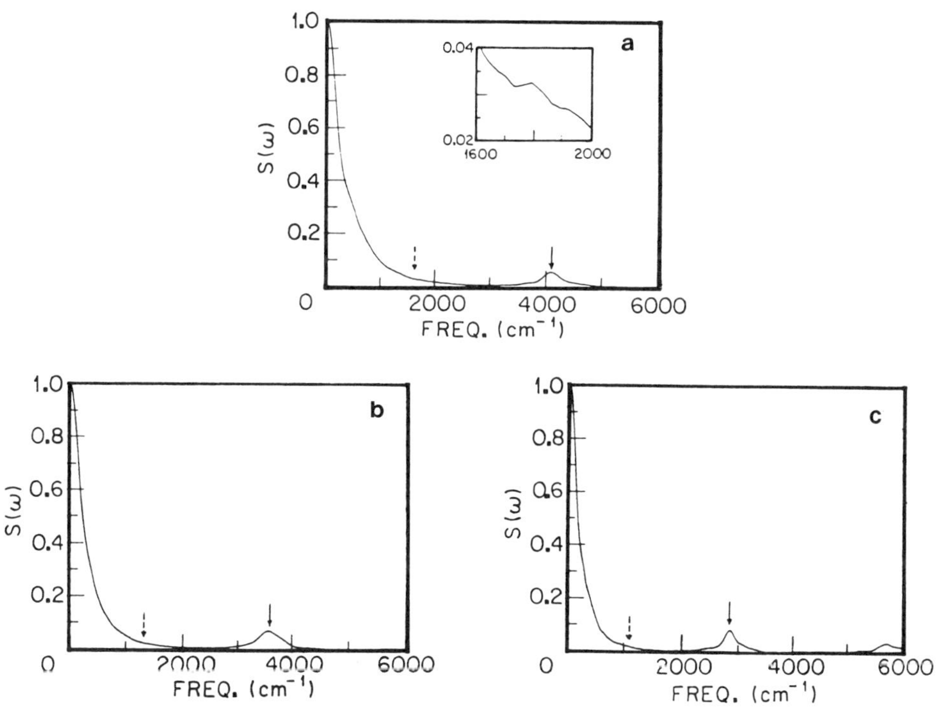

Fig. 3. (a) The classical spectral density S(w) for FH_2. The two arrows are discussed in the text. The ordinate is in arbitrary units and the abscissa in cm^{-1}. (b) Same for FDH (c) Same for FD_2

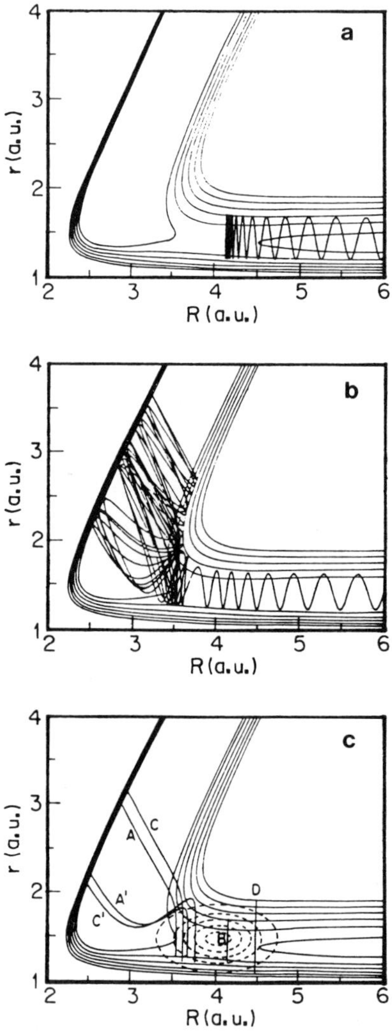
a
b
c
r (a.u.)
R (a.u.)
A
C
A'
C'
B
D

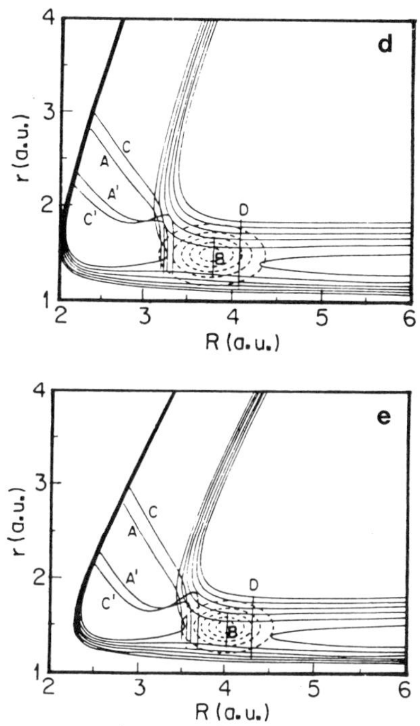

Figure 4 (a) A generic trajectory trapped in the channel at E=0.3256eV. the experimental energy of peak B (b) A generic trajectory trapped in the exchange region at E=0.2706eV, the experiment energy of peak A. (c) Two in-channel POs at thepeaks B and D experimental energies (E=0.3256 and 0.7876eV), respectively, and two H exchange POs at the experimental energies of peaks A and C (E = 0.2706 and 0.4476eV) for FH_2. The POs A,B,C, and D are single unstable, whereas the POs A' and C', at the same energies as A and C are double unstable. The values of the solid contours are the same as in Fig. 2. The dashed contours, showing the transition region, are at 10%, 30%, 50%, 70% and 90% of the center of the initial-state Gaussian. (d) Same as (c) for FDH, except that A'and C' in this figure are single unstable and the values of the potential contours are 0 .04, 0.16, 0.28, 0.40, 0.52, and 0.64 eV. (e) Same as (c) for FD_2. The potential contours are at 0.04, 0.14, 0.24, 0.34, 0.44 and 0.54 eV. The POs A' and C' are here double unstable.

the POs A and C in the deuterated cases actually explains why they do not appear in S(w) in Figs. 3(b) and 3(c) and why peak A disappears in the experiments on the deuterated species.

Can our assignment of peaks A and C being based on the H exchange POs and of peaks B and D being based on the in-channel POs be verified without any quantization? The answer is "yes" and the key lies in the isotope effect. As seen in Fig. 1, peak A disappears upon even a single deuteration and the B and D peaks move closer together. Table II shows as functions of isotope substitution w_{BD} taken from experiment and the w_{BD} of the in-channel PO at the energy half way between peaks B and D. The parallelism confirms the assignment of peaks B and D to the in-channel POs. As to the peaks A and C assignment, one notes in Fig. 4(d) drawn for FDH^- and Fig. 4(e) for FD_2^-, which are to be compared to Fig. 4(c) for FH_2^-, that the exchange PO has moved left and the transition region has, as usual with increasing mass, shrunk. As such, the exchange PO at the energy of peak A is at the edge of and poorly overlaps the transition region, and this correlates beautifully with the disappearance of peak A in Figs. 1(b) and 1(c) for FDH^- and FD_2^-, respectively. POC is seen to still lie in the transition region and peak C does not disappear.

Table 1. The energy (E), the action (S), the period (T), and the stability frequencies (w_2. and w_4) for FH_2. The energy is measured from the bottom of the H_2 asymptotic potential. The first two entries belong to the in-channel PO family and the rest to the hydrogen exchange PO family.

E (eV)	S(hartrees)	T(fs)	$w_2(cm^{-1})$	$w_3(cm^{-1})$
0.3142	0.500	8.38	406.1	(0.0, 350.3)
0.8035	1.501	8.59	463.5	(0.0, 307.4)
0.1358	3.991	20.3	390.6	(0.0, 549.1)
0.2475	4.532	19.8	361.0	(0.0, 544.7)
0.3471	5.005	19.5	344.5	(0.0, 527.2)
0.4265	5.378	19.3	334.3	(0.0, 510.7)
0.5605	6.002	19.2	319.8	(0.0, 481.5)

The fact that PO A is no longer in the transition region for the deuterated species at the energy of peak A also explains why in Figs 3(b) and 3(c) the low-frequency peak disappears. In FH_2 the $\rho(\mathbf{q}(0),\mathbf{q}(0))$ supplies trajectories that cover the energy range from peak A to peak C (thus, POs A and C) and w_{AC} is sampled at all these energies. With this w_{AC} still turns out to be hard to recognize with less than 1000 trajectories. Now for FDH and FD_2, $\rho(\mathbf{q}(0), \mathbf{p}(0)$ has a smaller unstable direction is along the reaction coordinate and clearly we have here a two-dimension near-threshold resonance associated with an adiabatic barrier.[41,42] Again the PO lies perpendicular to a barrier. Simple primitive quantization of the in-channel PO gives energies of 0.3394 and 0.8322 eV to be compared to those of 0.3256 eV for peak B and of 0.7876 eV for peak D (see Table II). In Table II, we have used zero-point energies shown in Fig. 1 or Ref. 22. Simple primitive quantization of A and C is not possible because of the double-well nature of the static potential cut along the exchange PO. We have not performed a uniform quantization and have relied on the A-C spacings and the isotope effect to verify our assignment. We do note, however, that the PO of peak A has an action that is consistent with Pollak and Wyatt's[16] assignment of a zero-node wave function over the arc of the H_2 channel and three nodes over the arc of the HF channel. Again we see that CMMA is able to extract the major dynamical features.

IV: Discussion

At this point the results here will be compared with some of the past theoretical works on the F + H_2 reactive scattering and the transition-state spectroscopy. Since over the history of this subject hypersurfaces have improved, computations have advanced from two to three dimensions and numerous interpretations have been reported, our discussion will favor comparison with 3D works on the more accurate surfaces. We shall divide the comparison into three parts. First of all there will be the comparison to exact calculations, almost all of which did not "assign' the resonances. In fact, most did not even perform time-delay[41] studies to distinguish resonance from background peaks. The second shall be comparisons to past periodic-orbit-based works. Lastly, some reference will be made to methods using global adiabatic separations.[14, 56]

Until very recently, all "exact" calculations except the work of Schatz, Bowman, and Kuppermann[43, 44] and that of Kress et al.[45] did not go to energies higher than peak B. Zhang and Miller[38] have recently published an exact calculation of the total spectra in Figs. 1(a) and 1(c); the so-called T5A surface was used. All recognize peak A in F_2^-. The works on F + HD (Ref. 46) and F + D_2 (Ref. 47) show peak A in the scattering as a small effect which is not inconsistent since the exchange PO is still there and the transition region is here irrelevant. Generally, some works use scattering results to compute either state-to-state reaction probabilities (P), cumulative reaction probabilities (CRP) for reactant channel to product channel, or the photodetachment spectra[38,39] itself. All works on F + H_2 and F + D_2 show peak B as a broad peak, but most do not suggest it as a resonance. The latest work[48] by Lynch et al., notes in the CRP two cusps which would be peaks in the energy derivative of the CRD, the so-called density of reactive states, which are clearly A and B. Unlike H + H_2 [49, 50] here both the in-channel and the exchange PO, which are called the quantized transition states (or threshold or quantized dynamical bottleneck) and trapped state, respectively, appear in the CRP. This is also clear from Pollak's early calculation of the CRP for colinear F + H_2.[51]

The majority of calculations on the surfaces 5SEC, M5, and T5a show peak B as a broad structure for all isotopes done.[44, 46,47] The calculation of Kress et al.,[45] additionally computes time delays and other measures of resonances. The problem in making a comparison of their work[45] with our work and the transition-state spectra is that except for peak A, the use of the T5A surface seems to cause the energies of the delayed states to shift just enough that we are unable to interpret their results as no dynamics or resonant topology is given. Again, 5SEC has no such problems for peaks A and B. The work of Schatz, Bowman, and Kuppermann[43] on F + H_2 uses the most primitive surface and obtains in the P, peak A and two other resonances at energies higher than TSS probes, at roughly $2w_{AC}$ and $4w_{AC}$. These are clearly exchange resonances and were assigned in terms of resonant periodic orbits (RPO) by Pollak and Child.[15] Peaks B, C, and D do not seem to show up. Whether this is due to the surface or to the fact that certain resonances are more or less prominent in different measures, i.e., CRP, P, TSS, etc., we cannot comment on.

The work of Zhang et al.[38] is certainly the most relevant for us. It shows directly by a good agreement with the TSS of FH_2^- and FD_2^- [Figs. 1(a) and 1(c)] the power of exact quantum calculations. It also shows, when compared to this work, the difficulty that exact calculation has in making assignments in the sense of Sec. I. To uncover the type of dynamics that we discuss, the wave function would have to be inspected for scars.[21] This may not be simple due to the wavefunction complexity induced by orthogonality requirements, due to problems associated with graphics in multidimension, and due to computer storage problems. The exact calculation, of course, does have the additional ability, as demonstrated in Ref. 38, to discern the contribution of each different resonant state A, B, C, or D to the final and initial asymptotic channels. This could be done classically but certainly not as well as quantum mechanically. Since this work agrees with the experiment, we will not compare our results to it, as we have already compared to the experiment. However, a qualitative point is noted, which is the disappearance of peak A in going from FH_2^- to FD_2^-. The experiment obtains no "A" for FD_2^-, while the calculation obtains a tiny residual of peak A. This fits beautifully with our results that show the exchange PO at the energy of A, which is just tangential to our limit of the transition region for the case of FD_2. We could not predict if it should appear or not but it is quite clear that, relative to FH_2^-, peak A in FD_2^- should be greatly reduced in size.

The PO analysis work in open systems and its relation to resonances goes back to the semiclassical work of Pollak and Child[15] and of Pollak[51] . For bound chaotic systems and quasicontinua very similar ideas were developed in the USC group.[1,19] Now clearly exchange POs belong to the RPO family and our in-channel orbits are (PODS) based POs. Pollak argues that in H + H_2,[52] the former are resonances and the latter are "barrier time delays" and that it would be better not to call the latter resonances. Such strong classification can have problems and require some caveats. First, peak C of F + H_2 does not fit the quantization scheme of Pollak and Child.[15] Their quantization scheme leads to the $2w_{AC}$ spacing as was seen in the scattering work of Schatz, Bowman, and Kuppermann. The $2w_{AC}$ spacing was assumed in studies by Pollak and Wyatt[16] and by Wyatt and coworkers.[17, 18]. The $2w_{AC}$ quantization goes back to the idea that the PODS mark the barriers behind which the resonance on the RPOs exist. Since peak A is below the so-called (0,3) PODS barrier and peak C, at a $1w_{AC}$ increase in energy, is above them, one assumed peak C did not exist. It does exist and is due to the local potentials around the exchange PO as are the "RPO" resonances. The difference is simply the RPOs are narrower (longer lived) because to decay they must escape two potential features, ie., the local and PODS barriers, while peak C type resonances are broad, not having the two PODS barrier to reflect its packet back toward the complex.

Are the PODS-based peaks resonances? Pollak would argue[52] that they should be better thought of as barrier caused time delays. This question of whether or not being on a barrier means it is a time delay and not a resonance, comes up in another problem that could be studied analytically, namely the hydrogen atom in an electric field. Here in the photoabsorption cross-section peaks or oscillations were seen to occur above the H-atom's ionization limit. These oscillations occurred with a frequency equal to that of a PO that went back and forth from the nucleus to the field wall. This periodic orbit was totally unstable in all directions. It was also adiabatically unstable. Yet, because the problem is separable, Grozdanov et al.[53] could solve analytically, by a second-order WKB method, for the S matrix and its poles which were found to correlate with the energies and widths of the oscillation. Also, when the classical limit of the expression for the complex pole was taken analytically, the action and stability frequency of the PO were recovered. In short, barrier POs sustain resonances. This calls into question the use of the name "resonant PO" which arose because the exchange PO was noted to cause resonances. The PODS PO supports resonances also. We prefer the names "exchange" PO for the old RPO and in-channel barrier PO or in-channel PO for the PODS states.

The message is clear, nonmotion based classifications do not really work well. Fortunately, CMMA does not depend on them. The CMMA and the phase-space structure yield for each system and experiment its own picture of the nuclear motions independent of the names given them.

Lastly, we comment on global adiabatic hyperspherical separations as was done for $F + H_2$ by Kuppermann.[54] In this work only the lowest resonance is studied, i.e., peak A for all isotopes. Does this study show the barrier resonances [Fig. 4(c)]? Note that even for in-channel motions, θ moves rapidly and R slowly. Certainly, this question has not been answered but it is clear that, in general, a single global separation will get resonances associated with some but not necessarily all important POs. Different coordinate separations need be used in different regions. For I + HI, it was shown[55] how different forms of the Jacobi coordinate in a stabilization calculation of resonant energies yield superior results to any single breakup.

In conclusion, we have presented the CMMA and have shown that for $F+H_2$ it is a simple, powerful method of analysis that is capable of extracting the motions and physics underlying transition-state and reactive scattering resonances. Again, these motions are the very ones that, when quantized and allowed to interfere, give the experimentally observed and/or theoretically computed structures in spectra and cross sections. The motions are the essence of the dynamics that chemists wish to extract from experimental data and/or theoretical simulations. As shown, it is fortunate that these motions can be extracted, for the most part, from relatively simple but judiciously designed classical mechanics.

ACKNOWLEDGEMENTS

H. S. T. thanks R. Pack and D. Neumark for pointing out that in spite of large-scale calculations and detailed experiments, no complete mechanical understanding of the FH_2 resonances existed. We thank Professor D.G. Truhlar for sending us the 5SEC potential, and Professor E. Pollak, Professor J.M. Gomez Llorente, and Professor S. Farantos for useful discussions. This work was funded by the NSF through Grants No. CHEM-8820886 and CHEM-5348155760.

REFERENCES

1. J.M. Gomez Llorente and H.S. Taylor, J. Chem. Phys., 91:953 (1989).
2. J.M. Gomez Llorente, J. Zakrzewski, K.C. Kulander, and H.S. Taylor, J. Chem. Phys., 91:1505 (1989).
3. O. Hahn, J.M. Gomez Llorente, and H.S. Taylor, J. Chem. Phys., 94:2608 (1991).
4. A. Carrington and R.A. Kennedy, J. Chem. Phys., 81:91 (1984).
5. M. Broyer, G. Delacretaz, G.-Q. Ni, R.L. Whetton, L.J. Wolf, and L. Woste, J. Chem. Phys., 90:4620 (1989); Phys. Rev. Lett., 62:2100 (1989).
6. A. Weaver and D.M. Neumark, Far. Dis. Chem. Soc., 91:5 (1991).
7. R. Schinke, K. Weide, B. Heumann, Far. Dis. Chem. Soc., 91:000 (1991).
8. K.C. Kulander, C. Cerjan, and A.E. Orel, J. Chem. Phys., 94:2571 (1991).
9. K. Yamanouchi, N. Ikeda, S. Tsuchiya, D. Jonas, J. Lindberg, G. Anderson, and R.W. Field, J. Chem. Phys.,95:6330 (1991).
10. S.F. Wu and R.D. Levine, Mol. Phys., 22:881 (1971).
11. P.G. Truhlar and A. Kuppermann, J. Chem. Phys., 56:2232 (1972).
12. G.C. Schatz and A. Kuppermann, J. Chem. Phys., 59:964 (1973).
13. G.C. Schatz, J.M.Bowman, and A. Kuppermann, J. Chem. Phys., 58:4023 (1973).
14. V.K. Babamov and A. Kuppermann, J. Chem. Phys. ,77:1891 (1982).
15. E. Pollak and M.S. Child, Chem. Phys., 60:23 (1981).
16. E. Pollak and R.E. Wyatt, J. Chem. Phys., 81: 1801 (1984).
17. C.C. Marston and R.E. Wyatt, J. Chem. Phys., 81:1819 (1984).
18. C.C. Marston and H.S. Taylor, Phys. Rev. A, 38:3732 (1988).
19. J. Zakrzewski and H.S. Taylor, Phys. Rev. A, 38:3732 (1988).
20. G. Hose, Y. Y. Bai, and H.S. Taylor, J. Chem. Phys., 80:4363 (1984).
21. E.J. Heller, Phys. Rev. Lett., 53:1515 (1984).
22. H.S. Taylor, Acct. Chem. Res., 22:263 (1989).
23. E.J. Heller, Acct. Chem. Res., 14:368 (1981).
24. W. H. Miller, J. Chem. Phys., 62:1899 (1975).
25. K. Stefanski and E. Pollak, J. Chem. Phys. 87:1079 (1987).
26. M.A. Sepulveda, R. Badii, and E. Pollak, Phys. Rev. Lett., 63:1226 (1989).
27. H. S. Taylor and A. U. Hazi, Phys. Rev. A1:1109 (1970).
28. J.H. Kim and G.S. Ezra, Proc. Adiatico Conf. Quantum Chaos, ed. H. Cerdeira and R.R. Ramaswamy (World Scientific, Hong Kong, 1991).
29. J. M. Gomez Llorente, F. Borondo, N. Berenguer, and R.M. Benito, C.P.L. in press 1992.
30. O. Hahn and H.S. Taylor, J.C.P. (1992) in press.
31. D.K. Bondi, J.N.L. Connor, J. Manz, and J. Romelt, J. Mol. Phys., 50:467 (1985).
32. D.M. Neumark, A.M. Wodtke, G.N. Robinson, C.C. Hayden, K. Shobatake, R.K. Sparkes, T.P. Schafer, and Y.T. Lee, J. Chem. Phys., 82:3067 (1985); D.M.

Neumark, A.M. Wodtke, G.N. Robinson, C.C. Hayden, K. Shobatake, R.K. Sparkes, T.P. Schafer, and Y.T. Lee, J. Chem. Phys., 82:3067 (1985).
33. R.B. Metz, T. Kitsopoulos, A. Waver, and D.M. Neumark, J. Chem. Phys., 88:1463 (1988).
34. G.C. Schatz, J. Phys. Chem., 94:6157 (1990); G.C. Schatz, J. Chem. Phys., 90:3582 (1989)
35. B. Gazdy and J.M. Bowman, J. Chem. Phys. 91:4615 (1989).
36. E. Pollak, in Theory of Chemical Reaction Dynamics, ed M. Baer (CRC Press, Boca Raton, Fl 1985) Vol. 3, P. 123.
37. J. H. Parker and G.C. Pimentel, J. Chem. Phys., 51:91 (1969).
38. J.Z.H. Zhang, W.H. Miller, A. Weaver, and D. Neumark, Chem. Phys. Lett, 182:283 (1991).
39. J.Z.H. Zhang and W.H. Miller, J. Chem. Phys. 92:1811 (1990).
40. G.C. Lynch, R. Steckler, D.W. Schwenke, A.J.C. Varandas, D.G. Truhlar, and B.C. Garret, J. Chem. Phys, 94:7136 (1991).
41. E. Pollak, M.S. Child, and P. Pechukas, J.Chem. Phys. 72:1669 (1980).
42. E. Pollak, Chem. Phys. 61:305 (1981).
43. G.C. Schatz, J.M. Bowman, and A. Kuppermann, J. Chem. Phys. 63:674 (1975).
44. G.C. Schatz, J.M. Bowman, and A. Kuppermann, J. Chem. Phys. 63:685 (1975).
45. J.D. Kress, G.A. Parker, Z Bacic, and R.T. Pack, J. Chem. Phys., 92:2344 (1990).
46. D.E. Manolopoulos, M. D'Mello, and R.E. Wyatt, J. Chem. Phys., 169:483 (1990).
47. M. D'Mello, D.E. Manolopoulos, and R.E. Wyatt, J. Chem. Phys., 168:113 (1990).
48. G.C. Lynch, P. Halrick, M. Zhao, D.G. Truhlar, C. Yu, and D.J. Kouri, and D.W. Schwenke, J. Chem. Phys. 94:7150 (1991).
49. D.C. Chatfield, R.S. Friedman, and D.G. Truhlar, Faraday Discuss, Chem. Soc., 91:289 (1991).
50. D.C. Chatfield, R.S. Friedman, D.G. Truhlar, B.C. Garret, and D.W. Schwenke, J. Am. Chem. Soc., 113:486 (1991).
51. E. Pollak, J. Chem. Phys., 75:443 (1981).
52. E. Pollak, Chem. Phys. Lett., 137:171 (1987).
53. T.P. Grosdanov, P.S. Kristic, M.J. Rakovic, and E.A. Solov'ev, Phys. Lett A., 132:262 (1988).
54. A. Kuppermann, in Potential Energy Surfaces and Dynamics Calculations, ed. D.G. Truhlar (Plenum, New York, 1981) p. 375.
55. B.B. Grayce and R.T. Skodje (unpublished).
56. J. Manz and J. Romelt, Chem. Phys. Lett., 77:172 (1981).
57. Quantum Chaos, ed. H. Cedeira and R. Ramaswamy (World Scientific, Hong Kong, 1991).

AN ANALYTIC DISCRETE VARIABLE REPRESENTATION

FOR THE COULOMB PROBLEM

R. V. Weaver*, J. T. Muckerman**, and T. Uzer*

*School of Physics, Georgia Institute of Technology
Atlanta, GA 30332-0430, USA
**Chemistry Department, Brookhaven National Laboratory
Upton, Long Island, NY 11973, USA

INTRODUCTION

Light and coworkers[1,2] have pioneered the use of so-called discrete variable representations (DVRs) in quantum mechanical problems. The basic idea of a DVR is applicable whenever two conditions are satisfied: (1) there is a basis set $\{\Psi_n(x);\ n=1,N\}$ in which to expand the wavefunction for a system, and (2) a (Gaussian) quadrature rule consisting of a set of quadrature points $\{x_n;\ n=1,N\}$ and weights $\{w_n;\ n=1,N\}$ is used to compute matrix elements in this basis.[1] Light and coworkers,[1,2] through use of a "continuous to discrete transformation" to define a DVR, have focused on the resulting set of points as the primary representation. While most of the work of Light's group has been applied to grid-method solutions to time-independent problems, the popular FFT (fast Fourier transform) method of Kosloff and Kosloff[3] for propagating wavepackets may be viewed as a discrete variable representation for time-dependent problems.

The types of Gaussian quadrature rules employed by Light and coworkers do not include the end points of the integration variable as quadrature points. This makes it difficult to apply arbitrary boundary conditions at the end points of the interval. Manolopoulos and Wyatt,[4] however, employed the Gauss-Lobatto quadrature rule, which explicitly includes the end points of the integration interval as quadrature points, in the context of a solution to the time-independent Schrödinger equation. Rather than focusing on the grid points, they constructed "Lobatto shape functions", as they termed the set of Lagrange polynomials defined by these quadrature points, which they used as the primary representation. These basis functions have by construction the Kronecker delta property at the quadrature points, and they can be differentiated analytically to evaluate the action of momentum operators on the wavefunction.

Lin and Muckerman[5] have extended the approach of Manolopoulos and Wyatt to time-dependent problems[6,7] and to the use of more general classes of Gaussian quadrature rules which include the end points of the integration interval in solving both the time-dependent and time-independent Schrödinger equations. The advantages of this analytical Discrete Variable Representation approach to time-dependent problems over the FFT method are: (1) it is not necessary to transform the wavefunction into momentum space at each step in the propagation; (2) there is no noise level limiting the accuracy of the evaluation of the action of either the kinetic or potential energy operators; (3) it is more convenient for the application of fixed-node boundary conditions at the end points; (4) we are at liberty to choose a set of basis functions that fit the potential so that fewer basis functions are needed; and (5) curvilinear coordinate systems, which have kinetic energy operators involving both coordinate and momentum operators in a single term, can be straightforwardly handled. The principal disadvantages of this approach are: (1) it scales as N^2 rather than N logN; however all the overhead of stuffing and unstuffing the input and output arrays of numerical FFT routines is totally avoided; and (2) the relevant quadrature rules tend to bunch the points near the end points which may be "wasteful", while the FFT method employs equally spaced points.

Muckerman[8] has developed a DVR method which retains the analytical property of the primary basis functions without being restricted to Gauss-Lobatto-type quadrature rules. His approach places the emphasis on incorporating convenient properties of the original basis set $\{\Psi_n(x); n=1,N\}$ into the primary basis of analytical "coordinate eigenfunctions." The idea is to choose an original basis appropriate for solutions to the Hamiltonian operator, e.g., incorporating the appropriate boundary conditions, without undue regard for the quadrature rule to be employed or the spacing of the grid points associated with it. What one then seeks is a basis of continuous coordinate eigenfunctions, $\{u_j; j=1,N\}$, and a quadrature rule consisting of a set of points $\{x_k; k=1,N\}$ and weights $\{w_k; k=1,N\}$ such that the $\{u_j(x)\}$ satisfy the Kronecker delta property, $u_j(x_k) = \delta_{jk}$, at the quadrature (grid) points.

In this article we construct an analytical DVR for the Coulomb problem which follows Muckerman's approach[8] for incorporating convenient properties of the original basis $\{\Psi_n\}$ into the basis used for the primary representation (the set of functions $\{u_n\}$). The description of time dependent ionization phenomena, including multiphoton ionization,[9,10] relies on propagating states in Coulomb (or Coulomb-like) potentials. The Coulomb potential differs from most interparticle potentials in atomic and molecular physics through its long range and its singularity at the origin. For this intrinsically multidimensional problem, it is often advantageous to use spherical coordinates[11], even though the kinetic energy becomes somewhat difficult to treat in these curvilinear coordinates. We have been interested in constructing a DVR for the unscreened Coulomb problem in connection with the suppression of multiphoton ionization[9] in superintense laser fields[10,12] ("stabilization"[13]). While the general problem involves three dimensions, for the purposes of illustration we develop here a two-dimensional model, which describes, for instance, laser-hydrogen atom interactions in which the radiation is linearly polarized so that m remains a good quantum number.[14]

GENERALITIES AND ONE-DIMENSIONAL CASES

We begin by specifying a basis set, which need not consist of orthogonal polynomials, that has certain desired properties such as fixed nodes at the end points of the coordinate interval. As above, we will denote this basis by $\{\Psi_n(x); n=1,N\}$. What we seek is a basis of "coordinate eigenfunctions" $\{u_j(x); j=1,N\}$, and a generalized quadrature rule consisting of a set of points $\{x_k; k=1,N\}$ and weights $\{w_k; k=1,N\}$ such that the $\{u_j(x)\}$ satisfy the Kronecker delta property

$$u_j(x_k) = \delta_{jk} \tag{1}$$

at the quadrature (grid) points. Let us choose a set of functions $\{\Psi_k\}$ which are particularly appropriate for our physical problem, and construct our unknown functions $\{u_j\}$ from them, according to

$$u_j(x) = \sum_{n=1}^{N} \Psi(x) \; (\Psi_n, u_j), \tag{2}$$

where (Ψ_n, u_j) denotes the integral

$$(\Psi_n, u_j) = \int \Psi_n^* \, u_j \, dx \; . \tag{3}$$

The overlap on the right can be approximated using our quadrature rule

$$(\Psi_n, u_j) \approx \sum_{k=1}^{N} w_k \; \Psi_n^*(x_k) \; u_j(x_k) \; , \tag{4}$$

and remembering requirement (1),

$$(\Psi_n, u_j) \cong w_j \; \Psi_n^*(x_j) \; . \tag{5}$$

Thus,

$$u_j \cong w_j \sum_{n=1}^{N} \Psi_n(x) \; \Psi_n^*(x_j) \; . \tag{6}$$

No matrix diagonalization is needed. This expression also emphasizes that the $u_j(x)$ are, apart from their Kronecker-δ property, defined in the entire interval. However, when integrations are needed, only their values at the grid points need to be known; thus they can be accessed from storage. Note that

$$(u_j, u_k) = \sum_{n=1}^{N} w_n \; u_j(x_n) \; u_k(x_n)$$

$$= \sum_{n=1}^{N} w_n \; \delta_{jn} \; \delta_{nk} = w_j \; \delta_{jk} \tag{7}$$

and therefore the functions

$$\Phi_j(x) = w_j^{1/2} \sum_{n=1}^{N} \Psi_n^*(x_j) \; \Psi_n(x) \tag{8}$$

form an orthonormal basis. Once these functions Φ_j have been determined, the propagation of a function is an easy matter.

Suppose the function F to be propagated is given in terms of the normalized coordinate eigenfunctions

$$\Phi_j(x) = w_j^{-1/2} u_j(x) \tag{9}$$

as

$$F(x,t) = \sum_{n=1}^{N} \Phi_n(x) \; (\Phi_n(x), \; F(x,t)) \; . \tag{10}$$

It can be shown that

$$F(x,t) = \sum_{n=1}^{N} u_n(x)\ F(x_n,t) \tag{11}$$

Then the m'th derivative is

$$\frac{d^m F(x,t)}{dx^m} = \sum_{n=1}^{N} F(x_n,t)\ u_n^{(m)}(x) \tag{12}$$

and multiple derivatives of u can be taken and stored at the grid points once and for all, since

$$\frac{d^m}{dx^m} u_n(x) \cong w_n \sum_{j=1}^{N} \Psi_n^{(m)}(x)\ \Psi_j^*(x_n) \ . \tag{13}$$

MULTIDIMENSIONAL PROBLEMS

Because of the paucity of multidimensional Gaussian integration formulas,[1] we simply use direct product bases. For each dimension in the problem, 3 sets of data must be supplied:

1. A set of N orthogonal polynomials selected from a complete set.
2. The N points for the Gaussian quadrature technique.
3. The corresponding N weights for the above points.

The procedure will be illustrated for two dimensions. With a view to later applications, let us call the two dimensions r and θ, and let time be a parameter. Let our coordinate eigenfunctions with the correct Kronecker δ property be denoted $u_j(r)$ and $v_k(\theta)$; there are N_r and N_θ of them, respectively. Then, by extension of Eq. (11) to two dimensions,

$$F(r,\theta,t) = \sum_{j=1}^{N_r} \sum_{k=1}^{N_\theta} F(r_j,\theta_k,t)\ u_j(r)\ v_k(\theta) \tag{14}$$

where r_j and θ_k are the Gaussian quadrature points in r and θ. Since we are using Gaussian integration, any function V of the coordinates (r,θ) (such as a potential energy) is diagonal in our basis, therefore

$$V(r,\theta,t)\ F(r,\theta,t)\Big|_{r_j,\theta_k} = V(r_j,\theta_k,t)\ F(r_j,\theta_k,t) \tag{15}$$

Furthermore, derivatives are easily computed:

$$\frac{\partial^2}{\partial r^2} F(r,\theta,t)\Big|_{r_j,\theta_k} = \left(\sum_{n=1}^{N_r} \sum_{\ell=1}^{N_\theta} F(r_n,\theta_\ell,t)\ u_n''(r)\ v(\theta)\right)\Bigg|_{r_j,\theta_k}$$

$$= \sum_{n=1}^{N_r} F(r_n,\theta_k,t)\ u_n''(r_j) \tag{16}$$

and again are quite simplified because they can be taken from a pre-stored matrix.

THE COULOMB PROBLEM

In spherical coordinates, the time-dependent Schrödinger equation for the Coulomb potential reads:

$$i\hbar \frac{\partial\Psi}{\partial t} = \frac{-\hbar^2}{2m}\left[\frac{1}{r^2}\frac{\partial}{\partial r} r^2 \frac{\partial\Psi}{\partial r} + \frac{1}{r^2 \sin\theta}\frac{\partial}{\partial\theta}\left(\sin\theta \frac{\partial\Psi}{\partial\theta}\right) + \frac{1}{r^2 \sin^2\theta}\frac{\partial^2\Psi}{\partial\phi^2}\right] - \frac{Z e^2}{r}\Psi \tag{17}$$

Using atomic units, taking Z=1, and defining $x \equiv \cos\theta$, $\overline{\Psi} \equiv r\Psi$, the equation becomes

$$i \frac{\partial}{\partial t}\overline{\Psi} = -\frac{1}{2}\left[\frac{\partial^2\overline{\Psi}}{\partial r^2} + \frac{1}{r^2}\left(-2x\frac{\partial\overline{\Psi}}{\partial x} + (1-x^2)\frac{\partial\overline{\Psi}}{\partial x} + \frac{1}{(1-x^2)}\frac{\partial^2\overline{\Psi}}{\partial\phi^2}\right)\right] - \frac{1}{r}\overline{\Psi} \quad . \tag{18}$$

We will build our two sets of coordinate eigenfunctions from

(a) A set of Sturmian functions constructed from generalized Laguerre polynomials $L_n^\alpha(x)$ for the radial coordinate,

(b) Legendre and associated Legendre polynomials (P_ℓ and P_ℓ^m) for the angular part (for m=0 and m≠0 cases).

Explicitly, our radial basis functions R_n are given by

$$R_n(\rho) = \frac{\Gamma(\alpha+n+1)}{\Gamma(n+1)} e^{-\rho/2} \rho^{\alpha/2} L_n^\alpha(\rho) \quad . \tag{19}$$

Note that these functions are similar to the radial solutions to the one-electron atom problem, which have $\rho = Zr/n$. In the case of the hydrogen atom solutions, the parameter α is given in terms of the angular momentum quantum number ℓ. In our calculations, it is an adjustable parameter. It is known that such a Sturmian basis set (namely one in which α is adjustable and the ρ does not depend on n) forms a complete set of discrete functions.[15,16] These functions are superpositions of the bound and continuum states of the hydrogen atom, and their overlaps with those more conventional basis states can be calculated analytically.[16]

Now define as our basis functions in r

$$R_n(r) = \frac{\Gamma(\alpha+n+1)}{\Gamma(n+1)} e^{-r/2} r^{\alpha/2} L_n^\alpha(r) \tag{20}$$

and our basis in $x = \cos\theta$ to be

$$\Theta_\ell(x) = \sqrt{\frac{(2\ell+1)}{2}\frac{(\ell-m)!}{(\ell+m)!}} P_\ell^m(x) \tag{21}$$

so our coordinate eigenfunctions u and v look as follows:

$$u_j(r) = w_j \sum_{k=1}^{N_r} R_k(r) R_k(r_j) \tag{22}$$

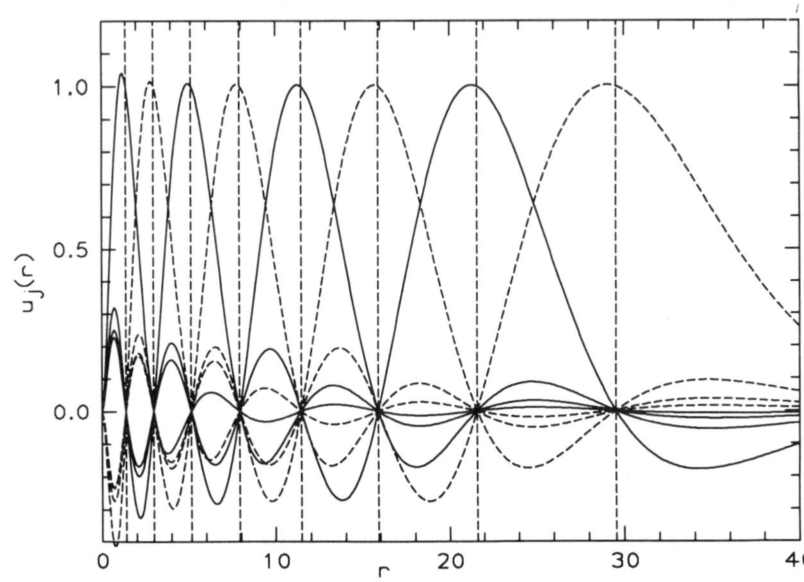

Figure 1. A DVR constructed from a basis of eight Sturmian functions with α=4. The "continuous coordinate eigenfunctions" with odd and even indices are shown as solid and dashed curves, respectively. The dashed vertical lines indicate the positions of the grid points. Note that the $u_j(r)$ exhibit the Kronecker delta property at the grid points.

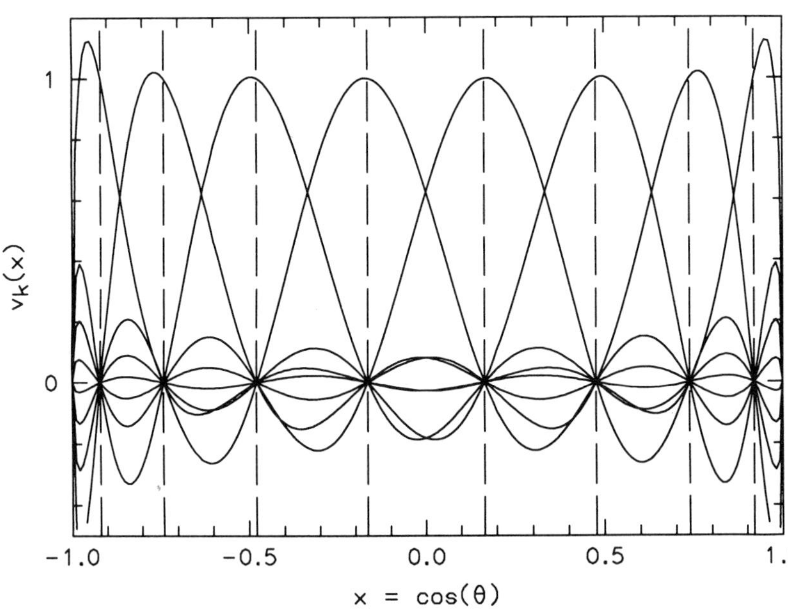

Figure 2. A DVR constructed from a basis of eight associated Legendre polynomials for m=1. The solid and dashed lines are as described in the caption to Fig. 1.

$$v_j(x) = y_j \sum_{k=1}^{N_\theta} \Theta_k(x)\ \Theta_k(x_j) \tag{23}$$

Here, $\{w_j\}$ and $\{y_j\}$ are weights that can be found from the forms of R and Θ. Recall that m is conserved. Thus, if m=1, these v's would be

$$v_j(x) = \frac{w_j}{2} \sum_{\ell=1}^{N_\theta} \frac{2\ell+1}{\ell(\ell+1)}\ P_\ell^1(x)\ P_\ell^1(x_j) \quad . \tag{24}$$

These functions are shown in Figures 1 and 2.

Now $\overline{\Psi}$ may be written as

$$\overline{\Psi} = \sum_{j=1}^{N_r} \sum_{k=1}^{N_\theta} \overline{\Psi}(r_j,x_k,t)\ u_j(r)\ v_k(x) \tag{25}$$

$$= \sum_{j=1}^{N_r} \sum_{k=1}^{N_\theta} \Psi(r_j,x_k,t)\ u_j(r)\ y_k \sum_{\ell=1}^{N_\theta} \phi_\ell(x)\ \phi_\ell(x_k) \quad . \tag{26}$$

Let us consider the angular momentum term

$$\frac{1}{r^2}\left(-2x\frac{\partial\overline{\Psi}}{\partial x} + (1-x^2)\frac{\partial^2\overline{\Psi}}{\partial x^2} - \frac{m^2\ \overline{\Psi}}{(1-x^2)} \right) \equiv -\frac{1}{r^2}\ L^2\ \overline{\Psi} \tag{27}$$

$$= \frac{1}{r^2} \sum_{j=1}^{N_r} \sum_{k=1}^{N_\theta} \overline{\Psi}(r_j,x_k,t)\ u_j(r)\quad L_k(x) \tag{28}$$

where

$$L_k(x) = -\ y_k \sum_{\ell=1}^{N_\theta} \ell(\ell+1)\ \phi_\ell(x)\ \phi_\ell(x_k) \quad . \tag{29}$$

Thus, at the grid point (r_a,x_b) the DVR form of the unscreened Coulomb problem becomes

$$i\frac{\partial}{\partial t}\ \overline{\Psi}(r_a,x_b,t) = \sum_{j=1}^{N_r}\sum_{k=1}^{N_\theta} \overline{\Psi}(r_j,x_k,t) \left[-\frac{1}{2}\left(u_j''(r_a)v_k(x_b) + \frac{L_k(x_b)}{r_a^2}\ u_j(r_a)\right) - \frac{1}{r_a}\ u_j(r_a)\ v_k(x_b) \right] \quad . \tag{30}$$

And if there is an external interaction (such as the coupling to a radiation field) $V_{ext}(r,\theta,t)$, then the fundamental equation to be propagated becomes

$$i\frac{\partial}{\partial t}\ \overline{\Psi}(r_a,x_b,t) = -\frac{1}{2}\sum_{j=1}^{N_r} \overline{\Psi}(r_j,x_b,t)\ u_j''(r_a)$$

$$-\frac{1}{2}\sum_{k=1}^{N_\theta} \overline{\Psi}(r_a,x_b,t)\ \frac{L_k(x_b)}{r_a^2}$$

$$- \frac{\overline{\Psi}(r_a,x_b,t)}{r_a} + V_{ext}(r_a,x_b,t)\ \overline{\Psi}(r_a,x_b,t) \tag{31}$$

which defines the wavefunction with the help of precomputed numbers for u", L, and V at the grid points.

Using the above formula we may calculate $\frac{\partial\Psi}{\partial t}$ at every grid point. We then use a simple point-slope method of propagation, i.e.,

$$\Psi(t+2\Delta t) = \Psi(t) + 2\Delta t\ \frac{\partial\Psi(t+\Delta t)}{\partial t} \tag{32}$$

so all we need to start the integration is $\Psi(t=0)$ and $\Psi(t=\Delta t)$. We can estimate $\Psi(\Delta t)$ using a second order Runge-Kutta method by first calculating

$$\Psi\left(\frac{\Delta t}{2}\right) \cong \Psi(0) + \frac{\Delta t}{2} \left.\frac{\partial\Psi}{\partial t}\right|_{t=0}$$

then

$$\Psi(\Delta t) \cong \Psi(0) + \Delta t \left.\frac{\partial\Psi}{\partial t}\right|_{t=\frac{\Delta t}{2}}$$

APPLICABILITY TO HYDROGEN ATOM RYDBERG STATES

In this section we test the applicability of the DVR based on Sturmian and Legendre functions as described above to the calculation of hydrogen atom Rydberg states. This exercise may be viewed as a necessary prelude to the use of this DVR in studying the dynamics of hydrogen atom Rydberg states in intense laser fields.[14] Our objective for this basis here is the accurate description of the energy levels of low-lying Rydberg states from a time-independent calculation using the Hamiltonian from the right hand side of (18) and the direct-product basis $\{u_j(r)\ v_k(x)\}$.

We employ 64 Sturmian functions with $\alpha = 4$ in the original radial basis, and 16 Legendre polynomials in the original angular basis for the case of m = 0 states of the hydrogen atom. This radial basis is most appropriate for d states, and should be especially poor for s states. The resulting DVR is a grid of 1024 (64x16) points. The Hamiltonian matrix element between each pair of direct-product basis functions is readily evaluated using the properties of the DVR basis at the grid points. No attempt was made to factor the Hamiltonian matrix according to different eigenvalues of angular momentum. When the resulting Hamiltonian matrix is then diagonalized, 149 of the 1024 eigenvalues are negative, i.e., correspond to bound states. The remaining 875 positive eigenvalues correspond to discrete states useful for describing the ionization continuum of the hydrogen atom.

Table 1 presents a comparison of the lowest 32 eigenvalues resulting from this calculation with the exact energy levels of the hydrogen atom. As expected, the DVR basis does not represent the 1s state well, but it does an excellent job of representing the low-lying Rydberg states up through n=7. These are, in fact, the most important states for treating the dynamics of, e.g., the $3d_0$ Rydberg state in an intense infrared laser field.

Table I. Comparison of lowest 32 hydrogen atom energy levels for m=0 computed with 1024 point (64 x 16) DVR basis with exact values.

n	exact levels	ns levels	np levels	nd levels	nf levels	ng levels	nh levels
1	-0.500000	-0.441136					
2	-0.125000	-0.117273	-0.125000				
3	-0.055556	-0.053229	-0.055556	-0.055556			
4	-0.031250	-0.030262	-0.031251	-0.031251	-0.031251		
5	-0.020000	-0.019500	-0.020008	-0.020006	-0.020005	-0.020002	
6	-0.013889	-0.013631	-0.013922	-0.013918	-0.013912	-0.013905	-0.013898
7	-0.010204		-0.010520	-0.010465	-0.010410	-0.010337	-0.010253
8	-0.007813		-0.010502	-0.010307	-0.010292	-0.010273	-0.010245
9	-0.006173		-0.010317				

ACKNOWLEDGEMENTS

This research was carried out at the Georgia Institute of Technology with the support of the U. S. National Science Foundation grant No. CHE89-22420, and at Brookhaven National Laboratory under Contract No. DE-AC02-76H00016 with the U. S. Department of Energy and supported by its Division of Chemical Sciences.

REFERENCES

1. J. C. Light, I. P. Hamilton and J. V. Lill, J. Chem. Phys. 82, 1400 (1985).
2. I. P. Hamilton and J. C. Light, J. Chem. Phys. 84, 306 (1986); Z. Bacic and J. C. Light, Annu. Rev. Phys. Chem. 40, 469 (1989); S. E. Choi and J. C. Light, J. Chem. Phys. 90, 2593 (1989).
3. D. Kosloff and R. Kosloff, J. Comp . Phys. 52, 35 (1983).
4. D. E. Manolopoulos and R. E. Wyatt, Chem. Phys. Lett. 152, 23 (1988).
5. F. J. Lin and J. T. Muckerman, Comp. Phys. Commun. 63, 538 (1991).
6. C. Leforestier, R. H. Bisseling, C. Cerjan, M. D. Feit, R. Friesner, A. Guldberg, A. Hammerlich, G. Jolicard, W. Karrlein, H.-D. Meyer, N. Lipkin, O. Roncero, and R. Kosloff, J. Comp. Phys. 94, 59 (1991).
7. C. Cerjan and K. C. Kulander, Comp. Phys. Commun. 63, 529 (1991).
8. J. T. Muckerman, Chem. Phys. Lett. 173, 200 (1990).
9. K. C. Kulander, Phys. Rev. A 35, 445 (1987).
10. C. Cerjan, J. Opt. Soc. Am. B 7, 680 (1990).
11. C. E. Dateo, V. Engel, R. Almeida, and H. Metiu, Comp. Phys. Commun. 63, 435 (1991); C. E. Dateo and H. Metiu, J. Chem. Phys. 95, 7392 (1991).
12. J. H. Eberly, J. Javanainen, and K. Rzazewski, Phys. Rep. 204, 331 (1991).
13. Q. Su and J. H. Eberly, J. Opt. Soc. Am. B 7, 564 (1990).
14. K. C. Kulander, K. J. Schaefer, and J. L. Krause, Phys. Rev. Lett. 66, 2601 (1991).
15. R. Blümel and U. Smilansky, Z. Phys. D 6, 83 (1987).
16. R. V. Jensen, S. M. Susskind, and M. M. Sanders, Phys. Rep. 201, 1 (1991).

NOTE ADDED IN PROOF

In their calculation of bound and quasi-bound states of ArHCl, S. E. Choi and J. C. Light (J. Chem. Phys. 92, 2129 (1990)) used Sturmian vibrational functions, the usefulness of which for describing stretching motions had been established by E. Yurtsever, O. Yilmaz and D. D. Shillady (Chem. Phys. Lett. 85, 111 (1982)). These Sturmians differ from ours [Eq. (20)] in their restriction of α to odd integers. As mentioned in our text, in our case, α is arbitrary and for the calculation in Table 1 was equal to 4.

A MECHANISM FOR ULTRAFAST DYNAMICS IN CURVE CROSSING SYSTEMS

J. Broeckhove, B. Feyen, L. Lathouwers and P. Van Leuven

Universitair Centrum Antwerpen
Departement Wiskunde en Informatica
Groenenborgerlaan 171, B2020 Antwerpen, Belgium

1 INTRODUCTION

Recent experimental femtosecond techniques have made it possible to perform time resolved analyses of intramolecular dynamics. The theoretical counterpart of such studies necessitates the solution of the time-dependent Schrödinger equation. Developments in numerical methods now allow the propagation of molecular wavepackets over long time intervals. Given the electronic potential surfaces and their interaction matrix elements, one can study the motion of the nuclei in the molecule to gain insight in the physical processes of intramolecular dynamics. In particular for diatomics, where the problem is essentially a one-dimensional one, computations can be performed with high accuracy. A particularly interesting situation is that of two potential curves with an avoided crossing. In that case the time evolution of the system can be described by following the two components of the wave function as they propagate on the potentials and transfer probability through the coupling which is effective in the crossing region. This transfer mechanism is sensitive to the topology of the potential curves. In this paper we investigate the essential features of such a mechanism and how they are influenced by peculiarities of the crossing topology.

2 SPLIT-OPERATOR METHOD

The numerical scheme we use to efficiently and accurately evaluate time evolution is based on the succesive application of the short time propagator, i.e. the time evolution $\psi_0 \rightarrow \psi_t = U(t)\psi_0$, generated by the quantum propagator

$$U(t) = \exp(-itH) = \exp(-it(T+V)) \tag{1}$$

Time-Dependent Quantum Molecular Dynamics, Edited by J. Broeckhove and L. Lathouwers, Plenum Press, New York, 1992

(we use $\hbar = 1$, T stands for kinetic energy and V for potential energy operator) is broken up into short time slices

$$U(t) = \prod_1^N U(\varepsilon), \quad t = \varepsilon N \tag{2}$$

with N large, ε small. This allows one to use an approximate short time propagator for $U(\varepsilon)$ which, in the "split operator scheme", is given by

$$U(\varepsilon) \approx \exp\left(-i\frac{\varepsilon}{2}V\right)\exp(-i\varepsilon T)\exp\left(-i\frac{\varepsilon}{2}V\right) \tag{3}$$

and is correct up to second order in ε [4]. The usefulness of the algorithm lies in the splitting of the V and T terms. The action of the first is evaluated while the wavefunction ψ is in a coordinate representation, i.e. defined by its values on a grid in coordinate space. For the second, one transforms the wavefunction to the momentum representation, using the discrete Fourier Transform associated with the grid. The T operator is multiplicative in that representation, so its action is easy to evaluate. Afterwards one transforms ψ back to the coordinate representation with an inverse Fourier Transform. This is a viable approach because of the availability of Fast Fourier Transform software that perform these transformations very efficiently.

The above scheme is readily applied to the nuclear dynamics on a single electronic potential energy surface. Fast Fourier Transform methods have also been extended to processes in which the non-adiabatic coupling between two electronic potential energy surfaces is active. In such situations the time evolution is more involved because the propagation along a surface competes and combines with the transitional dynamics between the two surfaces. We have proposed an improvement to the two-state split-operator formalism by using analytic exponentiation of two by two matrices [5]. This leads to a significant reduction in computational effort. It allows us to push the evolution to longer times and to investigate the dynamics of vibrational states in curve crossing systems, where the characteristic features develop on a much longer time scale than the collisional events considered in [6, 7]. We first outline our approach.

The Hamiltonian for a two-curve crossing system may be referred either to the adiabatic or to the diabatic electronic basis. In the latter we can write

$$H^d = T^d + V^d = \frac{1}{2m}\begin{pmatrix} -\frac{d^2}{dR^2} & 0 \\ 0 & -\frac{d^2}{dR^2} \end{pmatrix} + \begin{pmatrix} V_{11}(R) & V_{12}(R) \\ V_{21}(R) & V_{22}(R) \end{pmatrix} \tag{4}$$

for the Hamiltonian operator. Because the weak dependence of the diabatic electronic states on the nuclear coordinate R is neglected, the kinetic energy is diagonal in this basis.

In the adiabatic basis the potential energy matrix is diagonalized, with Born-Oppenheimer potentials on the diagonal, while the nuclear kinetic energy has R-dependent non-adiabatic coupling terms through the matrix element $g = \langle 1|\frac{d}{dR}|2\rangle$ between the two electronic states The transformation between both bases is a two-state rotation

$$\begin{pmatrix} \chi_1^a \\ \chi_2^a \end{pmatrix} = \begin{pmatrix} \cos\theta & \sin\theta \\ -\sin\theta & \cos\theta \end{pmatrix}\begin{pmatrix} \chi_1^d \\ \chi_2^d \end{pmatrix} \tag{5}$$

The diabatic kinetic energy and the adiabatic potential energy operators are diagonal and can readily be exponentiated. This observation has lead to the scheme proposed in [6], whereby the kinetic energy propagator is evaluated in the diabatic-and-momentum representation and the potential energy propagator is evaluated in the adiabatic-and-coordinate representation. This, however, necessitates two rotations (5) between adiabatic and diabatic states for every time step in addition to the Fourier transforms.

Table 1. Potential parameters (a.u.)

$T_1 = 0.475$	$D_1 = 0.248$	$\beta_1 = 1.602$	$R_1 = 2.116$
$T_2 = 0.472$	$D_2 = 0.169$	$\beta_2 = 0.673$	$R_2 = 2.731$
	$D_{12} = 0.004$	$\beta_{12} = 0.500$	$R_{12} = 2.299$

One can avoid these transformations by using direct analytical exponentiation for an hermitian two by two matrix as given in [2]

$$\exp\left(i\sum_0^3 v_j\sigma_j\right) = \exp(iv_0)\left[\sigma_0\cos|v| + i\left(\sum_1^3 \sigma_j v_j\right)\frac{\sin|v|}{|v|}\right] \tag{6}$$

$$|v| = \sqrt{\sum_1^3 v_j^2} \tag{7}$$

where the σ_j are the usual Pauli matrices. This formula can straightforwardly be applied to the exponentiation of the non-diagonal potential energy matrix in the diabatic representation by rewriting V^d as a sum of Pauli matrices

$$V^d = \frac{1}{2}[V_{11} + V_{22}]\sigma_0 + V_{12}\sigma_1 + \frac{1}{2}[V_{11} - V_{22}]\sigma_3 \tag{8}$$

This allows one to calculate the potential as well as the kinetic energy propagator in the diabatic representation and apply the split operator scheme without alternating between the adiabatic and diabatic representation.

3 VIBRATIONAL DYNAMICS

We shall start from the realistic case of the non-adiabatically coupled electronic states b', c' with ${}^1\Sigma_u^+$ symmetry in the N_2 molecule. In reference [3] the deperturbed spectroscopic constants for these states were derived using an iterative fitting scheme with Morse potentials. We have used these deperturbed potentials, V_{11} and V_{22}, in our calculations. Since physically speaking there can be no coupling at infinity, we have replaced the constant V_{12} by a Gaussian centered on the crossing point $R = R_{12}$ of the diabatic potentials. The diabatic coupling must have a weak dependence on R which implies that the width of this Gaussian is very large. We have used the following expressions for the potentials involved

$$V_{11} = T_1 + D_1[1 - \exp -\beta_1(R - R_1)]^2 \tag{9}$$

$$V_{22} = T_2 + D_2[1 - \exp -\beta_2(R - R_2)]^2 \tag{10}$$

$$V_{12} = D_{12}\exp -\beta_{12}(R - 12)^2 \tag{11}$$

The values of the parameters are given in table 1. The diabatic potentials with their energy levels are shown in figure 1.

We assume that initially a wave packet is created on the repulsive part of the inner diabatic well by a vertical or Franck-Condon excitation out of the $X\ {}^1\Sigma_g^+$ ground state. The data for the ground state were taken from [8]. To a good approximation the ground state wave function can be represented by a Gaussian wave packet

$$\chi_1^d(t=0) = (2\pi w^2)^{-\frac{1}{4}}\exp -\frac{(R - R_0)^2}{4w^2} \tag{12}$$

$$\chi_2^d(t=0) = 0 \tag{13}$$

with $R_0 = 2.074, w = .06$ in a.u.

The following scenario then develops for the time evolution of this system. The initial wave packet, created in state 1 by the Franck-Condon transition, has an energy in between the bottom of the well and the crossing region. By sliding down the potential wall the wave packet moves outwardly. When it enters the crossing region, the coupling becomes effective and transfers wave packet amplitude to the outer well. In this way the amplitude χ_2^d builds up. Because we are in an energy regime where no dissociation can take place, the wavepacket reflects from the outer potential wall and after a certain recurrence time the whole process repeats quasi periodically. Within the time interval of one recurrence period the details of the transfer mechanism, in particular the phase relation between the two components of the wavefunction show up as oscillatory behaviour.

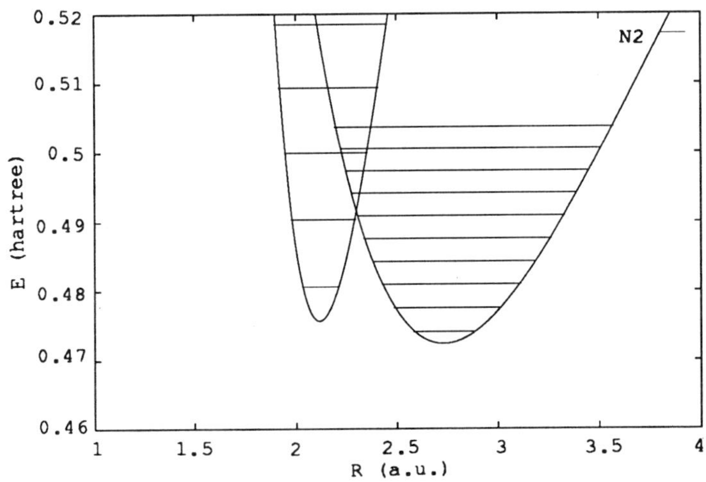

Figure 1. The diabatic potentials and energy levels of N_2

In order to study the transitional dynamics in the adiabatic picture, the probability to find the system in the lower adiabatic state and to the right of the crossing point has been suggested

$$P_1^a(R > R_c)(t) = \int_{R_c}^{+\infty} |\chi_1^a(R,t)|^2 dR$$

If the propagation is performed in the diabatic representation, the probability of occupying the second diabatic state suggests itself as an alternative :

$$P_2^d(t) = \int_{-\infty}^{+\infty} |\chi_2^d(R,t)|^2 dR$$

In all our curve crossing situations no significant difference was observed in the behaviour of the above two probabilities over a 1 ps time period. We have chosen to present the results for P_2^d.

In figure 2 we show the time evolution of P_2^d. We can distinguish three time scales : the overall recurrence period of the order of 100 fs, an oscillatory behaviour with period of the order of 15 fs and a further fluctuations on a still smaller time scale.

4 DEPENDENCE ON POTENTIAL TOPOLOGY

In order to study the stability of the probability transfer pattern, described in the previous section, against variations of the potential energy curve topology we have performed a series of time evolution calculations in which we varied the parameters of V_{22}.

Variation of T_2

By varying T_2 one shifts the right-hand diabatic potential up and down relative to the left hand one. In figure 3 we show P_{max}, the maximum value of P_2^d (in a time interval of 1 ps)

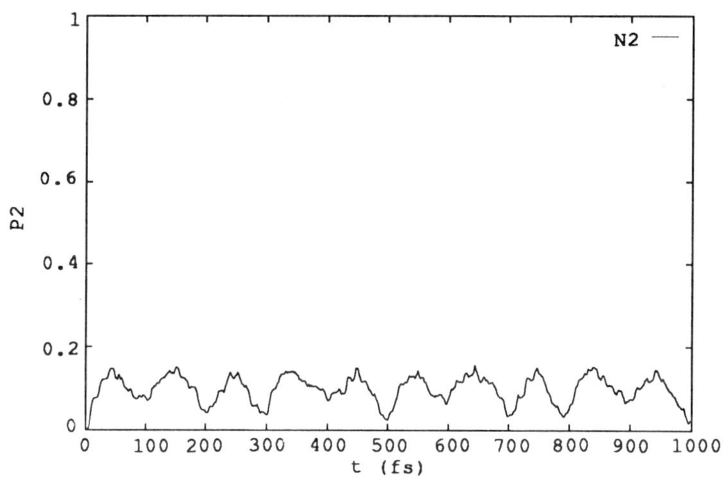

Figure 2. Time evolution of the occupation probability of state 2 for N_2

as T_2 ranges from 0.45 au to 0.50 au. There is an overall downward slope as T_2 increases. This is to be expected : as the right-hand side potential is lifted it becomes energetically more difficult to transfer probablity to it. However, the phenomenon is not a monotonic one. Superimposed on the general downward slope is a series of sharp maxima that mark the values of T_2 where the transfer mechanism is extremely effective. These resonant peaks are approximately equidistant with a spacing that corresponds to the vibrational frequency of the right-hand well. This suggests that the probability transfer depends strongly on the relative position of the energy levels of the two diabatic Morse wells.

In figure 4 we show the potentials with their respective vibrational levels for one of the resonant cases. We see that the situation is one in which each of the left-hand levels is quasi degenerate with a level on the right hand. Due to the approximate three-to-one ratio between the energy spacings in left- and right-hand wells, this situation is repeated each time T_2 is incremented by the harmonic frequency of the right-hand well.

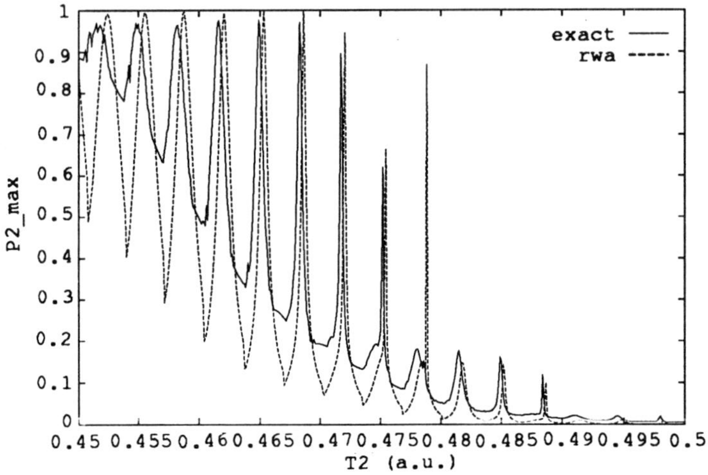

Figure 3. Maximum occupation probability of state 2 as a function of T_2

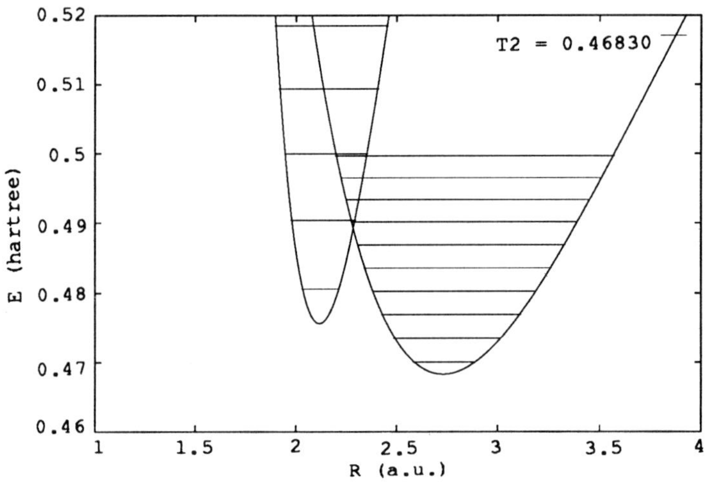

Figure 4. The diabatic potentials and energy levels of a resonant T_2-case

When the time evolution of P_2^d at such a resonance value of T_2 (figure 5) is compared with the reference case (figure 2) one notices that indeed P_{max} has increased but also the time scale has lengthened.

We conclude that the occurrence of near-degeneracies in the (diabatic) spectra is a controlling factor for the probablity transfer.

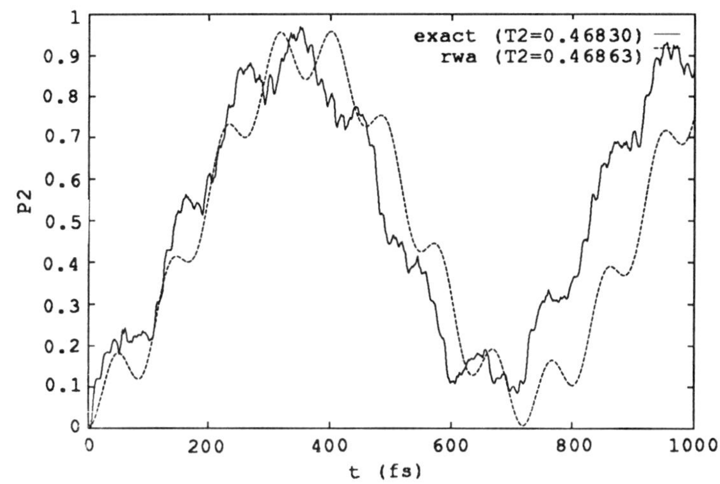

Figure 5. Time evolution of the occupation probability of state 2 for a resonant T_2-case

Variation of D_2

By varying D_2 one changes the dissociation energy of the right well, keeping the total number of bound states constant. In this way one stretches or compresses the vibrational spectrum of V_{22}. Clearly this procedure will also influence the degeneracy between pairs of levels of the left- and right-hand potential wells. In figure 6 we give the P_{max}, as defined before, versus D_2. Again there is a general downward trend as D_2 increases. When D_2 is small the V_{22} spectrum is dense and near-degeneracy for a number of levels is likely.

As before there are a number of peaks superimposed on the general trend. If we examine more closely the potential topology for a resonant case (figure 7), we observe a pair of levels whose classical turning points nearly coincide with the crossing point. such a situation can lead to large values of the coupling matrix element between the diabatic states and hence to large probablilty transfer even though the levels are not exactly degenerate.

Thus a second controlling factor for the build up of transfer probability is the overlap of left- and right-hand diabatic eigenfunctions. In figure 8 we show the time evolution of P_2^d at the D_2 value of figure 7. The same effects of increasing amplitude and time scales as in figure 5 are observed.

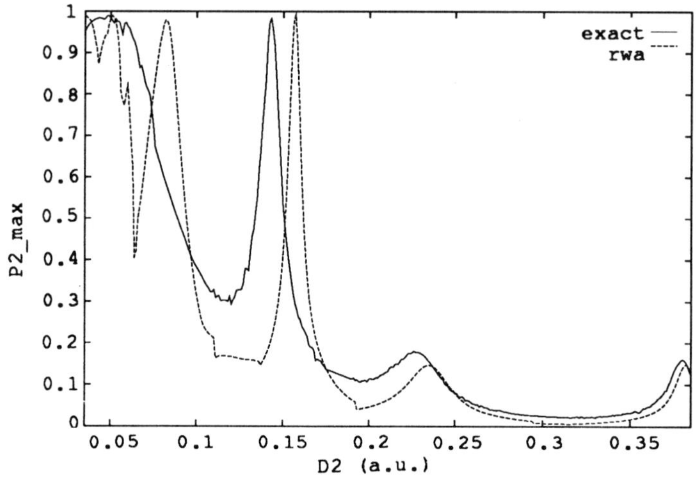

Figure 6. Maximum occupation probability of state 2 as a function of D_2

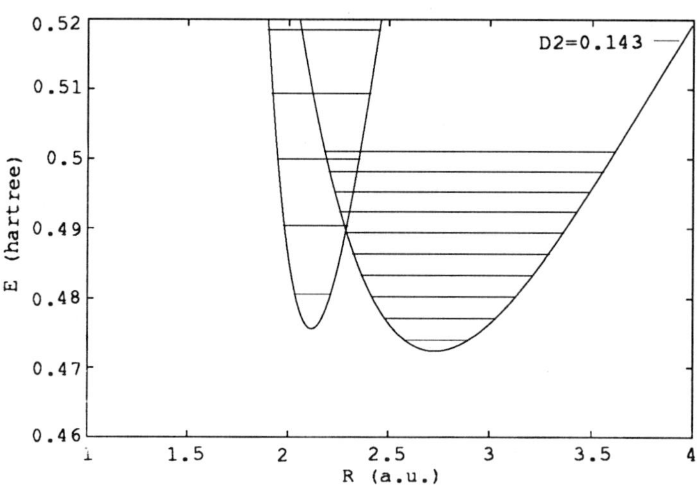

Figure 7. The diabatic potentials and energy levels of a resonant D_2-case

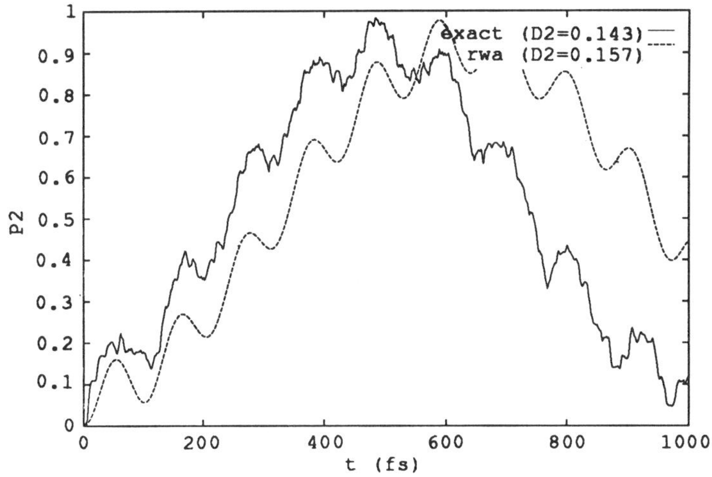

Figure 8. Time evolution of the occupation probability of state 2 for a resonant D_2-case

Variation of β_2

Because the harmonic frequency of the Morse potential is given by a combination of the parameters $\omega = \sqrt{D/m}\beta$, we expect the variation of β_2 to produce the same effects as the D_2 variation on a scale which is square root of the D_2 scale. Upon calculation we found this indeed to be the case (see figure 9).

From the above graphical material it is clear that the degeneracy pattern between the ground state and the first excited state of V_{11} and the spectrum of V_{22} strongly influences the transfer probablity. Since the initial state originates through a Franck-Condon excitation of the vibrational ground state of N_2, it is of interest to see how this state is distributed over the V_{11} spectrum. By performing a time evolution with the coupling between the diabatic states set equal to zero, we obtain from the auto-correlation function

$$A(t) =< \chi_0 \mid \chi_t > \tag{14}$$

where χ_0 is the intitial state, the Fourier transform

$$A(\omega) = \sum_n |< \chi_0 \mid \psi_n >|^2 \delta(\omega - \omega_n) \tag{15}$$

This yields the decomposition of the initial state in the eigenbasis of the left-hand well. The occupation probablities of χ_0 on the ground, first and second excited state respectively are about 85%, 14% and 1%. The final outcome of the transfer dynamics depends on the combination of these occupation probablities and the spectral properties of the topology.

5 MULTILEVEL ROTATING WAVE APPROXIMATION

The numerical experiments of the previous section indicate that an approximation scheme aimed at explaining the mechanism of the probablity transfer in curve crossing systems should focus on two features : the degeneracy of the (diabatic) spectra and the coupling matrix elements. this suggests that we apply a multilevel version of the familiar Rotating Wave approximation (RWA). We proceed as follows : we expand the exact time dependent diabatic components as

$$\chi_1^d(t) = \sum_m c_m^{(1)}(t)\psi_m^{(1)} \exp(-iE_m^{(1)}t) \tag{16}$$

$$\chi_2^d(t) = \sum_n c_n^{(2)}(t)\psi_n^{(2)} \exp(-iE_n^{(2)}t) \tag{17}$$

where $E_m^{(1)}$,$E_n^{(2)}$ and $\psi_m^{(1)}$,$\psi_n^{(2)}$ are the uncoupled diabatic spectrum and their respective eigenstates. Substitution in the time dependent Schrödinger equation yields the evolution equations for the coefficients

$$i\frac{dc_m^{(1)}(t)}{dt} = \sum_k c_k^{(2)}(t)V_{mk} \exp i\omega_{mk}t \tag{18}$$

$$i\frac{dc_n^{(2)}(t)}{dt} = \sum_l c_l^{(1)}(t)V_{ln} \exp i\omega_{ln}t \tag{19}$$

where $\omega_{mk} = E_m^{(1)} - E_k^{(2)}$ and $V_{mk} =< \psi_m^{(1)}|V_{12}|\psi_k^{(2)} >$ are the coupling matrix elements. We now apply the multilevel RWA (MRWA), i.e. we retain in (18) and (19) only the low frequency components. The contribution of a single quasi-degenerate pair (m, n) to the transfer probablity can then be expressed as [1]

$$P_{(m,n)}(t) = |c_n^{(2)}(t)|^2 = |c_m^{(1)}(0)|^2 \frac{V_{mn}^2}{\Omega_{mn}^2} \sin^2\Omega_{mn}t \tag{20}$$

where

$$\Omega_{mn} = \sqrt{\omega_{mn}^2/2 + V_{mn}^2} \tag{21}$$

The probablity of finding the wave packet in the right hand potential well can then be written

$$P_2^d(t) = \sum_n |c_n^{(2)}(t)|^2 \approx \sum_{(m,n)} P_{(m,n)}(t) \tag{22}$$

In the second step we have made the approximation of limiting the sum to the quasi degenerate pairs only. It is clear that the controlling factors that emerged from the numerical studies described above appear in our final formula. In the T_2 variation where quasi-degeneracy for all occupied levels can be achieved (i.e.$\Omega_{mn} \approx 0$) the V_{mn} don't matter much as far as determining P_{max}. One has $P_{mn}(t) = |c_m^{(1)}(0)|^2\sin^2(|V_{mn}|t)$ and the P_{max} peaks get very close to amplitude one. When such high degree of degeneracy cannot be achieved, as in the case of the D and β variations, the V_{mn} do matter. The formula also explaines the lenghtening of the dominant time scale in the P_2^d time evolution at resonance.

We now compare the results of the multilevel rotating wave approximation with the exact time evolutions over the whole parameter range of T_2 and D_2. Because the initial wave packet contains only three significant components, we restict the above sum to three terms corresponding to m=1, 2 and 3 and their nearest partners in the right hand spectrum. The comparison is shown on figures 3, 5, 6, 8 and 9. We can see that the overal resonance structure of P_{max} is rendered well. In the time evolution picture we distinguish the main recurrence profile which derives from the ground state pair while the modulation is due to the first excited state. The overall agreement is satisfactory to get a qualitative insight in the mechanism of the probablity transfer dynamics.

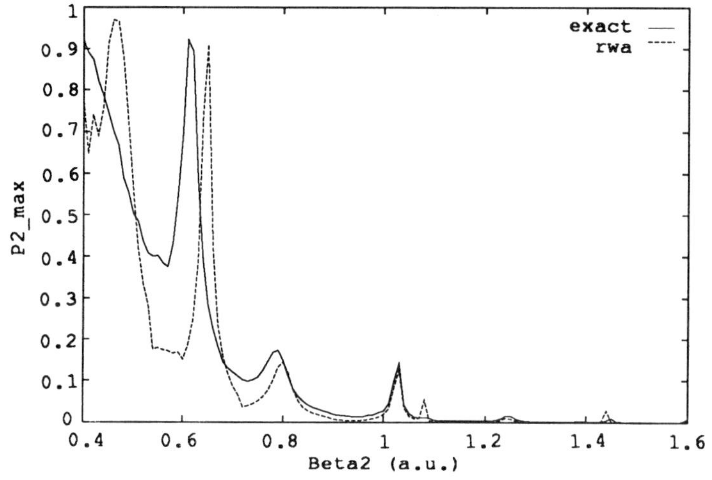

Figure 9. Transfer as a function of β_2

6 CONCLUSIONS

Analysis of the transfer probablity between two diabatic states in terms of the stationary spectral properties of the potentials shows that two factors govern the occurrence of superstrong transitions. On the one hand the topology of the crossing which can produce accidental degeneracies between the two spectra as well as strong coupling matrix elements and on the other hand the composition of the initial state. Maximum probablity transfer is obtained when the larger components in the expansion of the initial wave function on the eigenstates of the left hand potential are those for which there is a match with the levels of the right hand potential.

The rotating wave approximation, generalized to the case of two potentials instead of two levels, renders well the qualitative features of the dynamics.

References

[1] B.H.Bransden and J.J.Joachain. Introduction to Quantum Mechanics, Lomgman 1989, p.425

[2] R.Gilmore. Lie Groups, Lie algebras, and Some of Their Applications, J.Wiley and sons 1974

[3] D.Stahel, M.Leoni and K.Dressler. J.Mol.Spec., **77**:286, 1983

[4] J.A.Fleck, J.R.Morris and M.D.Feit. Appl.Phys., **10**:129, 1976

[5] J.Broeckhove, B.Feyen, L.Lathouwers, F.Arickx and P.Van Leuven. Chem.Phys.Lett., **174**:504, 1990

[6] R.Almeida and H.Metiu. Chem.Phys.Lett., **146**:47, 1988

[7] V.Engel and H.Metiu. J.Chem.Phys., **90**:6116, 1989

[8] K.P.Huber and G.Herzberg. Van Nostrand Reinhold, New York, 1983

TIME DEPENDENT QUANTUM MECHANICS WITH CORRELATED COHERENT STATES

Brian Weiner

Department of Physics
Pennsylvania State University
College Place, DuBois, PA 15801

Abstract

A formulation of Multi Configurational States (MCS), of the Complete Active Space (CAS) type, that describe correlation affects between electrons is formulated in terms of Vector Coherent States (VCS's). The Time Dependent Variational Principle is applied to these VCS's obtaining a description of the quantum evolution as a generalized classical dynamical system. Both approximate and *exact* quantum evolutions can be described in this fashion and equations of motion are obtained, which are expressed in terms of the reduced second order density matrix elements of CASMCS states. Spin adaptation of the states and of the equations of motion are formulated directly on the variables parameterizing the manifold of CASMCS's.

0. Introduction

In order to describe physical processes quantum mechanically one must invariably use an approximate form of Schrödinger's or Liouville's equation of motion for the state of the system. Approximations can be introduced in almost unlimited ways, their efficacy, however, must be judged against the criterion of whether they either give insight into the physical process or produce meaningful quantitative approximations to observable quantities, as well as to the practicalities of their use. Two basic philosophies control the choice of approximation method, (a) approximate the hamiltonian operator, i.e. the form and type of interactions involved in the evolution of the state of the system and (b) restrict the domain that the evolution equations refer to, i.e. constrain the form of the states. These approaches are not exclusive and are mostly used in tandem. These considerations are particularly appropriate when describing chemical reactions as it is important to take into account correlation between electrons in order to obtain accurate results, however this leads to equations of motion that are hard to solve and a balance must be sought between chemical accuracy and practical considerations.

One of the most promising approaches for approximating quantum evolution equations is to use the Time Dependent Variational Method (TDVP) /1/ coupled with the Coherent State (CS) description /2,3,4/ of a restricted domain for these equations. The standard CS method utilizes a domain formed by a manifold of states that are indexed by a left coset G/H

Time-Dependent Quantum Molecular Dynamics, Edited by J. Broeckhove
and L. Lathouwers, Plenum Press, New York, 1992

of a group G, where the group, G, has an irreducible representation as operators acting in the hilbert space, $\mathcal{H}$, of all pure states of the system and the subgroup, H, has a one dimensional representation in $\mathcal{H}$ i.e. there exist one dimensional H invariant subspaces in $\mathcal{H}$. States in $\mathcal{H}$ that produce one dimensional representations are called fiducial /2,3,4/ and the family of G/H related CS's are the set of states produced by the action of G/H on fiducial states. Subgroups, H, leaving the one dimensional subspace spanned by the fiducial state invariant are called isotropy subgroups for that state. In the study of systems of N electrons the group, G, has usually been chosen to be $U(\mathcal{H}^1)$, the unitary group of associated with the hilbert space, $\mathcal{H}^1$, describing the states of single electrons and the irreducible representation one considers is in the space, $\mathcal{H}^N$, describing the pure states of N electrons. The subgroup, H, is $U(\mathcal{H}_h^1)\otimes U(\mathcal{H}_p^1)$ where $\mathcal{H}_h^1$ is an N dimensional subspace (the hole space) associated with "occupied" one particle states in an Independent Particle State (IPS) that acts as the fiducial state $|\Phi\rangle \in \mathcal{H}^N$, which spans the one dimensional subspace, $\mathcal{H}_h^N$, of $\mathcal{H}^N$. The particle space, $\mathcal{H}_p^1$, which corresponds to "unoccupied" one particle states, is defined to be the orthogonal compliment of $\mathcal{H}_h^1$ in $\mathcal{H}^1$. The subgroup $U(\mathcal{H}_h^1)\otimes U(\mathcal{H}_p^1)$ is a maximal isotropy group for IPS's. The family of CS's produced in this case is known as Thouless Coherent States and form the complete manifold of IPS's contained in $\mathcal{H}^N$. These states do not, however, explicitly describe coulombic correlations between electrons, but rather the motion of single electrons in the mean field of the other electrons. In order to describe coulombic correlations explicitly one must use Multi Configurational States (MCS's). Care must however be exercised in the choice of configurations in order that chemical prejudices do not predetermine the outcome of calculations, to avoid this danger mathematically balanced MCS's can be chosen, such states are called Complete Active Space (CAS) MCS's in the quantum chemical literature.

The non linear manifold of Complete Active Space Multi Configurational States (CASMCS's) can be formulated in terms of Vector Coherent States (VCS) /5,6,7/, which are a generalization of the standard CS's. Instead of considering subgroups, H, that have one dimensional representations one considers subgroups that have multi dimensional irreducible representations in $\mathcal{H}^N$. These subgroups are then maximal isotropy groups for these subspaces, $\mathcal{H}_h^N$, (called coherent subspaces /7/) of dimension greater than one. The subspaces, $\mathcal{H}_h^N$, are produced by taking the Nth exterior power of a hole space of "active" spin orbitals, $\mathcal{H}_h^1$, which now has dimension greater than N. VCS's are generated by the action of G/H on $\mathcal{H}_h^N$ and are partially parameterized by this coset. The dimension of the VCS's correspond to the dimension of the invariant subspace and can be represented in component form, thus the name Vector Coherent State, by choosing a basis for the coherent subspace.

The general problem of reactive scattering of electrons and nuclei can be treated by a VCS description where either both nuclei and electrons are described by VCS's or just the electrons are and the nuclei are treated as classical point particles /8/. The trajectories in the parameter space obtained by solving the TDVP equations gives detailed information on the variation of the properties of the combined system as a whole and details of fluctuations of the parts. For instance amongst the properties that can be monitored are the change of electronic charge distributions during vibrational motion of the nuclei, formation and breaking of chemical bonds and the occurrence of resonances and meta stable states. However in order to analyze and understand reaction mechanisms and the spectra of the products and transient species produced it is important to identify states that refer to true or approximate symmetries. In this paper it is shown how to symmetry adapt CASMCS's in terms of the parameters describing the coset space G/H.

This manuscript is organized in the following manner:

0. Introduction.
1. Holomorphic Representations and Basic Vector Coherent State Theory.
2. Description of Group and Subspace Structure of $\mathcal{H}^N$.
3. The Time Dependent Variational Principle.
4. Time Evolution of Complete Active Space Multi Configurational States.
5. Spin Adaptation of Vector Coherent States.
6. Discussion.

1. Holomorphic Representations and Basic Vector Coherent State Theory

The cosets G/H form differentiable manifolds which can be given complex coordinate systems, $\mathbf{z}$, /9/ of dimension m and the space $\mathcal{H}^N$ can be shown to be isometrically isomorphic to a subspace of the direct sum of hilbert spaces of square integrable holomorphic functions of $\mathbf{z}$ with respect to a measure μ which reflects the antisymmetric nature of $\mathcal{H}^N$, i.e.

$$\mathcal{H}^N \rightarrowtail L_2(\mathbb{C}^m, d\mu_1)\oplus..\oplus L_2(\mathbb{C}^m, d\mu_N) \tag{1}$$

the above direct sum is isomorphic to a hilbert space, $L_2(\mathbb{C}^m, \mathcal{H}^N_h, \mu)$, of d-dimensional vector holomorphic functions defined on $\mathbb{C}^m$. The embedding eq.(1) is a consequence of the irreducibility of the unitary representation $\mathcal{U}$ of G in $\mathcal{H}^N$ which leads to the resolution of the identity

$$I = \int_{G/H} \sum_{k=1}^{d} \frac{|\mathbf{z},k\rangle\langle\mathbf{z},k|}{\langle\mathbf{z},k|\mathbf{z},k\rangle}\, d\mu(\mathbf{z}) = \int_{G/H} \sum_{k=1}^{d} |\mathbf{z},k\rangle\langle\mathbf{z},k|\, d\mu_k(\mathbf{z}) \tag{2}$$

where d is the dimension of $\mathcal{H}^N_h$, μ is a measure on G associated with the antisymmetric representation $\mathcal{U}$ and the states $|\mathbf{z},k\rangle$ are *unnormalized* with norms $\Big\| |\mathbf{z},k\rangle \Big\| = N_k(\mathbf{z})$ which are conveniently absorbed into the measure μ to form new measures $\mu_k = N_k^{-1}\mu$. The states $|\mathbf{z},k\rangle$ are defined by

$$N_k(\mathbf{z})^{-\frac{1}{2}}|\mathbf{z},k\rangle = \mathcal{U}(g(\mathbf{z}))|k\rangle = \mathcal{U}(c(\mathbf{z})h)|k\rangle = \mathcal{U}(c(\mathbf{z}))\mathcal{U}(h)|k\rangle$$
$$= \mathcal{U}(c(\mathbf{z}))\rho(h)|k\rangle = \mathcal{U}(c(\mathbf{z}))|k_h\rangle \tag{3}$$

for the factorization of G

$$g(\mathbf{z}) = c(\mathbf{z})h; \quad g(\mathbf{z})\in G,\ c(\mathbf{z})\in G/H, \text{ for a } \textit{fixed } h\in H \tag{4}$$

$\mathcal{H}^N_h$ is spanned by the orthonormal bases $\Big\{|k\rangle\Big\}$ and $\Big\{|k_h\rangle\Big\}$ and produces an irreducible representation ρ_0 of H, which is contained in ρ the reducible representation of H subduced in $\mathcal{H}^N$ by $\mathcal{U}$. The coset representatives c(**z**) depend on the fixed group element h which can be chosen to be the identity. Applying the resolution of the identity eq.(2) to states belonging to $\mathcal{H}^N$ one obtains

$$|\Psi\rangle = \int_{G/H} \sum_{k=1}^{d} |\mathbf{z},k\rangle\langle\mathbf{z},k|\Psi\rangle d\mu_k(\mathbf{z}) = \int_{G/H} \sum_{k=1}^{d} |\mathbf{z},k\rangle\Psi_k(\mathbf{z}^*)d\mu_k(\mathbf{z}) \tag{5}$$

where $\Psi_k(\mathbf{z}^*)=\langle\mathbf{z},k|\Psi\rangle$. The norm of the state $|\Psi\rangle$ is given by

$$\left\| |\Psi\rangle \right\|^2 = \sum_{k=1}^{d} \int_{G/H} |\Psi_k(\mathbf{z}^*)|^2 \, d\mu_k(\mathbf{z}) \tag{6}$$

and thus one has a mapping from $\mathcal{H}^N$ to a hilbert space, $\mathfrak{H}\subset L_2(\mathbb{C}^m,\mathcal{H}^N_h,\mu)$, of d dimensional vector holomorphic functions

$$|\Psi\rangle \hookrightarrow \Psi(\mathbf{z}^*) = \left(\Psi_1(\mathbf{z}^*),\ldots,\Psi_d(\mathbf{z}^*)\right) \tag{7}$$

which is a subspace of all functions defined on the coset G/H. A Vector Coherent State is defined to be

$$|\mathbf{z},\mathbf{d}\rangle = \sum_{k=1}^{d} |\mathbf{z},k\rangle d_k \tag{8}$$

Which by eq.(7) corresponds to a value of a holomorphic wavefunction at a point $\mathbf{z}^*$.

Operators, A, acting in $\mathcal{H}^N$ can be represented as operators in $\mathfrak{H}$ in two forms (a) as matrix integral operators

$$A|\Psi\rangle = \sum_{k,k'=1}^{d} \int_{G/H} |\mathbf{z},k\rangle\langle\mathbf{z},k|A\mathbf{z}',k'\rangle\langle\mathbf{z}',k'|\Psi\rangle d\mu_k(\mathbf{z})d\mu_{k'}(\mathbf{z}')$$

$$= \sum_{k,k'=1}^{d} \left[\int_{G/H} A_{kk'}(\mathbf{z},\mathbf{z}')\Psi_{k'}(\mathbf{z}^{*\prime})d\mu_{k'}(\mathbf{z}')\right] |\mathbf{z},k\rangle d\mu_k(\mathbf{z}) \tag{9}$$

or as (b) matrix differential operators

$$A|\Psi\rangle = \sum_{k=1}^{d} \int_{G/H} P^A_k\left(\mathbf{z}^*,\frac{\partial}{\partial\mathbf{z}^*}\right)\Psi_k(\mathbf{z}^*)|\mathbf{z},k\rangle d\mu_k(\mathbf{z}) \tag{10}$$

The form of the polynomials, P^A_k, depends on the group G, the coset G/H, the space $\mathcal{H}^N_h$ and the operator A /Appendix 1/.

Group theoretically the most important role of the subgroup H and the subspace $\mathcal{H}^N_h$ is to induce a representation, Γ, of G in the space $\mathfrak{H}$ which is defined by /10/

$$\Gamma(g)\Psi(\mathbf{z}^*) = \Psi\left(\sigma(g)^{-1}\mathbf{z}^*\right) \tag{11}$$

The symbol $\sigma(g)$ stands for the representation of the group G as automorphisms /Appendix 2/ of the complex space $z \in \mathbb{C}^m$ which parameterizes the coset G/H. The representations Γ depend on ρ_0, thus on $\mathcal{H}_h^N$, and H. The value of Ψ at a point z^* corresponds to VCS by

$$\Psi(z^*) \rightarrowtail\!\!\!\twoheadrightarrow \sum_{k=1}^{d} |z,k\rangle \Psi_k(z^*) \tag{12}$$

and each VCS corresponds to an element of $\mathcal{H}_h^N$ by

$$\mathcal{U}(c(z))^{-1} \sum_{k=1}^{d} |z,k\rangle \Psi_k(z^*) \rightarrowtail\!\!\!\twoheadrightarrow \sum_{k=1}^{d} |k\rangle \Psi_k(z^*) = \sum_{k=1}^{d} |k\rangle\langle z,k|\Psi\rangle \tag{13}$$

the last expression on the preceding line is called a VCS representative.

The full quantum mechanical equations of time evolution, i.e. Schrödingers or Liouvilles equation can expressed in the space $\mathfrak{H}$ and is thus a holomorphic representation of quantum theory.

2. Description of group and subspace structure of $\mathcal{H}^N$

The space of pure states of systems of N electrons can be constructed from one electron spaces $\mathcal{H}^1$ of pure one particle states called spin orbitals, e. g. in the Schrödinger representation as complex valued functions of position space and spin. The space $\mathcal{H}^1$ can be spanned by a bases set of orthonormal spin orbitals $\left\{|\varphi_i\rangle;\ 1\le i\le r\right\}$ from which a basis $\left\{|\varphi_{i_1}..\varphi_{i_N}\rangle;\ 1\le i_1\le..\le i_N\le r\right\}$ of $\mathcal{H}^N$ can be constructed by taking N fold antisymmetric tensor products of the spin orbitals. A fixed, but arbitrary, subset, $\left\{|\varphi_h\rangle;\ 1\le h\le r_1\right\}$ of the spin orbital basis can be chosen to produce the hole space, $\mathcal{H}_h^1$, and thus construct the subspace $\mathcal{H}_h^N$ of $\mathcal{H}^N$ called the active N particle space based on the spin orbitals $\left\{|\varphi_h\rangle;\ 1\le h\le r_1\right\}$. Vectors belonging to this space are CASMCS's described earlier. The decomposition

$$\mathcal{H}^1 = \mathcal{H}_h^1 \oplus \mathcal{H}_p^1;\quad 1 \le \text{Dimension}\left(\mathcal{H}_h^1\right) = r_1 \le r \tag{14}$$

is invariant to the subgroup $U(\mathcal{H}_h^1)\otimes U(\mathcal{H}_p^1)$ of $U(\mathcal{H}^1)$. This subgroup, which depends on r_1, reduces $\mathcal{H}^N$ into N+1 invariant subspaces, $\left\{V_\kappa^N,\ 1\le\kappa\le N+1\right\}$, which give rise to irreducible representations, ρ_κ, of $U(\mathcal{H}_h^1)\otimes U(\mathcal{H}_p^1)$. These subspaces are defined as

$$V_\kappa^N = \text{l.s.}\left\{|\varphi_{h_1}..\varphi_{h_{N-\kappa}}\varphi_{p_1}..\varphi_{p_\kappa}\rangle : \begin{matrix} 1 \le h_1\le\ldots\le h_{N-\kappa}\le r_1,\ 0\le\kappa\le N \\ r_1+1\le p_1\le..\le p_\kappa\le r \end{matrix}\right\} \tag{15}$$

where "l.s." denotes linear span of the basis set. The parameter, κ, designates the level of excitation of the subspace, commonly described as all κ-tuples out of the active space. The dimensions of the subspaces, V^N_κ, are $\binom{r_1}{N-\kappa}\binom{r-r_1}{\kappa}$ and the subduced representation ρ is a direct sum of the irreps ρ_κ, each appearing with frequency one in ρ.

The representation, $\mathcal{U}$, of $U(\mathcal{H}^1)$ and its lie algebra $u(\mathcal{H}^1)$ can be expressed in terms of the second quantized operators $\left\{a_i^\dagger, a_i;\ 1\le i\le r\right\}$, based on the spin orbitals $\left\{|\varphi_i\rangle;\ 1\le i\le r\right\}$ and defined by

$$a_i^\dagger|\phi\rangle = |\varphi_i\rangle;\quad \left[a_i, a_j^\dagger\right]_+ = \delta_{ij}\mathsf{I} \tag{16}$$

where "$\dagger$" denotes operator adjoint, $|\phi\rangle$ the zero particle vacuum, I the identity operator and $[\ ]_+$ the positive commutator. Using these operators the representation $\mathcal{U}(G/H)$ of the coset G/H can be expressed in Gauss factored form as the product of factors belonging to three subgroups contained in the complexification $GL(\mathcal{H}^1,\mathbb{C})$ of $U(\mathcal{H}^1)$

$$\mathcal{U}\left(c(\mathbf{z})\right) = \exp\left\{\mathcal{N}_+(\mathbf{z})\right\}\exp\left\{\mathcal{N}_0(\lambda(\mathbf{z}))\right\}\exp\left\{\mathcal{N}_-(\mathbf{z})\right\} \tag{17}$$

where

$$\mathcal{N}_0(\lambda(\mathbf{z})) = \sum_{1\le h_1,h_2\le r_1}\lambda^H_{h_1h_2}a^\dagger_{h_1}a_{h_2} + \sum_{r_1+1\le p_1,p_2\le r}\lambda^P_{p_1p_2}a^\dagger_{p_1}a_{p_2} = \mathcal{N}_0(\lambda(\mathbf{z}))^\dagger \tag{18}$$

Note that the operator $\exp\left\{\mathcal{N}_0(\lambda(\mathbf{z}))\right\}$ is *not* unitary but a positive operator as $\mathcal{N}_0(\lambda(\mathbf{z}))$ is hermitian *not* anti hermitian. The affect of this term is to maintain the normalization of states produced by the action of the operator eq.(17) and depends on both $\mathbf{z}^*$ and $\mathbf{z}$. The other infinitesimal generators in eq.(17) are defined as

$$\mathcal{N}_+(\mathbf{z}) = \sum_{\substack{1\le h\le r_1\\ r_1<p\le r}} z_{ph}a^\dagger_p a_h = -\mathcal{N}_-(\mathbf{z})^\dagger \tag{19}$$

again note that the operators $\mathcal{N}_+(\mathbf{z})$ and $\mathcal{N}_-(\mathbf{z})$ are *not* anti hermitian and thus the exponentials of these operators are *not* unitary. The operator $\mathcal{N}_-(\mathbf{z})$ annihilates the subspace V^N_0 i.e.

$$\mathcal{N}_-(\mathbf{z})|\Phi\rangle = 0\ \forall\ |\Phi\rangle\in V^N_0 \text{ and } \forall\mathbf{z}\in\mathbb{C}^m \tag{20}$$

while the operator $\mathcal{N}_+(\mathbf{z})$ is non zero on any vector in V^N_0.

An arbitrary CASMCS belonging to V^N_0 can expanded as

$$|\Phi\rangle = \sum_{1\le h_1<..<h_N\le r_1} d_{h_1..h_N}|\varphi_{h_1}..\varphi_{h_N}\rangle \tag{21}$$

The group $U(\mathcal{H}^1)$ acts on such states to produce variations of the spin orbitals $\left\{|\varphi_h\rangle;\ 1\le h\le r_1\right\}$

$$\mathcal{U}\left(c(\mathbf{z})h(\eta)\right)|\Phi\rangle = \exp\left\{\mathcal{N}_+(\mathbf{z})\right\}\exp\left\{\mathcal{N}_0(\lambda(\mathbf{z}))\right\}\exp\left\{i\mathcal{N}_0(\eta)\right\}|\Phi\rangle$$

$$= \sum_{1\le h_1<..<h_N\le r_1} d(\mathbf{z},\eta)_{h_1..h_N} \exp\left\{\mathcal{N}_+(\mathbf{z})\right\}|\varphi_{h_1}..\varphi_{h_N}\rangle = \sum_{1\le h_1<..<h_N\le r_1} d(\mathbf{z},\eta)_{h_1..h_N}\ |\varphi(\mathbf{z})_{h_1}..\varphi(\mathbf{z})_{h_N}\rangle \tag{22}$$

where the expansion coefficients $d(\mathbf{z},\eta)_{h_1..h_N}$ reflect the action of $\mathcal{U}(h(\eta))$ on $|\Phi\rangle$ and the variation of the norm of the states $|\varphi(\mathbf{z})_{h_1}..\varphi(\mathbf{z})_{h_N}\rangle$. The set of states expressed in eq.(22) form an $U(\mathcal{H}^1)$ orbit, for fixed $|\Psi\rangle$, in the non linear manifold of CASMCS's, $V_0^N(\mathbf{z})$. This latter manifold is produced by acting on V_0^N with an operator belonging to $\mathcal{U}(G/H)$, taking the linear closure of the states produced and then the set theoretic union of the resulting spaces as one ranges over all elements of $\mathcal{U}(G/H)$. The mathematical name for the structure so formed is a vector bundle. States belonging to $V_0^N(\mathbf{z})$ can be parameterized by the coset $U(\mathcal{H}^1)/U(\mathcal{H}^1_h)\otimes U(\mathcal{H}^1_p)$, thus by $\mathbf{z}$, and the linear expansion coefficients $d_{h_1..h_N}$ in the following manner

$$|\Phi(\mathbf{z},\mathbf{d})\rangle = \sum_{1\le h_1<..<h_N\le r_1} d_{h_1..h_N}\exp\left\{\mathcal{N}_+(\mathbf{z})\right\}|\varphi_{h_1}..\varphi_{h_N}\rangle$$

$$= \sum_{1\le h_1<..<h_N\le r_1} d_{h_1..h_N}\ |\varphi(\mathbf{z})_{h_1}..\varphi(\mathbf{z})_{h_N}\rangle \tag{23}$$

where $d_{h_1..h_N}$ are variable complex coefficients not depending on $\mathbf{z}\in\mathbb{C}^m$. A state $|\Phi(\mathbf{z},\mathbf{d})\rangle$ thus corresponds to a value of a holomorphic wavefunction, Ψ, at the point $\mathbf{z}^*$ given by

$$\Psi(\mathbf{z}^*) = \sum_{1\le h_1<..<h_N\le r_1} d_{h_1..h_N}\ |\varphi(\mathbf{z})_{h_1}..\varphi(\mathbf{z})_{h_N}\rangle = \sum_{1\le h_1<..<h_N\le r_1} \Psi_{h_1..h_N}(\mathbf{z}^*)\ |\varphi(\mathbf{z})_{h_1}..\varphi(\mathbf{z})_{h_N}\rangle \tag{24}$$

where the components $\Psi_{h_1..h_N}(\mathbf{z}^*)$ are set equal to $d_{h_1..h_N}$. Thus each set of expansion coefficients, $\mathbf{d}$, can be associated with a state belonging to $\mathfrak{H}$ and each CASMCS corresponds to the value of such a state at a particular $\mathbf{z}^*$ and therefore is a VCS. In mathematical terminology the states described by $\mathfrak{H}$ are cross sections of the vector bundle $V_0^N(\mathbf{z})$, and CASMCS's are values of the cross sections in the fibers of $V_0^N(\mathbf{z})$.

The manifold $V_0^N(\mathbf{z})$ is well suited to provide a restricted domain in which to approximate the time dependent Schrödinger equation for the electronic states of a system. This domain approaches the full domain in a balanced manner as $r_1 \longrightarrow r \longrightarrow \infty$ and it naturally describes an approximate time evolution as described in the following section.

3. The Time Dependent Variational Principle

The time dependent variational principle has been describe in detail in many places /1,8,9/, so only a brief outline will be presented here. Let ζ denote the coordinates of the domain of the Schrödinger equation, e.g. $\zeta = (\mathbf{z},\mathbf{d})$ for $V_0^N(\mathbf{z})$ in the preceding when we are considering the manifold of CASMCS's. The TDVP is derived from an action principle based on the Lagrangian

$$\mathcal{L}=\left\{\frac{i\hbar}{2}\left[\frac{\partial}{\partial t}-\frac{\partial}{\partial t'}\right]\langle\zeta(t')|\zeta(t)\rangle|_{t=t'}-\langle\zeta(t')|H\zeta(t)\rangle\right\}\langle\zeta(t')|\zeta(t)\rangle^{-1} \tag{25}$$

and action

$$A = \int\mathcal{L}\left(\zeta^*(t),\zeta(t')\right)dt \tag{26}$$

with the conditions $\delta A=0$ and the end points of the complex valued paths $\zeta(t)$ held fixed. This leads to the generalized classical equations of motion

$$\begin{aligned} i\sum C_{ij}\frac{\partial\zeta_j}{\partial t} &= \frac{\partial E}{\partial\zeta_i^*} \\ -i\sum C_{ij}^*\frac{\partial\zeta_j^*}{\partial t} &= \frac{\partial E}{\partial\zeta_i} \end{aligned} \tag{27}$$

and an independent equation for the overall phase, γ, of the state

$$\frac{d\gamma}{dt} = -\mathrm{Im}\left\{\sum \xi_i\frac{\partial\ln S}{\partial\xi_i}\right\} - E \tag{28}$$

with the metric and classical Hamiltonian functions defined as

$$\begin{aligned} C_{ij} &= \left.\frac{\partial^2\ln S}{\partial\zeta_i'\partial\zeta_j^*}\right|_{\zeta=\zeta'} \\ S &= S(\zeta'^*,\zeta) = \langle\zeta'|\zeta\rangle \\ E &= E(\zeta^*,\zeta) = \frac{\langle\zeta|H\zeta\rangle}{\langle\zeta|\zeta\rangle} \end{aligned} \tag{29}$$

where H is the quantum Hamiltonian operator. The matrix C is hermitian and the equation of motion eq.(27) can be written as

$$i\begin{pmatrix} C & 0 \\ 0 & -C^* \end{pmatrix}\begin{pmatrix} \frac{d\zeta}{dt} \\ \frac{d\zeta^*}{dt} \end{pmatrix} = \begin{pmatrix} \frac{\partial E}{\partial\zeta^*} \\ \frac{\partial E}{\partial\zeta} \end{pmatrix} \tag{30}$$

One can also define a Poisson bracket $\{,\}$ for two functions f and g of ζ and ζ^* by

$$\{f,g\} = \begin{pmatrix} \dfrac{\partial f}{\partial \zeta} & \dfrac{\partial f}{\partial \zeta^*} \end{pmatrix} \begin{pmatrix} -iC^{-1} & 0 \\ 0 & iC^{-1*} \end{pmatrix} \begin{pmatrix} \dfrac{\partial g}{\partial \zeta^*} \\ \dfrac{\partial g}{\partial \zeta} \end{pmatrix} \tag{31}$$

This Poisson bracket defines a symplectic structure on the manifold coordinatized by ζ and leads to a succinct expression of the time dependent Schrödinger equation restricted to states belonging to this manifold as a classical Hamiltonian system

$$\dot{\zeta} = \{\zeta^*, E\}, \quad \dot{\zeta}^* = \{\zeta^*, E\} \tag{32}$$

together with eq.(28) for the overall phase of the state. The time independent Schrödinger equation restricted to this manifold is obtained by setting the time derivatives to zero, which leads to a system of algebraic equations for the approximate stationary states, which are the solutions $\{\zeta_k\}$ of these equations.

4. Time Evolution of Complete Active Space Multi Configurational States

When the restricted domain of the Schrödinger equation is the CASMCS manifold $V_0^N(\mathbf{z})$ described in the preceding, the overlap kernel and classical Hamiltonian have the following form for the unnormalized states $|\Phi(\mathbf{z},\mathbf{d})\rangle$

$$S = S(\mathbf{z}'^*, \mathbf{d}'^*, \mathbf{z}, \mathbf{d}) = \langle \Phi(\mathbf{z}', \mathbf{d}') | \Phi(\mathbf{z}, \mathbf{d}) \rangle = \mathbf{d}'^\dagger \mathbf{S}(\mathbf{z}'^*, \mathbf{z}) \mathbf{d} \tag{33}$$

$$E = E(\mathbf{z}^*, \mathbf{d}^*, \mathbf{z}, \mathbf{d}) = \frac{\langle \Phi(\mathbf{z},\mathbf{d}) | H\Phi(\mathbf{z},\mathbf{d}) \rangle}{\langle \Phi(\mathbf{z},\mathbf{d}) | \Phi(\mathbf{z},\mathbf{d}) \rangle} = \frac{\mathbf{d}^\dagger \mathbf{E}(\mathbf{z}^*, \mathbf{z}) \mathbf{d}}{S} \tag{34}$$

Where the matrix elements of **S** and **E** are defined as

$$\mathrm{S}_{\kappa\kappa'}(\mathbf{z}^*, \mathbf{z}) = \langle \mathbf{z}, \kappa | \mathbf{z}, \kappa' \rangle; \quad \mathrm{E}_{\kappa\kappa'}(\mathbf{z}^*, \mathbf{z}) = \langle \mathbf{z}, \kappa | H\mathbf{z}, \kappa' \rangle \tag{35}$$

where $\kappa \equiv (h_1, .., h_N)$ for $1 \le h_1 < .. h_N \le r_1$. It useful to note that these matrix elements can be identified as the amplitudes of the states $|\mathbf{z},\kappa'\rangle$ and $H|\mathbf{z},\kappa\rangle$ in the holomorphic representation $\mathfrak{H}$. The electronic hamiltonian, H, can be written in terms of the second quantized operators, eq.(16), as

$$H = H_0 + \sum h_{ij} a_i^\dagger a_j + \sum V_{ijlk} a_i^\dagger a_j^\dagger a_k a_l \tag{36}$$

leading to the following expression for the "matrix elements" of the classical Hamiltonian

$$\begin{aligned} \mathrm{E}_{\kappa\kappa'}(\mathbf{z}^*, \mathbf{z}) = H_0 &+ \sum h_{ij} \langle \mathbf{z}, \kappa | a_i^\dagger a_j \mathbf{z}, \kappa' \rangle \\ &+ \sum V_{ijlk} \langle \mathbf{z}, \kappa | a_i^\dagger a_j^\dagger a_k a_l \mathbf{z}, \kappa' \rangle \end{aligned} \tag{37}$$

Where h_{ij} and V_{ijkl} are one particle kinetic energy and coulombic interaction integrals respectively. The terms $\left\{\langle z,\kappa|a_i^{\dagger}a_j z,\kappa'\rangle\right\}$ and $\left\{\langle z,\kappa|a_i^{\dagger}a_j^{\dagger}a_k a_l z,\kappa'\rangle\right\}$ define the matrix elements $\left\{D^{(1)}_{\kappa\kappa' ji}\right\}$, $\left\{D^{(2)}_{\kappa\kappa' lkij}\right\}$ of the first, $D^{(1)}_{\kappa\kappa'}(z^*,z)$, and second, $D^{(2)}_{\kappa\kappa'}(z^*,z)$, order *unnormalized* reduced transition density matrices respectively between IPS's. The action of any operator P(A) that is a polynomial in $\left\{a_h^{\dagger}a_p, a_p^{\dagger}a_h, a_h^{\dagger}a_{h'}, a_p^{\dagger}a_{p'}\right\}$ inside expectation values $\langle z,\kappa|P(A)z,\kappa'\rangle$ can be reformulated in terms of the affect of the differential operator $P^A\left(z^*,\dfrac{\partial}{\partial z^*}\right)$ on the amplitudes $S_{\kappa\kappa'}(z^*,z)$ $= \langle z,\kappa|z,\kappa'\rangle$, /see Appendix 1/, giving the following expressions for the *unnormalized* first order reduced transition density matrices elements:

$$\Gamma^{(1)}(z^*,d^*,z,d) = d^{\dagger}D^{(1)}_{\kappa\kappa'}(z^*,z)d \;;\; \Gamma^{(2)}(z^*,d^*,z,d) = d^{\dagger}D^{(2)}_{\kappa\kappa'}(z^*,z)d \tag{38}$$

$$\left(D^{(1)}_{\kappa\kappa'}(z^*,z)\right)_{ij} = \langle z,\kappa|a_j^{\dagger}a_i z,\kappa'\rangle \tag{39}$$

$$\langle z,\kappa|a_h^{\dagger}a_p z,\kappa'\rangle = \frac{\partial}{\partial z^*_{ph}}\langle z,\kappa|z,\kappa'\rangle \tag{40}$$

$$\langle z,\kappa|a_h^{\dagger}a_{h'}z,\kappa'\rangle = \left\{(1-\delta_{h'\kappa})\delta_{h\kappa} + \sum_p z^*_{ph'}\frac{\partial}{\partial z^*_{ph}}\right\}\langle z,\kappa|z,\kappa'\rangle \tag{41}$$

$$\langle z,\kappa|a_p^{\dagger}a_{p'}z,\kappa'\rangle = -\sum_h z^*_{ph}\frac{\partial}{\partial z^*_{p'h}}\langle z,\kappa|z,\kappa'\rangle \tag{42}$$

$$\langle z,\kappa|a_p^{\dagger}a_h z,\kappa'\rangle = \left\{\sum_{h'} z^*_{ph'}(1-\delta_{h\kappa})\delta_{h'\kappa} - \sum_{p'h'} z^*_{p'h}z^*_{ph'}\frac{\partial}{\partial z^*_{p'h'}}\right\}\langle z,\kappa|z,\kappa'\rangle \tag{43}$$

The *unnormalized* second order reduced transition density matrix elements can be obtained by expressing $\left\{a_i^{\dagger}a_j^{\dagger}a_k a_l\right\}$ in generalized anti normal order form $P_-(A)P_0(A)P_+(A)$ where the first term is a polynomial in $\left\{a_h^{\dagger}a_p\right\}$, the second term is a polynomial in $\left\{a_h^{\dagger}a_{h'}, a_p^{\dagger}a_{p'}\right\}$ and the third term is a polynomial in $\left\{a_p^{\dagger}a_h\right\}$ and iterating on the expressions eq.'s (40)-(43). The kernel on which these differentiations are carried out on is given by

$$\langle \mathbf{z},\kappa|\mathbf{z},\kappa'\rangle = S_{\kappa\kappa'}(\mathbf{z}^*,\mathbf{z}) = \det\left(\Lambda_{\kappa\kappa'}\right) \tag{44}$$

Where

$$\Lambda_{\kappa\kappa'} = J_{\kappa}^{\dagger}(I_{r_1} + \mathbf{z}^{\dagger}\mathbf{z})J_{\kappa'} \tag{45}$$

J_{κ} is a $r_1\times N$ matrix defined by

$$\left(J_{\kappa}\right)_{hj} = \delta_{hh_j};\ \kappa\equiv(h_1..h_N);\ 1\le j\le N;\ 1\le h\le r_1 \tag{46}$$

where I_{r_1} is a $r_1\times r_1$ unit matrix. The results of these differentiations on $S_{\kappa\kappa'}(\mathbf{z}^*,\mathbf{z})$ are /9/

$$D^{(2)}_{\kappa\kappa'}(\mathbf{z}^*,\mathbf{z}) = \begin{pmatrix} D_{hhhh} & D_{hhhp} & D_{hhph} & D_{hhpp} \\ \cdot & D_{hphp} & D_{hpph} & D_{hppp} \\ \cdot & \cdot & D_{phph} & D_{phph} \\ \cdot & \cdot & \cdot & D_{ppp} \end{pmatrix}$$

$$= \begin{pmatrix} I_{r_1}\otimes I_{r_1} \\ I_{r_1}\otimes\mathbf{z} \\ \mathbf{z}\otimes I_{r_1} \\ \mathbf{z}\otimes\mathbf{z} \end{pmatrix}\left(J_{\kappa'}\otimes J_{\kappa'}\,\mathrm{ad}^{(2)}(\Lambda_{\kappa\kappa'})J_{\kappa}^{\dagger}\otimes J_{\kappa}^{\dagger}\right)$$

$$\left(I_{r_1}\otimes I_{r_1},\ I_{r_1}\otimes\mathbf{z}^{\dagger},\ I_{r_1}\otimes\mathbf{z}^{\dagger},\ \mathbf{z}^{\dagger}\otimes\mathbf{z}^{\dagger}\right) \tag{47}$$

where $\mathrm{ad}^{(2)}(\Lambda_{\kappa\kappa'})$ is the second order classical adjugate matrix with matrix elements $\left(\mathrm{ad}^{(2)}(\Lambda_{\kappa\kappa'})\right)_{ijkl}$ that are determinants with sign factor $(-1)^{i+j+k+l+\delta_{k<i}+\delta_{l<j}}$, $\delta_{a<b}=1$ if $a<b$ and zero otherwise, obtained by deleting columns i and k and rows j and l in the matrix $\Lambda_{\kappa\kappa'}$.

The classical hamiltonian, E, and the symplectic structure $\{,\}$ can thus be formulated in terms of the derivatives up to second order of the overlap kernels, $S_{\kappa\kappa'}(\mathbf{z}^*,\mathbf{z})$. The equation of motion eq.(32) also requires first order derivatives of E which can also be obtained by using eq.'s (40)-(43) and thus include third order derivatives.

5. Spin Adaptation of Vector Coherent States

The group, SU(2), describes the internal spin properties of electronic states and can be used to label those states when relativistic corrections are not significant. This group is a subgroup of $U(\mathcal{H}^1)$ and of $U(\mathcal{H}^1_h)$, thus representations of these groups subduce representations of SU(2). The representation of the Lie algebra $su(2)$ of SU(2) subduced by $\mathcal{U}$

is generated by the basis $\left\{S_+-S_-,\ i(S_++S_-),\ S_0\right\}$ where

$$S_+ = \sum a^\dagger_{i+s}a_i;\ S_- = \sum a^\dagger_i a_{i+s};\ S_0 = \sum\left\{a^\dagger_i a_i - a^\dagger_{i+s}a_{i+s}\right\} \tag{48}$$

and we have ordered the spin orbital basis of $\mathcal{H}^1$ according to

$$|\varphi_i\rangle = |\chi_i\rangle|\alpha\rangle,\ 1\le i\le s;\qquad |\varphi_{i+s}\rangle = |\chi_i\rangle|\beta\rangle \tag{49}$$

and $\left\{|\chi_i\rangle;\ 1\le i\le s\right\}$ are an orthonormal set of spatial orbitals. The complexification of the Lie algebra $su(2)$ is the Lie algebra $sl(2,\mathbb{C})$ which is generated by the operators $\left\{S_+,S_0,S_-\right\}$, hence the representation, $\mathcal{U}$, of the group SU(2) also gives a representation of SL(2,ℂ). Elements of SL(2,ℂ) in the representation $\mathcal{U}$ can be parameterized as

$$\mathcal{U}\left(s(\xi)\right) = \exp\left\{\xi_-S_-\right\}\exp\left\{\xi_+S_+\right\}\exp\left\{\xi_0S_0\right\} \tag{50}$$

The group $U(\mathcal{H}^1)$ can be factored over the group SU(2) (as it is a subgroup) so that $U(\mathcal{H}^1) = \left[U(\mathcal{H}^1)/SU(2)\right]\left(SU(2)\right)$ and is actually equal to the group direct product $\left[U(\mathcal{H}^1)/SU(2)\right]\otimes\left(SU(2)\right)$. The representatives of the group $U(\mathcal{H}^1)/SU(2)$ in the representation $\mathcal{U}$ are produced by exponentiation of the spinless Lie algebra generators $\left\{A^\dagger_iA_j = a^\dagger_ia_j + a^\dagger_{i+s}a_{j+s};\ 1\le i,j\le s\right\}$ where $\dim(\mathcal{H}^1)=r=2s$. If one chooses a spin balanced $\mathcal{H}^1_h$ defined by

$\mathcal{H}^1_h=\text{l.s.}\left\{|\varphi_h\rangle,|\varphi_{h+s}\rangle;\ 1\le h\le s_1=\frac{r_1}{2}\right\}$ then this subspace carries a representation, Π, of SU(2) generated by the Lie algebra elements $\left\{a^\dagger_ha_{h'};\ 1\le h,h'\le r_1\right\}$. This representation is subduced by the representation, ρ_0, of $U(\mathcal{H}^1_h)\otimes U(\mathcal{H}^1_p)$ carried by $\mathcal{H}^N_h$ i.e.

$$\rho_0\big|_{SU(2)} = \Pi = \underset{s}{\oplus}\,\Pi_{N,S}\ ;\ \frac{N}{2}\ge S\ge\frac{1}{2}\ \textbf{(N odd) or 0 (N even)} \tag{51}$$

Using SU(2) the coset $U(\mathcal{H}^1)/U(\mathcal{H}^1_h)\otimes U(\mathcal{H}^1_p)$ can be factored as

$$\left[U(\mathcal{H}^1)/SU(2)\right]\otimes\left(SU(2)\right)/\left[\left[(U(\mathcal{H}^1_h)\otimes U(\mathcal{H}^1_p))/\left(SU(2)\otimes SU(2)\right)\right]\right.$$

$\left.\otimes\left(SU(2)\otimes SU(2)\right)\right]$ and each $\mathcal{U}(c(\mathbf{z})) = \mathcal{U}(c_0(\mathbf{Z}))\mathcal{U}(s(\xi)) = \mathcal{U}(s(\xi))\mathcal{U}(c_0(\mathbf{Z}))$ so that the action of the complexfication (transformations that do not conserve norm) of the above coset on $\mathcal{H}^N_h$ is

$$\mathcal{U}\left(c(\mathbf{z})\right)|\Phi\rangle=\exp\left\{\mathcal{N}_{+}(\mathbf{Z})\right\}\exp\left\{\xi_{-}S_{-}\right\}\exp\left\{\xi_{+}S_{+}\right\}\exp\left\{\xi_{0}S_{0}\right\}|\Phi\rangle \tag{52}$$

where

$$\mathcal{N}_{+}(\mathbf{Z}) = \sum_{\substack{1\le h\le s_1 \\ s_1<p\le s}} Z_{ph}A_p^{\dagger}A_h; \tag{53}$$

The states, $|\Phi(\mathbf{z})\rangle = \mathcal{U}(c(\mathbf{z}))|\Phi\rangle$, belonging to the fiber $V_0^N(\mathbf{z})$, can be expressed as $|\Phi(\mathbf{z})\rangle = |\Phi(\mathbf{Z},\xi)\rangle = \mathcal{U}(s(\xi))\mathcal{U}(c_0(\mathbf{Z}))|\Phi\rangle$. Now $\mathcal{U}(s(\xi))\mathcal{U}(c_0(\mathbf{Z})) = \mathcal{U}\left(s(\xi)c_0(\mathbf{Z})\right) = \mathcal{U}\left(c\left(\mathbf{z}(\mathbf{Z},\xi)\right)\right)$ where the variable $\mathbf{z}$ is related to the $s_1\times(s-s_1)$ matrix $\mathbf{Z}$ and the complex variable $\xi\in\mathbb{C}^3$ by /Appendix 2/

$$\mathbf{z} = \mathbf{z}(\mathbf{Z},\xi) = \left(A(\xi)\mathbf{Z} + B(\xi)\right)\left(C(\xi)\mathbf{Z} + D(\xi)\right)^{-1} \tag{54}$$

$$\mathbf{z}(\mathbf{Z},0) = \begin{pmatrix}\mathbf{Z} & \mathbf{0}\\ \mathbf{0} & \mathbf{Z}\end{pmatrix} \tag{55}$$

and $\begin{pmatrix}A(\xi) & B(\xi)\\ C(\xi) & D(\xi)\end{pmatrix}$ is the representation of $s(\xi)$ on $\mathcal{H}^1$ which can be shown to be /12/

$$\begin{pmatrix}A(\xi) & B(\xi)\\ C(\xi) & D(\xi)\end{pmatrix} = \begin{pmatrix}\exp\left(\frac{\xi_0}{2}\right)I_{s_1} & \exp\left(\frac{-\xi_0}{2}\right)\xi_+ I_{s_1}\\ \exp\left(\frac{\xi_0}{2}\right)\xi_- I_{s_1} & \exp\left(\frac{-\xi_0}{2}\right)\left(1+\xi_-\xi_+\right)I_{s_1}\end{pmatrix} \tag{56}$$

This parameterization can then be used to express components of the vector holomorphic wavefunction as $\Psi_\kappa(\mathbf{Z}(\xi)) = \langle\kappa,\mathbf{Z}(\xi)|\Psi\rangle$. In general the states $|\Phi(\mathbf{Z}(\xi))\rangle$ are not pure spin states i.e. not eigenstates of the casimir operator S^2 defined by

$$S^2 = S_-S_+ + S_0^2 + S_0 \tag{57}$$

nor even of the weight operator S_0. In order to produce pure spin states one must in general apply projection operators /11/

$$\mathcal{P}_{NSM_s} = \Theta_{NSM_s}\left(S^2\right)\Theta_{NSM_s}\left(S^2\right) \tag{58}$$

The factors in this operator are given by /12/

$$\Theta_{NSM_s}\left(S^2\right) = \sum_{\nu=0}^{\frac{N}{2}-S} a_{SM_s\nu}S_-^{S-M_s+\nu}S_+^{S-M_s+\nu} \tag{59}$$

$$\Theta_{NSM_s}\left[S_0\right] = \prod_{m\neq M_s} \frac{(S_0-m)}{(M_s-m)} = \sum_{j=0}^{N} b_{M_s j} S_0^j \tag{60}$$

The coefficients $\left\{a_{SM_s\nu}\right\}$ are defined as

$$a_{SM_s\nu} = \frac{(2S+1)(S+M_s)!(-1)^\nu}{(S-M_s)!\nu!(2S+\nu+1)!} \tag{61}$$

and $\left\{b_{Mj}\right\}$ are defined by eq.(60). These projection operators can be generated by differentiation of $U(s(\xi))$ as

$$\Theta_{NSM_s}\left[S^2\right] = \sum_{\nu=0}^{\frac{N}{2}-S} a_{SM_s\nu} \frac{\partial^{2(S-M_s+\nu)}}{\partial\xi_-^{(S-M_s+\nu)}\partial\xi_+^{(S-M_s+\nu)}} U(s(\xi))\Big|_{\xi_-=\xi_+=\xi_0=0} \tag{62}$$

$$\Theta_{NSM_s}\left[S_0\right] = \sum_{j=0}^{N} b_{M_s j} \frac{\partial^j}{\partial\xi_0^j} U\,(s(\xi))\Big|_{\xi_-=\xi_+=\xi_0=0} \tag{63}$$

Where the differential operators of order zero are to be interpreted as the identity operator. Spin adapted vector holomorphic wavefunctions, $\Psi_\kappa^{NSM_s}(\mathbf{z})$ can thus be obtained as

$$\sum_{\nu=0}^{\frac{N}{2}-S} a_{SM_s\nu} b_{M_s j} \frac{\partial^{2(S-M_s+\nu)+j}}{\partial\xi_-^{*(S-M_s+\nu)}\partial\xi_+^{*(S-M_s+\nu)}\partial\xi_0^{*j}} \Psi_\kappa(\mathbf{Z}(\xi))\Big|_{\xi_-^*=\xi_+^*=\xi_0^*=0} \tag{64}$$

If the hamiltonian is spin independent then the evolution equations of section 4 can be spin adapted by applying the differential operator eq.(64) on the overlap kernels $S_{\kappa\kappa'}$ defined in terms of the variables $\mathbf{Z}(\xi)$

$$\langle \mathbf{Z}(\xi),\kappa|\mathbf{Z}(\xi),\kappa'\rangle = S_{\kappa\kappa'}(\mathbf{Z}(\xi)^*,\mathbf{Z}(\xi)) = \det\left[J_\kappa^\dagger(I_{r_1}+\mathbf{Z}(\xi)^\dagger\mathbf{Z}(\xi))J_{\kappa'}\right] \tag{65}$$

and are special cases of vector holomorphic wavefunctions when $\Psi = \Phi_{\kappa'}$.

Only differentiations with respect to ξ^* need to be carried out as the projection operator $\mathcal{P}_{NSM_s}$ is idempotent and commutes with spin independent hamiltonians.

The method of producing spin adaptation described in this section depended on the fact that the spaces $\mathcal{H}^N$, $\mathcal{H}_h^N$ and $\mathcal{H}_p^N$ could be expressed as tensor products $\mathcal{H}_\gamma^N(\text{space})\otimes\mathcal{H}_\gamma^N(\text{spin})$, γ = h,p or "blank", thus leading to the factorizing of the groups $U(\mathcal{H}^1)$, $U(\mathcal{H}_h^1)$ and $U(\mathcal{H}_p^1)$ as direct products involving the internal spin group SU(2). Spatial symmetries such as point, space group or angular momentum can be handled by considering reducible representations of the appropriate group carried by $\mathcal{H}^{N}_{\gamma}$(space) and the associated decompositions of these subspaces into invariant subspaces with respect to these groups. Producing CASMCS's with prescribed spatial symmetries in terms of the parameterization $\mathbf{z}$ will be considered in more detail elsewhere.

6. Discussion

Using the TDVP with states belonging to the vector bundle $V^{N}_{0}(\mathbf{z})$, in other words the nonlinear manifold of CASMCS's, one can obtain approximate solutions of the Time Dependent Schrödinger Equation (TDSE) that are determined by the hamiltonian, initial state and the active space chosen. These solutions are paths $|\mathbf{z}(t)\rangle$ in $V^{N}_{0}(\mathbf{z})$. As the dimension r_1 of the one particle active space increases towards r, the dimension of the one particle space, the paths tend towards the exact solutions of the TDSE for bases sets of size r. As these bases become complete these solutions tend towards the *exact* solutions of the TDSE for the given hamiltonian, this thus determines a well defined sequence of approximations. If the hamiltonian is chosen to be approximate then a further set of sequential refinements can be overlayed as one "improves" the description of the interactions involved in the physical processes under study.

In order to study reactive scattering of atoms and molecules the CASMCS's described in this article are made to depend on nuclear positions and momentum via the dependence of the basis $\left\{|\varphi_k\rangle\right\}$ *of* the one electron space on these parameters, usually the bases are tied down to and move with each nuclear center in the system /8/. In particular for the reactive collision of two stable subsystems A and B, e.g. two stable molecules, the initial state for the evolution equations is a direct product of electronic stationary states of A and B calculated, within the same level of approximation, at a fixed large separation of A and B /13/. These stationary states can be found by solving the algebraic equations obtained by setting all time derivatives equal to zero in the evolution equations when each subsystem is separated by an arbitrary large distance. The state describing the evolution of the combined system is a function of nuclear positions, $\mathbf{R}(t)$, and nuclear momentums, $\mathbf{P}(t)$, as well as the electronic parameters $\mathbf{d}(t)$ and $\mathbf{z}(t)$. The properties of the combined system before collision, during collision and after collision can be analyzed in terms of various time dependent functions of these variables such as $\langle\Psi_0|\Psi\big(\mathbf{z}(t),\mathbf{d}(t),\mathbf{R}(t),\mathbf{P}(t)\big)\rangle$, which approximate the extinction coefficient $\langle\Psi_0|V(t)\Psi_0\rangle$ for the initial state, Ψ_0, and for the time variation, $Q(t)=\langle\Psi\big(\mathbf{z}(t),\mathbf{d}(t),\mathbf{R}(t),\mathbf{P}(t)\big)|Q\Psi\big(\mathbf{z}(t),\mathbf{d}(t),\mathbf{R}(t),\mathbf{P}(t)\big)\rangle$, of expectation values of observables Q. The poles of the Laplace-Fourier transform of the extinction coefficient give approximate stationary state energies and resonance information about the electronic, vibrational and rotational levels of the system during different periods of time during the evolution produced by the operator $V(t) = \exp\left\{-iHt\right\}$, based on the general nuclear-electron hamiltonian $H=H_n+H_e+H_{ne}$ that is a sum of three terms one describing the nuclei, one describing the electrons and one describing the electron-nuclear interaction. The Laplace-Fourier transform is given by

$$\left\langle \Psi_0 \middle| \left(\zeta I - H\right)^{-1} \Psi_0 \right\rangle \simeq \int_0^\infty \left\langle \Psi_0 \middle| \Psi\left(\mathbf{z}(t), \mathbf{d}(t), \mathbf{R}(t), \mathbf{P}(t)\right) \right\rangle e^{-i\zeta t} dt \tag{66}$$

for $\mathrm{Im}\zeta<0$. The values of the poles obtained reflect that a slice of the non linear manifold $V_0^N(\mathbf{z})$ is being sampled rather than of all of $\mathcal{H}^N$ during the evolution.

Acknowledgements

The author wishes to acknowledge the strong collaboration and multiple discussions with Eric Deumens and Yngve Öhrn on the subject matter of this contribution.

Bibliography

1. P. Kramer and C. Saraceno, *Geometry of the Timed Dependent Variational Principle in Quantum Mechanics* (Springer, New York, 1981).
2. J.R. Klauder and B. S. Skagerstam, *Coherent States, Applications in Physics and Mathematical Physics* (World Scientific, Singapore, 1985).
3. A. M. Perelomov, *Generalized Coherent States and their Applications* (Springer, New York, 1986).
4. W. M. Zhang, D. H. Feng and R. Gilmore, Rev. Mod. Phys. **62** (1990).
5. D. J. Rowe, G. Rosensteel and R. Gilmore, J. Math. Phys. **26**, 2787 (1985).
6. K. T. Hecht, *The Vector Coherent State Method and its Applications to Problems of Higher Symmetries* (Springer,New York, 1987)
7. D. J. Rowe, and J. Repka, J. Math. Phys. **32**, 2614 (1991).
8. E. Deumens and Y. Öhrn, J. Mol. Struct. **199**, 23 (1989).
9. E. Deumens, Y. Öhrn and B. Weiner J. Math. Phys. **32**, 1166, (1991).
10. R. Hermann, *Lie Groups for Physicists* (Benjamin, New York, 1966).
11. P. O. Löwdin, Rev. Mod. Phys. **36**, 966 (1964).
12. B. Weiner, E. Deumens and Y. Öhrn, J. Math. Phys. **32**, 2413, (1991).
13. The basis set superposition problem must of course be taken into account.

Appendix 1

The Gauss factorization decomposes a group element into a product that corresponds to a direct sum decomposition of the complexification, $\mathcal{G}^c$, of the Lie algebra, $\mathcal{G}$, of the group G into three subalgebras

$$\mathcal{G}^c = \mathcal{G}_+ \oplus \mathcal{G}_0 \oplus \mathcal{G}_- \tag{A1}$$

where the subalgebras $\mathcal{G}_+$ and $\mathcal{G}_-$ are nilpotent and

$$\left[\mathcal{G}_+, \mathcal{G}_0\right] \subset \mathcal{G}_+, \left[\mathcal{G}_-, \mathcal{G}_0\right] \subset \mathcal{G}_-, \left[\mathcal{G}_+, \mathcal{G}_-\right] \subset \mathcal{G}_0 \tag{A2}$$

The algebra $\mathcal{G}_- = \left(\mathcal{G}_+^\dagger\right)$ (we are implicitly working in a representation of the Lie algebra in an algebra of operators that contains the adjoint operation). Let $\left\{A_k^\dagger\right\}$, $\left\{\mathcal{A}_k\right\}$ and $\left\{A_k\right\}$ be bases for $\mathcal{G}_+$, $\mathcal{G}_0$ and $\mathcal{G}_-$ respectively then the relevant commutation relations defining $\mathcal{G}_+$ are

$$\left[A_i^\dagger, A_j^\dagger\right] = \sum s_{ij}^k A_k^\dagger \tag{A3}$$

where $\left\{s_{ij}^k\right\}$ are the structure coefficients of the Lie algebra. Those defining $\mathcal{G}_-$ which are the operator adjoint of eq.(A3)

$$\left[A_i, A_j\right] = \sum c_{ij}^k A_k \tag{A4}$$

where $s_{ij}^k = \left(c_{ji}^k\right)^*$. The commutation relations between $\mathcal{G}_0$ and $\mathcal{G}_-$ are defined by

$$\left[A_i, \mathcal{A}_j\right] = \sum \alpha_{ij}^k A_k \tag{A5}$$

and those between $\mathcal{G}_-$ and $\mathcal{G}_+$ defined by

$$\left[A_i, A_j^\dagger\right] = \sum \chi_{ij}^k \mathcal{A}_k \tag{A6}$$

We will use the operator identity

$$\mathrm{ad}(A)B = \left[A,B\right] = AB-BA \tag{A7}$$

$$\exp\left\{A\right\}B = \exp\left\{\mathrm{ad}(A)\right\}B\exp\left\{A\right\}. \tag{A8}$$

In order to evaluate components, $\langle\kappa,\mathbf{z}|B\Psi\rangle$, of holomorphic wavefunctions, $B\Psi(\mathbf{z})$, we need to consider the operator $A(\mathbf{z})^\dagger$

$$A(\mathbf{z})^\dagger = \sum z_i^* A_i \tag{A9}$$

and basis elements $A_k^\dagger$, $\mathcal{A}_k$, A_k of $\mathcal{G}_+$, $\mathcal{G}_0$ and $\mathcal{G}_-$ respectively. First we will consider $B = A_k$

$$\exp\left\{\mathrm{ad}(A(\mathbf{z})^\dagger)\right\}A_k = A_k + \left[A(\mathbf{z})^\dagger, A_k\right] + \frac{1}{2}\left[A(\mathbf{z})^\dagger, \left[A(\mathbf{z})^\dagger, A_k\right]\right] + .. \tag{A10}$$

$$= A_k + \sum z_i^*\left[A_i, A_k\right] + \frac{1}{2}\sum z_i^* z_m^*\left[A_m, \left[A_i, A_k\right]\right] + \ldots.. \tag{A11}$$

$$= A_k + \sum z_i^* c_{ik}^j A_j + \frac{1}{2}\sum z_i^* z_m^* c_{ml}^j c_{ik}^l A_j + \ldots.. \tag{A12}$$

$$= \sum\left\{\delta_{jk} + z_i^* c_{ik}^j + \frac{1}{2}\sum z_i^* z_m^* c_{ml}^j c_{ik}^l + \ldots\right\}A_j = \sum_j p_{kj}(\mathbf{z}^*, \mathbf{c})A_j \tag{A13}$$

The derivative of a component of a holomorphic wavefunction with respect to z_k^* is

$$\frac{\partial}{\partial z_k^*}\langle\kappa|\exp\left(\sum z_i^* A_i\right)\Psi\rangle = \langle\kappa|A_k\Psi\rangle + \frac{1}{2}\langle\kappa|A_k\left(\sum z_i^* A_i\right)\Psi\rangle$$

$$+ \frac{1}{2} \langle\kappa| \left[\sum z_i^* A_i\right] A_k \Psi\rangle + \ldots . \tag{A14}$$

$$= \langle\kappa|A_k\Psi\rangle + \langle\kappa|A_k\left[\sum z_i^* A_i\right]\Psi\rangle + \frac{1}{2} \langle\kappa| \left[\sum_{ij} z_i^* c_{ik}^j A_j\right]\Psi\rangle + \ldots \tag{A15}$$

$$= \langle\kappa|A_k \exp\left[\sum z_i^* A_i\right]\Psi\rangle + \frac{1}{2} \sum_{ij} z_i^* c_{ik}^j \langle\kappa|A_j \exp\left[\sum z_i^* A_i\right]\Psi\rangle + \ldots \tag{A16}$$

where we expanded the exponential, differentiated, used eq.(A4) on the third term and the collected powers of $A(\mathbf{z})^\dagger$ into an exponential again. One then obtains by inverting the series as a system of coupled equations that

$$\langle\kappa|A_k \exp\left[\sum z_i^* A_i\right]\Psi\rangle = \left[\frac{\partial}{\partial z_k^*} - \frac{1}{2} \sum_{ij} z_i^* c_{ik}^j \frac{\partial}{\partial z_j^*} + \ldots\right] \langle\mathbf{z},\kappa|\Psi\rangle \tag{A17}$$

$$= \left[\sum_i R_{ki}(\mathbf{z}^*,\mathbf{c}) \frac{\partial}{\partial z_i^*}\right] \langle\mathbf{z},\kappa|\Psi\rangle \tag{A18}$$

Hence the value of a component of a holomorphic wavefunction corresponding to $A_k|\Psi\rangle$ is given by

$$\langle\kappa|\exp\left\{A(\mathbf{z})^\dagger\right\} A_k\Psi\rangle = \langle\kappa|\exp\left\{\mathrm{ad}(A(\mathbf{z})^\dagger)\right\} A_k \exp\left\{A(\mathbf{z})^\dagger\right\}\Psi\rangle \tag{A19}$$

$$= \sum p_i(\mathbf{z}^*,\mathbf{c},A_k) \; \langle\kappa|A_i \exp\left\{A(\mathbf{z})^\dagger\right\}\Psi\rangle \tag{A20}$$

$$= \sum_{ij} p_{ki}(\mathbf{z}^*,\mathbf{c}) \; R_{ij}(\mathbf{z}^*,\mathbf{c}) \frac{\partial}{\partial z_j^*} \langle\kappa|\exp\left\{A(\mathbf{z})^\dagger\right\}\Psi\rangle \tag{A21}$$

Next we consider $B = \mathcal{A}_k$

$$\exp\left\{\mathrm{ad}(A(\mathbf{z})^\dagger)\right\}\mathcal{A}_k = \mathcal{A}_k + \left[A(\mathbf{z})^\dagger, \mathcal{A}_k\right] + \frac{1}{2}\left[A(\mathbf{z})^\dagger, \left[A(\mathbf{z})^\dagger, \mathcal{A}_k\right]\right] + \ldots \tag{A22}$$

$$= \mathcal{A}_k + \sum z_i^* \left[A_i, \mathcal{A}_k\right] + \frac{1}{2} \sum z_i^* z_m^* \left[A_m, \left[A_i, \mathcal{A}_k\right]\right] + \ldots\ldots \tag{A23}$$

$$= \mathcal{A}_k + \sum_j \left\{\sum z_i^* \alpha_{ik}^j + \frac{1}{2} \sum z_i^* z_m^* \alpha_{ik}^l c_{ml}^j + \ldots\right\} A_j \tag{A24}$$

$$= \mathcal{A}_k + \sum \left\{q_{kj}(\mathbf{z}^*,\mathbf{c},\alpha) A_j\right\} \tag{A25}$$

Thus the value of a component of a holomorphic wavefunction corresponding to $\mathcal{A}_k|\Psi\rangle$ is given by

$$\left\langle\kappa\left|\exp\left\{A(\mathbf{z})^{\dagger}\right\}A_k\Psi\right.\right\rangle = \left\langle\kappa\left|\exp\left\{\mathrm{ad}(A(\mathbf{z})^{\dagger})\right\}A_k\exp\left\{A(\mathbf{z})^{\dagger}\right\}\Psi\right.\right\rangle \tag{A26}$$

$$= \left\langle\kappa\left|A_k\exp\left\{A(\mathbf{z})^{\dagger}\right\}\Psi\right.\right\rangle + \sum q_{kj}(\mathbf{z}^*,\mathbf{c})\left\langle\kappa\left|A_j\exp\left\{A(\mathbf{z})^{\dagger}\right\}\Psi\right.\right\rangle \tag{A27}$$

$$= \left\langle\kappa\left|A_k\exp\left\{A(\mathbf{z})^{\dagger}\right\}\Psi\right.\right\rangle + \sum_{ij} q_{kj}(\mathbf{z}^*,\mathbf{c})R_{ji}(\mathbf{z}^*,\mathbf{c})\frac{\partial}{\partial z_i^*}\langle\kappa,\mathbf{z}|\Psi\rangle \tag{A28}$$

While if $B=A_k^{\dagger}$ one obtains

$$\exp\left\{\mathrm{ad}(A(\mathbf{z})^{\dagger})\right\}A_k^{\dagger} = A_k + \left[A(\mathbf{z})^{\dagger},A_k^{\dagger}\right] + \frac{1}{2}\left[A(\mathbf{z})^{\dagger},\left[A(\mathbf{z})^{\dagger},A_k^{\dagger}\right]\right] + \ldots \tag{A29}$$

$$= A_k + \sum z_i^*\left[A_i,A_k^{\dagger}\right] + \frac{1}{2}\sum z_i^*z_m^*\left[A_m,\left[A_i,A_k^{\dagger}\right]\right] + \ldots\ldots \tag{A30}$$

$$= A_k + \sum z_i^*\chi_{ik}^j A_j + \frac{1}{2}\sum z_i^*z_m^*\chi_{ik}^l\left[A_m,A_l\right] + \ldots\ldots \tag{A31}$$

$$= A_k + \sum z_i^*\chi_{ik}^j A_j + \frac{1}{2}\sum z_i^*z_m^*\chi_{ik}^l\alpha_{ml}^j A_j^{\dagger} + \ldots\ldots \tag{A32}$$

$$= A_k + A_k(\chi,\mathbf{z}) + \sum\left\{\frac{1}{2}z_i^*z_m^*\chi_{ik}^l\alpha_{ml}^j + \frac{1}{6}z_n^*z_m^*z_m^*\chi_{ik}^l\alpha_{ml}^{j'}c_{nj'}^j, +\ldots\right\}A_j \tag{A33}$$

$$= A_k + A_k(\chi,\mathbf{z}) + \sum\left\{h_{kj}(\mathbf{z}^*,\mathbf{c},\alpha,\chi)A_j\right\} \tag{A34}$$

and the value of a component of a holomorphic wavefunction corresponding to $A_k^{\dagger}|\Psi\rangle$ is given by

$$\left\langle\kappa\left|\exp\left\{A(\mathbf{z})^{\dagger}\right\}A_k^{\dagger}\Psi\right.\right\rangle = \left\langle\kappa\left|\exp\left\{\mathrm{ad}(A(\mathbf{z})^{\dagger})\right\}A_k^{\dagger}\exp\left\{A(\mathbf{z})^{\dagger}\right\}\Psi\right.\right\rangle \tag{A35}$$

$$= \left\langle\kappa\left|A_k^{\dagger}\exp\left\{A(\mathbf{z})^{\dagger}\right\}\Psi\right.\right\rangle + \left\langle\kappa\left|A_k(\chi,\mathbf{z})\exp\left\{A(\mathbf{z})^{\dagger}\right\}\Psi\right.\right\rangle$$

$$+ \sum h_{kj}(\mathbf{z}^*,\mathbf{c},\alpha,\chi)\left\langle\kappa\left|A_j\exp\left\{A(\mathbf{z})^{\dagger}\right\}\Psi\right.\right\rangle \tag{A36}$$

$$= \left\langle\kappa\left|A_k(\chi,\mathbf{z})\exp\left\{A(\mathbf{z})^{\dagger}\right\}\Psi\right.\right\rangle + \sum h_{kj}(\mathbf{z}^*,\mathbf{c},\alpha,\chi)\left\langle\kappa\left|A_j\exp\left\{A(\mathbf{z})^{\dagger}\right\}\Psi\right.\right\rangle \tag{A37}$$

$$= \left\langle\kappa\left|A_k(\chi,\mathbf{z})\exp\left\{A(\mathbf{z})^{\dagger}\right\}\Psi\right.\right\rangle +$$

$$\sum_{ij} h_{kj}(\mathbf{z}^*,\mathbf{c},\alpha,\chi)R_{ji}(\mathbf{z}^*,\mathbf{c})\frac{\partial}{\partial z_i^*}\left\langle\kappa\left|\exp\left\{A(\mathbf{z})^{\dagger}\right\}\Psi\right.\right\rangle \tag{A38}$$

In the Gauss decomposition of the group, $U(\mathcal{H}^1)$ based on the coset $U(\mathcal{H}^1)/U(\mathcal{H}^1_h)\otimes U(\mathcal{H}^1_p)$ the subgroups $\mathcal{G}_+$ and $\mathcal{G}_-$ are abelian, thus the structure constants **c=0** and in the representation $\mathcal{U}$

$$\left[a^\dagger_{h_1}a_{p_1},a^\dagger_{h_2}a_{p_2}\right] = \left[a^\dagger_{p_1}a_{h_1},a^\dagger_{p_2}a_{h_2}\right] = 0 \tag{A39}$$

leading to

$$R_{ki}(\mathbf{z}^*,\mathbf{c}) = \delta_{ki} \tag{A40}$$

The commutation relations between $\mathcal{G}_-$ and $\mathcal{G}_0$ are given by

$$\left[a^\dagger_{h'}a_{p'},a^\dagger_{h_1}a_{h_2}\right] = a^\dagger_{h'}a_{h_2}\delta_{h_1p'} - a^\dagger_{h_1}a_{p'}\delta_{h'h_2} = a^\dagger_{h_1}a_{p'}\delta_{h'h_2} \tag{A41}$$

$$\left[a^\dagger_{h'}a_{p'},a^\dagger_{p_1}a_{p_2}\right] = a^\dagger_{h'}a_{p_2}\delta_{p_1p'} - a^\dagger_{p_1}a_{p'}\delta_{h'p_2} = -a^\dagger_{h'}a_{p_2}\delta_{p'p_1} \tag{A42}$$

thus

$$\alpha^{hp}_{h'p'k_1k_2} = \delta_{hk_1}\delta_{pp'}\delta_{h'k_2} - \delta_{hh'}\delta_{pk_2}\delta_{p'k_1}. \tag{A43}$$

The commutation relations between $\mathcal{G}_-$ and $\mathcal{G}_+$ are given by

$$\left[a^\dagger_{h'}a_{p'},a^\dagger_{p_1}a_{h_1}\right] = a^\dagger_{h'}a_{h_1}\delta_{p_1p'} - a^\dagger_{p_1}a_{p'}\delta_{h'h_1} \tag{A44}$$

thus

$$\chi^{kk'}_{h'p'p_1h_1} = \delta_{kh'}\delta_{k'h_1}\delta_{p_1p'} - \delta_{kp_1}\delta_{k'p'}\delta_{h'h_1}. \tag{A45}$$

Thus eq. (A19) becomes for $A_k = a^\dagger_h a_p$

$$\left\langle\kappa\middle|\exp\left\{A(\mathbf{z})^\dagger\right\}a^\dagger_h a_p\Psi\right\rangle = \frac{\partial}{\partial z^*_{ph}}\left\langle\kappa\middle|\exp\left\{A(\mathbf{z})^\dagger\right\}\Psi\right\rangle \tag{A46}$$

After noting that

$$q_{k_1k_2hp}(\mathbf{z}^*,\mathbf{c},\alpha) = \sum_{p'h'} z^*_{p'h'}\alpha^{hp}_{p'h'k_1k_2} =$$

$$\sum_{p'h'} z^*_{p'h'}(\delta_{hk_1}\delta_{pp'}\delta_{h'k_2} - \delta_{hh'}\delta_{pk_2}\delta_{p'k_1}) \tag{A47}$$

The expression for $\langle\kappa,\mathbf{z}|a^\dagger_{h_1}a_{h_2}\Psi\rangle$ becomes

$$\left\langle\kappa\middle|\exp\left\{A(\mathbf{z})^\dagger\right\}a^\dagger_{h_1}a_{h_2}\Psi\right\rangle = \left\langle\kappa\middle|a^\dagger_{h_1}a_{h_2}\exp\left\{A(\mathbf{z})^\dagger\right\}\Psi\right\rangle +$$

$$\sum_{hp,p'h'} z^*_{p'h'}(\delta_{hh_1}\delta_{pp'}\delta_{h'h_2} - \delta_{hh'}\delta_{ph_2}\delta_{p'h_1})\left\langle\kappa\middle|a^\dagger_h a_p\exp\left\{A(\mathbf{z})^\dagger\right\}\Psi\right\rangle \tag{A48}$$

$$= \left\langle\kappa\middle|a^\dagger_{h_1}a_{h_2}\exp\left\{A(\mathbf{z})^\dagger\right\}\Psi\right\rangle + \sum_p z^*_{ph_2}\left\langle\kappa\middle|a^\dagger_{h_1}a_p\exp\left\{A(\mathbf{z})^\dagger\right\}\Psi\right\rangle \tag{A49}$$

$$= (1-\delta_{h_2\kappa})\delta_{h_1\kappa}\langle\kappa|\exp\left\{A(\mathbf{z})^\dagger\right\}\Psi\rangle + \sum_p z^*_{ph_2}\frac{\partial}{\partial z^*_{ph_1}}\langle\kappa|\exp\left\{A(\mathbf{z})^\dagger\right\}\Psi\rangle \tag{A50}$$

where $\kappa\equiv(h'_1,..h'_N)$ and kronecker delta functions indicate whether h_1 or h_2 belong to the configuration κ. The expression for $\langle\kappa,\mathbf{z}|a^\dagger_{p_1}a_{p_2}\Psi\rangle$ becomes

$$= \sum_{hp,p'h'} z^*_{p'h'}(\delta_{hp_1}\delta_{pp'}\delta_{h'p_2} - \delta_{hh'}\delta_{pp_2}\delta_{p'p_1})\ \langle\kappa|a^\dagger_h a_p\exp\left\{A(\mathbf{z})^\dagger\right\}\Psi\rangle \tag{A51}$$

$$= -\sum_h z^*_{p_1h}\frac{\partial}{\partial z^*_{p_1h}}\langle\kappa|\exp\left\{A(\mathbf{z})^\dagger\right\}\Psi\rangle \tag{A52}$$

In order to study the matrix element $\langle\kappa,\mathbf{z}|a^\dagger_p a_h\Psi\rangle$ we need first to consider

$$\sum_{p'h'h_1h_2} z^*_{p'h'}\chi^{h_1h_2}_{p'h'hp}a^\dagger_{h_1}a_{h_2} + \sum_{p'h'p_1p_2} z^*_{p'h'}\chi^{p_1p_2}_{p'h'hp}a^\dagger_{p_1}a_{p_2} \tag{A53}$$

which by eq.(A41)

$$= \sum_{h'} z^*_{ph'}a^\dagger_{h'}a_h - \sum_{p'} z^*_{p'h}a^\dagger_p a_{p'} \tag{A54}$$

and

$$\frac{1}{2}\sum z^*_{p'h'}z^*_{p''h''}\left[a^\dagger_{h'}a_{p''},\left[a^\dagger_{h''}a_{p''},a^\dagger_h a_p\right]\right] \tag{A55}$$

$$= -\sum_{p'h'} z^*_{p'h}z^*_{ph'}a^\dagger_{h'}a_{p'} \tag{A56}$$

Thus eq.(A35) becomes

$$\langle\kappa,\mathbf{z}|a^\dagger_h a_p\Psi\rangle = \sum_{h'} z^*_{ph'}\langle\kappa|a^\dagger_{h'}a_h\exp\left\{A(\mathbf{z})^\dagger\right\}\Psi\rangle +$$

$$- \sum_{p'h'} z^*_{p'h}z^*_{ph'}\langle\kappa|a^\dagger_{h'}a_{p'}\exp\left\{A(\mathbf{z})^\dagger\right\}\Psi\rangle \tag{A57}$$

$$= \sum_{h'} z^*_{ph'}(1-\delta_{h\kappa})\delta_{h'\kappa}\langle\kappa|\exp\left\{A(\mathbf{z})^\dagger\right\}\Psi\rangle$$

$$- \sum_{p'h'} z^*_{p'h}z^*_{ph'}\frac{\partial}{\partial z^*_{p'h'}}\langle\kappa|\exp\left\{A(\mathbf{z})^\dagger\right\}\Psi\rangle \tag{A58}$$

Appendix 2

The representation $\sigma(g)$ of the group element $g\in U(\mathcal{H}^1)$ on the coset space $U(\mathcal{H}^1)/U(\mathcal{H}^1_h)\otimes U(\mathcal{H}^1_p)$ is given by /7/

$$z' = \left(Az + B\right)\left(Cz + D\right)^{-1} \tag{B1}$$

where $g = \begin{pmatrix} A & B \\ C & D \end{pmatrix} \in U(\mathcal{H}^1)$ and is of dimension $r \times r$ for $r=2s$ and $\dim(\mathcal{H}^1) = r$. The submatrices are of dimension $A: r_1 \times (r-r_1)$; $B: (r-r_1) \times r_1$; $C: r_1 \times (r-r_1)$; $D: (r-r_1) \times r_1$.

CONTRIBUTORS

CERJAN C.
Lawrence Livermore Nat. Lab.
University of California
P.O.B. 808, L -296
Livermore, CA 94550
U.S.A.

CIMIRAGLIA R.
Instituto di Chimica Fysica
Univ .di Ferrara
Via Borsari 46
I-44100, Ferrara
ITALY

FARANTOS S.
Dept. of Chemistry & IESL FORTH
University of Crete
P.O. Box 1527
Heraklion, Crete 71110
CRETE GREECE

FARRELLY D.
Dept. of Chemistry
Utah State University
Logan, UT 84322-0300
U. S. A.

GRAY S.
Chemistry Division
Argonne National Laboratory
Argonne, IL 60439
U.S.A.

JOHNSON B.R.
Dept. of Chemistry
Rice University
P.O. Box 1892
Houston, Texas 77251
U.S.A.

KOPPEL H.
Theoretische Chemie
Universität Heidelberg
Im Neuenheimer Feld 253
D-6900 Heidelberg
GERMANY

KOSLOFF R.
Chemistry Dept.-The Fritz Haber Ctr
Hebrew University
91904 Jerusalem
ISRAEL

KOURI D.J.
Dept. of Chemistry
Univ. of Houston
Houston, TX 77204-5641
U.S.A.

KRAUSE J
Lawrence Livermore Nat Lab
Univ. of California
P.O. BOX 808
Livermore, Cal 94551-9900
U.S.A.

KULANDER K.C.
Lawrence Livermore Nat Lab
Univ. of California
P.O. BOX 808
Livermore, Cal 94551-9900
U.S.A.

KURUOGLU Z.
Dept. of Chemistry
Bilkent University
06533 Bilkent, Ankara
TURKEY

LEVINE R.D.
Chemistry Dept.-The Fritz Haber Ctr
Hebrew University
91904 Jerusalem
ISRAEL

LIGHT J.C.
James Franck Institute
The University of Chicago
5640 Ellis Avenue
Chicago, ILL 60637
U.S.A.

LINDERBERG J.
Dept. of Chemistry
Aarhus University
DK-8000 Aarhus C.
DENMARK

MAKRI N.
School of Chem. Sciences
Univ. of Illinois
505 S. Mathews Ave
URBANA, IL 61801
U.S.A.

MATHIES R.A.
Dept. of Chemistry
University of California
Berkeley, CA 94720
U.S.A.

MEYER H.D.
Theoretische Chemie
Universität Heidelberg
Im Neuenheimer Feld 253
D-6900 Heidelberg
GERMANY

MICHA D.
QTP, Williamson Hall 365
University of Florida
366 Williamson Hall
Gainesville, FL 32611
U.S.A.

MILLER W.H.
Dept. of Chem.-L. Berkeley Lab
Univ. of California
Berkeley, CA 94720
U.S.A.

OHRN Y.
QTP, Willamson Hall 363
University of Florida
WM 363, ,
Gainesville, FL 32611
U.S.A.

QUACK M.
Lab. für Phys. Chem.
ETH Zürich (Zentrum)
CH-8092 Zürich
SWITZERLAND

RIERA A.
Dept. de Quimica C-XIV
Univ. Autonoma de Madrid
Canto Blanco,
28049 Madrid
SPAIN

TANNOR D.
Dept. of Chemistry
Univ. of Notre Dame
Notre Dame, IN 46556
U.S.A.

TAYLOR H.S.
Dept. of Chemistry, SSC 704
Univ. of Southern California
Los Angeles 90089-0482
U.S.A.

UZER T.
School of Physics
Georgia Inst. of Technology
Atlanta, GA 30332-0430
U.S.A.

VAN LEUVEN P.
Dept. of Mathematics & Informatics
Univ. Antwerp (RUCA)
Groenenborgerlaan 171
B-2020 Antwerpen 2
BELGIUM

WEINER B.
Dept. of Physics
Pennsylvania State Univ.
College Place
Dubois, PA 15801
U.S.A.

INDEX